T0269178

Studies in Classification, Data Analysis, and Knowledge Organization

More information about this series at http://www.springer.com/series/1564

Adalbert F.X. Wilhelm • Hans A. Kestler
Editors

Analysis of Large and Complex Data

 Springer

Editors
Adalbert F.X. Wilhelm
Jacobs University Bremen
Bremen, Germany

Hans A. Kestler
Institute of Medical Systems Biology
Universität Ulm
Ulm, Germany

ISSN 1431-8814   ISSN 2198-3321   (electronic)
Studies in Classification, Data Analysis, and Knowledge Organization
ISBN 978-3-319-25224-7   ISBN 978-3-319-25226-1   (eBook)
DOI 10.1007/978-3-319-25226-1

Library of Congress Control Number: 2016930307

Springer Cham Heidelberg New York Dordrecht London

Springer International Publishing AG Switzerland is part of Springer Science+Business Media
(www.springer.com)

Foreword

Dear Scholars,

The world we live in is producing vast amounts of data everywhere and anytime. Wider use of the Internet with smartphones and tablets and increasing interconnection of equipment, vehicles and machines are swelling the data flow into a veritable flood of information. This flood of information, better known as "Big Data", is a valuable resource—if you know how to use it. Only efficient and intelligent analysis of Big Data can help us to understand linkages and to make better decisions on this basis. Its potential can be found in many areas: Evaluation of large volumes of data helps to improve medical care, to optimise use of natural resources, to increase our security or also to develop new products and services. We are only just beginning to exploit this treasure. This book is the outcome of the second European Conference on Data Analysis (ECDA) held in Bremen in 2014. The scientific programme of the conference covered a broad range of topics. Special emphasis was given to research on and development of innovative tools, techniques and strategies that address current challenges in the data analysis process. I warmly invite you to read this book in order to get a deep insight into the present state of research and into the pivotal areas of data analysis.

President of the Confederation of German Ingo Kramer
Employers' Association
(Bundesvereinigung der Deutschen
Arbeitgeberverbände – BDA)
Berlin, Germany
June 2015

Preface

The volume that you hold now in your hand or read electronically comprises the revised versions of selected papers presented during the European Conference on Data Analysis (ECDA 2014) and the Workshop on Classification and Subject Indexing in Library and Information Science (LIS' 2014). This second edition of the European Conference on Data Analysis was held at Jacobs University Bremen (Germany) under the patronage of Ingo Kramer, President of the Confederation of German Employers' Association (Bundesvereinigung der Deutschen Arbeitgeberverbände—BDA). The conference marked also the occasion of the 38th anniversary of the German Classification Society (GfKl). The conference was organised by the German Classification Society (GfKl) in cooperation with the Italian Statistical Society Classification and Data Analysis Group (SIS-Cladag), Vereniging voor Ordinatie en Classificatie (VOC), Sekcja Klasyfikacji i Analizy Danych PTS (SKAD) and the International Association for Statistical Computing (IASC).

In early July 2014, a total of 193 participants from 27 countries gathered at the beautiful campus of Jacobs University Bremen to listen to and critically discuss on 154 presentations including two plenary and eight semi-plenary keynote speeches, six invited symposia and a plenary panel discussion on "The Future of Publications in Classification and Data Sciences". Having most participants accommodated on campus allowed for additional discussions and mutual exchange of knowledge outside the conference presentations, and it created an excellent stimulus for fostering international collaborations and networks. With half of the participants coming from outside of Germany, the conference truly lived up to the idea of an international convention. The selection of keynote speakers from six European countries, as well as from China, Israel and the United States of America, is another indicator for the growing international network of researchers in this area.

The scientific programme extended across a broad range of sessions dealing with different aspects of the data analysis process. Quite a spectrum of application fields had been covered in the presentations showing the importance of a close interaction between theory and practice as well as between scientific disciplines. The members of the scientific programme committee under the lead of the Scientific

Programme Chair Hans A. Kestler stipulated and selected a truly inspiring and interdisciplinary programme bridging theory, methods and applications of data analysis in the following seven thematic areas:

1. Statistics and Data Analysis, organised by Claus Weihs, Francesco Mola, Roberto Rocci and Christian Hennig
2. Machine Learning and Knowledge Discovery, organised by Eyke Hüllermeier, Friedhelm Schwenker and Myra Spiliopoulou
3. Data Analysis in Marketing, organised by Józef Pociecha, Daniel Baier, Wolfgang Gaul and Reinhold Decker
4. Data Analysis in Finance and Economics, organised by Marlene Müller, Gregor Dorfleitner and Colin Vance
5. Data Analysis in Medicine and the Life Sciences, organised by Hans A. Kestler, Matthias Schmid, Iris Pigeot and Berthold Lausen
6. Data Analysis in the Social, Behavioural, and Health Care Sciences, organised by Ali Ünlü, Ingo Rohlfing, Karin Wolf-Ostermann and Jeroen K. Vermunt
7. Data Analysis in Interdisciplinary Domains, organised by Adalbert F.X. Wilhelm, Patrick Groenen, Sabine Krolak-Schwerdt, Frank Scholze and Andreas Geyer-Schulz
8. The Workshop Library and Information Science (LIS' 2014), organised by Frank Scholze

For each of these topics, a number of well-elaborated papers have been submitted for the proceedings volume after the conference took place. The 55 contributions that you find now in this volume have been accepted after a peer-reviewing process and provide a good representation of the topics covered at the conference. The contributions to the proceedings represent a diverse range of scientific disciplines, namely, Statistics, Psychology, Biology, Information Retrieval and Library Science, Archeology, Banking and Finance, Computer Science, Economics, Engineering, Geography, Geology, Linguistics and Musicology, Marketing, Mathematics, Medical and Health Sciences, Sociology and Educational Sciences. In all these disciplines, *Data Science* is a major unifying topic, which is reflected in papers that cover the meaningful extraction of knowledge from diverse data sources via structural, quantitative and statistical approaches. Examples are advances in classification and clustering and other pattern recognition methods. The explicit modelling of complex data in specific domains also includes the issues that come with *Big Data* in terms of numerical stability, set size and model learning or adaptation time and effort.

Empirical research in these fields requires the analysis of multiple data types. Even though underlying research questions and corresponding data emerge from most various areas, they often require similar statistical, structural or quantitative approaches for the analysis of data. The specific scientific impact of the post-conference volume concerns the presentation of methods, which may commonly be used for the analysis of data stemming from different domains and domain-specific research questions with the aim of solving the numerous domain-specific problems of data analysis on a theoretical as well as on a practical level, fostering their effective use for answering specific questions in various areas of application as well as evaluating alternative methods in the framework of applications.

Accordingly, the volume is organised with the following subsections:

Part I   Invited Papers
Part II   Big Data
Part III   Clustering
Part V   Regression and Other Statistical Techniques
Part VI   Applications
Part VII   Data Analysis in Marketing
Part VIII   Data Analysis in Finance
Part IX   Data Analysis in Medicine and Life Sciences
Part X   Data Analysis in Musicology
Part XI   Data Analysis in Interdisciplinary Domains
Part XII   Data Analysis in Social, Behavioural and Health Care Sciences
Part XIII   Data Analysis in Library Science

Organising this second ECDA conference required the coordination of many people and topics; dedicated colleagues and the great team of the Jacobs University Bremen made this possible. We would like to thank the area chairs and the LIS workshop chair for organising the areas during the conference, author recruitment and the evaluation of submissions. We are grateful to all reviewers: Daniel Baier, Andre Burkovski, Reinhold Decker, Gregor Dorfleitner, Axel Fürstberger, Wolfgang Gaul, Andreas Geyer-Schulz, Patrick Groenen, Christian Hennig, Eyke Hüllermeier, Johann Kraus, Sabine Krolak-Schwerdt, Berthold Lausen, Ludwig Lausser, Francesco Mola, Marlene Müller, Christoph Müssel, Magnus Pfeffer, Iris Pigeot, Józef Pociecha, Roberto Rocci, Ingo Rohlfing, Florian Schmid, Matthias Schmid, Frank Scholze, Friedhelm Schwenker, Myra Spiliopoulou, Eric Sträng, Ali Ünlü, Colin Vance, Jerome K. Vermunt, Claus Weihs, Heidrun Wiesenmüller and Karin Wolf-Ostermann.

Furthermore, we would like to thank Martina Bihn and Alice Blanck, Springer-Verlag, Heidelberg, for their support and dedication to the production of this volume. Last but not least, we would like to thank all participants of the ECDA 2014 conference for their interest and activities, which made the conference such a great interdisciplinary venue for scientific discussion.

Bremen, Germany   Adalbert F.X. Wilhelm
Ulm, Germany   Hans A. Kestler
July 2015

Acknowledgements

We gratefully acknowledge financial support by the Deutsche Forschungsgemeinschaft (DFG), the Bremen International Graduate School of Social Sciences
(BIGSSS), the Bankenverband Bremen e.V. and Springer-Verlag. We are very
thankful for Jacobs University Bremen for hosting the conference. We also like to
thank both Jacobs University Bremen and the Fritz-Lipmann Institute in Jena for
providing us with the necessary infrastructure and time to edit this volume.

Contents

Part I   Invited Papers

Latent Variables and Marketing Theory: The Paradigm Shift 3
Adam Sagan

Business Intelligence in the Context of Integrated Care Systems (ICS): Experiences from the ICS "Gesundes Kinzigtal" in Germany 17
Alexander Pimperl, Timo Schulte, and Helmut Hildebrandt

Clustering and a Dissimilarity Measure for Methadone Dosage Time Series 31
Chien-Ju Lin, Christian Hennig, and Chieh-Liang Huang

Linear Storage and Potentially Constant Time Hierarchical Clustering Using the Baire Metric and Random Spanning Paths 43
Fionn Murtagh and Pedro Contreras

Standard and Novel Model Selection Criteria in the Pairwise Likelihood Estimation of a Mixture Model for Ordinal Data 53
Monia Ranalli and Roberto Rocci

Part II   Big Data

Textual Information Localization and Retrieval in Document Images Based on Quadtree Decomposition 71
Cynthia Pitou and Jean Diatta

Selection Stability as a Means of Biomarker Discovery in Classification 79
Lyn-Rouven Schirra, Ludwig Lausser, and Hans A. Kestler

Active Multi-Instance Multi-Label Learning 91
Robert Retz and Friedhelm Schwenker

**Using Annotated Suffix Tree Similarity Measure for Text
Summarisation** 103
Maxim Yakovlev and Ekaterina Chernyak

**Big Data Classification: Aspects on Many Features and Many
Observations**... 113
Claus Weihs, Daniel Horn, and Bernd Bischl

Part III   Clustering

**Bottom-Up Variable Selection in Cluster Analysis Using
Bootstrapping: A Proposal** 125
Hans-Joachim Mucha and Hans-Georg Bartel

**A Comparison Study for Spectral, Ensemble
and Spectral-Mean Shift Clustering Approaches
for Interval-Valued Symbolic Data**... 137
Marcin Pełka

Supervised Pre-processings Are Useful for Supervised Clustering 147
Oumaima Alaoui Ismaili, Vincent Lemaire,
and Antoine Cornuéjols

Similarity Measures on Concept Lattices 159
Florent Domenach and George Portides

Part IV   Classification

**Multivariate Functional Regression Analysis with Application
to Classification Problems**.. 173
Tomasz Górecki, Mirosław Krzyśko, and Waldemar Wołyński

Incremental Generalized Canonical Correlation Analysis 185
Angelos Markos and Alfonso Iodice D'Enza

Evaluating the Necessity of a Triadic Distance Model 195
Atsuho Nakayama

Assessing the Reliability of a Multi-Class Classifier 207
Luca Frigau, Claudio Conversano, and Francesco Mola

Part V   Regression and Other Statistical Techniques

Reviewing Graphical Modelling of Multivariate Temporal Processes 221
Matthias Eckardt

The Weight of Penalty Optimization for Ridge Regression 231
Sri Utami Zuliana and Aris Perperoglou

**Monitoring a Dynamic Weighted Majority Method Based
on Datasets with Concept Drift** 241
Dhouha Mejri, Mohamed Limam, and Claus Weihs

Part VI   Applications

**Specialization in Smart Growth Sectors vs. Effects of Change
of Workforce Numbers in the European Union Regional Space** 253
Elżbieta Sobczak and Marcin Pełka

**Evaluation of the Individually Perceived Quality from Head-Up
Display Images Relating to Distortions** 265
Sonja Köppl, Markus Hellmann, Klaus Jostschulte,
and Christian Wöhler

**Minimizing Redundancy Among Genes Selected Based
on the Overlapping Analysis** 275
Osama Mahmoud, Andrew Harrison, Asma Gul, Zardad Khan,
Metodi V. Metodiev, and Berthold Lausen

**The Identification of Relations Between Smart Specialization
and Sensitivity to Crisis in the European Union Regions** 287
Beata Bal-Domańska

Part VII   Data Analysis in Marketing

**Market Oriented Product Design and Pricing: Effects
of Intra-Individual Varying Partworths** 301
Stephanie Löffler and Daniel Baier

The Use of Hybrid Predictive C&RT-Logit Models in Analytical CRM ... 311
Mariusz Łapczyński

Part VIII   Data Analysis in Finance

**Excess Takeover Premiums and Bidder Contests in Merger &
Acquisitions: New Methods for Determining Abnormal Offer Prices** 323
Wolfgang Bessler and Colin Schneck

**Firm-Specific Determinants on Dividend Changes: Insights
from Data Mining** 335
Karsten Luebke and Joachim Rojahn

**Selection of Balanced Structure Samples in Corporate
Bankruptcy Prediction** 345
Mateusz Baryła, Barbara Pawełek, and Józef Pociecha

**Facilitating Household Financial Plan Optimization
by Adjusting Time Range of Analysis to Life-Length Risk Aversion** 357
Radoslaw Pietrzyk and Pawel Rokita

Dynamic Aspects of Bankruptcy Prediction Logit Model
for Manufacturing Firms in Poland 369
Barbara Pawełek, Józef Pociecha, and Mateusz Baryła

Part IX   Data Analysis in Medicine and Life Sciences

Estimating Age- and Height-Specific Percentile Curves
for Children Using GAMLSS in the IDEFICS Study 385
Timm Intemann, Hermann Pohlabeln, Diana Herrmann, Wolfgang
Ahrens, and Iris Pigeot, on behalf of the IDEFICS consortium

An Ensemble of Optimal Trees for Class Membership
Probability Estimation 395
Zardad Khan, Asma Gul, Osama Mahmoud, Miftahuddin
Miftahuddin, Aris Perperoglou, Werner Adler, and Berthold Lausen

Ensemble of Subset of k-Nearest Neighbours Models for Class
Membership Probability Estimation 411
Asma Gul, Zardad Khan, Aris Perperoglou, Osama Mahmoud,
Miftahuddin Miftahuddin, Werner Adler, and Berthold Lausen

Part X   Data Analysis in Musicology

The Surprising Character of Music: A Search for Sparsity
in Music Evoked Body Movements 425
Denis Amelynck, Pieter-Jan Maes, Marc Leman,
and Jean-Pierre Martens

Comparing Audio Features and Playlist Statistics for Music
Classification 437
Igor Vatolkin, Geoffray Bonnin, and Dietmar Jannach

Duplicate Detection in Facsimile Scans of Early Printed Music 449
Christophe Rhodes, Tim Crawford, and Mark d'Inverno

Fast Model Based Optimization of Tone Onset Detection
by Instance Sampling 461
Nadja Bauer, Klaus Friedrichs, Bernd Bischl, and Claus Weihs

Recognition of Leitmotives in Richard Wagner's Music:
An Item Response Theory Approach 473
Daniel Müllensiefen, David Baker, Christophe Rhodes,
Tim Crawford, and Laurence Dreyfus

Part XI   Data Analysis in Interdisciplinary Domains

Optimization of a Simulation for Inhomogeneous Mineral
Subsoil Machining 487
Swetlana Herbrandt, Claus Weihs, Uwe Ligges, Manuel Ferreira,
Christian Rautert, Dirk Biermann, and Wolfgang Tillmann

Fast and Robust Isosurface Similarity Maps Extraction Using Quasi-Monte Carlo Approach 497
Alexey Fofonov and Lars Linsen

Analysis of ChIP-seq Data Via Bayesian Finite Mixture Models with a Non-parametric Component 507
Baba B. Alhaji, Hongsheng Dai, Yoshiko Hayashi, Veronica Vinciotti, Andrew Harrison, and Berthold Lausen

Information Theoretic Measures for Ant Colony Optimization 519
Gunnar Völkel, Markus Maucher, Christoph Müssel, Uwe Schöning, and Hans A. Kestler

A Signature Based Method for Fraud Detection on E-Commerce Scenarios 531
Orlando Belo, Gabriel Mota, and Joana Fernandes

Three-Way Clustering Problems in Regional Science 545
Małgorzata Markowska, Andrzej Sokołowski, and Danuta Strahl

Part XII   Data Analysis in Social, Behavioural and Health Care Sciences

CFA-MTMM Model in Comparative Analysis of 5-, 7-, 9-, and 11-point A/D Scales 553
Piotr Tarka

Biasing Effects of Non-Representative Samples of Quasi-Orders in the Assessment of Recovery Quality of IITA-Type Item Hierarchy Mining 563
Ali Ünlü and Martin Schrepp

Correlated Component Regression: Profiling Student Performances by Means of Background Characteristics 575
Bernhard Gschrey and Ali Ünlü

Analysing Psychological Data by Evolving Computational Models 587
Peter C.R. Lane, Peter D. Sozou, Fernand Gobet, and Mark Addis

Use of Panel Data Analysis for V4 Households Poverty Risk Prediction 599
Lukáš Sobíšek and Mária Stachová

Part XIII   Data Analysis in Library Science

Collaborative Literature Work in the Research Publication Process: The Cogeneration of Citation Networks as Example 611
Leon Otto Burkard and Andreas Geyer-Schulz

**Subject Indexing of Textbooks: Challenges in the Construction
of a Discovery System**... 621
Bianca Pramann, Jessica Drechsler, Esther Chen,
and Robert Strötgen

**The Ofness and Aboutness of Survey Data: Improved Indexing
of Social Science Questionnaires** 629
Tanja Friedrich and Pascal Siegers

**Subject Indexing for Author Name Disambiguation:
Opportunities and Challenges** 639
Cornelia Hedeler, Andreas Oskar Kempf, and Jan Steinberg

Index... 651

Contributors

Mark Addis Faculty of Arts, Design and Media, Birmingham City University, Birmingham, UK

Werner Adler Department of Biometry and Epidemiology, University of Erlangen-Nuremberg, Erlangen, Germany

Wolfgang Ahrens Leibniz Institute for Prevention Research and Epidemiology - BIPS GmbH, Bremen, Germany

Oumaima Alaoui Ismaili Orange Labs, Lannion Cedex, France

AgroParisTech, Paris, France

Denis Amelynck Ghent University, Gent, Belgium

Daniel Baier Institute of Business Administration and Economics, BTU Cottbus-Senftenberg, Senftenberg, Germany

David Baker Goldsmiths, University of London, London, UK

Beata Bal-Domańska Department of Regional Economics, Wroclaw University of Economics, Jelenia Góra, Poland

Hans-Georg Bartel Department of Chemistry, Humboldt University Berlin, Berlin, Germany

Mateusz Baryła Cracow University of Economics, Cracow, Poland

Nadja Bauer Chair of Computational Statistics, Faculty of Statistics, TU Dortmund, Dortmund, Germany

Orlando Belo University of Minho, Guimaraes, Portugal

Wolfgang Bessler Center for Finance and Banking, Justus-Liebig University Giessen, Giessen, Germany

Dirk Biermann Institute of Machining Technology, TU Dortmund, Dortmund, Germany

Bernd Bischl Department of Statistics, TU Dortmund University, Dortmund, Germany

Geoffray Bonnin LORIA, Nancy, France

Baba B. Alhaji Department of Mathematical Sciences, University of Essex, Colchester, UK

Leon Otto Burkard Information Services and Electronic Markets, Karlsruhe Institute of Technology, Karlsruhe, Germany

Esther Chen Georg Eckert Institute for International Textbook Research - Member of the Leibniz Association, Braunschweig, Germany

Max-Planck-Institut für Wissenschaftsgeschichte - Max Planck Institute for the History of Science, Berlin, Germany

Ekaterina Chernyak National Research University Higher School of Economics, Moscow, Russia

Pedro Contreras Thinking Safe Ltd., Egham, Surrey, UK

Claudio Conversano Department of Business and Economics, University of Cagliari, Cagliari, Italy

Antoine Cornuéjols AgroParisTech, Paris, France

Tim Crawford Goldsmiths, University of London, London, UK

Mark d'Inverno Goldsmiths, University of London, London, UK

Hongsheng Dai Department of Mathematical Sciences, University of Essex, Colchester, UK

Jean Diatta EA2525-LIM, University of Reunion Island, Ile de La Reunion, Saint Denis, France

Florent Domenach Computer Science Department, University of Nicosia, Nicosia, Cyprus

Jessica Drechsler Georg Eckert Institute for International Textbook Research - Member of the Leibniz Association, Braunschweig, Germany

Laurence Dreyfus University of Oxford, Oxford, UK

Matthias Eckardt Institute of Computer Science, Humboldt-Universität zu Berlin, Berlin, Germany

Joana Fernandes Farfetch, Braga, Portugal

Manuel Ferreira Institute of Materials Engineering, TU Dortmund, Dortmund, Germany

Alexey Fofonov Jacobs University Bremen, Bremen, Germany

Tanja Friedrich GESIS, Cologne, Germany

Klaus Friedrichs Chair of Computational Statistics, Faculty of Statistics, TU Dortmund, Dortmund, Germany

Luca Frigau Department of Business and Economics, University of Cagliari, Cagliari, Italy

Andreas Geyer-Schulz Information Services and Electronic Markets, Karlsruhe Institute of Technology, Karlsruhe, Germany

Fernand Gobet Department of Psychological Sciences, University of Liverpool, Liverpool, UK

Tomasz Górecki Faculty of Mathematics and Computer Science, Adam Mickiewicz University, Poznań, Poland

Bernhard Gschrey Chair for Methods in Empirical Educational Research, TUM School of Education and Centre for International Student Assessment (ZIB), TU München, Munich, Germany

Asma Gul Department of Mathematical Sciences, University of Essex, Colchester, UK

Department of Statistics, Shaheed Benazir Bhutto Women University Peshawar, Khyber Pukhtoonkhwa, Pakistan

Andrew Harrison Department of Mathematical Sciences, University of Essex, Colchester, UK

Yoshiko Hayashi Department of Mathematical Sciences, University of Essex, Colchester, UK

Cornelia Hedeler School of Computer Science, The University of Manchester, Manchester, UK

Markus Hellmann Daimler AG, Stuttgart, Germany

Christian Hennig Department of Statistical Science, University College London, London, UK

Swetlana Herbrandt Department of Statistics, TU Dortmund, Dortmund, Germany

Diana Herrmann Leibniz Institute for Prevention Research and Epidemiology - BIPS GmbH, Bremen, Germany

Helmut Hildebrandt OptiMedis AG, Hamburg, Germany

Daniel Horn Department of Statistics, TU Dortmund University, Dortmund, Germany

Chieh-Liang Huang Department of Psychiatry, China Medical University Hospital, Taichung City, Taiwan, ROC

Timm Intemann Leibniz Institute for Prevention Research and Epidemiology - BIPS GmbH, Bremen, Germany

Alfonso Iodice D'Enza Department of Economics and Law, Università di Cassino e del Lazio Meridionale, Cassino FR, Italy

Dietmar Jannach Department of Computer Science, TU Dortmund, Dortmund, Germany

Klaus Jostschulte Daimler AG, Stuttgart, Germany

Daimler Research & Development, Ulm, Germany

Andreas Oskar Kempf GESIS - Leibniz Institute for the Social Sciences, Cologne, Germany

Hans A. Kestler Institute of Medical Systems Biology, Universität Ulm, Ulm, Germany

Zardad Khan Department of Mathematical Sciences, University of Essex, Colchester, UK

Sonja Köppl Daimler AG, Stuttgart, Germany

Daimler Research & Development, Ulm, Germany

Mirosław Krzyśko Faculty of Mathematics and Computer Science, Adam Mickiewicz University, Poznań, Poland

Peter C.R. Lane School of Computer Science, University of Hertfordshire, Hertfordshire, UK

Mariusz Łapczyński Cracow University of Economics, Kraków, Poland

Berthold Lausen Department of Mathematical Sciences, University of Essex, Colchester, UK

Ludwig Lausser Core Unit Medical Systems Biology and Institute of Neural Information Processing, Ulm University, Ulm, Germany

Leibniz Institute for Age Research–Fritz Lipmann Institute, Jena, Germany

Vincent Lemaire Orange Labs, Lannion Cedex, France

Marc Leman Ghent University, Gent, Belgium

Uwe Ligges Department of Statistics, TU Dortmund, Dortmund, Germany

Mohamed Limam ISG, University of Tunis, Tunis, Tunisia

Dhofar University, Salalah, Oman

Chien-Ju Lin Department of Statistical Science, University College London, London, UK

MRC Biostatistics Unit, Cambridge, UK

Lars Linsen Jacobs University Bremen, Bremen, Germany

Stephanie Löffler Institute of Business Administration and Economics, BTU Cottbus-Senftenberg, Senftenberg, Germany

Karsten Luebke FOM Hochschule für Oekonomie und Management, Dortmund, Germany

Pieter-Jan Maes Ghent University, Gent, Belgium

Osama Mahmoud Department of Mathematical Sciences, University of Essex, Colchester, UK

Department of Applied Statistics, Helwan University, Cairo, Egypt

Angelos Markos Department of Primary Education, Democritus University of Thrace, Xanthi, Greece

Małgorzata Markowska Wroclaw University of Economics, Wroclaw, Poland

Jean-Pierre Martens Ghent University, Gent, Belgium

Markus Maucher Core Unit Medical Systems Biology, Ulm University, Ulm, Germany

Dhouha Mejri Technische Universität Dortmund, Dortmund, Germany

ISG, University of Tunis, Tunis, Tunisia

Metodi V. Metodiev School of Biological Sciences/Proteomics Unit, University of Essex, Colchester, UK

Miftahuddin Miftahuddin Department of Mathematical Sciences, University of Essex, Colchester, UK

Francesco Mola Department of Business and Economics, University of Cagliari, Cagliari, Italy

Gabriel Mota University of Minho, Guimaraes, Portugal

Hans-Joachim Mucha Weierstrass Institute for Applied Analysis and Stochastics (WIAS), Berlin, Germany

Daniel Müllensiefen Goldsmiths, University of London, London, UK

Fionn Murtagh Department of Computing, Goldsmiths University of London, London, UK

De Montfort University, Leicester, UK

Department of Computing and Mathematics, University of Derby, Derby, UK

Christoph Müssel Core Unit Medical Systems Biology, Ulm University, Ulm, Germany

Atsuho Nakayama Tokyo Metropolitan University, Tokyo, Japan

Barbara Pawełek Cracow University of Economics, Cracow, Poland

Marcin Pełka Department of Econometrics and Computer Science, Faculty of Economics, Management and Tourism, Wrocław University of Economics, Jelenia Góra, Poland

Aris Perperoglou Department of Mathematical Sciences, University of Essex, Colchester, UK

Radoslaw Pietrzyk Wroclaw University of Economics, Wroclaw, Poland

Iris Pigeot on behalf of the IDEFICS consortium Leibniz Institute for Prevention Research and Epidemiology - BIPS GmbH, Bremen, Germany

Alexander Pimperl OptiMedis AG, Hamburg, Germany

Cynthia Pitou EA2525-LIM, University of Reunion Island, Ile de La Reunion, Saint Denis, France

Józef Pociecha Cracow University of Economics, Cracow, Poland

Hermann Pohlabeln Leibniz Institute for Prevention Research and Epidemiology - BIPS GmbH, Bremen, Germany

George Portides Computer Science Department, University of Nicosia, Nicosia, Cyprus

Bianca Pramann Georg Eckert Institute for International Textbook Research - Member of the Leibniz Association, Braunschweig, Germany

Monia Ranalli Department of Statistics, The Pennsylvania State University, University Park, State College, PA, USA

Sapienza, University of Rome, Rome, Italy

Christian Rautert Institute of Machining Technology, TU Dortmund, Dortmund, Germany

Robert Retz Institute of Neural Information Processing, Ulm University, Ulm, Germany

Christophe Rhodes Goldsmiths, University of London, London, UK

Roberto Rocci IGF Department, University of Tor Vergata, Rome, Italy

Joachim Rojahn FOM Hochschule für Oekonomie und Management, Essen, Germany

Pawel Rokita Wroclaw University of Economics, Wroclaw, Poland

Adam Sagan Cracow University of Economics, Cracow, Poland

Lyn-Rouven Schirra Core Unit Medical Systems Biology and Institute of Neural Information Processing, Ulm University, Ulm, Germany

Institute of Number Theory and Probability Theory, Ulm University, Ulm, Germany

Colin Schneck Center for Finance and Banking, Justus-Liebig University Giessen, Giessen, Germany

Uwe Schöning Institute of Theoretical Computer Science, Ulm University, Ulm, Germany

Martin Schrepp SAP AG, Walldorf, Germany

Timo Schulte OptiMedis AG, Hamburg, Germany

Friedhelm Schwenker Institute of Neural Information Processing, Ulm University, Ulm, Germany

Pascal Siegers GESIS, Cologne, Germany

Elżbieta Sobczak Department of Regional Economics, Wrocław University of Economics, Jelenia Góra, Poland

Lukáš Sobíšek Faculty of Informatics and Statistics, University of Economics, Prague, Czech Republic

Andrzej Sokołowski Cracow University of Economics, Cracow, Poland

Peter D. Sozou Centre for Philosophy of Natural and Social Science, London School of Economics and Political Science, London, UK

Department of Psychological Sciences, University of Liverpool, Liverpool, UK

Mária Stachová Faculty of Economics, Matej Bel University, Banská Bystrica, Slovakia

Jan Steinberg GESIS - Leibniz Institute for the Social Sciences, Cologne, Germany

Danuta Strahl Wroclaw University of Economics, Wroclaw, Poland

Robert Strötgen Georg Eckert Institute for International Textbook Research, Braunschweig, Germany

Stiftung Wissenschaft und Politik - German Institute for International and Security Affairs, Berlin, Germany

Piotr Tarka Department of Marketing Research, Poznan University of Economics, Poznan, Poland

Wolfgang Tillmann Institute of Materials Engineering, TU Dortmund, Dortmund, Germany

Ali Ünlü Chair for Methods in Empirical Educational Research, TUM School of Education and Centre for International Student Assessment (ZIB), TU München, Munich, Germany

Igor Vatolkin Department of Computer Science, TU Dortmund, Dortmund, Germany

Veronica Vinciotti Department of Mathematical Sciences, University of Essex, Colchester, UK

Gunnar Völkel Institute of Theoretical Computer Science and Core Unit Medical Systems Biology, Ulm University, Ulm, Germany

Claus Weihs Chair of Computational Statistics, Faculty of Statistics, TU Dortmund, Dortmund, Germany

Waldemar Wołyński Faculty of Mathematics and Computer Science, Adam Mickiewicz University, Poznań, Poland

Christian Wöhler TU Dortmund, Dortmund Germany

Maxim Yakovlev National Research University Higher School of Economics, Moscow, Russia

Sri Utami Zuliana Department of Mathematical Sciences, University of Essex, Colchester, UK

Part I
Invited Papers

Latent Variables and Marketing Theory: The Paradigm Shift

Adam Sagan

Abstract An extensive discussion concerning formal, empirical, and ontological status of latent variables in psychological literature concerns the distinction between the realist and anti-realist positions within the classical test theory and item response theory (IRT) psychometric traditions in measurement of latent variables (Measurement 6:25–53, 2008; Salzberger and Koller, J Bus Res 66:1307–1317, 2013). However, this bi-polar view seems to be too distant from the perspectives of schools of thought in the marketing discipline and actual developments of measurement models in specific fields of marketing research. An extensive discussion concerning the reflective–formative latent variables dilemma and relational status of constructs in the contemporary marketing opens space for the redefinition of the nature and role of latent variables in marketing science. The aim of the paper is to outline the interlink between theoretical schools within marketing discipline and contemporary discussion concerning the nature and use of latent variables in marketing.

1 Introduction

Latent variables play an important role in testing substantive theories in various fields of social sciences including psychology, sociology, and marketing.

As many authors stress, in a formal sense, there is nothing special about latent variables (Bartholomew et al. 2011) but usually, they form theory-free measurement models in which only the structural part is strongly embedded in theoretical assumptions. Therefore, the measurement models in marketing and social science are less theory oriented and the theoretical framework of measurement part receives usually less attention than the structural part of SEM model.

However, the measurement model and latent variable conceptualization are also connected to theoretical assumptions and stances. This issue in model-building process is often neglected, taking the assumptions that latent variable definition and

A. Sagan (✉)
Cracow University of Economics, Rakowicka 27, 31-510 Cracow, Poland
e-mail: sagana@uek.krakow.pl

© Springer International Publishing Switzerland 2016
A.F.X. Wilhelm, H.A. Kestler (eds.), *Analysis of Large and Complex Data*, Studies in Classification, Data Analysis, and Knowledge Organization,
DOI 10.1007/978-3-319-25226-1_1

operationalization is a theory-free process based solely on statistical and numerical assumptions.

Theoretical assumptions of the latent variables and measurement models are based not only on so called small-m-methodology (assumptions of statistical model, sampling size, identification rules, etc.), but also on big-M-methodology connected with the existing paradigms, meta-theoretical assumptions and schools of thought in a given discipline (McCloskey 1983).

This problem is important in managerial marketing, where latent variables are regarded almost exclusively as a tool-box with the avoidance of discussion about theoretical assumptions and schools of thought in marketing in the context of big-M-methodology.

2   Marketing Constructs and Latent Variables Models

2.1   The Evolution of Marketing Theory

One of the earliest definitions of marketing says that "Marketing is buying and selling activities." This definition was published in *Miss Parloa's New Cookbook and Marketing Guide* around 1880 (Shaw and Jones 2005). The domain of marketing discipline, its constructs, theories, and methodologies is presented in Hunt's three dichotomies model (1976) who tried to summarize the major aspects of the marketing field of research. He depicted three basic criteria that define marketing as a scientific discipline: (1) research objects (profit and non-profit organizations), (2) the level of analysis (micro–macromarketing perspectives), and (3) research objectives (positive and normative marketing). Within this framework, the dominant schools of thought in marketing can be located. Three-dichotomy model helps us to understand the self-identity of marketing and provides a broader context of marketing discipline.

In the evolution of marketing discipline several schools and research paradigms have emerged (Jones et al. 2010). Figure 1 presents the dominant schools of thought within the marketing discipline.

The variety of approaches can be summarized in three dominant views (paradigms).

1. The cognitive view, based on the realist paradigm that underlines the problems of causality, functional relations between marketing concepts and, referring to Hunt, takes macro and positive perspective on marketing. Historically, the cognitive view is close to famous Bartels' question in the origin era of scientific marketing—"can marketing be a science?" (Bartels 1951).

   The cognitive view encompassed many schools of thought like distributional, macro-marketing and social exchange, system school, and information processing theory (IPT) in the field of consumer research.

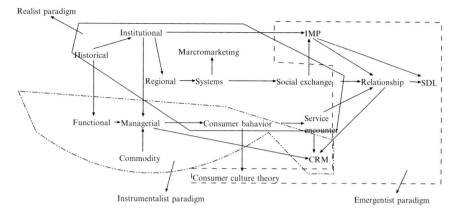

Fig. 1 Marketing schools and research paradigms

2. The behavioral view, which is rooted in a predictive-instrumental approach and "toolbox" analogy of marketing. This approach, mostly popularized and developed by P. Kotler and J. McCarthy textbooks, is more micro and normative oriented. The main schools of thought are managerial, functional, CRM and behavioral perspective models (BPM) in the consumer theory. This view of marketing is labelled also as "mainstream" or "transactional."
3. The relationship marketing school opens the perspective for contextual, service- and process-oriented view on marketing as a business domain integrator (Vargo and Lusch 2006). Within this perspective several schools are present i.e. the neo-institutional, Nordic school of relationship marketing, Interactive Marketing and Purchasing (IMP) group and Service-Dominant Logic approach.

2.2 Domains of Marketing Constructs

The variety of marketing schools outlined above implies many different method-ological orientations, ways of construct definition, and approaches to model build-ing. To simplify this view we can distinguish between two dimensions: (1) the type of causal explanation (prediction–postdiction) and (2) the nature of marketing constructs (descriptive–relational). The first dimension represents the way of causal explanation. In predictive, experimental, and nomothetic approach the objective is to predict future (new) phenomena based on the present (existing) data. In postdictive, historical and idiographic approach, one can explain the past and originated causes on the basis of present data. The second dimension deals with the nature of marketing variables. Descriptive variables are subject-oriented and are used to describe the "positional" characteristics of subjects under study, like socio-demographic, attitudes, preferences, and values. Relational variables in marketing describe the results of interactions, joint actions of the partners (mutual trust,

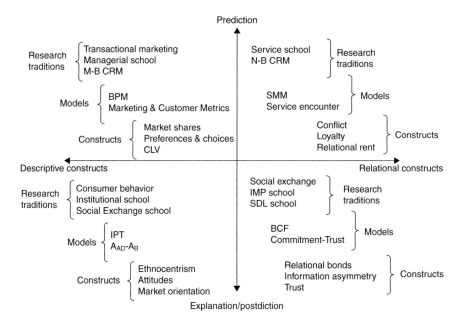

Fig. 2 Marketing constructs and research domain

information asymmetry, loyalty, etc.). In the classification depicted in Fig. 2, the four basic methodological stances of marketing models can be identified.

The first (upper left) represents the predictive models based on descriptive marketing constructs. This is the mainstream of managerial marketing, and market-based relationship management (M-B RM). The BPM, consumer metrics indices are defined to measure variables like market shares, preferences and choices, consumer lifetime values (CLV), etc.

The second (lower left) is dominant in marketing explanatory models of consumer behavior, social exchange, and institutional schools. Models of attitudes toward the ad and brand ($A_{AD} - A_B$), IPT-based models involve marketing constructs like values, attitudes, ethnocentrism, market orientation, etc.

The third (upper right) involves relational variables and predictive models in marketing. The services school, network-based relationship marketing (N-BRM) obey many specific models in the area of social media marketing (SMM), the service encounter approaches that try to predict relationship variables like conflict level, loyalty, or relational rent.

Fourth quadrant (lower right) represents explanatory models with relational variables. Social exchange school, IMP Group, and Service Dominant Logic schools develop the business cluster formation (BCF) models or commitment-trust theories to model the formal, social and structural bonds, information asymmetry, relational norms, or mutual trust of business partners.

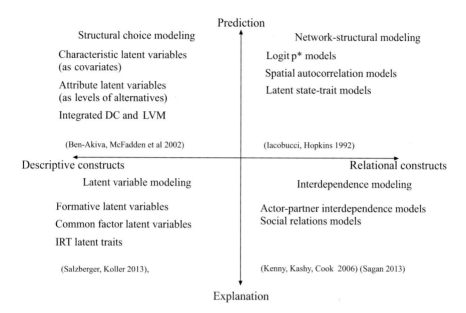

Fig. 3 Latent variables and research domains

All of the constructs above involve the concept of latent variables for model building within the substantial theory. The lack of understanding of the methodological assumptions underlying theoretical concepts may lead to misuse of measurement models of latent variables within particular marketing paradigms .

Figure 3 shows four areas of the application of latent variable models in marketing domains in fourfold classification from Fig. 2. The first quadrant (A: descriptive–predictive) represents the structural choice modeling with latent response variables (with probit or logit parameterization) in discrete choice models (DC) and integrated latent variable and choice modeling.

The second quadrant (B: descriptive–explanatory) shows construct-oriented latent variable models with reflective [CFA and item response theory (IRT)] indicators.

The third quadrant (C: relational–predictive) includes network-structural models with latent variables and latent state-trait models of consumer satisfaction and service encounter variables.

The fourth quadrant (D: relational–explanatory) stresses the importance of explanatory models of interdependences in buyer–seller interactions, formation of ego networks, and network competencies in business clusters.

3 Latent Variables in Marketing

3.1 Three Types of Latent Variables

Application of latent variables in marketing can be illustrated by two contradictory quotations. On one hand "SEM rarely receives any attention in marketing models textbooks" (Steenkamp and Baumgartner 2000). On the other, "It is probably difficult to find an issue of a major marketing journal in which SEM is not used in at least one of the articles" (Steenkamp and Baumgartner 2000). This contradictory view refers to philosophies of use of latent variables in marketing modeling.

Operational definition philosophy assumes one-to-one correspondence between a theoretical construct and its measurement. Manifest variables seem to be perfect measures of the underlying marketing constructs without measurement error and no errors-in-variables.

Partial interpretation philosophy requires multiple operationalizations of the underlying construct by individually imperfect but collectively reliable and valid measures. The observed variables are regarded as the imperfect measures of the underlying constructs with measurement errors. Latent variables that are variables when the inference from data structure to variable structure is prone to error (Borsboom 2008) allows for controlling the bias in model parameters and their standard errors (type II error) with the presence of masking effect of R^2 stability in both error-free and error-bound models.

As Bollen (2002) stresses, it is impossible to date the first use of latent variables that capture a wide variety of statistical concepts, including random effects, missing data, sources of variation in hierarchical data, finite mixtures, latent classes, and clusters (Muthén 2003; Bentler and Huang 2014). Formal definitions of latent variables involve true score (expected value) and common cause interpretation, and are defined as "random variables with no sample realizations, and about which we make distributional assumptions" (Bollen 2002), "variables in measurement model where the number of independent variables is greater than manifest one," "variables in the equations if the equations cannot be manipulated so as to express these variables as a function of manifest variables only" (Bentler 1982; Bollen 2002), "random variables whose realized values are hidden" (Skrondal and Rabe-Hesketh 2007). The diversity of measurement models for latent variables in marketing theory involves at least three approaches to use of latent variables in marketing research:

1. Common factor latent variables. Common factor (with parallel Likert-type items) and IRT (with cumulative Guttman-type items) models are rooted in behavior domain theory (BDT) together with the realist causal theory of measurement (CTM). In BDT, constructs are conceptualized in terms of domains of behavior, and item responses are considered as samples from this domain. Conceptualization of latent variables uses the common cause rule of causality and local independence assumption of indicators from the latent variable.

In CTM, the constructs refer to common causes that underlie a set of item responses, so that people respond to items differently because they have a different construct score (Borsboom 2008; Borsboom et al. 2004).

2. Composites. The latent construct depends on a constructivist, operationalist, or instrumentalist interpretation of latent variables (Borsboom et al. 2003). The operational definition treats composites as weighted linear combination of its indicators (Diamantopoulos et al. 2008; Williams et al. 2003). In marketing, the models with composites are popular in instrumentalist paradigm and for prediction rather than explanation of marketing phenomena (CRM, BPM, service encounter, SMM, etc.).

3. Formative latent variables. They differ from the composites having disturbance term reflected unexplained residuals in measurement model. In composites it is assumed that linear combination of indicators is a perfect measure of the latent variable without an error.

However, the use of formative latent variables have both strong opponents and proponents in marketing and statistical literature. Bollen and Lennox argue that in isolation, the formative measurement model is statistically underidentified (Bollen and Lennox 1991) and can only be estimated if it is placed within a larger model that incorporates consequences of the latent variable in question (Bollen 1989; Cadogan and Lee 2013). A necessary condition for identifying the disturbance term is that the latent variable emits at least two paths to other constructs measured with reflective indicators (2+ rule of identification) (MacCallum and Browne 1993).

With respect to dependent formative latent variables, Rigdon (2014) underlines that any distinction between a "measurement model" and a "structural model" is only a mental convenience, and not a mathematical reality because of necessary misspecification of the structural or measurement part of the whole model and the impossibility of co-existence of the formative measurement model and the structural one (if the structural part of the model is correct, then the measurement model is misspecified and vice versa Rigdon 2014).

The intermediate solution that combines two approaches above is the MIMIC model which consists of both reflective and formative indicators. Causality effect between latent variables and indicators is linked conceptually to Markov causal condition, instead of the common cause and local independence rule mentioned above.

4. Relational latent variables. They support the view on marketing constructs as emerging from the interactions or relationships between partners (mutual trust, loyalty, etc.), whose roles are often interchangeable (i.e. consumers and producers as value co-creators). This view broadens the psychometric and operational tradition in marketing and takes into account the interactive nature of marketing variables and the Service Dominant Logic idea of value co-creation.

Conceptually, these variables are close to common factors that represent constructs of both parties that form the dyadic or network structure as a unit of analysis with a dependency between subjects and correlated errors. Those kinds of models are in the family of actor–partner interdependence models (APIM). Also, the network structures and network indices (i.e. centrality, betweenness, in-degree, or out-degree) may be explained by the latent traits of both actors in logit p^* models or network-structural models with latent variables.

Contextual and situation specific marketing constructs are measured by the estimation of latent variable models with fixed and random effects like latent state-trait models (LS-T), multilevel actor–partner interdependence (APIM) with random loadings.

3.2 Reflective: Formative Dichotomy in Comparative Analysis

The direction of causal flow between a construct and its indicators evokes the problem of separation between these two entities, and raises the questions whether the measurement items are separate entities from the latent variable. In the realist perspective and reflective measurement, the indicators y (dependent variables) vary as a consequence of the variation in the latent variable ξ (independent variable) and their independence is represented by error terms. Causality effect between latent variables and indicators is linked to the common cause and the local independence rules.

In the constructivist approach and formative measurement, the latent variable η (dependent variable) varies as a consequence of the variation in its indicators x and their independence is represented by the disturbance term. In case of FLV the causality effect between indicators and latent variables is not directly measured or controlled.

To summarize the broad discussion concerning the conceptualization of latent variables in marketing as reflective or formative, many authors (Borsboom et al. 2003; Ping 2004; Wilcox et al. 2008) question the status of formatively measured constructs as latent variables in marketing. From practical point of view, reflective model is sometimes used for masking the problem with collinearity of formative indicators and formative one is adopted because of the problems with reliability of reflective items.

On the other hand, Diamantopoulos (2008) argues that both reflective and formative measurement models have a place in research, and neither formative nor reflective measurement is inherently wrong or right.

In order to illustrate the reflective–formative dichotomy we use the database satisfaction in *plspm* library of R package (Sanchez et al. 2008). The sample consists of 250 clients of a credit institution in Spain. The 10-point Likert scaled items were used to measure several customer satisfaction constructs like image (IMA—reputation, trustworthiness, seriousness, and caring about customer needs), customer expectation (EXP—products and services provided and expectations for the overall quality), (QUA—reliable products and services, range of products and services, and overall perceived quality), benefits (VAL—beneficial services and products, valuable investments, quality relative to price, and price relative to quality), overall satisfaction (SAT—overall rating of satisfaction, fulfillment of expectations, satisfaction relative to other banks, and performance relative to customer's ideal bank), and loyalty (LOY—propensity to choose the same bank again, propensity to switch to another bank, intention to recommend the bank to friends, and a sense of loyalty). Taking into account the semantic relations between constructs and indicators, most of items seem to be formative rather than reflective.

In SEM model (maximum likelihood estimation), the reflective indicators were used. PLS-PM model included formative indicators (model B) and internal weights estimated by path weighting. In both cases the 1000 bootstrap samples were used during the estimation of loadings and path parameters. Table 1 depicts the loadings for measurement (outer) models of SEM and PLS-PM. In measurement parts of the

Table 1 SEM and PLS-PM outer models

Latent variables	Manifest variables	SEM estimates	Confidence interval	PLS-PM estimates	Confidence interval
IMA	I1	1.16	0.84–1.41	0.57	0.39–0.74
	I2	1.67	1.44–1.93	0.84	0.73–0.92
	I3	1.52	1.27–1.77	0.87	0.77–0.93
	I4	0.81	0.59–1.05	0.50	0.29–0.68
	I5	1.30	1.00–1.67	0.80	0.69–0.89
LOY	L1	1.33	1.03–1.63	0.93	0.83–0.98
	L2	0.71	0.46–0.96	0.58	0.37–0.73
	L3	1.34	1.03–1.64	0.92	0.84–0.97
	L4	0.78	0.46–1.10	0.50	0.29–0.72
SAT	S1	0.47	0.20–0.74	0.96	0.92–0.98
	S2	0.24	0.20–0.71	0.92	0.86–0.96
	S3	0.36	0.14–0.58	0.74	0.60–0.82
	S4	0.36	0.14–0.59	0.79	0.67–0.88
VAL	V1	0.71	0.39–1.04	0.91	0.86–0.95
	V2	0.52	0.24–0.80	0.77	0.62–0.88
	V3	0.52	0.25–0.79	0.66	0.51–0.77
	V4	0.68	0.37–0.98	0.85	0.77–0.91
QUA	Q1	0.19	0.00–0.38	0.78	0.63–0.88
	Q2	0.24	0.01–0.47	0.90	0.84–0.94
	Q3	0.21	0.00–0.41	0.74	0.61–0.84
	Q4	0.17	0.00–0.36	0.81	0.70–0.90
	Q5	0.21	0.00–0.41	0.81	0.72–0.88
EXP	E1	0.99	0.73–1.24	0.71	0.53–0.83
	E2	1.28	1.05–1.50	0.88	0.77–0.93
	E3	1.08	0.84–1.32	0.70	0.54–0.80
	E4	0.86	0.65–1.08	0.78	0.65–0.87
	E5	1.27	1.03–1.50	0.80	0.70–0.87

model, the SEM parameters, have in comparison to PLS-PM, broader bootstrapped confidence intervals and higher point estimates of the parameters (see also Ringle et al. 2012). We can see that formative PLS-PM indicators have a higher precision of estimates, but loadings are attenuated due to unknown reliability of the latent variables (as composites). On the other hand, the SEM latent variables account for unreliability of reflective indicators and the factor loadings have a higher but partly insignificant values, indicating smaller type I error probability.

Table 2 presents the path coefficients for SEM and PLS-PM models and R^2 for dependent latent variables.

Similarly to the outer model, in the inner model the precision of the estimates (sampling variance) is lower for PLS-PM. Also, some of the path coefficients are insignificant in SEM model. However, the R^2 determination coefficients are generally higher in SEM than PLS-PM which indicates a greater explanatory power.

This comparative analysis shows the bias and sampling variance trade-off effects between these two models. PLS-PM are used in the instrumentalist paradigm (functional school, managerial marketing, and CRM) where latent variables represent the effect of data reduction and are nothing more than the empirical content of its indicators. Predictive power of such models benefits with a smaller sampling variance of estimates and a greater precision of the measurement of parameters' values.

SEM models are used within the realist paradigm in marketing (institutional, macromarketing, system schools) and are characterized by the higher explanatory

Table 2 SEM and PLS-PM inner models

Parameter	Paths and variables	SEM estimates	Confidence interval	PLS-PM estimates	Confidence interval
Paths	IMA → EXP	0.90	0.63–1.13	0.60	0.51–0.70
	IMA → LOY	0.35	0.11–0.60	0.23	0.10–0.39
	IMA → SAT	0.65	0.15–2.50	0.20	0.10–0.32
	EXP → QUA	5.88	2.09–7.67	0.85	0.81–0.89
	EXP → VAL	−3.22	−6.87–2.00	0.15	0.02–0.30
	EXP → SAT	−2.73	−12.37–7.18	0.00	−0.12–0.13
	QUA → VAL	0.82	−0.09–1.92	0.65	0.49–0.78
	QUA → SAT	0.38	−3.04–2.54	0.08	−0.08–0.26
	VAL → SAT	1.47	0.70–6.44	0.62	0.47–0.76
	SAT → LOY	0.28	0.09–0.46	0.56	0.42–0.69
R^2	EXP	0.45		0.37	
	QUA	0.98		0.72	
	VAL	0.84		0.61	
	SAT	0.93		0.73	
	LOY	0.65		0.55	
GFI/GoF		0.756		0.609	

power, smaller bias of estimates, and account for unreliability of the indicators (correction for attenuation of the parameters).

4 Measurement Models Life Cycles

The diversity of the use of latent variables in marketing makes it almost impossible to outline the evolution of latent variable concepts in the field of marketing. However, one can identify at least five basic steps in the development of latent variable models in marketing research.

The first stage (1960–1980) is based on classic psychometric approach (classical test theory) with the exploratory factor analysis (or PCA) as a basic model for estimation of latent variables. The validity and reliability were assessed by MTMM approach and Cronbach's α. The second stage (1980–1995) is a remarkable growth of CFA and classic SEM models and model-based methods of validity (correlated-traits correlated methods or correlated-traits correlated uniqueness) and reliability analysis (Jöreskog and McDonald's rho). The third era (1995-) represents the popularity of FLV and PLS-PM models in marketing and consumer behavior. Fourth stage (2000-) opens the growing importance of IRT and Rasch (Birnbaum) scaling in marketing. And the last developments (2005-) involve SEM/PLS models in the situation of population heterogeneity (multilevel CFA, models with random loadings, FIMIX-PLS, and REBUS-PLS).

Figure 4 presents the popularity of particular approaches in a selected marketing journals over the last 10 years. Undoubtedly, the major approach in marketing literature is the confirmatory factor analysis with reflective items (RCFA), which is the leading method used in marketing journals [Journal of Marketing (JM), Journal of Marketing Research (JMR), International Journal of Research in Marketing (IJRM), and Journal of Academy of Marketing Science (JAMS)].

However, many journals promote also partial least squares approach with formative indicators (FPLS). Marketing Information Systems Quarterly (MISQ), Academy of Marketing Science Review (AMSR), Journal of Marketing Targeting and Positioning (JMTP), Journal of Business Research (JBR), and Long Run Planning (LRP) among others, provide the platform for interesting (and often emotional) discussions concerning latent variables with formative indicators.

The IRT and Rasch modeling have also growing importance in marketing literature (Salzberger 2009; Salzberger and Koller 2013). JBR and JAMS are the representative journals for this approach.

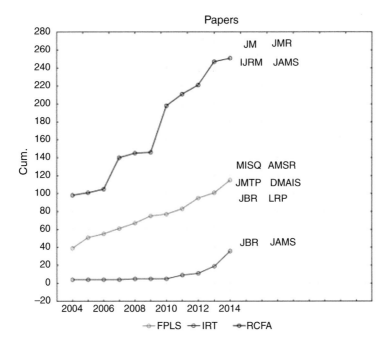

Fig. 4 Latent variables in selected marketing journals

5 Summary

Operational definitions of latent variables are strongly related to the underlying marketing paradigms. In last two decades the changing of marketing paradigms (with respect to Service Dominant Logic) has opened space for the analysis of state-based and dyadic/network relations among marketing actors.

Hence, the reference for the use of latent variables is rather a "descriptive–relational" dimension than "reflective–formative" for the measurement of the constructs within marketing theories. The use of latent variables in marketing is more oriented toward the measurement of interaction and shared values of dyadic/network entities as a unit of analysis. The growing acceptance of PLS-PM and IRT approaches in marketing discipline confirms the (Bollen 2002) claim that "there is no right or wrong definition of latent variables. It is more a question of finding the definition that is most useful and that corresponds to a common understanding of what should be considered latent variables."

Understanding the statuses of the latent variables and schools of thought in marketing may help to improve both the use of latent variables within discipline and the effectiveness of communication between scholars.

References

Bartels, R. (1951). Can marketing be a science? *Journal of Marketing, 15*, 319–328.

Bartholomew, D. J., Knott, M., & Moustaki, I. (2011). *Latent variable models and factor analysis: A unified approach*. Chichester: Wiley.

Ben-Akiva, M., Mcfadden, D., Train, K., Walker, J., Bhat, C., Bierlaire, M., et al. (2002). Hybrid choice models: Progress and challenges. *Marketing Letters, 13*(3), 163–175.

Bentler, P. M. (1982). Linear systems with multiple levels and types of latent variables. In K. G. Jöreskog & H. Wold (Eds.), *Systems under indirect observation: Causality, structure, prediction* (Part I, pp. 101–130). Amsterdam: North-Holland.

Bentler, P. M., & Huang, W. (2014). On components, latent variables, PLS and simple methods: Reactions to Rigdon's rethinking of PLS. *Long Range Planning, 47*, 138–145.

Bollen, K. (1989). *Structural equations with latent variables*. New York: Wiley.

Bollen, K. (2002). Latent variables in psychology and the social sciences. *Annual Review of Psychology, 53*, 605–634.

Bollen, K., & Lennox, R. (1991). Conventional wisdom on measurement: A structural equation perspective. *Psychological Bulletin, 100*, 305–314.

Borsboom, D. (2008). Latent variable theory. *Measurement, 6*, 25–53.

Borsboom, D., Mellenbergh, G. J., & Van Heerden, J. (2003). The theoretical status of latent variables. *Psycholological Review, 110*, 203–219.

Borsboom, D., Mellenbergh, G. J., & Van Heerden, J. (2004). The concept of validity. *Psychological Review, 111*(4), 1061–1071.

Cadogan, J. W., & Lee, N. (2013). Improper use of endogenous formative variables. *Journal of Business Research, 66*, 233–241.

Diamantopoulos, A. (2008). Formative indicators: Introduction to the special issue. *Journal of Business Research, 61*, 1201–1202.

Diamantopoulos, A., Riefler, P., & Roth, K. P. (2008). Advancing formative measurement models. *Journal of Business Research, 61*, 1203–1218.

Hunt, S. (1976). The nature and scope of marketing. *Journal of Marketing, 40*, 17–28.

Iacobucci, D., & Hopkins, N. (1992). Modeling dyadic interactions and networks in marketing. *Journal of Marketing Research, 29*, 5–17.

Jones, D. G. B., Shaw, E. H., & Mclean P. A. (2010). The modern schools of marketing thought. In P. Maclaran, M. Saren, B. Stern, & M. Tadajewski (Eds.), *The SAGE handbook of marketing theory* (pp. 42–58). London: SAGE.

Kenny, D. A., Kashy, D. A., & Cook, W. L. (2006). *Dyadic data analysis*. New York: Guilford Press.

MacCallum, R., & Browne, M. (1993). The use of causal indicators in covariance structure models: Some practical issues. *Psychological Bulletin, 114*(3), 533–541.

Mccloskey, D. (1983). The rhetoric of economics. *Journal of Economic Literature, 21*, 481–517.

Muthén, B. (2003). Beyond SEM: General latent variable modeling. *Behaviormetrika, 29*, 81–117.

Ping, R. A., Jr. (2004). On assuring valid measurement for theoretical models using survey data. *Journal of Business Research, 57*, 125–141.

Rigdon, E. (2014). Comment on "improper use of endogenous formative variables". *Journal of Business Research, 67*, 2800–2802.

Ringle, C. M., Sarstedt, M., & Strabub, D. W. (2012). A critical look at the use of PLS-SEM in MIS quarterly. *MIS Quarterly, 36*(1), 3–14.

Sagan, A. (2013). Market research and preference data. In M. A. Scott, J. S. Simonoff, & B. D. Marx (Eds.), *The SAGE handbook of multilevel modeling* (pp. 581–598). Los Angeles/London/Singapore: Sage.

Salzberger, T. (2009). *Measurement in marketing research: An alternative framework*. Cheltenham: Edward Elgar.

Salzberger, T., & Koller, M. (2013). Towards a new paradigm of measurement in marketing. *Journal of Business Research, 66*, 1307–1317.

Sanchez, G., Trinchera, L., & Russolillo, G. (2008). Tools for partial least squares path modeling (PLS-PM). Accessed September 21, 2014, http://cran.r-project.org/web/packages/plspm/plspm.pdf

Shaw, E. H., & Jones, D. G. B. (2005). A history of schools of marketing thought. *Marketing Theory, 3*(3), 239–281.

Skrondal, A., & Rabe-Hesketh, S. (2007). Latent variable modelling: A survey. *Scandinavian Journal of Statistics, 34*, 712–745.

Steenkamp, J.-B., & Baumgartner, H. (2000). On the use of structural equation models for marketing modeling. *International Journal of Research in Marketing, 17*, 195–202.

Wilcox, J. B., Howell, R. D., & Breivik, E. (2008). Questions about formative measurement. *Journal of Business Research, 61*, 1219–1228.

Williams, L. J., Edwards, J. R., & Vanderberg, R. J. (2003). Recent advances in causal modeling methods for organizational and management. *Journal of Management, 29*(6), 903–936.

Vargo, S. L., & Lusch, R. F. (2006). Evolving to a new dominant logic for marketing. In R. F. Lusch, & S. L. Vargo (Eds.), *The service dominant logic of marketing. Dialog, debate and directions*. New York: Sharpe.

Business Intelligence in the Context of Integrated Care Systems (ICS): Experiences from the ICS "Gesundes Kinzigtal" in Germany

Alexander Pimperl, Timo Schulte, and Helmut Hildebrandt

Abstract Patients generate various data with every contact to the health care system. In integrated care systems (ICS) these fragmented patient data sets of the various health care players can be connected. Business intelligence (BI) technologies are seen as valuable tools to gain insights and value from these huge volumes of data. However so far there are just sparse experiences about BI used in the integrated care (IC) context. Therefore the aim of this article is to describe how a BI solution can be implemented practically in an ICS and what challenges have to be met. By the example of a BI best practice model—the ICS Gesundes Kinzigtal—it will be shown that data from various data sources can be linked in a Data Warehouse, prepared, enriched and used for management support via a BI front-end: starting with the project preparation and development via the ongoing project management up to a final evaluation. Benefits for patients, care providers, the ICS management company and health insurers will be characterised as well as the most crucial lessons learned specified.

1 Background

Integrated Care (IC)—the networking of different healthcare providers—has had an increased presence on the German political agenda in the last decade (SVR 2012). Also internationally integrated care initiatives get political attention (Stein et al. 2013; Sun et al. 2014). These efforts can be seen as part of the governmental attempt to address a variety of problems observed in their healthcare systems resulting from institutional fragmentation. These problems include gaps in information exchange, difficulties in coordination between different levels of care, lack of coordination between sectors, poor financial incentives for unquantifiable service expansions as

A. Pimperl (✉)
OptiMedis AG, Borsteler Chaussee 53, 22453 Hamburg, Germany
e-mail: a.pimperl@optimedis.de

T. Schulte • H. Hildebrandt
OptiMedis AG, Hamburg, Germany
e-mail: t.schulte@optimedis.de; h.hildebrandt@optimedis.de

© Springer International Publishing Switzerland 2016
A.F.X. Wilhelm, H.A. Kestler (eds.), *Analysis of Large and Complex Data*, Studies in Classification, Data Analysis, and Knowledge Organization,
DOI 10.1007/978-3-319-25226-1_2

17

well as missing common goals and values (Amelung et al. 2012). These problems are associated with unnecessary risks for patients, such as: double examinations with X-rays (Schonfeld et al. 2011), untuned and inappropriate medication resulting in avoidable hospital admissions and deaths (Schrappe 2005). It has to be assumed that these problems will be further enhanced in the future with the high and ever-increasing number of patients with chronic and psychological conditions (Amelung et al. 2008).

Effective Information and Communication Technology (ICT) is seen as a vital tool to reduce this burden (Hammersley et al. 2006). ICT ensures in IC that the correct information is exchanged and understood, while at the same time laying the foundation for stable cooperation and changes (Janus and Amelung 2005).

However, the establishment of an ICT structure alone may likely be insufficient. Data which could be generated by such an ICT structure must in addition be processed in a way that it can trigger a continuous learning and improvement process (Kupersmith et al. 2007; Vijayaraghavan 2011). "Business Intelligence" (BI) is often used as generic term for this: "Business Intelligence is the process of transforming data into information and, through discovery, into knowledge" (Behme 1996).

In Germany most IC approaches are in practice still a long way from BI solutions. A study of practice networks, showed, for instance, that only a very few networks have a common IT architecture or strategy, let alone a BI system (Purucker et al. 2009). One of the leading networks is Gesundes Kinzigtal (GK). The OptiMedis AG, the shareholder and management partner of the Gesundes Kinzigtal (GK) GmbH, was awarded with the "BARC Best Practice Award for Business Intelligence and Data Management 2013," for the BI system it runs for GK. The award was granted by a jury of experts and 300 visitors from the BARC Business Intelligence Congress in 2013 (BARC 2013).

The regional healthcare network GK was established in 2006 by two statutory health insurers (SHI)—AOK Baden-Württemberg (AOK BW) and LKK Baden-Württemberg (LKK BW)—and the GK GmbH. This company is two-thirds owned by the Medizinisches Qualitätsnetz Ärzteinitiative Kinzigtal e.V. (MQNK—a network of physicians in the region) and one-third by the management and investment company OptiMedis AG. It is the only population-based IC-contract in Germany that took on joint economic and medical responsibility for all indications and health service sectors (excluding dental health) in a long term contract (9 years) with sufficient investment and which is thoroughly scientifically evaluated compared to standard care (Siegel et al. 2014). The IC-contract includes approximately 33,000 insured persons, of both SHI, living in the region of Kinzigtal. This is approximately half of the whole population living in this region (Hermann et al. 2006). Compensation of GK by cooperating SHIs is based on performance. Only when the quality of medical care is at least as good or better than, and simultaneously less expensive than, similar treatment and results for comparable insured persons of similar age, gender and health, does GK receive appropriate remuneration (Hildebrandt et al. 2010). Besides the improvement in the health of the population and the economic efficiency of healthcare, improving the patient's experience represents the third substantial goal of the IC agreement. This goal triangle of

GK resembles the Triple Aim developed by the IHI Institute (Berwick et al. 2008; Hildebrandt et al. 2012). To achieve these objectives GK coordinates care processes across sectors, implements its own disease management and prevention programs, but also makes extensive use of other existing disease management programs. In addition, it concludes on-site contracts with service providers for additional services and compensation, integrates sports and social clubs, social services and self-help organisations and conducts healthcare research studies as well as regular controlling (Hildebrandt et al. 2010). A fundamental requirement for both of the last-named tasks is efficient networking through ICT infrastructure with an adequate BI system built onto that. In the following sections the BI system in GK is discussed in more detail, starting from the data sources through the data processing and the Data Warehouse to the analytical capabilities and application in a continuous improvement process. Finally, challenges and lessons learned are outlined.

2   Transforming Data into Information: Data Sources, Data Processing and Data Warehouse of Gesundes Kinzigtal

As a first step, in a BI system data have to be collected and transformed into information (cf. hierarchical process in Fig. 1,[1]) whereas for this article the term data relates to discerned elements (Liebowitz 2006). Data is raw and can exist in any form useful, such as structured data (spreadsheet holding diagnoses data coded with ICD10) or not easily useful, i.e. unstructured data (clinical findings as free text) (Ackoff 1989 as cited in Riley and Delic 2010). In the context of the integrated care systems (ICS) GK there is a multitude of data sources used by OptiMedis in the Data Warehouse (for further details on the data sources and the Data Warehouse cf. Pimperl et al. 2014a). The main data source are the monthly updated insured person-based, but pseudonymised,[2] claims data from the two cooperating SHI for their 33,000 people living in the Kinzigtal region. These include data starting in 2003 for all care sectors, e.g. hospital claims including principle and secondary diagnoses, operations, fee schedule numbers and diagnoses by outpatient providers as well as drug remedy and aid prescriptions and much more (additional information on SHI data can be found at: GKV Spitzenverband 2013).

---

[1]This article follows the theoretical model of Ackoff (1989) as cited in Riley and Delic (2010) to define data, information and knowledge and bring them in a hierarchical connection. However, there are various perspectives on this field. For an in-depth discussion, see, e.g., Zins (2007).

[2]In terms of data protection law this actually concerns semi-anonymous data, since there is no way for OptiMedis and GK to resolve the unique pseudonyms related to insured persons. The pseudonyms are generated by the SHIs, and are not disclosed.

This SHI data is supplemented by further pseudonym-linked information from service providers contracted by the GK GmbH as well as documentation of the local management company GK. This includes, inter alia, secondary claims and medical data (e.g. laboratory or cytology results) extracted directly from the physicians' electronic medical records (EMRs), electronic documentation for disease management programs (eDMP) as well as billing or treatment pathway documentation[3] from the standardised IT solution for medical networking, CGM-NET,[4] used by the GK GmbH. The advantages of this data is that it is more readily and quickly accessible and also includes additional care relevant information. Above that, also primary data (e.g. from patient satisfaction surveys Stoessel et al. 2013), external catalog data (e.g. ICD, OPS, ATC) and external comparative data, such as risk structure equalisation scheme (=Morbi-RSA) allocations or quality report findings, are used.

This diverse data is integrated via various ETL (extract, transform, load) processes in the Core Data Warehouse (MS SQL Server). Thereby data are brought into relational connections, standardised (e.g. standardised unique pseudonyms for insured parties and service providers etc. across all data sources) and subjected to quality checks, cleansing and normalisation. After this stage, data are, via additional ETL processes, prepared for the analytical database. Amongst others, for instance,

- insurance selections,
- risk adjustment methods like propensity score and exact matching,
- relative time references (i.e. for example what happened a quarter prior or post enrolment on a Disease Management Program),
- model and scenario calculations (e.g. risk structure equalisation schemes calculations, disease-related expense attribution models) and
- predictive modelling approaches

are implemented (for further details see Pimperl et al. 2014a,b; Schulte et al. 2012). Subsequently the data is processed in a Health Data Analytics- and Cost-Accounting-OLAP Cube.[5] Once the data is patterned in that way and can be given meaning—this meaning can, but does not have to be useful—it becomes information (Ackoff 1989 as cited in Riley and Delic 2010; Liebowitz 2006) and can therefore be used in the Deltamaster BI front-end for various analyses and reports (cf. Fig. 1).

---

[3]In the Gesundes Gewicht program, for example, the following are recorded: weight, girth, blood pressure, fasting blood glucose, standard blood glucose, etc.

[4]For more information see OptiMedis AG (2014).

[5]OLAP is short for On-Line Analytical Processing and is often used as a synonym for multidimensional data analyses. More detailed explanations can be found in Azevedo et al. (2009).

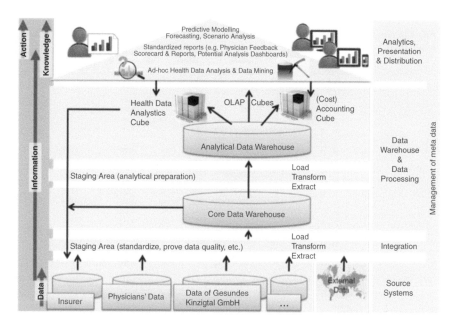

Fig. 1 Schematic overview of the OptiMedis BI system (cf. Pimperl et al. 2014a)

3 From Information to Knowledge to Action: Applications of the BI-System in Gesundes Kinzigtal

Information is factual (per definition of Kahaner 1997). It is "data that has been given meaning by way of relational connection" (Ackoff 1989 as cited in Riley and Delic 2010), such as in the Data Warehouse of OptiMedis. On its own this "meaning", as Ackoff further states, "can be useful, but does not have to be." Information pieces have to be filtered, distilled, analysed and synthesised in a collection that is intended to be useful and can be acted upon (Ackoff 1989 as cited in Riley and Delic 2010). Then, by adding insights and experience, information becomes knowledge (Liebowitz 2006), respectively (business) intelligence (Kahaner 1997), and may initiate actions. For that purpose the whole design of the BI system has been embedded in the organisational and management structure of the ICS GK. It supports the entire Plan-Do-Study-Act (PDSA)-management cycle, from setting goals, through creating necessary conditions to measuring performance, followed by taking action for further improvement and then repeat the model to anchor sustainable change (Furman and Caplan 2007).

In the planning phase, analyses are used to generate the initial findings required to determine the burden of disease, from a medical/epidemiological as well as an economic perspective, in the relevant population. The goal is to identify the needs and opportunities for intervention and then create priorities. Exemplary analyses used in the project preparation by GK project managers as well as the doctors on site in the project group, normally established for that reason, are shown in Fig. 2.

Age and gender distribution: patients with heart failure vs. standard Gesundes Kinzigtal population

2012 age group	female insurants total %	patients with diagnosis %		male patients with diagnosis %	insurants overall %
0 - 4	3,4%	0,1%		0,1%	4,1%
5 - 9	4,1%	0,3%		0,3%	4,6%
10 - 14	4,4%	0,8%		1,0%	5,1%
15 - 19	4,5%	2,3%		2,7%	5,5%
20 - 24	5,4%	4,0%		4,2%	6,4%
25 - 29	5,1%	3,8%		4,5%	5,8%
30 - 34	5,0%	4,7%		5,6%	5,2%
35 - 39	5,1%	5,4%		5,9%	5,3%
40 - 44	6,3%	6,6%		8,5%	7,2%
45 - 49	8,1%	9,0%		11,0%	8,5%
50 - 54	7,4%	9,3%		10,2%	7,8%
55 - 59	6,4%	9,7%		10,1%	7,0%
60 - 64	5,8%	8,2%		10,0%	6,2%
65 - 69	4,2%	4,7%		5,3%	4,0%
70 - 74	6,9%	10,0%		7,9%	6,0%
75 - 79	6,7%	9,2%		6,3%	5,4%
80 - 84	5,1%	6,0%		4,1%	3,7%
85 - 89	3,9%	4,3%		1,7%	1,6%
90 - 94	1,8%	1,4%		0,5%	0,5%
95 - 99	0,3%	0,1%		0,0%	0,1%
100 - 104	0,1%	0,0%		0,0%	0,0%
Sum	**16.137**	**4.148**		**2.982**	**15.243**
Ø-Age	47,74	55,91		52,20	43,59
Ø-Life exp.	83,65	77,84		76,32	77,22
prevalence		25,7%		19,6%	
incidence		3,6%		3,7%	

Health care cost distribution per patient: patients with heart failure vs. standard Gesundes Kinzigtal population

	2012	Ø Patients 2005 - 2012	Ø Population 2005 - 2012	expansion rate 2011 to 2012 %	alteration rate 2005 to 2012 %
⊞ Patients with diagnosis total	7.130	7.097	26.904	-0,1%	11,3%
Patientes with diagnosis (daily basis)	7.051,4	7.025,9	26.369,7	-0,1%	10,8%
sickness benefit	147,91	153,85	78,36	84,3%	2,3%
treatment expenses (incl. dialysis)	607,64	594,45	471,88	-1,5%	7,5%
hospital expense	1.126,34	982,50	805,54	6,8%	18,6%
rehab./cure expense	78,07	76,32	52,00	4,7%	-12,2%
other services payment amount	266,40	301,44	284,11	-2,1%	-24,3%
drug costs pre-tax	633,91	589,24	496,02	3,5%	14,3%
⊟ overall costs per patient	2.860,27	2.697,79	2.187,90	5,5%	7,7%
allocation per patient	2.655,82	2.466,38	2.129,54	3,8%	19,7%
⊟ contribution margin per patient	-204,45	-231,41	-58,36	-35,5%	53,2%

Drill-Down

Top 10 hospital diagnoses of heart failure patients

	2012 patients with principal diagnosis	patients with secondary diagnosis	medical expenses cases total	Ø Patients 2005 - 2012 patients with principal diagnosis	patients with secondary diagnosis	medical expenses cases total
⊞ M48 # Other spondylopathies	62	96	339.697	30	26	146.051
⊞ I50 # Heat failure	60	165	208.166	31	45	112.459
⊞ I48 # Atrial fibrillation and flutter	43	189	122.504	27	58	67.352
⊞ M54 # Dorsalgia	42	241	89.142	24	59	49.331
⊞ I10 # Essential (primary) hypertension	37	627	49.028	24	213	34.531
⊞ I20 # Angina pectoris	37	103	124.335	28	36	101.521
⊞ M51 # Other intervertebral disc disorders	36	73	99.429	30	29	95.300
⊞ I63 # Cerebral infarction	35	45	165.598	16	13	76.832
⊞ M17 # Gonarthrosis	33	58	233.517	32	27	213.730
⊞ R55 # Syncope and collapse	30	89	62.396	21	27	40.710

Fig. 2 Schematic overview of the OptiMedis BI system (cf. Pimperl et al. 2014a)

It illustrates a simple overview of the age and gender distribution of patients with heart failure in comparison to the standard GK population (= all 33,000 insured persons, of both AOK and LKK BW, living in the region of Kinzigtal). Via difference-bar-charts the difference between the heart failure patients and the standard GK population are plotted. In this way significant differences of the selected patient group can be easily spotted and the most relevant age groups for an intervention can be identified. To complete the picture, also mean age, life expectancy and incidence and prevalence are reported. Further, the health care cost distribution, the allocations from the Morbi-RSA (= revenues of the SHIs in Germany) and resulting contribution margins per patient for people with heart failure are also analysed in comparison to the standard GK population. The development over a period of 5 years is visually represented in small inline column charts (sparklines) directly next to the actual value. A trend arrow indicates if there is a significant increase or decrease. The exact alteration rate is also calculated for 1 year and 5 years. Since in this case hospital costs are high, a drill down on the hospital sector is shown (top 10 hospital diagnoses of heart failure patients) to further analyse comorbidities leading to a hospital case. Using OLAP-technology such an analysis could also easily be applied to other diseases, such as depression or hypertension, since only the selection on the medical indication would have to be changed. The analysis displayed in Fig. 2 will automatically be calculated for the new disease selected because of the underlying connections within the data in the OLAP-cube. If the joint decision is made to implement a special systematic support program for patients with a special condition, as it has happened, for example, for the disease heart failure shown above in Fig. 2 (Schmitt et al. 2011), then this program is also regularly evaluated (OptiMedis AG 2014). Control-group based approaches (exact matching, propensity score matching) are used for that purpose to reduce bias. These evaluations are carried out for the support, preventive and health promotion programs for specific conditions, such as heart failure (OptiMedis AG 2012) or osteoporosis (Schulte and Fichtner 2013), as well as for all enrolled insured persons as a whole (Schulte et al. 2012).

Besides this a comparative benchmarking approach, based on a combination[6] of Donabedian's (2005) (hierarchical) structure-process-outcome quality framework and Kaplan and Norton's balanced scorecard approach (Kaplan and Norton 2001, 2004), is used to stimulate continuous improvement in the network providers.[7] Performance measures for the evaluation are thereby grouped into the following dimensions:

- Structure, divided into the subdimensions "Learning and Innovation" and "Patient Structure" (illustrated e.g. by the indicators: participation in quality circles, age and morbidity structure indicators of patients/doctors),
- Processes (e.g. review of quality of medical care measures—amongst others guideline orientation, implementation of contracts as well as economical

---

[6]Cf. also similar approach in Gröbner (2007).

[7]For further details see Pimperl et al. (2013, 2014a).

process measures like economically sound prescriptions: me-too/generic quotas) and

- Healthcare outcomes, divided into the Triple Aim dimensions (operationalised e.g. via mortality ratio, years of potential life lost, contribution margin, patient experience).

This comparative performance management approach was and is continuously being rolled out across the depth and breadth of the network, from network management level to the individual network partners (GPs, specialists, hospitals, pharmacies etc.). An example of such a report (health services cockpit for a GP practice) is shown in Fig. 3.

Third quarter of 2013	**Quality indicators and key figures**		**Your practice (practice 8)**	**Ø-LP- GP's**	**Ø-NLP- GP's**	**Min/ Max GP**
3. Outcomes: Which impacts have interventions on medical and finacial outcomes and patient satisfaction?						
3.1 Economical outcomes	Allocation (Morbi-RSA) per patient		1.021,11 →	914,19	834,46	1.115,86
	- Total costs per patient		826,54	917,89	841,14	668,74
	= Contribution margin per patient		194,56	-3,70	-6,68	215,30
3.2 Health outcomes	Hospital cases per 1.000 patients (risk-adj.)		68,01	91,39	93,99	59,41
	Decedents % (risk-adj. mortality)		0,00 %	0,43 %	0,32 %	0,00 %
	Patients with osteoporosis & fracture %		1,8 %	1,3 %	1,3 %	0,0 %
3.3 Patient satisfaction	Impression of practice very good - exc. %		66,7%	61,0%		83,3%
	Med. treatment very good - exc. %		52,8%	53,0%		79,2%
	Recommendation likely - certain %		85,2%	84,6%		95,6%
2. Process - What must we excel at?			↑		↑	
2.1 Diagnostic quality	Unspecified diagnoses %		26,3 % →	27,7 %	34,3 %	17,0 %
	Suspected diagnoses %		1,8 %	1,4 %	1,6 %	0,8 %
2.2 Utilization	Patients >= 35 with health-check-up %		9,1 %	8,0 %	7,8 %	12,8 %
	Patients incapable of working %		27,2 %	25,3 %	26,8 %	18,1 %
	Lenght of incapacity for work		2,71	2,48	2,74	1,76
2.3 Improvement of Medication	Generic quota		92,2 %	88,5 %	87,0 %	92,2 %
	Pat. with heart-fail. & guideline prescr. %		72,7 %	71,5 %	68,8 %	84,6 %
	Patients >= 65 with pot. inad. med. (PRISCUS) %		14,4 % →	11,6 %	11,2 %	5,6 %
	Patients >= 65 with inad. med. (FORTA D) %		10,2 %	9,0 %	9,9 %	5,5 %
1. Structur - What is the target population? Where can we			↑		↑	
1.1 Patient structure						
1.1.1 Age, gender, etc.	Ø-Number of patients		481,0	480,9	326,1	934,0
	Ø-Age		57,88	55,31	52,96	54,2
	Female %		57,6 %	56,3 %	55,7 %	67,8 %
	Patients capable of work %		53,6 %	58,1 %	59,2 %	75,7 %
	Patients dependent on care %		8,7 % →	8,3 %	7,7 %	4,2 %
1.1.2 Morbidity	Ø-Charlson-comorbidity-score		2,15 →	1,37	1,26	0,75
	Regional GP-risk-score (Ø = 1,00)		1,16	1,04	0,95	0,81
1.1.3 Enrollment	Participants Integrated Care %		86,5 %	58,5 %	10,7 %	86,5 %
	Participants Disease Management Programs %		71,0 %	54,9 %	34,4 %	80,1 %
1.2 Learning & innovation	Participation in quality circles (Ø = 1,00)		1,50	1,00	-	4,00

Fig. 3 Health services cockpit for the GP practice (sample export of an overview dashboard from the Deltamaster Business Intelligence Suite)

These quarterly reports are sent to all GPs cooperating with the ICS GK. Each practice can see their results in comparison to those of their colleagues in the ICS (LP) as well as with practices in the region not contracted to GK, respectively not participating in the ICS (NLP). Also the minimum and maximum value for a measure is represented as a benchmark and every indicator has sparklines (small inline charts) showing the development over time as well as trend arrows indicating significant increases or decreases. A notation concept is also included: For example the colours blue, red and grey are used to indicate that a value of an indicator should be kept high (=blue), low (=red) or if the measure has just a general information character (=grey), for instance. An example for a red-coloured quality indicator that should be kept low is patients ≥65 with inadequate prescriptions in terms of the FORTA D classification (Kuhn-Thiel et al. 2014). The concrete evaluation is nevertheless left to the doctor concerned. For this purpose predefined detailed reports for each indicator, through which the doctor can browse when necessary, are also included: i.e. detailed benchmarking reports, or a list of medications that have been prescribed to patients of the practice and which are, in terms of the FORTA D list, potentially inadequate. To support the transformation of the knowledge gainable through the health services cockpits to actions of improvement along with the electronic dispatch of the feedback reports to the network providers, these are integrated, in a variety of ways, in the GK management routine. For example, they are utilised in the monthly Medical Advisory Board meetings, in project groups and quality circles (e.g. the ICS drug commission), for regular practice visits by employees of GK, at annual meetings between network management and network partners as well as for preparing selected extracts in the internal newsletter and during training and general meetings.

4   Lessons Learned and Outlook

Looking back at the more than 9-year history of the development and continuous improvement of the BI system, there are lessons learned which could be relevant to other ICS.[8] A critical success factor is the commitment of the independent network partners (doctors, SHIs etc.). These, with their data and engagement, are forming the basis for establishing a data-based management process. At GK close trustful cooperation in an organisational-contractual framework (heterarchic network) is seen as a fundamental prerequisite for the commitment and involvement of network members. First, network members have—besides their intrinsic motivation to deliver good medicine—on the basis of the contractual arrangement as partners in GK coupled with the success-oriented remuneration model of the SHIs, a reinforced desire to learn from the data and initiate a continuous process of improvement. On the other hand, from the start the content in the BI system has been developed in conjunction

---

[8]Cf. also book contribution in German on this topic from Pimperl et al. (2014a).

with the network members. This has created in turn confidence and trust, as well as identification with the BI project. In addition, a simple, unified and well-structured report design[9] is essential to ensure acceptance. Detailed information on feedback reports is even actively requested by some service providers now. For the majority, however, it is still necessary to combine electronic distribution with concrete face-to-face meetings in connection with the practice visits by employees of GK, project groups etc. Furthermore the BI system must be extended to include health science logic (cf., e.g. propensity score matching for risk adjustment), in order to generate valid analyses. This must be done in such a way so as to make standardised and automated analysis and reporting possible, since the evaluation of production can otherwise not be commercially viable. Simultaneously the BI System must maintain the flexibility necessary for ad-hoc analyses. This can be ensured by using the sophisticated models already on the database level combined with a high-performance OLAP system. Securing the quality, completeness and actuality of the data is a major challenge. To ensure the best possible results here, on one hand a variety of cleaning and projection modelling needs to be integrated in the Data Warehouse while, on the other hand, new and further developments (e.g. the networking software CGM-NET, developed with CGM) need to be implemented in the source systems. And there is still great future potential. To date GK has only really tapped into data from the medical service providers/partners in addition to the claims date delivered by the SHIs and the own data collected. Yet, data extracted directly from other partners, such as hospitals, pharmacies and physiotherapists etc., could increase the speed of availability, just as additional information not yet in the claims data could add important value to understanding and improving healthcare processes. Over and above this, for example, a medical training centre operated by GK is in the planning stages. In this centre a wide range of data for training development and the state of health of the insured will be generated by the training equipment, mobile apps and other devices, which could then also be made available to all participants in the treatment process. On the whole the quick provision of accurate, complete and quality-assured data is just a first step. The greatest added value will only be seen with the enrichment and combination of the data with other information already in the BI system, from which decision support at the right time in the workflow, for the right actor, can be generated. Naturally the highest priority in such a BI solution is also the protection of the sensitive data of the insured persons being treated. It is important here to weigh the potential benefits against the risk. In Germany to date there is also much which has not been clearly defined by the law, creating some challenges for the IC projects. Often sensible applications are blocked because of data protection, as unclear statutory regulations are, out of a need to be cautious, applied more restrictively by data protection authorities. Applications such as predictive modelling, in which the risk of patients before the potential occurrence of severe health events (hospitalisation,

---

[9]Reduction in complexity with simultaneous density of information, e.g. graphic table, sparklines, notation etc. (Bissantz 2010; Gerths and Hichert 2011).

for example) can be assessed, with patients then being enrolled on appropriate disease management programs, are currently possible in principle in GK, but require substantial effort and have therefore not been applied fully. On the other hand, the data protection topic in healthcare is, correctly so, a particularly sensitive issue, and could—in the event of abuse—severely discredit IC. At GK the protagonists, i.e., besides management and, amongst others, a specially contracted external and legally qualified data protection supervisor are therefore in very close consultation with the various authorities in Baden-Württemberg and at federal level. This has already led to legislative improvements as a result of federal initiatives. Even if this isn't the right place for an intensive cost–benefit ratio analysis of the investment effort, since 2006, in the Data Warehouse and the evaluation routine,[10] it can still be stated that the very high levels of investment have nonetheless paid off, and that without this investment the ever-growing seven figure savings contract results produced by GK would not have happened. Above all, without this database, GK would not have gained the ability to implement their own results, and would not be in a fair negotiation position with SHIs. A business which depends on goodwill and blind acceptance of the evaluation of their contract partners, is not really an independent business, nor is it an equal partner. To conclude it should be noted that a variety of data-based management opportunities already exist under the current conditions. It is assumed that their full exploitation—and even more so their development into a comprehensive BI system—including mobile and patient-generated health data, real-time decision support systems and sophisticated predictive management approaches—will still require some time. The goal remains that the management of ICS must continue to evolve from pure empiricism to data-based evidence.

References

Ackoff, R. L. (1989). From data to wisdom. *Journal of Applied Systems Analysis, 16*, 3–9.

Amelung, V., Wolf, S., & Hildebrandt, H. (2012). Integrated care in Germany–A stony but necessary road! *International Journal of Integrated Care, 12*, 1. Accessed August 23, 2014, http://www.ijic.org/index.php/ijic/article/view/URN%3ANBN%3ANL%3AUI%3A10-1-112901

Amelung, V. E., Sydow, J., & Windeler, A. (2008). Vernetzung im Gesundheitswesen im Spannungsfeld von Wettbewerb und Kooperation. In V. E. Amelung, J. Sydow, & A. Windeler (Eds.), *Vernetzung im Gesundheitswesen. Wettbewerb und Kooperation* (pp. 9–24). Unterschleißheim/Stuttgart: Kohlhammer.

Azevedo, P. (2009). *Business Intelligence und Reporting mit Microsoft SQL Server 2008. OLAP, data mining, analysis services, reporting services und integration services mit SQL Server 2008.* Unterschleißheim: Gabler.

Behme, W. (1996). Business Intelligence als Baustein des Geschäftserfolgs. In H. Mucksch & W. Behme (Eds.), *Das Data-Warehouse-Konzept. Architektur – Datenmodelle – Anwendungen* (pp. 27–46). Wiesbaden.

Berwick, D. M., Nolan, T. W., & Whittington, J. (2008). The triple aim: Care, health, and cost. *Health Affairs (Project Hope), 27*(3), 759–769.

---

[10]For further details see Reime et al. (2014).

Bissantz, N. (2010). *Bella berät: 75 Regeln für bessere Visualisierung*. Nürnberg: Bissantz & Company.

Business Application Research Center (BARC). (2013). Merck und OptiMedis gewinnen den BARC best practice award 2013. Accessed on 14 September 14, 2014, http://www.barc.de/content/news/merck-und-optimedis-gewinnen-den-barc-best-practice-award-2013

Donabedian, A. (2005). Evaluating the quality of medical care. *Milbank Quarterly, 83*(4), 691–729.

Furman, C., & Caplan, R. (2007). Applying the Toyota production system: Using a patient safety alert system to reduce error. *Joint Commission Journal on Quality and Patient Safety/Joint Commission Resources, 33*(7), 376–386.

Gerths, H., & Hichert, R. (2011). *Professionelle Geschäftsdiagramme nach den SUCCESS-Regeln gestalten*, 1st edn. Freiburg im Breisgau: Haufe.

GKV Spitzenverband. (2013). GKV-Datenaustausch. Elektronischer Datenaustausch in der gesetzlichen Krankenversicherung. Accessed on September 10, 2014, http://www.gkv- datenaustausch.de/

Gröbner, M. (2007). Controlling mit Kennzahlen in vernetzten Versorgungsstrukturen des Gesundheitswesens. Dissertation, Munich. Accessed on August 3, 2014, http://ub.unibw-muenchen.de/dissertationen/ediss/groebner-martin/inhalt.pdf

Hammersley, V. S., Morris, C. J., Rodgers, S., Cantrill, J. A., & Avery, A. J. (2006). Applying preventable drug-related morbidity indicators to the electronic patient record in UK primary care: Methodological development. *Journal of Clinical Pharmacy and Therapeutics, 31*(3), 223–229.

Hermann, C., Hildebrandt, H., Richter-Reichhelm, M., Schwartz, W. F., & Witzenrath, W. (2006). Das Modell "Gesundes Kinzigtal". Managementgesellschaft organisiert Integrierte Versorgung einer definierten Population auf Basis eines Einsparcontractings. *Gesundheits- und Sozialpolitik, 5*(6), 11–29.

Hildebrandt, H., Hermann, C., Knittel, R., Richter-Reichhelm, M., Siegel, A., & Witzenrath, W. (2010). Gesundes Kinzigtal integrated care: Improving population health by a shared health gain approach and a shared savings contract. *International Journal of Integrated Care, 10*. Accessed on September 11, 2014, http://www.ijic.org/index.php/ijic/article/view/ 539/1050

Hildebrandt, H., Schulte, T., & Stunder, B. (2012). Triple aim in Kinzigtal. Improving population health, integrating health care and reducing costs of carelessons for the UK? *Journal of Integrated Care, 20*(4), 205–222.

Janus, K., & Amelung, V. E. (2005). Integrated health care delivery based on transaction cost economics. Experiences from California and cross-national implications. In T. G. Savage, A. J. Chilingerian, & M. F. Powell (Eds.), *International health care management. Bd. 5. Advances in health care management.* Greenwich: JAI Press.

Kahaner, L. (1997). *Competitive intelligence: How to gather analyze and use information to move your business to the top.* New York: Touchstone.

Kaplan, R. S., & Norton, D. P. (2001). *The strategy-focused organization: How balanced scorecard companies thrive in the new business environment.* Boston, MA: Harvard Business School Press.

Kaplan, R. S., & Norton, D. P. (2004). *Strategy maps: Converting intangible as-sets into tangible outcomes.* Boston, MA: Harvard Business School Press.

Kuhn-Thiel, A. M., Weiß, C., & Wehling, M. (2014). Consensus validation of the FORTA (Fit fOR The Aged) list: A clinical tool for increasing the appropriateness of pharmacotherapy in the elderly. *Drugs & Aging, 31*(2), 131–140.

Kupersmith, J., Francis, J., Kerr, E., Krein, S., Pogach, L., & Kolodner, R. M., et al. (2007). Advancing evidence-based care for diabetes: Lessons from the veterans health administration. *Health Affairs (Project Hope), 26*(2), w156–w168.

Liebowitz, J. (2006). *Strategic intelligence: Business intelligence, competitive intelligence, and knowledge management.* Boca Raton, FL: Auerbach Publications.

OptiMedis AG. (2012). Gesundes Kinzigtal: Programmblatt zu Starkes Herz. Accessed on September 16, 2014, http://de.scribd.com/doc/152716963/Gesundes-Kinzigtal-Programmblatt-zu-Starkes-Herz

OptiMedis AG. (2014). Gesundheitsprogramme. Gesundheitsprogramme werden regelmäßig evaluiert. Accessed on August 8, 2014, http://www.optimedis.de/gesundheitsnutzen/gesundheitsprogramme

Pimperl, A., Schulte, T., Daxer, C., Roth, M., & Hildebrandt, H. (2013). Balanced scorecard-Ansatz: Case study Gesundes Kinzigtal. *Monitor Versorgungsforschung, 6*(1), 26–30.

Pimperl, A., Dittmann, B., Fischer, A., Schulte, T., Wendel, P., & Hildebrandt, H. (2014a). Wie aus Daten Wert entsteht: Erfahrungen aus dem Integrierten Versorgungssystem "Gesundes Kinzigtal". In P. Langkafel (Ed.), *Big data in der Medizin und Gesundheitswirtschaft: Diagnose, Therapie, Nebenwirkungen.* Neckar, Heidelberg: med-hochzwei Verlag.

Pimperl, A., Schreyögg, J., Rothgang, H., Busse, R., Glaeske, G., & Hildebrandt, H. (2014b). Ökonomische Erfolgsmessung von integrierten Versorgungsnetzen – Gütekriterien, Herausforderungen, Best-Practice-Modell. *Gesundheitswesen.* Accessed on November 12, 2014, https://www.thieme-connect.de/products/ejournals/html/10.1055/s-0034- 1381988

Purucker, J., Schicker, G., Böhm, M., & Bodendorf, F. (2009): *Praxisnetz-Studie 2009. Management – Prozesse – Informationstechnologie. Status quo, Trends und Herausforderungen.* Nürnberg: Lehrstuhl Wirtschaftsinformatik II, Universität Erlangen-Nürnberg.

Reime, B., Kardel, U., Melle, C., Roth, M., Auel, M., & Hildebrand, H. (2014). From agreement to realization: Six years of investment in integrated eCare in Kinzigtal. In I. Meyer, S. Müller, & L. Kubitschke (Eds.), *Achieving effective integrated e-care beyond the Silos.* Hershey, PA: IGI Global.

Riley, J., & Delic, K. A. (2010). Enterprise knowledge clouds: Applications and solutions. In B. Furht & A. Escalante (Eds.), *Handbook of cloud computing* (pp. 437–452). Boston, MA: Springer.

Sachverständigenrat Zur Begutachtung Der Entwicklung Im Gesundheitswesen – SVR. (2012). Sondergutachten 2012 des Sachverständigenrates zur Begutachtung der Entwicklung im Gesundheitswesen. Wettbewerb an der Schnittstelle zwischen ambulanter und stationärer Gesundheitsversorgung. Accessed on August 17, 2014, http://www.svr-gesundheit.de/ fileadmin/user_upload/Gutachten/2012/GA2012_Langfassung.pdf

Schmitt, G., Hildebrandt, H., Roth, M., Auel, M., Deschler, T., & Witzenrath, W. (2011). 'Starkes Herz'/strong heart: Integrated health care for patients with history of heart failure in the Kinzigtal region, a rural area in south of Germany. *International Journal of Integrated Care, 11,* 7. Accessed on August 6, 2014, http://www.ijic.org/index.php/ ijic/article/view/URN%3ANBN%3ANL%3 AUI%3A10-1-101546

Schonfeld, S. J., Lee, C., & Berrington De Gonzales, A. (2011). Medical exposure to radiation and thyroid cancer. *Clinical Oncology (Royal College of Radiologists (Great Britain)), 23*(4), 244–250.

Schrappe, M. (2005). Patientensicherheit und Risikomanagement. *Medizinische Klinik, 100*(8), 478–485.

Schulte, T., & Fichtner, F. (2013). *Erste Ergebnisse der internen Evaluation des Programms Starke Muskeln – Feste Knochen für Osteoporose-Patienten aus dem IV-Projekt Gesundes Kinzigtal.* Berlin: German Medical Science GMS Publishing House.

Schulte, T., Pimperl, A., Dittmann, B., Wendel, P., & Hildebrandt, H. (2012). Drei Dimensionen im internen Vergleich: Akzeptanz, Ergebnisqualität und Wirtschaftlichkeit der Integrierten Versorgung Gesundes Kinzigtal. Accessed on August 19, 2014, http://www.optimedis.de/ images/docs/aktuelles/121026_drei_dimensionen.pdf

Siegel, S., et al. (2014). Utilization dynamics of an integrated care system in Germany: Morbidity, age, and sex distribution of Gesundes Kinzigtal integrated care's membership in 2006–2008. In C. Janssen, E. Swart, & T. von Lengerke (Eds.), *Health care utilization in Germany: Theory, methodology, and results* (pp. 321—335). New York: Springer.

Stein, V., Barbazza, E. S., Tello, J., & Kluge, H. (2013). Towards people-centred health services delivery: A frame-work for action for the World Health Organisation (WHO) European region. *International Journal of Integrated Care, 13,* 4. Accessed on August 12, 2014, http://www.ijic.org/index.php/ijic/article/view/ URN%3ANBN%3ANL%3AUI%3A10-1-114766

Stoessel, U., Siegel, A., Zerpies, E., & Körner, M. (2013). Integrierte Versorgung Gesundes Kinzigtal – Erste Ergebnisse einer Mitgliederbefragung. *Das Gesundheitswesen, 75*(08/09), A265.

Sun, X., Tang, W., Ye, T., Zhang, Y., Bo, W., & Zhang, L. (2014). A comprehensive bibliometric analysis and literature review. *International Journal of Integrated Care, 14*, 2. Accessed on August 23, 2014, http://www.ijic.org/index.php/ijic/article/view/URN%3ANBN%3ANL%3AUI%3A10-1-114784

Vijayaraghavan, V. (2011). *Disruptive innovation in integrated care delivery systems.* Mountain View, CA: Innosight Institute. Accessed September 4, 2014, Downloaded on 2014-03-31, www.christenseninstitute.org/wp-content/uploads/2013/04/Disruptive-innovation-in-integrated-care-delivery-systems.pdf

Zins, C. (2007). Conceptual approaches for defining data, information, and knowledge. *Journal of the American Society for Information Science and Technology, 58*(4), 479–493.

Clustering and a Dissimilarity Measure for Methadone Dosage Time Series

Chien-Ju Lin, Christian Hennig, and Chieh-Liang Huang

Abstract In this work we analyse data for 314 participants of a methadone study over 180 days. Dosages in milligram were converted for better interpretability to seven categories in which six categories have an ordinal scale for representing dosages and one category for missing dosages. We develop a dissimilarity measure and cluster the time series using "partitioning around medoids" (PAM). The dissimilarity measure is based on assessing the interpretative dissimilarity between categories. It quantifies the structure of the categories which is partly categorical, partly ordinal and also involves quantitative information. The principle behind the measure can be used for other applications as well, in which there is more information about the meaning of categories than just that they are "ordinal" or "categorical".

1 Introduction

Heroin is an expensive and highly addictive drug. Heroin-dependent individuals who aim at overcoming their addiction are offered a methadone maintenance therapy (MMT) for many years. The main purpose of the MMT is not to help them to achieve abstinence but to minimize the harm associated with the use of heroin. The idea of MMT is to let drug users reduce the use of heroin by addicting to methadone and then to quit the use of methadone. The effect of methadone lasts 24 hours and consequently it has to be taken on a daily basis. To date there is no clear principle

C.-J. Lin (✉)
Department of Statistical Science, University College London, London, UK

MRC Biostatistics Unit, Cambridge, UK
e-mail: chienju@mrc-bsu.cam.ac.uk; chien-ju.lin.10@ucl.ac.uk

C. Hennig
Department of Statistical Science, University College London, London, UK
e-mail: c.hennig@ucl.ac.uk

C.-L. Huang
Department of Psychiatry, China Medical University Hospital, Taichung City, Taiwan, ROC
e-mail: psyche.hcl@gmail.com

© Springer International Publishing Switzerland 2016   31
A.F.X. Wilhelm, H.A. Kestler (eds.), *Analysis of Large and Complex Data*, Studies in Classification, Data Analysis, and Knowledge Organization,
DOI 10.1007/978-3-319-25226-1_3

for the determination of the methadone dosage. Physicians prescribe dosages based on their own intuition.

Research has been done on daily methadone dosage taken by participants. Strain et al. (1993) studied treatment retentions and illicit drugs use and found that low dose of methadone (≤ 20 mg) may improve retention but were inadequate for suppressing illicit drug use. Langendam et al. (1998) observed that participants requested to stay at a lower dosage (≤ 60 mg) because of fear of double addiction. Bellin et al. (1999) studied associations between criminal activity and methadone dosage and found drug users on a high dose (≥ 60 mg) were less likely to return to jail. Murray et al. (2008) found that methadone dosage might be a response to misery. Peles et al. (2007) reported that the major risk factors for depression were female gender and high dose (> 120 mg). Gossop et al. (2000) applied the K-Means clustering method with four groups to a 1 year follow-up study. Two groups showed substantial reductions in their illicit drug use and criminality. They concluded that in a certain group MMT was appropriate.

Ideally, drug users are expected to reduce the use of heroin by addicting to methadone and then to quit use of methadone. The dosages should consequently have a pattern in which they go up at the beginning of the treatment and later go down. This would indicate detoxification. Physicians think that participants with such a dosage pattern and a high attendance rate most likely will have a positive outcome. Therefore, our objective is to develop a method to divide participants into groups according to their behaviour, that is, patterns of daily methadone dosage, and then find the differences between the groups. By clustering, we can study the association between dosage patterns and demographic factors, the degrees of addictions and retention of MMT. Also, the dosage patterns provide the possibility of developing a guideline for prescribing a proper methadone dosage.

The problems of clustering the participants in our study are the fluctuations of dosages and missing dosages. First of all, some participants who abused heroin while receiving the MMT did not need the full dosages indicated on their prescriptions to accommodate their addictions. In fact, they took a combination of drug and methadone in order for their addictions to be satisfied, so it was not guaranteed that the observed methadone dosages represented detoxication. Secondly, missing dosages were not missing at random. They were recorded as zeroes but zero is not normally a proper description of their state of addiction. We take account of these issues and propose to categorize dosages for alleviating the fluctuations of observed dosages and for keeping the sequences of missing dosages. Also, we propose a new dissimilarity ("p-dissimilarity") that quantifies the structure of the categories and involves quantitative information. The dosage patterns will then be represented by sequences of categories.

2   Data

2.1   Plain Dosage Data

Daily dosages in milligram for 314 participants who received MMT between 01 January 2007 and 31 December 2008 were collected. These participants were selected from a larger study using the criterion that they had not left the study before the completion of 180 days and that they had at least 70 % nonzero records of taking methadone. One month is often regarded as the minimum length of receiving methadone treatment. Participants who stay in MMT for 6 months are considered to be candidates who can achieve abstinence. We considered data over 180 days. Normally, participants got weekly prescriptions. They would occasionally have multiple prescriptions but only one record of dosage taken on a single day. Besides, there was a chance that participants abused drugs, so their demand for daily methadone differed. Participants were allowed to take a dosage that was lower than the prescribed dosage to avoid overdosing. Their behaviour of abusing drugs was reflected by fluctuations in their dosage taken records. By and large, following a weekly prescription, a participant took methadone daily for a period of 7 days. Many participants dropped out and later returned to the treatment, or missed their treatment on a number of days. This resulted in many "zero dosage" records. However, their addiction to drugs was not zero, and these records were therefore treated as "missing values".

2.2   Category-Ordered Data

There are two problems with the plain dosages data. Firstly, some degrees of dosage fluctuation are not meaningful. We take the assessment of the physician, reflected in the prescription, as being more meaningful of the participant's state of addiction than occasional drug use or mood changes that may have led them to select slightly lower dosages on specific days. However, daily dosages are still of some informative value because not for every dosage taken there is a unique prescription explaining it; sometimes participants had more than one prescription from more than one physician to choose from. In principle, if no drugs are abused, methadone taken by participants should show long sequences of stability. Secondly, missing dosages are to be taken into account. These missing dosages usually have a specific meaning, namely that the participant did not show up for receiving methadone. Technically, these events are recorded as zero, but they could have various meanings. In some occasions there was no change to the status of the participant compared to surrounding days at which methadone was taken; the participant was just unreliable or felt so well on the day that they did not believe that they needed methadone. However, in some cases, particularly if there were longer absences, missing values point to more severe problems of the participant, or a tendency to leave the study, or illicit drug use. In any case, this cannot be properly reflected by the value "zero". In this paper, we treat missing values as a specific category. Lin (2014) also carried out experiments with imputation. A sensible scheme for imputation is difficult to define, because it should depend on the non-missing values surrounding the missing days, and also on length of periods of missingness.

Therefore we constructed a new data format by categorizing daily dosage. In our study, cut points for categorizing dosages were defined by the physician. He suggested that dosage in the range of 20 mg could be considered virtually the same. This meant that the qualitative difference between two dosages in the same interval could be treated as irrelevant. This corresponds to the fact that physicians often used multiples of 20 mg in their prescriptions. We defined six categories for dosages smaller than or equal to 20 mg, 21–40 mg, 41–60 mg, 61–80 mg, 81–100 mg and greater than 100 mg, recoded as 1,..., 6 (Likert-coding). Another category was "missing (zero)". The resulting dataset is called "category-ordered data". This minimized the implications of irrelevant daily fluctuations and outliers (although there were no really extreme ones), and reflected the interpretation of the dosages by the physicians.

To explore the uncategorized and the categorized datasets, a heatplot is used. This is a technique to represent data by colour. Each horizontal line in a heatplot represents the data of each participant. Figure 1 shows the heatplots of plain dosage and of the category-ordered data. Each horizontal line represents records of a participant from day 1 to day 180. In the graph, 314 participants are ordered by the average of their dosages. The colour spectrums of dosage and that of category are displayed below the heatplots. We observe that most dosage records in the first week are in category 1, as the initial prescription dosage for participants, most of

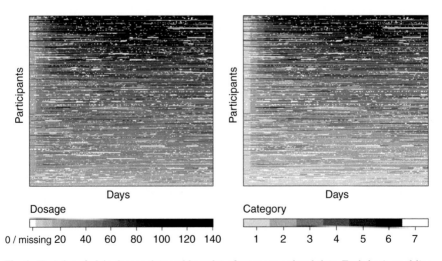

Fig. 1 Heatplot of plain dosage data and heatplot of category-ordered data. Each *horizontal line* represents records of a participant from day 1 to day 180. The 314 participants are ordered by the average of their dosages

which have no previous experience of the MMT, is 20 mg. Subsequently, the colours of dosage start to change, reflecting the fact that the doctors started adjusting the dosage. Also, it can be seen that most of these movements from category to category go to the next nearest category.

3 The New p-Dissimilarity Measure

3.1 Motivation

We define a new dissimilarity measure called "p-dissimilarity" for the category-ordered data by summing up daily dissimilarities between categories. In Lin (2014) there is a discussion of dissimilarity measures that take into account the time series structure, but as far as such measures are already in the literature, they cannot be easily adapted to our data structure (such as methods related to fitting autoregressive models to time series with continuous data) or they seem inappropriate such as "time warping" (Berndt and Clifford 1994) because the absolute length of periods of stability is very meaningful in the context of methadone therapy whereas such lengths are treated as unimportant in time warping.

We will follow the philosophy outlined in Hennig and Hausdorf (2006), according to which a dissimilarity measure should formalize the "interpretative distance" between objects according to knowledge of the subject matter.

A specific feature of the "interpretative distance" for the data at hand is that similarity between participants is mainly governed by periods in which they are on the same dosage (category). The distinction whether categories on a day are the same or different is more important than how different they are given that they are different, because changing categories even between neighbouring categories is interpreted as indicating a substantial change in the condition of the participant. We will define a dissimilarity function that assigns a quantitative value to distances between neighbouring categories and categories further apart in a concave monotonic fashion, i.e. further categories are further away, according to the dissimilarity, but the increase of the distance becomes smaller moving further away from a category and its neighbours. This implies information that can be seen as stronger than ordinal; note, however, that any method for defining distances between ordinal categories amounts to imposing a quantitative effective distance between them. This may be governed by the distribution of the data (as when midranks are assigned or latent normality is assumed) or by the meaning of the categories and the context, which we prefer. Our dissimilarity will involve a tuning constant p, which will tune the level of information between treating the categories as purely categorical (for $p = 0$, the dissimilarity will only count how often participants are in different categories) and treating them as quantitative and equidistant ($p = 1$). The "missing value" category will be treated in a specific way, as having the same dissimilarity from all other categories.

An example for the above is that if data for three participants (A,B,C) for 7 days are [1, 1, 1, 1, 1, 2, 2], [1, 1, 4, 1, 1, 2, 2] and [1, 2, 1, 2, 1, 3, 2], we want to define a dissimilarity measure that treats A and B as more similar than A and C, so the effective distance between 1 and 4 between A and B on day 3 should not dominate the fact that A and C differ on three different days.

3.2   Definition of the p-Dissimilarity

Let $D(\cdot, \cdot)$ denote the dissimilarity between participants and $d(\cdot, \cdot)$ denote the dissimilarity between categories. Let $x_{it} \in \Theta = \{1, \ldots, \theta, \theta + 1\}$ be the category-ordered data for the participant i on the tth day since they joined the MMT. θ is the number of Likert-coded dosage categories, and missing values are coded as $\theta + 1$. The p-dissimilarity between participants i and i' with a category for missing values is defined by

$$d(x, y) = \delta(x, y)(1 - p^{\alpha(x,y)}) + (1 - \delta(x, y))(1 - p^{\beta}), \tag{1}$$

$$D(i, i') = \sum_{t=1}^{T} d(x_{it}, x_{i't}). \tag{2}$$

where $\delta(x, y) = 1(x \leq \theta, y \leq \theta)$, an indicator that neither x nor y are missing, $0 < p < 1$, $\alpha(x, y) = |x - y|$, the difference between the Likert codes of the categories, and $1 < \beta < (\theta - 1)$.

The constant p tunes the dissimilarity between categories. The p can be interpreted as a switch between data being treated as categorical and linear in the Likert codes. For $x \neq y$ and both non-missing, it can be shown that $d(x, y) = (1 - p) \sum_{l=0}^{|x-y|-1} p^{l}$. Therefore d is monotonic in $|x - y|$ and concave, see Fig. 2. For

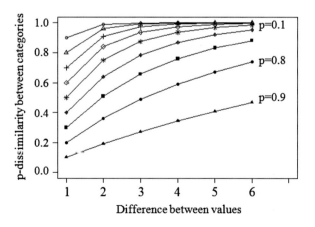

Fig. 2 p-dissimilarity between categories

p-dissimilarity between categories

p=0.1

p=0.8

p=0.9

Difference between values

$p \approx 1$ this is almost linear in $\alpha(x_{it}, x_{i't})$, whereas for $p \approx 0$ this is close to 1 for all nonzero differences between the Likert codes.

The dissimilarity between any category and a missing value is $(1-p)^{\beta}$. Missing values can have very different meanings as explained above. Therefore missing values were not treated as particularly close to any specific category, and the constant of $(1-p)^{\beta}$ was even applied between two missing values, implying that there could be a nonzero dissimilarity between two participants with the same values on all 180 days if this included missing values. Also it means that in general the p-dissimilarity violates the triangle inequality (although this is not the case if no missing values occur). Note that Hennig and Hausdorf (2006) argue that fulfilling the triangle inequality is not in itself a virtue of a dissimilarity measure, but only if there are subject matter reasons why it should be fulfilled for the "interpretative distance" between objects.

The parameter β tunes the dissimilarity involving missing values compared to the distances between non-missing values. $\beta = 1$ means that missing values are treated as if they were neighbouring to any category.

For practical application, the parameters p and β need to be specified. p was specified by subject matter considerations. Given the arguments before, it is clear that p should neither be very close to 1 nor very close to zero, because the very motivation for the p-dissimilarity is that a compromise between these extremes is attempted. Guided by medical considerations, we chose $p = 0.6$, for which $|x - y| = 2$ leads to a dissimilarity already very close to the maximum value, i.e. a difference of two dosage categories between participants is already implied to be very substantial, but $d(x, y)$ with $|x - y| = 2$ is still considerably larger than with $|x - y| = 1$. β was chosen as 1.42, which was the average of all $|x - y|$ occurring in the dataset between different participants on the same day with $\delta(x, y) = 1$, so missing values were treated as "in average distance to everything".

Arguments for choosing these parameters can only be imprecise. Lin (2014) carried out sensitivity analyses using other values of p and β, showing that changing p and β to values that can be seen as having a similar interpretation does not affect the clustering below much.

4 Clustering of the Category-Ordered Data

We apply the PAM clustering method (Kaufman and Rousseeuw 1990) with five clusters and the p-dissimilarity with $(p = 0.6, \beta = 1.42)$ to the category-ordered data. The clustering is then related with further information about the participants. In Lin (2014), various clustering methods (PAM, complete, average and single linkage clustering) (Gordon 1999) have been compared for this dataset regarding the average silhouette width (Kaufman and Rousseeuw 1990) and the prediction strength (Tibshirani and Walther 2005), which resulted in the choice of PAM with five clusters.

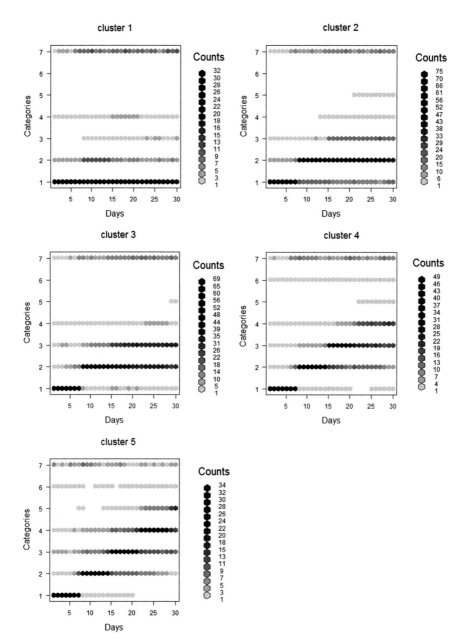

Fig. 3 Frequency of the categories from day 1 to day 30 for the five clusters. Each *horizontal line* represents the number of participants in each of the seven categories from day 1 to day 30. The colour designates the number of participants

Several one-way ANOVAs and χ^2-tests were run to see whether characteristics of the participants varied across clusters. There is not enough evidence to conclude that the mean participant ages differ between the five clusters (p-value is 0.084). The result of the ANOVA test about the cluster-wise mean ages of heroin onset is borderline significant (p-value is 0.062), there could be some effect of age of heroin onset. χ^2 tests show that there is not enough evidence to conclude that there exists a relationship between clusters with respect to gender ($p = 0.377$), education ($p = 0.996$), marital status ($p = 0.429$) and occupation, which is a binary variable indicating whether the participant is occupied or not ($p = 0.310$). Figure 3 shows the frequency of the categories from day 1 to day 30 for the five clusters. The y-axis indicates category and the x-axis indicates days. The colour designates frequency. We observe the following: (1) There is an upward trend in categories over time, particular in cluster 5. (2) Cluster 1 seems to have more missing values than cluster 5. For convenience, we define the pattern of detoxification in three stages. Stage I represents that the methadone dosage goes up, stage II represents the dosage staying stable, and stage III represents the dosage going down. We attempt to summarize the pattern of detoxification for each cluster by the (rough) time point on which the three stages are observed to start: Cluster 2 (day 1-40-100), Cluster 3 (day 1-80-140), cluster 4 (1-100-150), cluster 5 (1-100-150). Also, the majority of patients in cluster 4 and cluster 5 have their dosages in high categories. This means that participants who are highly addicted to heroin might take longer to finish the detoxification process.

5 Discussion

We selected a meaningful sample with 314 participants. We took account of the weekly prescriptions, fluctuations in dosage taken records and patterns of missing dosages, and defined the category-ordered data. The final clusters were obtained by using the PAM method with five clusters and the p-dissimilarity. Unfortunately, without the data of whether participants achieve abstinence or not, we cannot understand the relationship between treatments and final outcomes. Despite the fact that none of the five clusters could very easily be distinguished in terms of, say, their demographics, the sequences of categories for the five clusters were clinically useful. We found that the heroin onset age might have an influence on the patterns of detoxification. Participants with low addiction reduced the use of heroin by addicting to methadone in the first month and attempted to reduce/quit the use of methadone by the third month. As for participants with high addictions, few attempted to reduce the use of methadone up until the fifth month and most required more time to finish the detoxification process.

Lin (2014) presents more cluster validation and discussion. She found that the data could not be significantly distinguished from a Markov model without clustering structure, which means that the observed patterns cannot be safely be assumed to correspond to a "real" clustering of methadone participants. She also computed a clustering based on raw data (instead of category-ordered data), again with PAM and five clusters. The adjusted Rand index between this clustering and the one presented here was 0.54. In any case the clustering can be used for helping physician's decision making, because they give a simple summary of the complex range of existing dosage patterns.

The p-dissimilarity is based on assessing the interpretative dissimilarity between categories and focused more on sequence of constancy and less on sudden changes in categories. This was used to measure dissimilarity between the 180-day time series of the participants. It implements concepts of variables the categories of which cannot properly be classified as purely categorical or ordinal, and can be used for incomplete data. It could be applied in wider areas of application where researchers have a quantitative idea about the interpretative distance between categories, which could be between a categorical concept in which all differences between categories have the same distance, and a Likert-scaling concept with linearity in the Likert codes. See Hennig and Liao (2013) about related ideas for quantifying distances between categories.

References

Bellin, E., Wesson, J., Tomasino, V., Nolan, J., Glick, A. J. & Oquendo, S. (1999). High dose methadone reduces criminal recidivism in opiate addicts. *Addiction Research, 7*, 19–29.
Berndt, D., & Clifford J. (1994) Using dynamic time warping to find patterns in time series. *AAAI-94 Workshop on Knowledge Discovery in Databases* (pp. 229–248).
Gordon, A. D. (1999). *Classification*. New York: Chapman and Hall.
Gossop, M., Marsden, J., Stewart, D., & Rolfe, A. (2000). Patterns of improvement after methadone Treatment: 1 Year follow-up results from the national treatment outcome research study. *Drug Alcohol Abuse, 60*, 275–286.
Hennig, C., & Hausdorf, B. (2006). Design of Dissimilarity Measures: a New Dissimilarity Measure between Species Distribution Ranges. In V. Batagelj, H.-H. Bock, A. Ferligoj, & A. Žiberna (Eds.), *Data science and classification* (pp. 29–37). Berlin: Springer.
Hennig, C., & Liao, T. F. (2013). How to find an appropriate clustering for mixed type variables with application to socioeconomic stratification (with discussion). *Journal of the Royal Statistical Science, Series C (Applied Statistics), 62*, 309–369.
Kaufman, L., & Rousseuw, P. J.(1990). *Finding groups in data: an introduction to cluster analysis*. New Jersey: Wiley
Langendam, M., Van Haastrecht, H., Brussel, G., Van Den hoek, A., Coutinho, R., & Van Ameijden, E. (1998). Research report differentiation in the Amsterdam dispensing circuit: determinants of methadone dosage and site of methadone prescription. *Addiction, 93*, 61–72.
Lin, C. J. (2014) *A Pattern-Clustering Method for Longitudinal data - Heroin Users Receiving Methadone*. PhD thesis, Department of Statistical Science, University College London.
Murray, H., Mchugh, R., Behar, E., & Pratt, E. (2008). Personality factors associated with methadone maintenance dose. *The American Journal of Drug and Alcohol Abuse, 34*, 634–641.
Peles, E., Schreibera, S., Naumovskya, Y., & Adelsona, M. (2007). Depression in methadone maintenance treatment patients: Rate and risk factors. *Affective Disorders, 99*, 213–220.

Strain, E. C., Stitzer, M. L., Liebson, I. A., & Bigelow, G. E. (1993). Methadone dose and treatment outcome. *Drug and Alcohol Dependence, 33*, 105–117.

Tibshirani, R., & Walther, G. (2005). Cluster validation by prediction strength. *Journal of Computational and Graphical Statistics, 14*, 511–528.

Linear Storage and Potentially Constant Time Hierarchical Clustering Using the Baire Metric and Random Spanning Paths

Fionn Murtagh and Pedro Contreras

Abstract We study how random projections can be used with large data sets in order (1) to cluster the data using a fast, binning approach which is characterized in terms of direct inducing of a hierarchy through use of the Baire metric; and (2) based on clusters found, selecting subsets of the original data for further analysis. In this work, we focus on random projection that is used for processing high dimensional data. A random projection, outputting a random permutation of the observation set, provides a random spanning path. We show how a spanning path relates to contiguity- or adjacency-constrained clustering. We study performance properties of hierarchical clustering constructed from random spanning paths, and we introduce a novel visualization of the results.

1 Introduction

In our current era of Big Data, and given the central importance of hierarchical clustering for so many application domains, there is a need to improve computationally on standard quadratic time algorithms (i.e. $O(n^2)$ for n observation vectors, see, e.g., Murtagh 1985). In Murtagh (2004), we even discuss constant time hierarchical clustering, which presupposes that our data is, naturally or otherwise, embedded in an ultrametric topological space. In this article, we take further our work in Contreras and Murtagh (2012) and Murtagh et al. (2008). In those works, we demonstrated the effectiveness of linear computational time hierarchical clustering, using a range of examples, including from astronomy and chemistry.

F. Murtagh (✉)
Department of Computing, Goldsmiths University of London, London SE14 6NW, UK

De Montfort University, Leicester, UK

Department of Computing and Mathematics, University of Derby, Derby, UK
e-mail: fmurtagh@acm.org

P. Contreras
Thinking Safe Ltd., Egham, Surrey TW20 0EX, UK
e-mail: pedro.contreras@acm.org; pedro@cs.rhul.ac.uk

© Springer International Publishing Switzerland 2016   43
A.F.X. Wilhelm, H.A. Kestler (eds.), *Analysis of Large and Complex Data*, Studies
in Classification, Data Analysis, and Knowledge Organization,
DOI 10.1007/978-3-319-25226-1_4

In particular we focus on very high dimensional data. We have demonstrated in many clustering case studies that random projection can work very well indeed. Random projection is a first stage of the processing, which allows both computationally efficient and demonstrably effective hierarchical clustering, using the Baire metric (Murtagh et al. 2008; Contreras and Murtagh 2012). The Baire metric is simultaneously an ultrametric. In this article, we further develop the theory and the practice of random projection in very high dimensional spaces. We are seeking a computationally efficient clustering method for massive (large n, number of rows), very high dimensional, very sparse data. Massive high dimensional data are typically sparse (i.e. containing many non-presence or 0 terms).

To help the reader to reproduce our results, some ancillary material, including R code, is available at http://www.multiresolutions.com/HiClBaireRanSpanPaths

2 Data

We present our methodology using a case study. We used the textual content of 34,352 research funding proposals, that were submitted to, and evaluated by, a research funding agency in the years 2012–2013. We refer to these proposals as proposals or documents. Because it permits search, and basic clustering, we used the Apache Solr software (Solr 2013), which is based on the Apache Lucene indexing software. Clustering in Solr is nearest neighbour-based, and is termed MLT, "more like this". Similarity scores between pairs of documents, based on textual content, are produced. Murtagh (2013) provides a short description of the MLT similarity. (Murtagh 2013, is available with this article's ancillary material.) Our documents were indexed by Solr, and MLT similarity coefficients were generated for the top 100 matching proposals. A selection of 10,317 of these proposals constituted the set that was studied. Our major aim in this work was prototyping our approach, based on the results provided by Solr. The R sparse matrix format (Matrix Market 2013) was used for subsequent R processing. The maximum MLT score (i.e. similarity coefficient value) was 3.218811. In matrix terms, we have 10,317 proposals (rows) crossed by 34,352 proposals (columns). Non-zero values accounted for 0.2854 % of the elements of this matrix.

Figure 1 serves to describe the properties of this data: a somewhat skewed Gaussian marginal distribution for the proposals, and a power law for the similar or matching MLT proposals. In Murtagh et al. (2008), we also find such Gaussian and power law behaviour for high dimensional chemical data.

In this article, we will seek to cluster the 10,317 proposals, using their similarities with the fuller set of 34,352 proposals as features. Justification for this feature space perspective, rather than directly using the MLT similarities, is that MLT similarities are asymmetric.

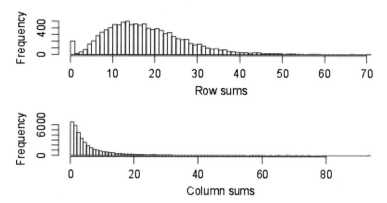

Fig. 1 Marginal distributions by row and by column. Numbers of rows, columns: 10,317, 34,352

3 Random Projection and Use for Clustering

For a discussion of random projections used for clustering, and also description and use of the Baire distance, see Contreras and Murtagh (2012). See also Sect. 4 below. Given our input data, i.e. a cloud of points in k-dimensional space, conventional random projection uses a random valued linear mapping in order to yield a much reduced dimensionality space: $f : \mathbb{R}^k \to \mathbb{R}^\ell$ where $\ell \ll k$. In our approach we are not seeking to use this ℓ-dimensional subspace, but rather we take a consensus ranked set of positions from the values of points on the ℓ axes. For this, we use a set of ℓ random projections, each onto a one-dimensional subspace. Therefore we consider ℓ random axes. The theory underpinning this, relative to conventional (Kaski 1998) random projection, is provided in Murtagh and Contreras (2015).

To deal with variability of outcomes in random projections used for clustering, Fern and Brodley (2003) project to a random subspace, apply a Gaussian mixture model, using expectation maximization, then an ensemble-based data aggregation matrix collects interrelationship information, which is submitted to an agglomerative hierarchical clustering. In Boutsidis et al. (2010), a random projection subspace is Shown to provide bounds on k-means clustering properties. The objective in Kaski (1998) is to determine the subspace of best metric fit to the original space.

For Urruty et al. (2007): "We begin by clustering the points of each of the selected uni-dimensional projections." And: "in the second phase we refine the clustering by using two processes: bimodulation and cluster expansion." (The former term introduced by those authors is for cluster specification using multiple random projections; and the latter term used by those authors uses hyper-rectangles to find the largest density cluster.) In this article, we develop in particular the early phase of clustering on uni-dimensional projections, and we relate such clustering to the Baire hierarchical clustering. Our objective is to develop a fast multiresolution hashing approach to clustering, rather than the optimal fit of proximity relations in $\mathbb{R}^\ell$, relative to proximity relations in $\mathbb{R}^k$.

4   Baire Clustering of a Random Spanning Path

4.1   Random Spanning Paths

Consider a random projection into a one-dimensional subspace, i.e. onto a random axis, of our set of documents. Such a random projection defines a permutation of the object set. It thereby defines a random spanning path. Spanning paths are useful and beneficial for data analysis. An optimal (i.e. minimum summed weight) spanning path has been used as an alternative to a minimal spanning tree (Murtagh 1985, ch. 4). The spanning path is the solution of the travelling salesman problem (Murtagh 1985, ch. 1). Braunstein et al. (2007) consider bounds for random path lengths relative to the optimal path length in the case of Erdős–Rényi and scale-free networks.

4.2   Inducing a Hierarchy through Endowing the Data with the Baire Metric

Our algorithm is as follows. Determine a random projection of our data. Induce a Baire hierarchy, using a regular 10-way tree. At level 1, the clusters will be labelled by 0, 1, 2, ... 9. At level 2, the labelling is 00, 01, ... 99. Full details of the Baire metric, and ultrametric, that endows the data with a hierarchy, is described in Contreras and Murtagh (2012). Our data values are univariate. Without loss of generality, take our values as being bounded by 0 and 1. An immediate consequence of the Baire metric is that, at level 1, all values that start with 0.3 will be in the same cluster; as will all values that start with 0.4; and so on for the 10 clusters at level 1. The Baire metric is a longest common prefix metric.

A random projection onto a one-dimensional axis provides a view of the relationships in the data, and hence a view of the clustering properties. See Contreras and Murtagh (2012). The random projected values are found to be quite similar in their interrelationships for different random vectors. We demonstrate this below. We determine the consensus or majority set of neighbourhood relationships from a sufficiently large set of random projections.

Now consider a given random projection. We determine a partition into clusters of the observables, following projection onto the random vector. A set of partitions can be sought, with their clusters ordered by inclusion.

Traditional approaches to clustering use pairwise dissimilarities, between adjacent clusters of points. (In partitioning, k-means takes a set of cluster centres and stepwise refines this set of cluster centres, together with their cluster assignments. Hierarchical clustering determines, stepwise, the smallest set of dissimilarities and agglomerates the associated pair of clusters.)

A direct reading of a partition is the alternative pursued here. Let the distance defined between adjacent clusters be a *p*-adic or *m*-adic distance (where typically

p refers to a prime number, and m refers to a non-prime integer). We define a cluster by an m-adic ball: $U_r(a) = \{x : |x-a|_m \leq r\}$. A Baire distance is associated with an ultrametric (a distance defined on a tree, rather than the real number line). Balls are either disjoint or are ordered by inclusion. It follows that for given r a partition is defined. For a set of values of r the set of associated partitions have clusters that are hierarchically structured, i.e. the associated set of clusters is a partially ordered set.

To address variability in results furnished by different random projections, we adopt the following approach: first, determine a stable, mean random projection. Then use it as the basis for a Baire clustering.

Parenthetically, let us address a comment sometimes made in regard to the m-adic distance used by us here. (We use $m = 10$; p-adic distances, where p is a prime, lead to an alternative to the real number system.) Consider two real measurements with values $2.99999\ldots$ and $3.0000\ldots$. These would be mapped onto different clusters in our approach. The following remark is however an appropriate one here: "two points on a complex protein may be close in Euclidean space but distant in terms of chemical reaction propensity" (Manton et al. 2008, pp. 81–82). In other words, if our digits have some form of inherent meaning, then it may well be fully appropriate to consider very similar real values to be quite separate and distinct.

4.3 Stability of Random Spanning Path

In Fig. 2 we assess the convergence, based on the first random projection, and the successive means of 2, 3, 4, ..., 98 random projections. The squared error is between the mean of these random projections, each normalized by its maximum value, and the mean of the 99 random projections, also normalized. We note the fast and stable (although not uniform) convergence. (The R code carrying out this processing is available on the web site containing our ancillary material.)

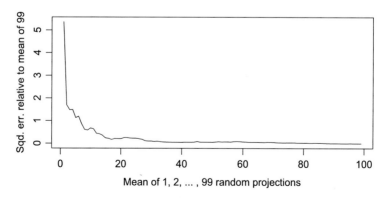

Fig. 2 Squared error of the mean of 1, 2, 3, ..., 98 random projections, relative to the mean of 99 random projections

4.4   Random Spanning Paths are Highly Correlated

In the case of random projection sets (mean of 99 realizations, sorted), for reproducibility we set the initialization seeds. For seeds 1471 and 3189, we had a correlation coefficient of 0.9999919.

Another random projection set (mean of 99 realizations, sorted) was generated with seed 7448. The correlations with the first two random projections were 0.9999937 and 0.9999905.

A further random projection was generated (seed 8914), and correlations with the first three sets were: 0.9999933, 0.9999898, 0.9999933. We conclude that using a given random projection set, in our work here resulting from 99 realizations, this is a fully sufficient basis for further cluster analysis.

In Murtagh and Contreras (2015) we consider theoretical properties of one-dimensional random projection in very high dimensional spaces.

5   Applying the Baire Distance to Obtain the Hierarchical Clustering

5.1   The Baire Metric and Ultrametric

As we have noted, the Baire distance is a longest common prefix distance which is also an ultrametric, or distance defined on a tree. In a given random projection, we can read off clusters using their Baire distance properties. Consider four adjacent, in rank order, projected values: 3.493297, 3.493731, 3.499185, 3.499410. The maximum value found for this particular random vector was 35.21912. We fix, in this instance, the projected values to be 8 digit values (viz., 2 digits in the integer part, and 6 digits in the fractional part, with zero padding if necessary). We define the Baire distance, with base 10, as 10 to the negative power of the last common, shared, digit.

The first two of our projected values above have Baire distance equal to 10^{-4} (because they share these digits: 3.493). The second two of our projected values above have Baire distance equal to 10^{-4}. The Baire distance between the second and third of our projected values above is 10^{-3}. The first and the fourth of our projected values have this same Baire distance, 10^{-3}.

Having defined the Baire distance between projected values, we next consider the Baire distance between clusters of projected values. Consistent with our consideration of adjacency of projected values, a cluster is a segment or succession of adjacent values. A singleton cluster is a single projected value. By considering the agglomeration of adjacent values 3.493297, 3.493731 at Baire distance 10^{-4}, and furthermore the adjacent values 3.499185, 3.499410 also at Baire distance 10^{-4}, we have the agglomeration of these two clusters, or segments, at Baire distance 10^{-3}, since the digits 3.49 are shared.

Computational analysis is as follows. For a random projection, we have the product of a (sparse) $k \times \ell$ matrix and a vector. Before taking sparsity into account, this gives $O(k\ell)$ time. The mean of a fixed number of random projections requires $O(k)$ time. The potentially linear Baire distance clustering comes from reading the mean random projection values, with assignment of each in turn to cluster nodes in the Baire hierarchy. In this way, a linked list of cluster (or node in the Baire hierarchy) members is built up.

5.2 A Theorem Ensuing from the Baire Ultrametric

The agglomeration of clusters takes a cluster or segment of ordered values $(x_{lo} \ldots x_{hi})$ to be agglomerated with a cluster of ordered values $(y_{lo} \ldots y_{hi})$. Based on adjacency in the clustering of random projections, and our definition of Baire distance between clusters, we have the following, where d_B is the Baire distance: $d_B(x_{lo}, y_{hi}) = d_B(x_{hi}, y_{lo})$.

If the two adjacent clusters are labelled c_x and c_y, then $\max\{d_B(i, j) \mid i \in c_x, j \in c_y\} = d_B(x_{lo}, y_{hi})$. Call this Baire distance $d_{max}(c_x, c_y)$. Similarly, $\min\{d_B(i, j) \mid i \in c_x, j \in c_y\} = d_B(x_{hi}, y_{lo})$. Call this Baire distance $d_{min}(c_x, c_y)$. What we availed of here was: $x_{lo} < x_{hi} < y_{lo} < y_{hi}$. From the foregoing description, the following theorem holds.

Theorem for Baire distance, d: $d_{min}(c_x, c_y) = d_{max}(c_x, c_y)$ for all contiguous clusters, c_x, c_y.

To show this, we start with singleton clusters, and the definition of the Baire distance, d. Following cluster formation, the cardinalities of the clusters will grow. By induction this theorem is extended to clusters c_x, c_y of any cardinality. A simple example ensues from the 4 points, together with their projected values, that were discussed above in Sect. 5.1. Given the terms "single link" and "complete link", as used in traditional hierarchical clustering, this theorem establishes that single and complete link agglomerative criteria are identical. This finding is consistent with having endowed our data not just with a metric, but with an ultrametric.

It has been noted how a random projection on a one-dimensional subspace is a random spanning path. This also establishes a contiguity or adjacency relationship between all points that we are analysing. So our hierarchical clustering can also be considered as a contiguity-constrained hierarchical clustering.

In Murtagh (1985) two contiguity-constrained hierarchical clustering algorithms were discussed. Proofs were provided that both would guarantee that no inversions could arise in the hierarchy, that is, there could be no non-monotonic change in cluster criterion value. One algorithm, also developed by other authors, Ferligoj and Batagelj (1982) and Legendre and Legendre (2012), was contiguity constrained complete link clustering: the pairwise most distant set of (by requirement, contiguous) cluster members determines the inter-cluster dissimilarity: $d_{max}(c_x, x_y)$. The other contiguity-constrained hierarchical clustering was single link, where inter-cluster dissimilarity is defined as the pairwise closest set of cluster members ($\min\{d_{ij}|i \in c_1, j \in c_2\}$), subject to the contiguity constraint.

By virtue of the theorem above, for all adjacent and agglomerable clusters c_x, c_y, $d_{\min}(c_x, c_y) = d_{\max}(c_x, c_y)$, we also have that the above described contiguity-constrained complete link and the contiguity-constrained single link hierarchical clustering methods are identical. This holds because of the Baire distance.

These perspectives add to the importance, in practice and in its theoretical foundations, of the theorem for the Baire distance.

5.3 Visualization of Baire Hierarchy

Using a regular 10-way tree, Fig. 3 shows a Baire hierarchy with nodes colour-coded (rainbow colour lookup table used), and with the root (a single colour, were it shown), comprising all clusters, to the bottom. The terminals of the 8-level tree are at the top. Ancillary material for this article, as noted in the "Introduction", is available. The R code used for Fig. 3 is listed there, and the code for the subsequent analysis of clusters extracted from the hierarchy.

The first Baire layer of clusters, displayed as the bottom level in Fig. 3, was found to have 10 clusters. (8 are very evident, visually.) The next Baire layer has 87

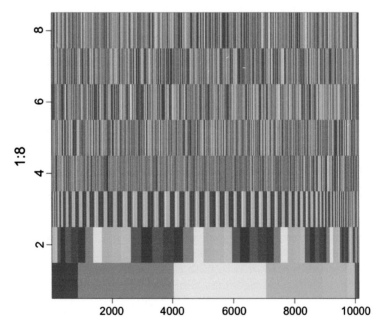

Fig. 3 Means of 99 random projections. Abscissa: the 10,118 (non-empty) documents are sorted (by random projection value). Ordinate: each of 8 digits comprising random projection values

clusters, and the third Baire layer has 671 clusters. See our ancillary material for a study of the clusters at layers 1 and 2.

6 Conclusions

In Contreras and Murtagh (2012), there is reporting on analysis of clusters found using the methodology developed here (in application domains that include astronomy and chemistry) and there is comparison with other, alternative processing approaches.

We can state that our work is oriented towards inter-cluster analysis, rather than intra-cluster analysis. That is to say, we want candidate observation classes, and furthermore we seek to be selective about what we derive from the data, in order to carry on to further use of the selected, derived clusters. Such overall processing is very suitable for big data analytics. The theorem stated in Sect. 5.2 points to the major importance of the Baire viewpoint. Further theoretical results are presented in Murtagh and Contreras (2015).

Acknowledgements We are grateful to Paul Morris for initial discussions related to this work.

References

Boutsidis, C., Zouzias, A., & Drineas, P. (2010). Random projections for k-Means clustering. *Advances in Neural Information Processing Systems, 23*(iii), 298–306.

Braunstein, L. A., Zhenhua W. U., Chen, Y., Buldyrev, S. V., Kalisky, T., Sreenivasan, S., Cohen, R., López, E., Havlin, S., & Stanley, H. E. (2007). Optimal path and minimal spanning trees in random weighted networks. *International Journal of Bifurcation and Chaos, 17* (7), 2215–2255.

Contreras, P., & Murtagh, F. (2012). Fast, linear time hierarchical clustering using the baire metric. *Journal of Classification, 29*, 118–143.

Ferligoj, A., & Batagelj, V. (1982). Clustering with relational constraint. *Psychometrika, 47*, 413–426.

Fern, X. Z., Brodley, C. E. (2003). Random projection for high dimensional data clustering: A cluster ensemble approach. In T. Fawcett & N. Mishra (Eds.), *Proceedings 20th International Conference on Machine Learning* (pp. 186–193).

Kaski, S. (1998). Dimensionality reduction by random mapping: Fast similarity computation for clustering. In *IJCNN'98, IEEE International Joint Conference on Neural Networks* (Vol. 1, pp. 413–418).

Legendre, P., & Legendre, L. (2012). *Numerical ecology* (3rd ed.). Amsterdam: Elsevier.

Manton, K. G., Huang, H. & Xiliang G. U. (2008). Chapter 3 - Molecular basis of CNS aging, frailty, fitness and longevity: A Model based on cellular energetic. In J. P. Tsai (Ed.), *Leading-edge cognitive disorders research*, New York: Nova Science, Hauppauge.

Matrix Market (2013). Matrix market exchange formats, http://math.nist.gov/MatrixMarket/formats.html

Murtagh, F. (1985). *Multidimensional clustering algorithms*. Heidelberg and Vienna: Physica-Verlag.

Murtagh, F. (2004). On ultrametricity, data coding, and computation. *Journal of Classification, 21*, 167–184.

Murtagh, F. (2013). MoreLikeThis and Scoring in Solr, report, 4 pp., 26 May 2013. http://www.multiresolutions.com/HiClBaireRanSpanPaths

Murtagh, F., & Contreras, P. (2015). Constant time search and retrieval in massive data with linear time and space setup, through randomly projected piling and sparse p-adic coding, article in preparation.

Murtagh, F., Downs, G., & Contreras, P. (2008). Hierarchical clustering of massive, high dimensional data sets by exploiting ultrametric embedding. *SIAM Journal of Scientific Computing, 30*, 707–730.

Solr (2013). Solr, Apache Lucene based search server, http://lucene.apache.org/solr

Urruty, T., Djeraba, C., & Simovici, D. A. (2007). Clustering by random projections, *Advances in data mining. Theoretical aspects and applications lecture notes in computer science* (Vol. 4597, pp. 107–119).

Standard and Novel Model Selection Criteria in the Pairwise Likelihood Estimation of a Mixture Model for Ordinal Data

Monia Ranalli and Roberto Rocci

Abstract The model selection in a mixture setting was extensively studied in literature in order to assess the number of components. There exist different classes of criteria; we focus on those penalizing the log-likelihood with a penalty term, that accounts for model complexity. However, a full likelihood is not always computationally feasible. To overcome this issue, the likelihood is replaced with a surrogate objective function. Thus, a question arises naturally: how the use of a surrogate objective function affects the definition of model selection criteria? The model selection and the model estimation are distinct issues. Even if it is not possible to establish a cause and effect relationship between them, they are linked to each other by the likelihood. In both cases, we need to approximate the likelihood; to this purpose, it is computationally efficient to use the same surrogate function. The aim of this paper is not to provide an exhaustive survey of model selection, but to show the main used criteria in a *standard* mixture setting and how they can be adapted to a *non-standard* context. In the last decade two criteria based on the observed composite likelihood were introduced. Here, we propose some new extensions of the standard criteria based on the expected complete log-likelihood to the non-standard context of a pairwise likelihood approach. The main advantage is a less demanding and more stable estimation. Finally, a simulation study is conducted to test and compare the performances of the proposed criteria with those existing in literature. As discussed in detail in Sect. 7, the novel criteria work very well in all scenarios considered.

M. Ranalli (✉)
Department of Statistics, The Pennsylvania State University, University Park,
State College, PA, USA

Sapienza, University of Rome, Rome, Italy
e-mail: monia.ranalli@psu.edu

R. Rocci
Department of Economics and Finance, University of Tor Vergata, Rome, Italy
e-mail: roberto.rocci@uniroma2.it

© Springer International Publishing Switzerland 2016
A.F.X. Wilhelm, H.A. Kestler (eds.), *Analysis of Large and Complex Data*, Studies in Classification, Data Analysis, and Knowledge Organization,
DOI 10.1007/978-3-319-25226-1_5

1   Introduction

The aim of statistical model selection is to pick the model which represents the best approximation of reality shown in the observed data. Model selection procedures are needed almost everywhere: for example, in a linear regression context, the focus is on variable selection; in cluster analysis, it is on the number of clusters and so on. An extensive literature exists about assessing the number of components for a finite mixture model. Different approaches were proposed, such as the bootstrap technique, the directional score functions, the nonparametric methods and the penalized likelihoods. In the sequel we focus on the last approach that balances accuracy, represented by the log-likelihood, with parsimony, represented by the penalty term whose function is to account for model complexity. One of the requirements is to adopt a full likelihood approach to obtain the maximized likelihood value. However, there exist situations in which the likelihood is cumbersome (for example, it involves multidimensional integrals) and thus the maximum likelihood estimation is not computationally feasible; for example, it happens in the mixture model for ordinal data proposed by Ranalli and Rocci (2014). They suggested to adopt a pairwise likelihood approach to estimate the model. Even if it is not possible to establish a cause and effect relationship between model estimation and model selection, they are definitely linked to each other by the likelihood. In other words, in both cases, we need to approximate the likelihood; to this purpose, it is computationally efficient to use the same surrogate function, i.e. the pairwise likelihood. It follows that the chosen surrogate function has some consequences on the definition of the model selection criteria. In fact, each sub-likelihood (a likelihood composed of a pair of variables) is a true likelihood, but overall the pairwise likelihood (the product of all sub-likelihoods) is not. Thus, the penalized criteria developed within a full likelihood approach should be adapted to this framework. In Sect. 2 we summarize the main features of the mixture model for ordinal data. In Sect. 3 we show the main used criteria in a *standard* mixture setting, distinguishing two classes of criteria: those based on the observed log-likelihood (Sect. 4) and those based on the complete log-likelihood (Sect. 5). Some extensions to a *non-standard* mixture setting are presented in Sect. 6: there we present both the existing and the novel criteria. In Sect. 7, a simulation study is conducted aimed at testing and comparing the performances of the criteria in a pairwise likelihood context. Finally some concluding remarks are pointed out in the last section.

2   Mixture Model for Ordinal Data

In this section we describe briefly the model proposed by Ranalli and Rocci (2014). Let $x_1, x_2, \ldots, x_P$ be ordinal variables and $c_i = 1, \ldots, C_i$ the associated categories for $i = 1, 2, \ldots, P$. There are $R = \prod_{i=1}^{P} C_i$ possible response patterns $\mathbf{x}_r = (x_1 = c_1, x_2 = c_2, \ldots, x_P = c_P)$, with $r = 1, \ldots, R$. The ordinal variables are assumed to be generated by thresholding $\mathbf{y} = \sum_g^G p_g \boldsymbol{\phi}(\boldsymbol{\mu}_g, \boldsymbol{\Sigma}_g)$ that is a multivariate continuous random variable distributed as a finite mixture of Gaussians (FMG). The link between $\mathbf{x}$ and $\mathbf{y}$ is expressed by a threshold model defined as

$$x_i = c_i \Leftrightarrow \gamma_{c_i-1}^{(i)} \le y_i < \gamma_{c_i}^{(i)}. \tag{1}$$

Let $\boldsymbol{\psi} = \{p_1, \ldots, p_G, \boldsymbol{\mu}_1, \ldots, \boldsymbol{\mu}_G, \boldsymbol{\Sigma}_1, \ldots, \boldsymbol{\Sigma}_G, \boldsymbol{\gamma}\}$ be the set of model parameters. The probability of response pattern $\mathbf{x}_r$ is given by

$$Pr(\mathbf{x}_r; \boldsymbol{\psi}) = \sum_{g=1}^{G} p_g \int_{\gamma_{c_1-1}^{(1)}}^{\gamma_{c_1}^{(1)}} \cdots \int_{\gamma_{c_P-1}^{(P)}}^{\gamma_{c_P}^{(P)}} \boldsymbol{\phi}(\mathbf{y}; \boldsymbol{\mu}_g, \boldsymbol{\Sigma}_g) d\mathbf{y} = \sum_{g=1}^{G} p_g \pi_r(\boldsymbol{\mu}_g, \boldsymbol{\Sigma}_g, \boldsymbol{\gamma}) \tag{2}$$

where $\pi_r(\boldsymbol{\mu}_g, \boldsymbol{\Sigma}_g, \boldsymbol{\gamma})$ is the probability of response pattern $\mathbf{x}_r$ in cluster g. Thus, for a random i.i.d. sample of size N, $\mathbf{X}$, the log-likelihood is

$$\ell(\boldsymbol{\psi}; \mathbf{X}) = \sum_{r=1}^{R} n_r \log \left[\sum_{g=1}^{G} p_g \pi_r(\boldsymbol{\mu}_g, \boldsymbol{\Sigma}_g, \boldsymbol{\gamma}) \right], \tag{3}$$

where n_r is the observed sample frequency of response pattern $\mathbf{x}_r$ and $\sum_{r=1}^{R} n_r = N$. In the sequel we only mention the relevant literature; we point the reader to the introduction of Ranalli and Rocci (2014) for a wider picture on the existing literature; some applications to real data can be found in Everitt and Merette (1990) and Ranalli and Rocci (2014).

A similar model is proposed by Everitt (1988) who introduces a mixture model for mixed data (see also Everitt and Merette 1990). The joint distribution of the variables is a homoscedastic FMG where some variables are observed as ordinal. In particular, the ordinal variables are seen as generated by thresholding some marginals of the joint FMG with different thresholds in each component. The same framework is used by Lubke and Neale (2008). Their model is specified for ordinal variables that are generated by thresholding an heteroschedastic mixture of Gaussians, whose covariance matrices are reparametrized using a simplified dependence structure (as it occurs in a factor analysis model). In all models estimation is carried out by full maximum likelihood. Nevertheless there is always a multidimensional integral, whose dimension depends on the number of observed variables. Its numerical computation is time consuming and becomes infeasible when more than 4 or 5 variables are involved. For this reason, in Ranalli and Rocci (2014), the authors propose to estimate the model within the EM framework maximizing the pairwise log-likelihood, i.e. a composite log-likelihood given by the sum of all possible log-likelihoods based on the bivariate marginals. In general, the pairwise maximum likelihood estimators have been proven to be consistent, asymptotically unbiased and normally distributed. They are usually less efficient than the full maximum likelihood estimators, but in many cases the loss in efficiency is very small or almost null (Lindsay 1988; Varin et al. 2011). In formulas the pairwise log-likelihood is of the form

$$pl(\boldsymbol{\psi}; \mathbf{X}) = \sum_{i=1}^{P-1} \sum_{j=i+1}^{P} \ell(\boldsymbol{\psi}; (x_i, x_j))$$

$$= \sum_{i=1}^{P-1} \sum_{j=i+1}^{P} \sum_{c_i=1}^{C_i} \sum_{c_j=1}^{C_j} n_{c_i c_j}^{(ij)} \log \left[\sum_{g=1}^{G} p_g \pi_{c_i c_j}^{(ij)} (\boldsymbol{\mu}_g, \boldsymbol{\Sigma}_g, \boldsymbol{\gamma}) \right], \qquad (4)$$

where $n_{c_i c_j}^{(ij)}$ is the observed joint frequency in category c_i and c_j for variables x_i and x_j, respectively, while $\pi_{c_i c_j}^{(ij)}(\boldsymbol{\mu}_g, \boldsymbol{\Sigma}_g, \boldsymbol{\gamma})$ is the corresponding probability obtained by integrating the bivariate marginal (y_i, y_j) of the normal distribution with parameters $(\boldsymbol{\mu}_g, \boldsymbol{\Sigma}_g)$ between their threshold parameters. It is clear that the pairwise approach is feasible as it requires only the evaluation of integrals on bivariate normal distributions, regardless of the number of observed or latent variables $\mathbf{y}$. Nevertheless the estimation of all parameters is carried out simultaneously. As regards the classification, in Ranalli and Rocci (2014) it has been suggested to use an iterative proportional fitting algorithm based on the pairwise posterior probabilities obtained as output of the pairwise EM algorithm in order to approximate the joint posterior probabilities. For identification reasons, a component is fixed as a reference group; thus, its mean vector is set to $\mathbf{0}$ and its variances to 1. On one hand, the model proposed in Ranalli and Rocci (2014) can be seen as a particular case of Everitt's proposal (Everitt 1988). In fact here only categorical ordinal variables are considered and the identifiability constraint is reformulated such that means and covariance matrices can be computed. On the other hand, it is more flexible, since each component has its own mean vector and covariance matrix and it is computationally more efficient to be estimated, since a pairwise likelihood approach is suggested. Indeed in Everitt, the means and the variances of the latent variables are fixed to zero and one, respectively; the correlations are invariant and only the thresholds are free to change over the components. In comparison with the proposal of Lubke and Neale (2008), it is computationally feasible regardless of the number of variables involved. Moreover, it is able to cover the true partition and the true parameters (even if the accuracy depends on the sample size). For more details see Ranalli and Rocci (2014).

3   Model Selection in a Mixture Setting

In order to estimate the mixture model, the selection of the number of components G is needed. In some applications, it is specified a priori through the available information; while in most of them, G has to be inferred from the data. A way of dealing with this problem may be to use the likelihood ratio statistic test. However, in a mixture model context, the asymptotic distribution of the Wilks statistics $-2 \log \lambda$ does not have the usual distribution of chi-squared with degrees of freedom equal to the number of independent constraints that are imposed to obtain the null hypothesis, since the regularity conditions do not hold. In literature several results about the asymptotic distribution of the likelihood ratio statistic, or some modified versions, have been provided. Unfortunately they hold only in particular cases. In general the problem is sorted out by using the bootstrap technique (Mclachlan 1987), even if, in practice, this approach could become infeasible due to the heavy computation complexity required. Beside this, we found other approaches: the proposal of Lindsay (1983) according to which the number of components is chosen based on the values assumed by the directional score functions; the nonparametric methods or the criteria based on method of moments. Furthermore, there exists the wide class of information criteria based on log-likelihood function with simple penalties accounting for model complexity. In the sequel, we focus on the penalized likelihoods distinguishing two classes of criteria: those based on the observed log-likelihood and those based on the expected complete log-likelihood.

4 Model Selection Criteria Based on the Observed Log-Likelihood

In this section we briefly review the more known information criteria based on the observed log-likelihood. These were developed by different theories and goals, but, algebraically, all of them share the same principle: choosing the model with the best penalized log-likelihood. Specific definitions of the penalty term lead to different criteria. Historically the first information criteria (AIC) was introduced by Akaike (1973). The best fitted model is chosen by minimizing

$$\text{AIC} = -2\ell(\hat{\boldsymbol{\psi}}) + 2d, \tag{5}$$

where $\ell(\hat{\boldsymbol{\psi}})$ represents the observed log-likelihood, while d is the number of independent parameters. It is based on the concept of minimizing the expected Kullback–Leibler divergence between the likelihood under the fitted model and the unknown true likelihood that generated data (see, e.g., Sawa 1978; Sugiura 1978). Furthermore, it was proved to be asymptotically efficient (Shibata 1980), but not consistent. In addition, especially when the sample size is small, AIC leads to overfitting. To overcome the inconsistency of AIC, two further criteria, developed from a Bayesian point of view, have been proposed by Akaike (1978) and Schwarz (1978), called BIC and SIC, Bayesian Information Criterion and Schwarz Information Criterion, respectively. Even if they are called differently, they were introduced about the same time and they are equivalent. The idea is to select the model with the highest posterior probability: the posterior probability can be approximated by a Taylor expansion, whose the first two terms represent the likelihood for the model considered and the model complexity, respectively. According to BIC, the model is chosen by minimizing

$$\text{BIC} = -2\ell(\hat{\boldsymbol{\psi}}) + d\log(N), \tag{6}$$

where N is the sample size. Comparing $2d$ with $d\log(N)$, we note the BIC penalizes the model complexity more heavily than AIC. Thus, it corrects the tendency of AIC to fit too many components; however, if the sample size is small, BIC may select too few components. Finally, Leroux (1992) has shown that AIC and BIC consistently do not underestimate the true number of components G, while Keribin (2000) has shown that BIC is a consistent criterion to estimate G.

4.1 Some Variants

Several variants of these model selection criteria have been proposed; these aim at penalizing heavily the model complexity. Here we mention some of them:

$$\text{AIC}^3 = -2\ell(\hat{\boldsymbol{\psi}}) + 3d$$

introduced by Bozdogan (1983). He shows that the so-called *magic number* 2 in the original definition of AIC is not adequate for the mixture model. We refer the reader to Bozdogan (1993) and references therein for more details. Liang et al. (1992) proposed

$$\text{BIC}^2 = -2\ell(\hat{\boldsymbol{\psi}}) + 2d\log(N)$$

and

$$\text{BIC}^5 = -2\ell(\hat{\boldsymbol{\psi}}) + 5d\log(N),$$

where the penalty terms have been specified on the basis of their experimental results.

5 Model Selection Criteria Based on the Expected Complete Log-Likelihood

In this section we survey some criteria developed specifically in a mixture model context. They have been derived from the expected complete log-likelihood that is usually considered to build the EM algorithm. As noted by Hathaway (1986), the log-likelihood of a mixture model can be written as

$$\ell(\boldsymbol{\psi}; \mathbf{x}) = \ell_c(\boldsymbol{\psi}; \mathbf{x}, \mathbf{z}) - \sum_{n=1}^{N}\sum_{g=1}^{G} p_{ng} \log \frac{p_g f(\mathbf{x}_n; \boldsymbol{\psi}_g)}{\sum_{h=1}^{G} p_h f(\mathbf{x}_n; \boldsymbol{\psi}_h)}$$

$$= \ell_c(\boldsymbol{\psi}; \mathbf{x}, \mathbf{z}) - \sum_{n=1}^{N}\sum_{g=1}^{G} p_{ng} \log(p_{ng}); \tag{7}$$

where ℓ_c is the conditional expectation of the complete log-likelihood given the observed data, p_{ng} is the posterior probability of gth component, given the nth observation, while the second term is known as *entropy* of the fuzzy classification obtained in the E-step of the EM algorithm, $EN(\mathbf{p})$. Biernacki and Govaert (1997) introduced the Classification Likelihood Criterion (CLC), which selects as the best model that one minimizing

$$\text{CLC} = -2\ell_c(\hat{\boldsymbol{\psi}}) = -2\ell(\hat{\boldsymbol{\psi}}) + 2EN(\hat{\mathbf{p}}), \tag{8}$$

where $(\hat{\cdot})$ indicates that we have replaced the unknown parameters with the corresponding maximum likelihood estimates. On the other hand, Banfield and Raftery (1993) suggested a modified version of BIC. This criterion is called approximate weight of evidence (AWE) and it is given by

$$\text{AWE} = -2\ell_c(\hat{\boldsymbol{\psi}}) + 2d(0.5 + \log N) = -2\ell(\hat{\boldsymbol{\psi}}) + 2EN(\hat{\mathbf{p}}) + 2d(0.5 + \log N). \tag{9}$$

When the mixture components are well separated, then $\ell_c(\hat{\boldsymbol{\psi}}) \approx \ell(\hat{\boldsymbol{\psi}})$ and thus AWE is expected to be similar to BIC. Furthermore, an analogous criterium to BIC, namely the integrated classification likelihood (ICL), was introduced by Biernacki et al. (2000),

$$\text{ICL} = -2\ell_c(\hat{\boldsymbol{\psi}}) + d\log N = -2\ell(\hat{\boldsymbol{\psi}}) + 2EN(\hat{p}) + d\log N. \tag{10}$$

BIC and ICL have the same structure, but they differ in the objective penalized log-likelihood. The former considers the observed log-likelihood, while the latter the expected complete log-likelihood.

The second equality in Eqs. (8)–(10) clarifies the role of *EN*: if we express CLC, AWE or ICL in terms of the observed likelihood, *EN* works as penalty term accounting for the fuzziness of the classification. How severely *EN* penalizes the log-likelihood depends on how well separated the fitted components are: if the components are well separated it takes values close to 0 (the minimum value), otherwise it will take larger values. Finally, Celeux and Soromenho (1996) proposed to use the normalized estimated entropy $EN(\hat{p})$ (NEC) as a criterion in its own right for selecting the number of components G,

$$\text{NEC} = \frac{\ell(\hat{\psi}) - \ell_c(\hat{\psi})}{\ell(\hat{\psi}) - \ell_1(\hat{\psi})} = \frac{EN(\hat{p})}{\ell(\hat{\psi}) - \ell_1(\hat{\psi})}, \qquad (11)$$

where $\ell_1(\hat{\psi})$ denotes the likelihood for a unicomponent mixture model. In this regard, Biernacki et al. (1999) introduced a procedure to choose the number of components: if $g = 1$, NEC takes value 1; to chose $g > 1$ rather than $g = 1$, NEC should take values less than 1.

6 Extensions of the Standard Criteria

As said in the Introduction, it is not always possible to estimate a model through a full maximum likelihood approach. In Sect. 2 we gave an illustrative example where a pairwise likelihood approach was adopted. In the last decade, criteria based on the observed composite likelihood were developed. However, analogous criteria to the expected complete log-likelihood do not exist. In the sequel, these are introduced for an experimental purpose.

6.1 Existing Criteria Based on the Observed Composite Log-Likelihood

The criteria based on the observed composite likelihood have a penalty term that was already known in model selection literature. This was introduced in a mis-specified model framework. These criteria are based on the following intuition: if the fitted model is different from that one generated data (in other words the fitted model is mis-specified), then, they need a correction for the bias introduced. A generalized version of AIC was introduced by Takeuchi (1976) with the criterion called Takeuchi Information Criterion (TIC), according to which the model is chosen minimizing

$$\text{TIC} = -2\ell(\hat{\psi}) + 2\text{tr}\left\{\hat{\mathbf{H}}^{-1}(\hat{\psi})\hat{\mathbf{V}}(\hat{\psi})\right\}, \qquad (12)$$

where

$$\hat{\mathbf{H}}(\hat{\psi}) = -\frac{1}{N}\sum_{i=1}^{N}\nabla^2 \log f(\mathbf{x}_i; \hat{\psi})$$

and

$$\hat{\mathbf{V}}(\hat{\boldsymbol{\psi}}) = \frac{1}{N} \sum_{i=1}^{N} \left(\nabla \log f(\mathbf{x}_i; \hat{\boldsymbol{\psi}}) \right) \left(\nabla \log f(\mathbf{x}_i; \hat{\boldsymbol{\psi}}) \right)'.$$

It has the same form as AIC if the model is true; in this case $\hat{\mathbf{H}}(\hat{\boldsymbol{\psi}})^{-1}$ tends to be equal to $\hat{\mathbf{V}}(\hat{\boldsymbol{\psi}})$ and the trace reduces to the number of estimated parameters (i.e. d). The two criteria based on composite likelihood share the same underlying assumption as TIC, since each term of the composite likelihood is a true likelihood, but overall it is not. Furthermore, the sub-likelihoods are correlated and thus $\hat{\mathbf{H}}(\hat{\boldsymbol{\psi}})^{-1} \neq \hat{\mathbf{V}}(\hat{\boldsymbol{\psi}})$. These criteria are the Composite AIC (C-AIC) and Composite BIC (C-BIC) introduced by Varin and Vidoni (2005) and Gao and Song (2010), respectively. They combine the goodness-of-fit for a given model (minus twice the composite log-likelihood — here, we consider the pairwise log-likelihood) and the penalty term for the model complexity,

$$\text{C-AIC} = -2p\ell(\hat{\boldsymbol{\psi}}) + 2\text{tr}\left(\hat{\boldsymbol{H}}^{-1} \hat{\boldsymbol{V}} \right), \tag{13}$$

$$\text{C-BIC} = -2p\ell(\hat{\boldsymbol{\psi}}) + \log N \text{tr}\left(\hat{\boldsymbol{H}}^{-1} \hat{\boldsymbol{V}} \right), \tag{14}$$

where $\mathbf{H}$ is the sensitivity matrix, $\mathbf{H} = E(-\nabla^2 p\ell(\boldsymbol{\psi}; \mathbf{x}))$ while $\mathbf{V}$ is the variability matrix (the covariance matrix of the score vector), $\mathbf{V} = \text{Var}(\nabla p\ell(\boldsymbol{\psi}; \mathbf{x}))$. They have the same structure of AIC and BIC; the only difference is the way used to count the number of parameters; the identity $\mathbf{H} = \mathbf{V}$ does not hold, since the likelihood components are not independent (in contrast to the full likelihood theory). Referring to the model presented in Sect. 2, sample estimates of $\mathbf{H}$ and $\mathbf{V}$ are

$$\hat{\boldsymbol{H}} = -\frac{1}{N} \sum_{r=1}^{R} n_r \nabla^2 p\ell(\hat{\boldsymbol{\psi}}; \mathbf{x}_r)$$

and

$$\hat{\boldsymbol{V}} = \frac{1}{N} \sum_{r=1}^{R} n_r (\nabla p\ell(\hat{\boldsymbol{\psi}}; \mathbf{x}_r))(\nabla p\ell(\hat{\boldsymbol{\psi}}; \mathbf{x}_r))'.$$

The main drawback of C-AIC and C-BIC is the computation of the matrices $\hat{\boldsymbol{H}}$ and $\hat{\mathbf{V}}$. In the following simulation study, in order to obtain the empirical estimates of the sensitivity and variability matrices, we have used a numerical approximation technique. More precisely, the derivatives are estimated by finite differences. As regards the variability matrix, a covariance matrix of the score function has been estimated for each response pattern. Referring to the model presented in the second section, computationally speaking the variability matrix has been obtained by multiplying a matrix including the score functions for each response pattern times a diagonal matrix with the frequencies n_r on the main diagonal times the first matrix transposed. As regards the sensitivity matrix, we know from the theoretical results of the pairwise that each sub-likelihood (i.e. each component of the pairwise likelihood) is a true likelihood; this means that the second Bartlett's identity holds. This allows us to estimate the sensitivity matrix in the same fashion as before. However in this case the diagonal matrix has the frequencies $n_{x_i x_j}$ on the main diagonal and the score functions refer to each response pattern for each pair of variables. Finally, the trace is obtained by summing the generalized eigenvalues of the two matrices, i.e. by solving the equation $\hat{\mathbf{V}}\mathbf{x} = \lambda\hat{\mathbf{H}}\mathbf{x}$. This allows to avoid inverting the sensitivity matrix, that may be imprecise and unstable.

6.2 Novel Criteria Based on the Expected Complete Composite Log-Likelihood

The computational complexity of the existing criteria based on composite log-likelihood, gave us motivation to investigate the behaviour of criteria based on the *expected complete* composite log-likelihood. We focus on the pairwise log-likelihood, but all the following extensions can be easily generalized to any other composite likelihood. The key point is to show that the principle of the fuzzy log-likelihood is also true for a pairwise log-likelihood, that is reported in the Appendix. This justifies the extension of all criteria based on the complete log-likelihood to the complete pairwise log-likelihood without violating any assumption about the specification of the model, in principle. Furthermore, they have the appealing feature to be estimable completely from the output of the pairwise EM algorithm. In the following simulation study, we explore more in depth the empirical behaviour of CLC, AWE, ICL and NEC by replacing the expected complete log-likelihood in (8)–(11) with the expected complete pairwise log-likelihood. We indicate these novel criteria with p-CLC, p-AWE, p-ICL and p-NEC, respectively. However, to be more precise, the principle of the fuzzy pairwise log-likelihood justifies only partially some of these extensions. In fact, as regards AWE and ICL, they require the specification of d; in the simulation study it was set equal to the sum of the number of estimated parameters in each term of the pairwise log-likelihood. However, this choice was made from a practical point of view and a honed choice should be considered in future.

7 Simulation Study

Here, we consider the model presented in Sect. 2 to conduct a simulation study aimed at comparing the performances of the model selection criteria presented previously. Eleven criteria, namely AIC, AIC^3, BIC, BIC^2, BIC^5, C-AIC, C-BIC, p-CLC, p-AWE, p-ICL and p-NEC were included in the study. We did not include CLC, AWE, ICL, NEC and TIC since the main aim was to see if the novel criteria, that are easier to estimate, have competitive performances to C-AIC and C-BIC. Nevertheless, a naive extension of the standard criteria, AIC, BIC and their variants have been included as benchmarks.

They were computed using the maximum pairwise log-likelihood instead of the full maximum likelihood. As regards the number of parameters d, it was specified as the sum of the number of parameters in each sub-likelihood. We simulated 250 samples from a latent mixture of Gaussians in eight different scenarios considering three different experimental factors: the sample size ($N = 500, 2500$), the thresholds (equidistant or non-equidistant) and the separation between clusters (well separated or non well separated). In the last case, the separation between clusters is given by distance between mixture component means. In each scenario, we simulated three ordinal variables with six categories by thresholding a two-component mixture with $p_1 = 0.3$, $\boldsymbol{\mu}_1 = [0, 0, 0]$ and

$$
\boldsymbol{\Sigma}_1 = \begin{bmatrix} 1 & -0.5 & -0.6 \\ -0.5 & 1 & -0.3 \\ -0.6 & -0.3 & 1 \end{bmatrix}, \boldsymbol{\Sigma}_2 = \begin{bmatrix} 2.9 & 1.9 & 2.8 \\ 1.9 & 1.3 & 2.0 \\ 2.8 & 2.0 & 4.2 \end{bmatrix}.
$$

The remaining parameters vary with the scenario, as shown in Table 1. For each of the 2000 simulated datasets, we have fitted different number of components, $G = 1, 2, 3$. A criterion based on the Aitken acceleration has been used as a convergence criteria (Böhning et al. 1994) for the EM algorithm. It was stopped when the increase in the asymptotic estimate log-likelihood between two consecutive steps was less than $\epsilon = 10^{-2}$, i.e. when $| \ell_{i+1}^{\infty} - \ell_i^{\infty} | < \epsilon$, where

$$
\ell_{i+1}^{\infty} = \ell_i + \frac{\ell_{i+1} - \ell_i}{1 - c_i} \text{ and } c_i = \frac{\ell_{i+1} - \ell_i}{\ell_i - \ell_{i-1}}.
$$

To carry out this simulation study, we wrote our own Matlab code. The results in Table 2 suggest that the naive AIC, BIC and their variants are not reliable at all. As expected, their performances are worse than those shown by any other criteria considered, since they are not taking into account the dependencies between the pairwise components. Moreover, it is surprising to know that sometimes when the

Table 1 True values of the mixture mean vectors and thresholds under different scenarios

Groups well separated and equidistant thresholds and non-equidistant thresholds	
$\boldsymbol{\mu}_2 = [3, 2, -4]$	
$\boldsymbol{\gamma} = [-1.2, -0.6, 0, 0.6, 1.2]$	$\boldsymbol{\gamma} = [-1.5, -0.5, 0.25, 0.75, 1]$
Groups non well separated and equidistant thresholds and non-equidistant thresholds	
$\boldsymbol{\mu}_2 = [1.5, 1, -1.5]$	
$\boldsymbol{\gamma} = [-1.2, -0.6, 0, 0.6, 1.2]$	$\boldsymbol{\gamma} = [-1.5, -0.5, 0.25, 0.75, 1]$

Table 2 Percentages of times out of the 250 samples for which the 11 criteria considered select the true number of components ($G = 2$) by different scenarios

Scenario	C-AIC	C-BIC	AIC	AIC^3	BIC	BIC^2	BIC^5	p-CLC	p-AWE	p-ICL	p-NEC
Well-separated groups Equidistant thresholds $N = 500$	98.4	100.0	44.4	44.4	44.4	44.4	44.4	100.0	98.0	100.0	100.0
Well-separated groups Equidistant thresholds $N = 2500$	99.6	100.0	62.8	62.8	62.8	62.8	62.8	100.0	100.0	100.0	100.0
Well-separated groups Non-equidistant thresholds $N = 500$	97.6	99.2	23.2	23.2	23.2	23.2	23.2	98.8	97.6	99.2	99.2
Well-separated groups Non-equidistant thresholds $N = 2500$	98.8	100.0	46.8	46.8	46.8	46.8	46.8	100.0	99.6	100.0	100.0
Non well-separated groups Equidistant thresholds $N = 500$	98.8	100.0	44.0	44.0	44.0	44.0	44.0	100.0	98.8	100.0	100.0
Non well-separated groups Equidistant thresholds $N = 2500$	99.6	100.0	68.4	68.4	68.4	68.4	68.4	100.0	100.0	100.0	100.0
Non well-separated groups Non-equidistant thresholds $N = 500$	97.6	100.0	18.4	18.4	18.4	18.4	18.4	99.6	98.4	100.0	99.6
Non well-separated groups Non-equidistant thresholds $N = 2500$	97.6	100.0	21.6	21.6	21.6	21.6	21.6	100.0	98.8	100.0	100.0

groups are not well-separated AIC and BIC work better compared to the cases in which the groups are well separated. However, a kind of consistency is presented, since their performances improve as the sample size increases. As regards C-AIC and C-BIC, the percentage of selecting the true number of components increases with the sample size. There is no significant difference between the scenarios with equidistant/non-equidistant thresholds and well-/non well-separated groups. However, C-BIC seems to work slightly better. Finally, all the criteria based on the expected complete pairwise log-likelihood show very competitive results; in some cases they behave even better than C-AIC. Once again, as the sample size increases they improve their performances, while the other two experimental factors (equidistant/non-equidistant thresholds and well/non well separated) do not affect them.

It is worth making some further observations. Firstly, compared to C-AIC and C-BIC they can be estimated more easily: their estimation is less demanding and more stable. Indeed they can be obtained completely from the output of the pairwise EM algorithm without requiring any derivative estimations. Secondly, even if, so far, there is no theoretical justification to equal d to the sum of the number of estimated parameters in each sub-likelihood, its weight as penalty does not seem to affect the performances of the criteria (differently from AIC, BIC and their variants).

8   Conclusions

In this paper we have briefly surveyed the more known model selection criteria used within a mixture framework, focusing on the model proposed by Ranalli and Rocci (2014) as motivating example. The estimation method used belongs to the composite likelihood; this has some consequences on the construction of the model selection criteria. We have outlined three different groups of criteria discussing their strengths and weaknesses: those based on the observed log-likelihood, those based on the expected complete log-likelihood and the extensions developed for a composite likelihood setting. In the latter case, we have introduced some novel extensions of the standard criteria (p-CLC, p-AWE, p-ICL and p-NEC). Then the differences in performances have been illustrated through a simulation study. It gave an interesting result: the novel criteria based on the expected complete pairwise log-likelihood work very well in all scenarios considered; they sometimes behave even better than C-AIC and C-BIC, i.e. the two main criteria developed within a composite likelihood framework. Thus, it seems reasonable to use these criteria; moreover, they are more stable numerically, since they do not require any derivative estimations and they can be easily obtained from the output of the EM-like algorithm. Although there still remains some undone work, for example, justifying the choice of equalling d to the sum of the number of estimated parameters in each sub-likelihood, the results seem to be very promising.

Appendix

Maximizing the observed pairwise log-likelihood is equivalent to maximize the fuzzy classification pairwise log-likelihood. This partially justifies the behaviour of the criteria based on the expected complete pairwise log-likelihood. In this appendix we derive the pairwise EN term. This is useful to two things: if we define the pairwise EN, the criteria based on the expected complete pairwise log-likelihood can be seen as the observed pairwise likelihood penalized by the pairwise EN term. Moreover, it gives us an idea about the separation between the mixture components.

$$
p\ell(\boldsymbol{\psi}; \mathbf{X}) = \sum_{i=1}^{P-1} \sum_{j=i+1}^{P} \sum_{c_i=1}^{C_i} \sum_{c_j=1}^{C_j} n_{c_i c_j}^{(ij)} \log \left[\sum_{h=1}^{G} p_h \pi_{c_i c_j}^{(ij)} (\boldsymbol{\mu}_h, \boldsymbol{\Sigma}_h, \boldsymbol{\gamma}) \right]
$$

$$
= \sum_{i=1}^{P-1} \sum_{j=i+1}^{P} \sum_{c_i=1}^{C_i} \sum_{c_j=1}^{C_j} n_{c_i c_j}^{(ij)} \sum_{g=1}^{G} p_{c_i c_j; g}^{(ij)} \log \left[\sum_{h=1}^{G} p_h \pi_{c_i c_j}^{(ij)} (\boldsymbol{\mu}_h, \boldsymbol{\Sigma}_h, \boldsymbol{\gamma}) \right]
$$

$$
= \sum_{i=1}^{P-1} \sum_{j=i+1}^{P} \sum_{c_i=1}^{C_i} \sum_{c_j=1}^{C_j} n_{c_i c_j}^{(ij)} \sum_{g=1}^{G} p_{c_i c_j; g}^{(ij)} \log \left[\sum_{h=1}^{G} p_h \pi_{c_i c_j}^{(ij)} (\boldsymbol{\mu}_h, \boldsymbol{\Sigma}_h, \boldsymbol{\gamma}) \right]
$$

$$
+ \sum_{i=1}^{P-1} \sum_{j=i+1}^{P} \sum_{c_i=1}^{C_i} \sum_{c_j=1}^{C_j} n_{c_i c_j}^{(ij)} \sum_{g=1}^{G} p_{c_i c_j; g}^{(ij)} \log \left[p_g \pi_{c_i c_j}^{(ij)} (\boldsymbol{\mu}_g, \boldsymbol{\Sigma}_g, \boldsymbol{\gamma}) \right]
$$

$$
- \sum_{i=1}^{P-1} \sum_{j=i+1}^{P} \sum_{c_i=1}^{C_i} \sum_{c_j=1}^{C_j} n_{c_i c_j}^{(ij)} \sum_{g=1}^{G} p_{c_i c_j; g}^{(ij)} \log \left[p_g \pi_{c_i c_j}^{(ij)} (\boldsymbol{\mu}_g, \boldsymbol{\Sigma}_g, \boldsymbol{\gamma}) \right]
$$

$$
= \sum_{i=1}^{P-1} \sum_{j=i+1}^{P} \sum_{c_i=1}^{C_i} \sum_{c_j=1}^{C_j} n_{c_i c_j}^{(ij)} \sum_{g=1}^{G} p_{c_i c_j; g}^{(ij)} \log \left[p_g \pi_{c_i c_j}^{(ij)} (\boldsymbol{\mu}_g, \boldsymbol{\Sigma}_g, \boldsymbol{\gamma}) \right]
$$

$$
- \sum_{i=1}^{P-1} \sum_{j=i+1}^{P} \sum_{c_i=1}^{C_i} \sum_{c_j=1}^{C_j} n_{c_i c_j}^{(ij)} \sum_{g=1}^{G} p_{c_i c_j; g}^{(ij)} \left[\log \left(p_g \pi_{c_i c_j}^{(ij)} (\boldsymbol{\mu}_g, \boldsymbol{\Sigma}_g, \boldsymbol{\gamma}) \right) \right.
$$

$$
\left. - \log \sum_{h=1}^{G} p_h \pi_{c_i c_j}^{(ij)} (\boldsymbol{\mu}_h, \boldsymbol{\Sigma}_h, \boldsymbol{\gamma}) \right]
$$

$$
= \sum_{i=1}^{P-1} \sum_{j=i+1}^{P} \sum_{c_i=1}^{C_j} \sum_{c_j=1}^{C_j} n_{c_i c_j}^{(ij)} \sum_{g=1}^{G} p_{c_i c_j; g}^{(ij)} \log \left[p_g \pi_{c_i c_j}^{(ij)} (\boldsymbol{\mu}_g, \boldsymbol{\Sigma}_g, \boldsymbol{\gamma}) \right]
$$

$$
-\sum_{i=1}^{P-1}\sum_{j=i+1}^{P}\sum_{c_i=1}^{C_i}\sum_{c_j=1}^{C_j}n_{c_ic_j}^{(ij)}\sum_{g=1}^{G}p_{c_ic_j;g}^{(ij)}\log\frac{p_g\pi_{c_ic_j}^{(ij)}(\mu_g,\Sigma_g,\gamma)}{\sum_{h=1}^{G}p_h\pi_{c_ic_j}^{(ij)}(\mu_h,\Sigma_h,\gamma)}
$$
$$
= p\ell_c(\psi;\mathbf{X}) - EN_{pl}(\mathbf{p})
$$

References

Akaike, H. (1973). Information theory and an extension of the maximum likelihood principle. In *Second International Symposium on Information Theory*. Akademinai Kiado (pp. 267–281).

Akaike, H. (1978). A bayesian analysis of the minimum AIC procedure. *Annals of the Institute of Statistical Mathematics, 30*(1), 9–14.

Banfield, J. D., & Raftery, A. E. (1993). Model-based Gaussian and non-Gaussian clustering. *Biometrics, 49*(3), 803–821.

Biernacki, C., Celeux, G., & Govaert, G. (1999). An improvement of the NEC criterion for assessing the number of clusters in a mixture model. *Pattern Recognition Letters, 20*(3), 267–272.

Biernacki, C., Celeux, G., & Govaert, G. (2000). Assessing a mixture model for clustering with the integrated completed likelihood. *IEEE Transactions on Pattern Analysis and Machine Intelligence, 22*(7), 719–725.

Biernacki, C., & Govaert, G. (1997). Using the classification likelihood to choose the number of clusters. *Computing Science and Statistics, 29*, 451–457.

Böhning, D., Dietz, E., Schaub, R., Schlattmann, P., & Lindsay, B. G. (1994). The distribution of the likelihood ratio for mixtures of densities from the one-parameter exponential family. *Annals of the Institute of Statistical Mathematics, 46*(2), 373–388.

Bozdogan, H. (1983). Determining the Number of Component Clusters in the Standard Multivariate Normal Mixture Model Using Model-Selection Criteria. No. UIC/DQM/A83-1. Illinois Univ at Chicago Circle Dept of Quantitative Methods.

Bozdogan, H. (1993). Choosing the number of component clusters in the mixture-model using a new informational complexity criterion of the inverse-Fisher information matrix. In O. Opitz, B. Lausen, & R. Klar *Information and classification. Studies in classification, data analysis and knowledge organization* (pp. 40–54). Berlin, Heidelberg: Springer.

Celeux, G., & Soromenho, G. (1996). An entropy criterion for assessing the number of clusters in a mixture model. *Journal of classification 13*(2), 195–212.

Everitt, B. (1988). A finite mixture model for the clustering of mixed-mode data. *Statistics & Probability Letters, 6*(5), 305–309.

Everitt, B., & Merette, C. (1990). The clustering of mixed-mode data: a comparison of possible approaches. *Journal of Applied Statistics, 17*(3), 283–297.

Gao, X., & Song, P. X. K. (2010). Composite likelihood bayesian information criteria for model selection in high-dimensional data. *Journal of the American Statistical Association, 105*(492), 1531–1540.

Hathaway, R. J. (1986). Another interpretation of the EM algorithm for mixture distributions. *Statistics and Probability Letters, 4*(2), 53–56.

Keribin, C. (2000). Consistent estimation of the order of mixture models. *Sankhya: The Indian Journal of Statistics, Series A, 62*, 49–66.

Leroux, B. G. (1992). Consistent estimation of a mixing distribution. *The Annals of Statistics, 20*(3), 1350–1360.

Liang, Z., Jaszczak, R. J., & Coleman, R. E. (1992). Parameter estimation of finite mixtures using the EM algorithm and information criteria with application to medical image processing. *IEEE Transactions on Nuclear Science, 39*(4), 1126–1133.

Lindsay, B. G. (1983). Efficiency of the conditional score in a mixture setting. *The Annals of Statistics, 11*, 486–497.

Lindsay, B. G. (1988). Composite likelihood methods. *Contemporary Mathematics, 80*, 221–239.

Lubke, G., & Neale, M. (2008). Distinguishing between latent classes and continuous factors with categorical outcomes: Class invariance of parameters of factor mixture models. *Multivariate Behavioral Research, 43*(4), 592–620.

Mclachlan, G. J. (1987): On bootstrapping the likelihood ratio test statistic for the number of components in a normal mixture. *Applied Statistics, 36*, 318–324.

Ranalli, M., & Rocci, R. (2014). Mixture models for ordinal data: a pairwise likelihood approach. *Statistics and Computing.* doi: 10.1007/s11222-014-9543-4.

Sawa, T. (1978). Information criteria for discriminating among alternative regression models. *Econometrica: Journal of the Econometric Society, 46*, 1273–1291.

Schwarz, G. et al. (1978). Estimating the dimension of a model. *The Annals of Statistics, 6*(2), 461–464.

Shibata, R. (1980). Asymptotically efficient selection of the order of the model for estimating parameters of a linear process. *The Annals of Statistics, 8*, 147–164.

Sugiura, N. (1978). Further analysts of the data by Akaike's information criterion and the finite corrections. *Communications in Statistics-Theory and Methods, 7*(1), 13–26.

Takeuchi, K. (1976). Distribution of informational statistics and a criterion of model fitting. *Mathematical Sciences, 153*, 1, 12–18.

Varin, C., Reid, N., & Firth, D. (2011). An overview of composite likelihood methods. *Statistica Sinica, 21*(1), 1–41.

Varin, C., & Vidoni, P. (2005) A note on composite likelihood inference and model selection. *Biometrika, 92*(3), 519–528.

Part II
Big Data

Textual Information Localization and Retrieval in Document Images Based on Quadtree Decomposition

Cynthia Pitou and Jean Diatta

Abstract Textual information extraction is a challenging issue in Information Retrieval. Two main approaches are commonly distinguished: texture-based and region-based. In this paper, we propose a method guided by the quadtree decomposition. The principle of the method is to recursively decompose regions of a document image is four equal regions, starting from the image of the whole document. At each step of the decomposition process an OCR engine is used for retrieving a given textual information from the obtained regions. Experiments on real invoice data provide promising results.

1 Introduction

Document dematerialization consists in the conversion of documents into digital contents, using optical character recognition (OCR), automatic document recognition, or automatic document reading tools. These tools have some limitations including:

- dependence to the nature of the documents: invoices, recipes, novels, forms, etc.;
- low capacity to recognize handwritten documents;
- low capacity to circumscribe given information.

Much research efforts have been performed in Information Extraction (IE) for Information Retrieval (IR) (Jacobs 2014). The task of Information Extraction is to identify a predefined set of concepts in a specific domain, ignoring other irrelevant information, where a domain consists of a corpus of texts together with a clearly specified information need (Piskorski and Yangarber 2013). According to Sumathi

C. Pitou (✉) • J. Diatta
EA2525-LIM, University of Reunion Island, Ile de La Reunion, Saint Denis, France
e-mail: cpitou@gaa.fr; jean.diatta@univ-reunion.fr

© Springer International Publishing Switzerland 2016
A.F.X. Wilhelm, H.A. Kestler (eds.), *Analysis of Large and Complex Data*, Studies in Classification, Data Analysis, and Knowledge Organization,
DOI 10.1007/978-3-319-25226-1_6

et al. (2012a) Text Information Extraction (TIE) in images as a sub-part of IE is concerned with extracting the relevant text data from a collection of images. Sumathi et al. (2012b) identified five stages in TIE process:

- Text detection: determine presence of text in images;
- Text localization: determine the location of text in images and generate boxes around text;
- Text tracking: reduce the processing time for text localization;
- Text extraction: separate text from the background of images;
- Text enhancement: improve the quality (resolution, noise,...) of extracted text.

A 6th stage can be added to the previous stages:

- Text recognition: transform extracted text images into plain text.

Different types of images can be distinguished in the literature: document image, scene text image, caption text image. Document image is a single-page produced from a scanner, a fax machine, or by converting an electronic document into an image format (JPEG or TIFF). Scene text image is natural scene which contains text. Caption text image is image with artificial embedded text. In this paper we are exclusively concerned with document images. In TIE system two main approaches are considered in the literature:

- the region-based approach (Ying et al. 2008): this approach is based on color or gray scale features of text regions in images.
- the texture-based approach (Wei et al. 2009; Ying et al. 2006): this approach is based on textural features (direction, intensity, regularity, alignment,...) of text regions in images.

Text localization as a part of TIE is a challenging domain mainly due to the font non-uniformity, styles, and quality variability for a given type of document (Emmanouilidis et al. 2009). Emmanouilidis et al. (2009) proposed a text localization approach for binarized printed document images. In the area of TIE, their approach is a mix of text detection and text localization according to the definition of text localization given above. Indeed, they try to detect text regions and include them in rectangular bounding boxes. Moreover, they work on document images such as pages of scientific journals which contain text and scenes.

Another issue in text localization is the location variability of the text sought, even if the corresponding textual information is inherent to the document type under consideration. For example, invoices must include the company identifier, but, the location of this information may vary depending on the company. This is most likely due to the lack of a standard in invoices' layout. The present paper is concerned with both of the localization and the retrieval of inherent textual information from document images. More precisely, we focus on how to define a rectangular box around a given textual information. To achieve that, we propose an approach based on the quadtree decomposition. The principle of the method is to recursively decompose regions of a document image in four equal regions, starting from the image of the whole document. At each step of the decomposition process an OCR engine is used for recognizing and extracting a given textual information from the obtained regions. Section 2 describes the quadtree decomposition. Section 3 presents the proposed textual information localization algorithm. Experiments on real invoices data are described in Sect. 4 followed by a conclusion.

2 Quadtree Decomposition

The quadtree decomposition, developed by Finkel and Bentley (1974), is a method of hierarchical cell decomposition of an environment that can be represented as a two dimensional map. A quadtree is a tree data structure in which each internal node has exactly four children. A quadtree is shown in Fig. 1. The decomposition starts by dividing a two dimensional space into four equal regions. Then, each region may be recursively divided into four equal regions. One or more criteria may be set in order to decide whether or not a region should be divided.

The quadtree decomposition is widely used in the literature for image processing and image retrieval. To mention a few: in Dagher and Taleb (2014) the quadtree decomposition is applied to an image in association with a wavelet transform for image denoizing. Ramanathan et al. (2011) proposed an image retrieval technique which takes into account the spatial occurrence of a visual word in an image along with the co-occurrence of other visual words in a predefined region of the image obtained by a quadtree decomposition of the image. Other applications of the quadtree decomposition or quadtree structure usage such as image segmentation (Minaee et al. 2014), video coding (Yuan et al. 2012), document analysis (Gatos 2014) can be found in the literature.

In the present paper, we used quadtree decomposition to localize and retrieve textual information from document images. Document images are divided into regions in which we try to extract given textual information. We mentioned in the introduction that OCR tools have limitations such as the difficulty to recognize efficiently different types of documents depending on the complexity of their layout in particular. We observed a better efficiency of the used OCR engine on sub-regions of images than on the full images. Indeed, we observed that the decomposition of images leads a reduction of noise in the images. The main interest of the decomposition is that it allows the OCR engine to benefit from a zoom

Fig. 1 A quadtree

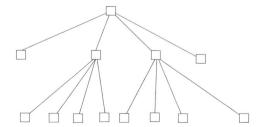

effect (applied to increasingly small regions) which enhances character recognition efficiency.

The following section describes the proposed algorithm for textual information extraction. To the best of our knowledge the quadtree decomposition has never been used for this purpose.

3   The Textual Information Localization Algorithm

We have mentioned earlier weaknesses of OCR tools which recognize only partially or incorrectly textual information contained in document images. Moreover these tools do not allow extraction of reusable (by a computer program) information. Indeed, OCR engines are designed to extract the whole mass of textual information without distinguishing the relevant ones from the others. We observed that when OCR is applied on full images the result is a block of plain text corresponding to all of textual information contained in the images. But a part of this retrieved textual information is often noise, compared to the desired information. The aim of the presented algorithm is to localize and extract these information without noise.

Let I be a document image from which we want to locate and extract a given textual information T. The image I may be obtained from a device for acquiring digital images such as a scanner. To locate and extract a given textual information from I we use a regular expression E describing T. For example, the regular expression to find the net amount of an invoice can be written "(\\bTOTAL\\b) ?\\d+[,]\\d+[€]?." Regular expressions are useful to find specific strings within a text. To extract all kinds of textual information we use Tesseract OCR, an open source OCR engine. Tesseract OCR engine was originally developed by HP Labs between 1985 and 1995. It is now owned by Google. Tesseract is released under the Apache License 2.0 and is available on SourceForge.net.

The quadtree decomposition that we adopt is to recursively subdivide I into four equal rectangular regions. An OCR is performed on each region to extract the contained text. We have two stopping criteria:

1. A text element which matches with a regular expression is discovered in a region.
2. A fixed maximum number of decompositions, say e, is reached.

The algorithm we propose has two main steps applied each in a breadth-first way:

1. Region decomposition: divide the current region into four equal regions.
2. Textual information retrieval: run OCR and retrieve any string that matches with the regular expression E associated with the sought textual information T.

The algorithm starts by applying step 1 to the whole document image. Step 1 is applied to a subsequently obtained region only if no string matching with E is found in that region and the maximum number e of division is not reached. Step 2 is applied to all the regions obtained from the decomposition of a given region while a string matching with E is not found.

Each region containing a sought textual information is described by a vector which is composed of dimensions (w,h) of the region, its location (x,y coordinates of the upper left corner point) and a list l containing the extracted string S. The content of the vector is stored in an XML file. This vector is used as a descriptor of the region. To deal with possible division of I into regions that would separate textual content and disperse them into several parts in different regions, we add a positive constant s to the height and the width of each region (Fig. 2). The constant s we used in our experiment are fixed empirically.

The algorithm can easily be generalized to extract a list $L = (T_1, T_2, \ldots T_n)$ of information and a list $M = (E_1, E_2, \ldots E_n)$ of regular expressions. In this case, the algorithm consists in run OCR and try to extract each element of L successively in a region. The algorithm is stopped if all elements of L are extracted or if the maximum number of decomposition is reached. A region containing at least one extracted string is not decomposed further.

Fig. 2 Example of division into four regions

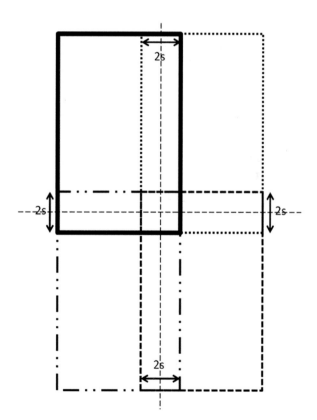

4   Experimentation

We experimented our method on a corpus of 65 scanned invoices. These invoices are issued by 10 service providers. These service providers are divided into four categories: towing companies, auto repair, taxi, and car rental, thus have heterogeneous layouts. Table 1 shows the original distribution of the invoices.

For this experiment, we have developed a JAVA program implementing the proposed algorithm. The program takes as input a directory containing all the files of the corpus and output XML files (one file per document). Each XML files contains a list of region's descriptor. An example of XML file obtained is given in Fig. 3.

Line 2 to line 17: description of the set of regions which were considered containing sought textual information by the algorithm.
Line 3 to line 10: description of region 1 obtained by dividing the whole image into four regions. Line 6 to line 9: the details of textual information extracted in region 1.
Line 7: string extracted from region 1. This string corresponds to the sought information: "invoice date."
Line 8: a second string extracted from region 1 corresponding to the sought information: "article reference."

Table 1 Original distribution of the corpus of invoices

	nb of service providers	nb of invoices
Towing	3	23
Auto repair	2	12
Car rental	3	18
Taxi	2	12
Total	10	65

```
1<invoice f-234>
2 <regions>
3 <region1>
4 <origine x=0,y=0/>
5 <dimension width=1024.0,height=1368.0/>
6 <information>
7 <invoiceDate> 13-04-11 </invoiceDate>
8 <articleReference> R 1104 0753 X04 </articleReference>
9 </information>
10 </region1>
11 <region4>
12 <origine x=0,y=0/>
13 <information>
14 <subtotal> 107.00 </subtotal>
15 </information>
16 </region4>
17 </regions>
18</invoice f-234>
```

Fig. 3 Invoice XML file example

The experiment is to try to extract a list L of 5 information from invoice images: subtotal, article reference, company identifier, invoice date, invoice identifier. We have performed two types of extraction:

1. extraction from whole images,
2. extraction by applying our algorithm implementing the quadtree decomposition.

The same OCR engine and the same regular expressions are used for both extractions. Also, the list of items recognize are stored in XML files. We considered two measures of evaluation:

• The rate of information correctly extracted.
• The rate of information correctly recognized.

The evaluation of the rate of information correctly extracted is based on the counting of how many elements of L have been extracted. An information is considered extracted if there is a stored string which matches with the corresponding regular expression. In another side, an information is considered correctly recognize if the stored string corresponds exactly to the visual information contained in the image. For example, let "net amount" be a sought information and assume that the net amount is 107 €in the original invoice. The regular expression of the net amount can be written "(\\bTOTAL\\b) ?\\d+[,]\\d+[€]?." During the processing of the algorithm, the string "101 €" may be extracted by the OCR engine. Such a string is considered correctly extracted because it matches with the regular expression. Indeed, it corresponds to a string which is a sequence of digits separated by a comma, preceded by the string TOTAL or not and followed by the euro symbol or not. However, the extracted string can't be considered correctly recognized because in the original invoice the textual information "107 €" appears instead of the extracted information "101 €." This can happen due to the weaknesses of the OCR engine.

We obtained a similar number of extracted information for the two types of extraction. However we have obtained a higher recognition rate for extraction from images divided by our algorithm. Indeed, the recognition rate of the selected OCR engine is of 83 % for whole invoices images, and of 97 % for decomposed invoice images. The results are presented in Table 2.

Table 2 Results

	Extraction rate (%)	Recognition rate (%)
Whole images	4.5	83
Decomposed images	4.7	97

5 Conclusion

In this paper, we presented a method to localize and retrieve textual information such as invoice's net amount or invoice's company ID. Experimental results show that the proposed decomposition approach has good potential to significantly improve the correctness of the retrieved textual information.

References

Dagher, I., & Taleb, C. (2014). Image denoising using fourth order wiener filter with wavelet quadtree decomposition. *Journal of Electrical and Computer Engineering, 2014*, 9.

Emmanouilidis, C., Batsalas, C., & Papamarkos, N. (2009). Development and evaluation of text localization techniques based on structural texture features and neural classifiers. In *10th International Conference on Document Analysis and Recognition* (pp. 1270–1274).

Finkel, R. A., & Bentley, J. L. (1974). Quad trees: A data structure for retrieval on composite keys. *Acta Informatica, 4*, 11–9.

Gatos, B. G. (2014). Imaging techniques in document analysis processes. In *Handbook of document image processing and recognition* (Vol. 1, pp. 73–131). London: Springer-Verlag

Jacobs, P. S. (2014). *Text-based intelligent systems: Current research and practice in information extraction and retrieval*. New York: Psychology Press.

Minaee, S., Yu, H., & Wang, Y. (2014). A robust regression approach for background/Foreground segmentation. arXiv preprint arXiv: 1412.5126.

Piskorski, J., & Yangarber, R. (2013). Information extraction: past, present and future. In *Multi-source, multilingual information extraction and summarization* (pp. 23–49). Berlin Heidelberg: Springer-Verlag

Ramanathan, V., Mishra, S., & Mitra, P. (2011). Quadtree decomposition based extended vector space model for image retrieval. *2011 IEEE Workshop on Applications of Computer Vision (WACV)*, pp. 139–144.

Sumathi, C. P., Santhanam, T., & Gayathri, D. (2012a). A survey on various approaches of text extraction in images. *International Journal of Computer Science & Engineering Survey, 3*(4), 27–42.

Sumathi, C. P., Santhanam, T., Priya, N. (2012b). Techniques and challenges of automatic text extraction in complex images: a survey. *Journal of Theoretical and Applied Information Technology, 35*(2), 225–235.

Wei, L., Lefebvre, S., Kwatra, V., Turk, G., (2009). State of the art in example-based texture synthesis. *Eurographics 2009, State of the Art Report, EG-STAR* (pp. 93–117).

Ying, L., Dengsheng, Z., & Guojun, L. (2008). Region based image retrieval with high-level semantics using decision tree learning. *Pattern Recognition, 41*, 2554–2570.

Ying, L., Zhang, G., & Wei-Ying, M. (2006). Study on texture feature extraction in region-based image retrieval system. In *International Multimedia Modelling Conference* (pp. 264–271).

Yuan, Y., Kim, I. K., Zheng, X., Liu, L., Cao, X., Lee, S., & Park, J. H. (2012). Quadtree based nonsquare block structure for inter frame coding in high efficiency video coding. *IEEE Transactions on Circuits and Systems for Video Technology, 22*(12), 1707–1719.

Selection Stability as a Means of Biomarker Discovery in Classification

Lyn-Rouven Schirra, Ludwig Lausser, and Hans A. Kestler

Abstract Diagnostic models for gene expression profiles need to operate on sparse collections of high-dimensional samples. In this context, highly accurate and low-dimensional decision rules with interpretable signatures are of great importance. Feature selection processes are essential for fulfilling these design criteria. They select small subsets of highly informative features that can be starting points for new biological hypotheses and experiments. In this work we present an empirical study on purely data-driven selection algorithms. These "standalone" selectors do not incorporate information about the utilized classification model into their criteria. We examine these methods regarding their selection stability. These classifier independent measures are used to distinguish subgroups of algorithms and to identify valuable filter/classifier combinations for the application on microarray datasets.

1 Introduction

A standard problem in the classification of biological data is the interpretation of high-dimensional gene expression profiles originating from high-throughput

L.-R. Schirra
Core Unit Medical Systems Biology and Institute of Neural Information Processing,
Ulm University, 89069 Ulm, Germany

Institute of Number Theory and Probability Theory, Ulm University, 89069 Ulm, Germany
e-mail: lyn-rouven.schirra@uni-ulm.de

L. Lausser
Core Unit Medical Systems Biology and Institute of Neural Information Processing,
Ulm University, Ulm, Germany

Bioinformatics and Systemsbiology, Leibniz Institute on Aging - Fritz Lipmann Institute,
07745 Jena, Germany
e-mail: ludwig.lausser@uni-ulm.de

H.A. Kestler (✉)
Institute of Medical Systems Biology, Universität Ulm, Ulm, Germany
e-mail: hans.kestler@uni-ulm.de; hkestler@fli-leibniz.de

© Springer International Publishing Switzerland 2016   79
A.F.X. Wilhelm, H.A. Kestler (eds.), *Analysis of Large and Complex Data*, Studies
in Classification, Data Analysis, and Knowledge Organization,
DOI 10.1007/978-3-319-25226-1_7

technologies such as microarrays or deep sequencing. Designed for screening, these profiles measure thousands of gene expression levels reflecting a wide range of biological processes. Combined with a small sample size, gene expression profiles are a typical example for high-dimensional datasets of low cardinality ($n \gg m$).

For such data, feature selection methods are often incorporated in the training of a classification model. These procedures reduce the high-dimensional profiles to low-dimensional signatures that can be used for prediction (Armstrong et al. 2001; Golub et al. 1999). Feature selection can improve both the accuracy and the interpretability of a classification model. While removing noisy or unrelated gene expression levels can eliminate distracting influences, the selection of highly informative genes, so-called key players, can focus the researchers' attention on biologically relevant processes. Another, more technical, reason might be the switch to a new experimental platform. In this case, the high-dimensional profile of a screening platform is replaced by a smaller signature of genes that are measured in a more accurate way. Here, the signature is constrained to a fixed number of measurements.

In this work we analyze the influence of different feature selection strategies on the task of classifying gene expression profiles from microarray experiments. We try to identify valuable feature selection /classifier combinations that lead to an improved classification performance in contrast to a standalone classification algorithm. We especially focus on a selection stability as a classifier independent measurement for the quality of a feature selection algorithm. In this empirical study, we compare 23×13 feature selection and classifier combinations on 9 microarray datasets in 10×10 cross-validation experiments.

2 Methods

The task of a classifier is to categorize objects into semantically meaningful classes, $y \in \mathcal{Y}$. This process is based on vectors of measurements, $\mathbf{x} = (x^{(1)}, \ldots, x^{(n)})^T \in \mathcal{X} \subseteq \mathbb{R}^n$. It can be modeled as a function $c : \mathcal{X} \to \mathcal{Y}$. The performance of a classifier can be evaluated on a test set of samples, $\mathcal{T} = \{(\mathbf{x}_j, y_j)\}_{j=1}^m$, by determining its accuracy in predicting the test set's labels.

$$A_{(c)} = \frac{1}{|\mathcal{T}|} \sum_{(\mathbf{x}, y) \in \mathcal{T}} \mathbb{I}_{[c(\mathbf{x})=y]} \qquad (1)$$

The classifier is typically unknown a priori. In that case, the classification model has to be trained according to a training set of labeled samples, $\mathcal{L} = \{(\mathbf{x}_j, y_j)\}_{j=1}^{m'}, \mathcal{L} \cap \mathcal{T} = \emptyset$,

$$\mathcal{C} \times \mathcal{L} \xrightarrow{\text{train}} c_{\mathcal{L}} \in \mathcal{C}. \qquad (2)$$

Here $\mathcal{C}$ denotes a function or concept class that describes structural properties that should be fulfilled by the classifier.

2.1   Feature Selection

The training process of a classifier can include a feature selection step in which the set of measurements is reduced to a small subset. Only these measurements will influence the predictions of the trained classification model. This process can give hints on the importance of the single measurements. Formally, feature selection can be seen as a function that maps from a training set $\mathscr{L}$ and a concept class $\mathscr{C}$ to an ordered and repetition-free index vector,

$$f : \mathscr{C} \times \mathscr{L} \xrightarrow{\text{select}} \mathbf{i} \in \mathscr{I} = \{\mathbf{i} \in \mathbb{N}^{\hat{n} \leq n} | i_k < i_{k-1}, 1 \leq i_k \leq n\}. \tag{3}$$

The derived index vector $\mathbf{i} = (i_1, \ldots, i_{\hat{n}})^T$ will be utilized to map input patterns $\mathbf{x}$ to a lower-dimensional representation $\mathbf{x}^{(\mathbf{i})} = (x^{(i_1)}, \ldots, x^{(i_{\hat{n}})})^T$. We will additionally use $\hat{\mathbf{i}}$ to denote the set of selected indices $\hat{\mathbf{i}} = \{i_k\}_{k=1}^{\hat{n}}$ for a fixed selection size $\hat{n}$.

Feature selection methods can be separated into *model-based* and purely *data-driven* algorithms (Saeys et al. 2007). Model-based algorithms incorporate knowledge about the utilized concept class $\mathscr{C}$ into the selection process. They can be further subdivided into *wrapper* methods and *embedded* methods. A wrapper algorithm constructs a feature set in an internal evaluation loop in which possible candidate gene sets are tested and modified in experiments with the chosen type of classifier (Kohavi and John 1997). An embedded algorithm is originally designed as a learning algorithm for a sparse classification model, which derives its decision according to a small number of feature evaluations (Blum and Langley 1997). These measurements afterwards can be extracted and utilized as a feature set for other classification algorithms.

In this paper we will mainly focus on purely data-driven feature selection methods which do not take into account any information about the concept class $\mathscr{C}$ of a classifier. These methods are often called *filters* in the literature (Guyon and Elisseeff 2003). They can be applied as an independent preprocessing step before adapting the final classification model. Most filters are *univariate* feature selectors, which validate the single measurements separately. They can further be distinguished in *supervised* filters (Liu and Motoda 2007), which utilize knowledge about the training labels (e.g., association measures) and *unsupervised* filters (Varshavsky et al. 2006), which only rely on characteristics of the feature values (e.g., dispersion measures).

2.2   Evaluation of Classification Models

For a dataset of low cardinality, a classification algorithm is typically evaluated in resampling experiments such as the $K \times L$ cross-validation (Bishop 2006). The dataset $\mathscr{S}$ is split into L folds of roughly equal size. Each fold serves once as a test set $\mathscr{T}$, while the other $L - 1$ folds are used to train the classifier (training set $\mathscr{L}$).

This procedure is repeated in K runs on independent permutations of the samples. The accuracy of a classifier can then be estimated by

$$A_{cv} = \frac{1}{K|\mathscr{S}|} \sum_{k=1}^{K} \sum_{l=1}^{L} \sum_{(\mathbf{x},y)\in\mathscr{T}_{kl}} \mathbb{I}_{[c\mathscr{L}_{kl}(\mathbf{x})=y]}. \tag{4}$$

In our evaluations, the cross-validation experiment will be the basic setting for characterizing filters and filter/classifier combinations.

2.2.1 Stability

An important characteristic of a feature selection method is its selection stability in a resampling experiment. Informative features should be selected in almost all single experiments while uninformative feature should only be detected in rare cases. If this is the case, a feature selection procedure is called stable and unstable otherwise. Lausser et al. (2013) introduced a stability score S_{stab} for measuring the stability in resampling experiments with fixed feature set sizes $\hat{n}$,

$$S_{\text{stab}} = \frac{\hat{n}^{-1}}{(KL)^2} \sum_{i=1}^{KL} i^2 a_{\hat{n}}^{(i)}, \quad \text{with} \quad a_{\hat{n}}^{(j)} = \sum_{i=1}^{n} \mathbb{I}_{[s_{\hat{n}}^{(i)}=j]}, \quad \text{and} \quad s_{\hat{n}}^{(i)} = \sum_{k=1}^{K} \sum_{l=1}^{L} \mathbb{I}_{[i\in\hat{\mathbf{i}}_{kl}]}. \tag{5}$$

Here, $a_{\hat{n}}^{(j)}$ denotes the number of features that are selected j times and $s_{\hat{n}}^{(i)}$ is the selection frequency of the ith feature. In our experiments, we will address the question how exactly the stability of a feature selection method is related to the changes in accuracy of a subsequent classifier.

2.2.2 Gain in Accuracy and Variability

The influence of a feature selection method on a classification model can be analyzed by the resulting gain in accuracy. In our experiments, we have chosen to compare the accuracy of a classifier that is trained on all available features to a classifier that incorporates a feature selection process in its training,

$$\text{Gain}_{\text{acc}} = A_{cv} - A_{cv}^{f,\hat{n}}. \tag{6}$$

Here f denotes the chosen feature selection algorithm and $\hat{n}$ denotes the selection size. In a similar way, the differences in the variability of the achieved accuracies can be analyzed. As a measure the standard deviations of the accuracies over the runs of one cross-validation experiment are compared. The corresponding score will be denoted by Gain_{sd}.

3   Experimental Setup

We conducted series of 10×10 cross-validation experiments on 9 microarray datasets (Table 1). We applied 23 feature selection methods (Table 2) as pre-processing steps to 13 classifiers (Table 3). For each selector, experiments with

Table 1 Description of the utilized microarray datasets

Nr.	Dataset	Identifier	Features (n)	Samples (m)		Pathology
d_1:	Armstrong et al. (2001)	ARM	12,582	72	(48/24)	Leukemia
d_2:	Bittner et al. (2000)	BIT	8067	38	(19/19)	Melanoma
d_3:	Golub et al. (1999)	GOL	7129	72	(25/47)	Leukemia
d_4:	Gordon et al. (2002)	GOR	12,533	181	(150/31)	Lung cancer
d_5:	Notterman et al. (2001)	NOT	7457	36	(18/18)	Colorectal cancer
d_6:	Pomeroy et al. (2002)	POM	7129	34	(25/9)	Medulloblastoma
d_7:	SHIPP et al. (2002)	SHI	7129	77	(58/19)	B-cell lymphoma
d_8:	Singh (2002)	SIN	12,600	102	(50/52)	Prostate cancer
d_9:	West et al. (2001)	WES	7129	62	(25/37)	Breast cancer

Table 2 List of utilized feature selection methods

Nr.	Feature selection criterion	Id.
Unsupervised filters (univariate)		
f_{1-3}:	Variance	VAR, VAR*, VAR**
f_{4-6}:	Interquartile range	IQR, IQR*,IQR**
f_{7-9}:	Median absolute deviation	MAD, MAD*, MAD**
f_{10}:	Range	RAN
Supervised filters (univariate)		
f_{11}:	Pearson correlation	COR$_P$
f_{12}:	Spearman correlation	COR$_S$
f_{13-15}:	Support vector weight	SVF, SVF*, SVF**
f_{16}:	Misclassification index	IMP$_M$
f_{17}:	Gini index	IMP$_G$
f_{18}:	Entropy index	IMP$_E$
f_{19}:	Threshold number of misclassification	TNOM
f_{20}:	Signal-to-noise ratio	SNR
f_{21}:	Nearest shrunken centroid	NSC
f_{22}:	Area under ROC curve	ROC
Supervised filter (multivariate)		
f_{23}:	Relief algorithm	REL
*	Rescaling I	$\dot{x}_j^{(i)} = \dfrac{x_j^{(i)} - x_{min}^{(i)}}{x_{max}^{(i)} - x_{min}^{(i)}}$
**	Rescaling II	$\ddot{x}_j^{(i)} = \dfrac{x_j^{(i)} - \bar{X}^{(i)}}{sd(X^{(i)})}$

varying selection sizes $\hat{n} \in \{5, 10, \ldots, 200\}$ were conducted. All cross-validation experiments on one dataset utilized the same list of folds. They were performed with the TunePareto software (Müssel et al. 2012).

Our experiments comprise filters based on unsupervised dispersion measures (VAR, IQR, MAD, RAN) (Upton and Cook 2002) as well as supervised measures based on the class labels of the samples (Table 2). Among the supervised criteria we utilize filters based on correlation (COR_P, COR_S) (Upton and Cook 2002), signal-to-noise ratio (SNR) (Guyon 2006), and impurity measures (IMP_M, IMP_G, IMP_E) (Breiman et al. 1984). Furthermore measures based on the evaluation of basic classification algorithms (SVF, TNOM, ROC) were used (Vapnik 1998, Ben-Dor et al. (2000), Fawcett (2006)). Also we utilize more complex criteria such as the multivariate Relief algorithm (REL) by Kira and Rendell (1992) or the Nearest Shrunken Centroid algorithm (NSC) by Tibshirani et al. (2002).

For classification, we utilized three groups of algorithms (Table 3), prototype-based classifiers (kNN, NCC, RPS, PAM), linear classifiers (PER, $SVM_{r,l}$, STC, FCC), and hierarchical classifiers (RF). The accuracies of standalone classifiers (without feature selection) are given in Table 4.

Table 3 List of utilized classification algorithms

Nr.	Classification algorithm	Id.	Parameter
$c_{1,3}$:	k Nearest neighbors (Fix and Hodges 1951)	kNN	$k \in \{1, 3, 5\}$
c_4:	Representative prototypes set (Lausser et al. 2012)	RPS	
c_5:	Nearest centroid	NCC	
c_6:	Prediction analysis for microarrays (Tibshirani et al. 2002)	PAM	steps = 30
c_7:	Single threshold classifier (Kestler et al. 2011)	STC	
c_8:	Fold change classifier (Lausser and Kestler 2014)	FCC	
c_{9-11}:	Linear support vector machine (Vapnik 1998) (r-reg., l-loss, see Abe 2010)	$SVM_{r,l}$	$C = 1$ $(r, l) \in \{(1, 2), (2, 1), (2, 2)\}$
c_{12}:	Perceptron (Rosenblatt 1958)	PER	iter. = 1000
c_{13}:	Random forest (Breiman 2001)	RF	trees = 100

Table 4 Accuracies of standalone classifiers (without feature selection)

	kNN			RPS	NCC	PAM	STC	FCC	$SVM_{r,l}$			PER	RF
	1	3	5						1,2	2,1	2,2		
Median	87.8	86.6	89.1	93.2	78.8	92.9	86.5	88.5	94.0	96.6	96.6	92.5	95.8
IQR	14.7	18.4	14.1	14.4	29.4	4.2	6.1	10.7	10.3	9.7	9.9	10.7	8.5

The median accuracy (%) and the interquartile range over all datasets are reported

4 Results

Figure 1a shows the median gain in accuracy ($Gain_{acc}$) of a classifier and feature selection combination (over all datasets and selection sizes). It can be observed that the accuracy of prototype-based classifiers can be improved by supervised filtering methods. The highest $Gain_{acc}$ is achieved by the NCC but also prototype-based classifiers with a higher standalone accuracy such as the RPS have improved. An exception is the PAM classifier that additionally utilizes its own embedded feature selection.

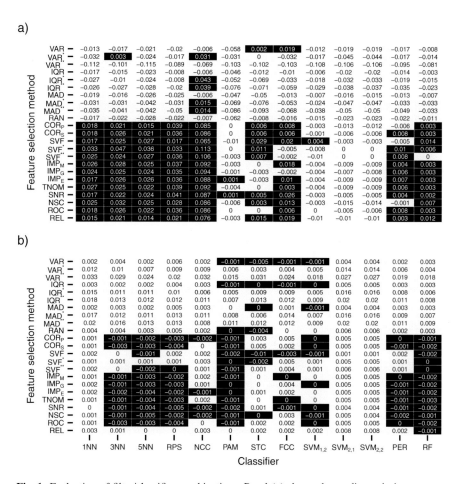

Fig. 1 Evaluation of filter/classifier combinations. Panel (**a**) shows the median gain in accuracy in comparison to a standalone classifier (over all datasets and selection sizes). Positive values (improved accuracies) are marked in *black*. Panel (**b**) shows the median gain in standard deviation. Here, negative values (reduced standard deviations) are highlighted

Two major patterns can be seen for the linear classifiers. No positive $Gain_{acc}$ is achieved for the large margin classifiers $SVM_{2,1}$ and $SVM_{2,2}$. As an exception, $SVM_{1,2}$, which intrinsically uses feature selection, has been improved by VAR. The other linear classifiers can all be improved by supervised filters. This set especially includes the STC and FCC classifiers. These algorithms are designed for operating on one or two single features. The $Gain_{acc}$ of RF is also improved by supervised filters.

The median gain in standard deviation ($Gain_{sd}$) is given in Fig. 1b. From the filters that operate on rescaled features (*,**), only SVF* and SVF** achieve a negative $Gain_{sd}$. $Gain_{sd}$ is positive for all feature selection algorithms for the classifiers 1NN, $SVM_{2,1}$, and $SVM_{2,2}$. For NCC the $Gain_{sd}$ is negative for NSC, SNR, IMP_E. The other prototype-based classifiers investigated show a negative $Gain_{sd}$ for all supervised filters despite REL and SVF. For the linear classifiers $SVM_{1,2}$, FCC, and STC, a negative $Gain_{sd}$ can be achieved by unsupervised filters.

The feature selection algorithms were characterized by their median stability score (S_{stab}) over all datasets and selection sizes (Fig. 2). A complete-linkage clustering was used to group the algorithms. The four top-level clusters (from most stable to most unstable) were structured as follows. The most stable subgroup consists of the unsupervised filter that does not operate on rescaled features (VAR, RAN, MAD, IQR). The second cluster comprises the univariate supervised filters. The third cluster mainly consists of the unsupervised filters that operate on rescaled features (VAR*, VAR**, MAD*, MAD**, IQR*, IQR**). The third cluster also includes the multivariate supervised REL. The most unstable cluster consists of only one member (VAR**). Comparing the median S_{stab} of a filter/classifier combination to its median $Gain_{acc}$ and median $Gain_{sd}$, a positive correlation (Pearson) of 0.513

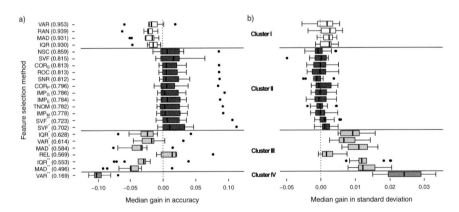

Fig. 2 Relation of stability and accuracy. The median $Gain_{acc}$ (Panel **a**) and median $Gain_{sd}$ (Panel **b**) over all datasets and selection sizes are shown. The gains achieved for one filter (over all classifiers) are summarized in one box plot. The box plots are sorted according to their stability scores S_{stab} (row labels). The different shades of *gray* indicate the four major clusters of a complete-linkage clustering of the stability scores

(to $Gain_{acc}$, $p < 2.2 \cdot 10^{-16}$) and a negative correlation of -0.777 (to $Gain_{sd}$, $p < 2.2 \cdot 10^{-16}$) were observed.

5 Discussion and Conclusion

Our study does not claim to be exhaustive. We focused on purely data-driven feature selectors, "standalone" techniques, which rely neither on domain knowledge nor on specific classification models. These algorithms may be seen as basic building blocks for a wide range of algorithms.

In our experiments a median gain in accuracy was mainly observed for supervised selection methods. Only the nearest centroid classifier could be improved by several unsupervised criteria. This may be due to the chosen selection sizes of less than 200 features. While unsupervised methods are mainly applied for removing a bulk of uninformative features, supervised methods are utilized for constructing relatively small signatures of informative features.

The selection frequency is an important indicator for the biological relevance of a feature. The results suggest that it has a positive influence on the gain in accuracy and reduces the classifiers standard deviation. Nevertheless, a high selection stability is no guarantee for a good classification result. This can be observed for the unsupervised dispersion measures which lead to a more stable feature selection than the supervised criteria. Being susceptible to the scaling of a feature these measures probably incorporate unrelated information into a feature selection. Feature-wise rescaling leads to a decreased selection stability for all analyzed unsupervised and supervised criteria but in many cases also decreased the gain in accuracy.

From the analyzed classifiers, prototype-based algorithms were more easily improved by an unspecific choice of a (supervised) selection strategy. Their median gain in accuracy was also larger than for other classifiers. An exception is the NSC classifier which additionally utilizes its own embedded feature selection strategy. It could only be improved in rare cases. A similar behavior could be observed for the L1-regularized support vector machine, while the random forest could be slightly improved by supervised selection criteria. Those classifiers which intrinsically operate on one or two selected features could also be improved by a supervised preselection. It is especially interesting to see that the single threshold classifier can be improved by the multivariate criterion of the REL. The median accuracies of the large margin classifiers were not improved.

In conclusion our experiments indicate that feature selection can improve the accuracy of classifiers that were developed for low-dimensional data, such as most prototype-based methods, to the level of algorithms that are more directly designed for operating in high-dimensional spaces. Thus feature selection methods can increase the repertoire of classification models suitable for gene expression profiles. Showing a comparable accuracy, such filter/classifier combinations might be preferable due to their higher interpretability.

Acknowledgements The research leading to these results has received funding from the European Community's Seventh Framework Programme (FP7/2007–2013) under grant agreement n°602783 to HAK, the German Research Foundation (DFG, SFB 1074 project Z1 to HAK), and the Federal Ministry of Education and Research (BMBF, Gerontosys II, Forschungskern SyStaR, project ID 0315894A to HAK).

References

Abe, S. (2010). *Support vector machines for pattern classification.* London: Springer.

Armstrong, S. A., Staunton, J. E., Silverman, L. B., Pieters, R., den Boer, M. L., Minden, M. D., et al. (2001). MLL translocations specify a distinct gene expression profile that distinguishes a unique leukemia. *Nature Genetics, 30,* 41–47.

Ben-Dor, A., Bruhn, L., Friedman, N., Nachman, I., Schummer, M., & Yakhini, Z. (2000). Tissue classification with gene expression profiles. *Journal of Computational Biology, 7*(3–4), 559–583.

Bishop, C. M. (2006). *Pattern Recognition and Machine Learning.* New York: Springer.

Bittner, M., Meltzer, P., Chen, Y., Jiang, Y., Seftor, E., Hendrix, M., et al. (2000). Molecular classification of cutaneous malignant melanoma by gene expression profiling. *Nature, 406,* 536–540.

Blum, A. L., & Langley, P. (1997). Selection of relevant features and examples in machine learning. *Artificial Intelligence, 97*(1–2), 245–271.

Breiman, L. (2001). Random forests. *Machine Learning, 45*(1), 5–32.

Breiman, L., Friedman, J., Olshen, R., & Stone, C. (1984). *Classification and regression trees.* Belmont, CA: Wadsworth.

Fawcett, T. (2006). An introduction to ROC analysis. *Pattern Recognition Letters, 27*(8), 861–874.

Fix, E., & Hodges, J. L. (1951). Discriminatory analysis: Nonparametric discrimination: consistency properties. *USAF School of Aviation Medicine, Randolf Fields, Tech. Rep. Project 21-49-004, Report Number 4.*

Golub, T. R., Slonim, D. K., Tamayo, P., Huard, C., Gaasenbeek, M., Mesirov, J. P., et al. (1999). Molecular classification of cancer: Class discovery and class prediction by gene expression monitoring. *Science, 286,* 531–537.

Gordon, G. J., Jensen, R. V., Hsiao, L. L., Gullans, S. R., Blumenstock, J. E., Ramaswamy, S., et al. (2002). Translation of microarray data into clinically relevant cancer diagnostic tests using gene expression ratios in lung cancer and mesothelioma. *Cancer Research, 62*(17), 4963–4967.

Guyon, I. (2006). *Feature extraction: Foundations and applications.* Heidelberg: Springer.

Guyon, I., & Elisseeff, A. (2003). An introduction to variable and feature selection. *The Journal of Machine Learning Research, 3,* 1157–1182.

Kestler, H. A., Lausser, L., Lindner, W., & Palm, G. (2011). On the fusion of threshold classifiers for categorization and dimensionality reduction. *Computational Statistics, 26*(2), 321–340.

Kira, K., & Rendell, L. A. (1992). The feature selection problem: Traditional methods and a new algorithm. In *Proceedings of the Tenth National Conference on Artificial Intelligence* (pp. 129–134). Menlo Park, CA: AAAI Press.

Kohavi, R., & John, G. (1997). Wrappers for feature subset selection. *Artificial Intelligence, 97*(1–2), 273–324.

Lausser, L., & Kestler, H. A. (2014). Fold change classifiers for the analysis of gene expression profiles. In W. Gaul, A. Geyer-Schulz, Y. Baba & A. Okada (Eds.), *German-Japanese interchange of data analysis results* (pp. 193–202). New York: Springer.

Lausser, L., Müssel, C., & Kestler, H. A. (2012). Representative prototype sets for data characterization and classification. In N. Mana, F. Schwenker & E. Trentin (Eds.), *Artificial neural networks in pattern recognition* (pp. 36–47). Berlin/Heidelberg: Springer.

Lausser, L., Müssel, C., Maucher, M., & Kestler, H. A. (2013). Measuring and visualizing the stability of biomarker selection techniques. *Computational Statistics, 28*, 51–65.

Liu, H., & Motoda, H. (2007). *Computational methods of feature selection*. Boca Raton: Chapman & Hall/CRC.

Müssel, C., Lausser, L., Maucher, M., & Kestler, H. A. (2012). Multi-objective parameter selection for classifiers. *Journal of Statistical Software, 46*(5), 1–27.

Notterman, D. A., Alon, U., Sierk, A. J., & Levine, A. J. (2001). Transcriptional gene expression profiles of colorectal adenoma, adenocarcinoma, and normal tissue examined by oligonucleotide arrays. *Cancer Research, 61*, 3124–3130.

Pomeroy, S. L., Tamayo, P., Gaasenbeek, M., Sturla, L. M., Angelo, M., McLaughlin, M. E., et al. (2002). Prediction of central nervous system embryonal tumour outcome based on gene expression. *Nature, 415*(6870), 436–442.

Rosenblatt, F. (1958). The perceptron: A probabilistic model for information storage and organization in the brain. *Psychological Review, 65*(6), 386.

Saeys, Y., Inza, I., & Larrañaga, P. (2007). A review of feature selection techniques in bioinformatics. *Bioinformatics, 23*, 2507–2517.

Shipp, M. A., Ross, K. N., Tamayo, P., Weng, A. P., Kutok, J. L., Aguiar, R. C. T., et al. (2002). Diffuse large B-cell lymphoma outcome prediction by gene-expression profiling and supervised machine learning. *Nature Medicine, 8*(1), 68–74.

Singh, D., Febbo, P. G., Ross K., Jackson, D. G., Manola J., Ladd C., et al. (2002). Gene expression correlates of clinical prostate cancer behavior. *Cancer Cell, 1*(2), 203–209.

Tibshirani, R., Hastie, T., Narasimhan, B., & Chu, G. (2002). Diagnosis of multiple cancer types by shrunken centroids of gene expression. *Proceedings of the National Academy of Sciences, 99*(10), 6567–6572.

Upton, G., & Cook, I. (2002). *A dictionary of statistics*. New York: Oxford University Press.

Vapnik, V. (1998). *Statistical learning theory*. New York: Wiley.

Varshavsky, R., Gottlieb, A., Linial, M., & Horn, D. (2006). Novel unsupervised feature filtering of biological data. *Bioinformatics, 22*(14), e507–e513.

West, M., Blanchette, C., Dressman, H., Huang, E., Ishida, S., Spang, R., et al. (2001). Predicting the clinical status of human breast cancer by using gene expression profiles. *Proceedings of the National Academy of Sciences, 98*(20), 11462–11467.

Active Multi-Instance Multi-Label Learning

Robert Retz and Friedhelm Schwenker

Abstract Multi-instance multi-label learning (MIML) introduced by Zhou and Zhang is a comparatively new framework in machine learning with two special characteristics: Firstly, each instance is represented by a set of feature vectors (a bag of instances), and secondly, bags of instances may belong to many classes (a Multi-Label). Thus, an MIML classifier receives a bag of instances and produces a Multi-Label. For classifier training, the training set is also of this MIML structure. Labeling a data set is always cost-intensive, especially in an MIMIL framework. In order to reduce the labeling costs it is important to restructure the annotation process in such a way that the most informative examples are labeled in the beginning, and less or non-informative data more to the end of the annotation phase. Active learning is a possible approach to tackle this kind of problems in this work we focus on the MIMLSVM algorithm in combination with the k-Medoids clustering algorithm to transform the Multi-Instance to a Single-Instance representation. For the clustering distance measure we consider variants of the Hausdorff distance, namely Median- and Average-Based Hausdorff distance. Finally, active learning strategies derived from the single-instance scenario have been investigated in the MIML setting and evaluated on a benchmark data set.

1 Multi-Instance Multi-Label Learning

In the standard case of supervised learning a classifier c has to learn a mapping function between objects, represented as feature vectors x_i, and their related labels y_i. Therefore the classifier gets a set of training examples L. This contains examples which consist of x_i and its label y_i, $L = \{(x_1, y_1), \dots, (x_n, y_n)\}$. After the training the classifier has to identify the labels of given objects, $c(x_u) = y_u$. This standard supervised learning case can be described as single-instance single-label learning,

R. Retz • F. Schwenker (✉)

Institute of Neural Information Processing, Ulm University, 89069 Ulm, Germany

e-mail: robert.retz@uni-ulm.de; friedhelm.schwenker@uni-ulm.de

© Springer International Publishing Switzerland 2016   91

A.F.X. Wilhelm, H.A. Kestler (eds.), *Analysis of Large and Complex Data*, Studies in Classification, Data Analysis, and Knowledge Organization,

DOI 10.1007/978-3-319-25226-1_8

short SISL. In this framework every object is represented by a single feature vector called instance and is related to a single label.

Dietterich et al. (1997) introduced a multiple-instance representation to extend SISL. They designed multi-instance single-label learning (MISL) to solve a drug activity prediction problem. The goal in this scenario is to predict the degree of binding of a drug molecule with a larger target molecule. Every drug molecule has multiple shapes which are crucial for its binding ability. A drug molecule has a good binding, if at least one of its shapes has a good binding to the target molecule. In MISL an object is represented by a set of instances $X_i = \{x_{i1}, \ldots, x_{im_i}\}$, a so-called bag. A bag is related to a label if at least one of its containing instances has this label, it is not known which or how many of the instances have the label. Every bag can only have one label. The training set in MISL is given as: $L = \{(X_1, y_1), \ldots, (X_n, y_n)\}$. The advantage of MISL is that learning tasks can be modelled in their "natural" design. In addition a more precise identification of the crucial parts of an object, respectively its instances, is possible compared with a single-instance representation.

Another expansion of SISL was introduced by Mccallum (1999) called single-instance multi-label learning (SIML), to solve a text classification task. In this framework every object is represented by a single instance x_i, but can have more than one correct label, $Y_i = \{y_{i1}, \ldots, y_{il_i}\}$. In this case the training set is defined as: $L = \{(x_1, Y_1), \ldots, (x_n, Y_n)\}$. There are two intuitive ways to transform Multi-Label data into a Single-Label representation. The first possibility is to train a separate classifier for every occurring label. This method is used by the MLSVM algorithm from Boutell et al. (2004). The problem of this approach is that possible correlations between the labels are not considered. The second transformation approach is to create a new label for each occurring label combination. The problem is that for certain label combinations only a few training examples exist. SIML makes it possible to deal with multiple correct labels, which occur in many learning tasks.

The learning framework used in this paper is multi-instance multi-label learning (MIML). MIML was introduced by Zhou and Zhang (2007) and combines MISL and SIML. More precise, this framework uses the bag representation of objects, $X_i = \{x_{i1}, \ldots, x_{im_i}\}$, here each bag can have more than one label, $Y_i = \{y_{i1}, \ldots, y_{il_i}\}$. According to that the training set is: $L = \{(X_1, Y_1), \ldots, (X_n, Y_n)\}$. The combination of both learning frameworks inherits not only the advantages but also the disadvantages. As in MISL the mapping between the instances of a bag and its related label is not known, additionally every bag and instance can have more than one label as in SIML. However it is possible to use ideas and algorithms of the predecessor learning frameworks to design approaches on MIML.

2   Distances in Multi- to Single-Instance Transformation

The core of an active learning approach is, besides the selection strategy, the used base classifier. Therefore the performance and results are heavily dependent on the chosen classifier. This section discusses the improvement of the MIMLSVM algorithm of Zhou and Zhang (2007), which is used in the active learning algorithm discussed in this paper in Sect. 3.

   An approach to solve complex problems is to transform them into simpler problems. This general approach is used by the MIMLSVM and MIMLBoost algorithms of Zhou and Zhang (2007) where the MIML problem is transformed into an SIML and MISL problem, respectively. In this paper the transformation used by MIMLSVM is examined. The idea is to map the Multi-Instance data into a Single-Instance representation, which in turn is given to an SIML algorithm. In the standard version of MIMLSVM the MLSVM algorithm is used. The transformation process itself is divided into two parts. The first one calculates the k most representative data points $M_1, \ldots, M_k$, so-called medoids, from the training data L with a k-Medoids algorithm using a cluster distance. Afterwards the cluster distance between bag $X_i \in L$ and every medoid $m_j \in M$ is computed and $x_i' = (d(X_i, M_1), \ldots, d(X_i, M_k))$ is used as new instance representation. The Multi-Label representation of the data is preserved. After the transformation the training set is: $L = \{(x_1', Y_1), \ldots, (x_n', Y_n)\}$. The resulting data highly depends on the used cluster distance. Furthermore the classification performance of MIMLSVM is influenced by the selected distance. The origin MIMLSVM algorithm uses the Hausdorff distance d_h, defined in Eq. (1).

$$d_h(A, B) = \max \left\{ \max_{a \in A} \min_{b \in B} d(a, b), \ \max_{b \in B} \min_{a \in A} d(a, b) \right\} \tag{1}$$

here d is the Euclidean distance. Variations of the Hausdorff distance are used in the MISL algorithms. Bayesian-kNN and Citation-kNN were proposed by Wang and Zucker (2000) and Zhang and Zhou (2009) proposed an averaged Hausdorff distance for their multiple-instance clustering algorithm BAMIC.

2.1   Average-Based Hausdorff Distances

Based on the averaged Hausdorff distance by Zhang and Zhou (2009), defined in Eq. (2), extensions of the Hausdorff distance are introduced in this paper. Here the key idea is to incorporate more geometric information in the distance measure that is supposed to be preserved during the transformation process and to improve the classification.

$$d_{\mathrm{avgH}}(A, B) = \frac{1}{|A| + |B|} \cdot \left(\sum_{a \in A} \min_{b \in B} d(a, b) + \sum_{b \in B} \min_{a \in A} d(a, b) \right) \tag{2}$$

On the basis of the averaged Hausdorff distance other so-called Average-Based Hausdorff distances are introduced. There are many ways to include the average function to the Hausdorff distance. The following four are used in this paper:

$$d_{\text{avgH}_A}(A, B) = \frac{1}{2} \cdot \left(\max_{a \in A} \min_{b \in B} d(a, b) + \max_{b \in B} \min_{a \in A} d(a, b) \right) \tag{3}$$

$$d_{\text{avgH}_B}(A, B) = \frac{1}{2} \cdot \left(\frac{1}{|A|} \cdot \sum_{a \in A} \min_{b \in B} d(a, b) + \frac{1}{|B|} \cdot \sum_{b \in B} \min_{a \in A} d(a, b) \right) \tag{4}$$

$$d_{\text{avgH}_C}(A, B) = \max \left\{ \frac{1}{|A|} \cdot \sum_{a \in A} \min_{b \in B} d(a, b), \ \frac{1}{|B|} \cdot \sum_{b \in B} \min_{a \in A} d(a, b) \right\} \tag{5}$$

$$d_{\text{avgH}_D}(A, B) = \min \left\{ \frac{1}{|A|} \cdot \sum_{a \in A} \min_{b \in B} d(a, b), \ \frac{1}{|B|} \cdot \sum_{b \in B} \min_{a \in A} d(a, b) \right\} \tag{6}$$

In order to select the most promising approach of the introduced distances, they are empirically compared on two data sets in two experiments. The experiment settings are described in Sect. 2.3.

2.2 Median-Based Hausdorff Distances

In addition to the averaged Hausdorff distances in Sect. 2.1 three Median-Based Hausdorff distances are examined:

$$d_{\text{medH}_A}(A, B) = \max \left\{ \text{med}_{a \in A} \left\{ \min_{b \in B} d(a, b) \right\}, \ \text{med}_{b \in B} \left\{ \min_{a \in A} d(a, b) \right\} \right\} \tag{7}$$

$$d_{\text{medH}_B}(A, B) = \frac{1}{2} \cdot \left(\text{med}_{a \in A} \left\{ \min_{b \in B} d(a, b) \right\} + \text{med}_{b \in B} \left\{ \min_{a \in A} d(a, b) \right\} \right) \tag{8}$$

$$d_{\text{medH}_C}(A, B) = \min \left\{ \text{med}_{a \in A} \left\{ \min_{b \in B} d(a, b) \right\}, \ \text{med}_{b \in B} \left\{ \min_{a \in A} d(a, b) \right\} \right\} \tag{9}$$

The different Median-Based Hausdorff distances were empirically compared to select the most promising one. The detailed experiment settings are described in Sect. 2.3.

2.3 Experiment

The empirical comparison of Average-Based, Median-Based, and original Hausdorff distances measures has been done on two different MIML data sets. As kernel function of the SVM in the MIMLSVM algorithm the Gaussian radial basis function $K(x, x') = \exp\left(-\frac{||x-x'||_2^2}{2 \cdot \sigma^2}\right)$ has been applied.

The two used data sets are MIMLText, a text classification task also called MIML-Reuters from Zhou and Zhang (2007), and MIMLImage, the data set on which MIML was originally introduced. Both consist of 2000 examples with seven and five possible labels. MIMLText has an average bag size of 3.56 and a Multi-Label amount of 15 % whereas MIMLImage has an average bag size of 9 with 22 % of Multi-Label examples.

In the first experiment the size of the training set L is varied between (200 : 200 : 1800) examples. Thereby every classifier gets the same splits. The second experiment varies the amount of multiple labels in the data from 5 % to 75 % in steps of 5 %. Therefore the data set is split into single and multiple label examples, thereafter they are randomly mixed until the Multi-Label amount reaches 75 %. Afterwards the Multi-Label examples are replaced randomly by unused Single-Label examples to achieve the other label amounts. On each amount step a Leave-one-out cross-validation on a tenfold is processed. Both experiments are repeated 50 times on random splits. Before the experiments are executed a grid search was performed to optimize the parameter of every MIMLSVM and distance combination on the data sets. The k value, the number of medoids, is set to 20 %. The results of the experiments with varying training set size are shown in Fig. 1 (top). Here the best cluster distances of each type as defined in Sect. 2 have been selected. Both plots show that the Median- and Average-Based approaches are able to achieve better results than the original Hausdorff distance. In both experiments d_{medH_B} performed very well. The Average-Based Hausdorff distances showed a good performance too, d_{avgH_B} on MIMLImage and d_{avgH_D} on MIMLText. But depending on the data set, different variants were selected as the best one in the previous experiments. The experiment with different multiple label amounts have lead to the results shown in Fig. 1 (bottom). Like before the median-based variant B [see Eq. (8)] achieved good and stable results in the experiments on both data sets. The average-based in turn showed similar results, but as in the other experiment setting none of them could achieve good results on both data sets. On MIMLImage, d_{avgH_B} and d_{avgH_C} were selected and both of them performed similar to d_{medH_B}. Equally d_{avgH_B} performed well on the MIMLText data set and is comparable with d_{medH_B}. Experiments show that the introduced distances are able to improve the performance of the MIMLSVM algorithm. Especially Median-Based Hausdorff distances, particularly d_{medH_B}, performed well in the evaluation. In addition it can be said that the median approaches perform more stable.

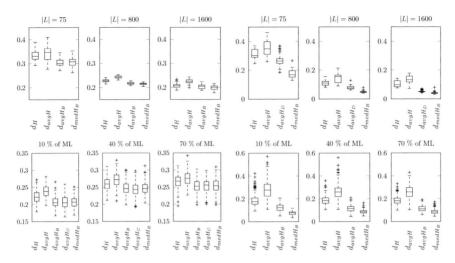

Fig. 1 Boxplot of the Hamming loss values (*down arrow*) achieved in the varying multi-label amount experiment (*top*) and the varying training set experiment (*bottom*) on MIMLImage (*left*) and MIMLText (*right*). Three specific points are given to display the progress along the increasing training set sizes and the number of examples with multiple labels

3   Active Multi-Instance Multi-Label Learning

Active learning is an example of a partially supervised learning algorithm, see Schwenker and Trentin (2014), and is applied in learning scenarios with expensive labeling costs, to reduce the amount of training data needed to achieve a good classification performance. The most common type of active learning is Pool-Based Selective Sampling, see Settles (2009). In this scenario two pools of data are given. One containing of already labeled data, another one containing unlabeled examples. Because of the complex structure of MIML, especially the multiple label part, the generation of labeled data is expensive. That is why the combination of active learning and MIML is a promising field of research. Active learning is already used in SIML, first by Li et al. (2004), and MISL, first by Settles et al. (2008).

There are two options to design the Multi-Instance part of MIML in active learning. Both variants can be found in active learning approaches made on MISL. The question is if instances or complete bags are queried. If instance level querying is used a very accurate information is given to the active learning agent. This approach is used by Settles et al. (2008). In the base design of MISL the classifier has to identify the relation between instances in a bag and the given label. A precise knowledge of the instance to label mapping would lead to a strong information gain for the classifier. If additionally the Multi-Label design of data in MIML is considered another problem arises. It is only possible to add a bag into the labeled example pool if all of its labels are known. This is only possible if every containing instance and their related labels are known or if every possible label is given through a known subset of the instances. The other, and mostly used, possibility is to query on the bag level, like Fu and Yin (2011). The difficulty using bag level active learning is to score the bags from the unlabeled pool. More precise, the question is how to combine the, for example Uncertainty Sampling (Settles 2009), scores of the different instances to a meaningful bag score. This problem gets even more difficult if the Multi-Label design is additionally used like in MIML. In this case the score of a bag or instance has to represent not only its score related to a single label but rather has to combine the scores of all labels. Another decision to be made is if after querying an instance or bag all related labels are revealed or only a single one. The resulting problems are similar to the one of instance versus bag level querying. An advantage of instance querying and single label revelation is that the labeling costs are reduced. Because usually it is simpler to find the labels of only one instance than of a complete bag and it is simpler to find only one label of a given example than to find all of them, respectively. Additionally the handling of the incomplete labeled examples has to be solved. In this paper active learning with complete label revelation on the bag level is used. This makes it possible to use Pool-Based Selective Sampling. The problem of the bag score is solved through the use of MIMLSVM as the base classifier. As mentioned before this algorithm transforms the Multi-Instance into a Single-Instance representation with the result that the instances of a bag are already combined into a single vector, and only the score of this vector has to be calculated.

3.1 Active Single-Instance Multi-Label Learning on MIML Data

The use of MIMLSVM makes it possible to use SIML active learning strategies on MIML data sets. The question is whether the transformation process of MIMLSVM conserves enough vital information. Therefore different active learning strategies are executed on an active learning experiment and later compared with the active learning approach, proposed in Sect. 3.2, for the experiment settings see Sect. 3.3.

The first and simplest active learning strategy which is used in this paper is BinMin (BM), see Brinker (2005). It simply sets the score of an instance to the minimum uncertainty value of all possible values. After the scores of all possible query candidates are computed, the one with the lowest score is selected. The score is calculated with: $score(x) = \min_{i=1,...,l} (|f^i(x)|)$, whereas $f^i(x)$ is the classifier value of the instance x for the label l_i. The next two approaches are called mean max loss (MML) and max loss (ML), see Li et al. (2004). The key idea behind both strategies is to select the instance which leads to the highest decrease of expected loss. The fourth strategy is called maximum loss reduction with maximal confidence (MMC), see Yang et al. (2009). In MMC a Logistic Regression approach is used to predict the number of labels m of an unlabeled instance. Afterwards the so-called most confident vector $Y' = (y'_1, \ldots y'_l)$ is generated, where l is the number of all possible labels. Therefore the values at the positions of the m labels with the highest classification probability are set to 1. The other positions are set to 0. In other words Y' specifies the most probable labels of an unlabeled instance. The score is calculated with the following equation: $score(x) = \sum_{j=1}^{l} \frac{1-y'_j \cdot f^j(x)}{2}$. The instance to query is the one with the highest score. These four active learning strategies are executed in an experiment to test if they can perform effective on an MIML data set using MIMLSVM as their base classifier.

3.2 MidSelect

The active learning strategy introduced in this paper is called MidSelect (MS), which uses MIMLSVM as its base classifier. The idea of this strategy is to consider all label scores of an instance in its overall query score calculation. Therefore the score is represented by the median value of all label scores of an instance. The selection strategy itself is an SIML active learning approach, which only has to handle the combination of the single label based scores. These label scores are based on uncertainty sampling. The idea of Uncertainty Sampling is that an example about whose label membership the classifier is most uncertain, is the most informative example and therefore it has to be queried. Another requirement on MidSelect is that it has to be simple like for example BinMin. A problem of BinMin is that only one label score is taken into account. As a result possible valuable query candidates are not selected because their minimum label score is slightly higher than the one of the chosen one. To overcome this problem an adapted median function is used in MidSelect to combine the single label scores. This adapted median, $median_{min}$, chooses the minimum absolute value of the lower and upper median if an even count of labels is given. This is necessary because as label score the classifier output values are used. Among these output values a value near zeros means a high uncertainty, a high positive or low negative value means a high certainty that the candidate has the label or not. The adaptation helps to overcome what would occur if the average of the lower and upper median were used. Otherwise it could be possible that a candidate with very similar but high valued lower and upper medians would be selected instead of a candidate with different but low valued lower and upper medians. The selection function of MidSelect is defined in Eq. (10). There $f(X) = (f^{(1)}(X), \ldots, f^{(n)}(X))$ defines the output values of MIMLSVM given the bag X with n as the number of labels.

$$\text{MidSelect}(Il) = \underset{X \in U}{\arg\min} \mid median_{min}(f(X))\mid \qquad (10)$$

Another extended approach of MidSelect is given in Eq. (11). This extended MidSelect (EMS) tries to measure the label score values of the two sets divided by the median value, to get a more precise measurement of the label score distribution.

This is done by the function $d_{HS}(x) = |\operatorname{med}(x_{low}) - \operatorname{med}(x_{high})|$, where $x_{low} = (x_1, \ldots, x_{\frac{n+1}{2}-1})$ and $x_{low} = (x_{\frac{n+1}{2}+1}, \ldots, x_n)$ are the two half spaces in the case that n is odd. Otherwise if n is even $x_{low} = (x_1, \ldots, x_{\lfloor \frac{n+1}{2} \rfloor})$ and $x_{low} = (x_{\lceil \frac{n+1}{2} \rceil}, \ldots, x_n)$.

$$\text{ExtMidSelect}(U) = \arg\min_{X \in U} |\operatorname{median}_{min}(f(X))| \cdot d_{HS}(c(X)) \tag{11}$$

3.3 Experiment

The experiment of SIML active learning strategies on an MIML data set and the experiment to compare the MidSelect approaches with different SIML approaches use the same setting. In both experiments MIMLSVM is used as the base classifier. The basic active learning algorithm used in combination with the different selection strategies in the experiments of this paper is outlined in Fig. 2. The used cluster distance is d_{medH_B} from Eq. (8), because it showed the best results in Sect. 2. As the experiment data set MIMLImage is used, introduced in Sect. 2.3, which seems to be more difficult compared with MIMLText. For each experiment run, the data is split randomly into the labeled Pool L and the unlabeled Pool U. Thereby L contains 200 examples at the beginning of each run. Each query strategy starts with the same L, U and the same MIMLSVM classifier trained with L. In every active learning iteration each query strategy chooses two bags from U to query. These bags are labeled and added to L and removed from U. Afterwards the classifier is trained with L and its classification performance is evaluated on U. Overall 50 experiment runs are executed. The parameter of MIMLSVM are the same used in Sect. 2.3.

The results of the experiments are shown as boxplots in Fig. 3. The left plot shows the results of the different SIML active learning strategies on MIMLImage.

Algorithm
1: Train MIMLSVM with L
2: Classify each bag X_u of U with trained MIMLSVM to achieve $f(X_u)$
3: Calculate $score(X_u)$ with $f(X_u)$
4: Select most valuable bag X_v of U according to a ML selection strategy
5: Request all labels Y_v of X_v
6: Remove X_v from U, add (X_v, Y_v) to L
7: goto 1: until iteration limit is reached

Fig. 2 The pseudo code of the generic active learning algorithm used in the experiments to compare the different multi-label selection strategies. As base classifier MIMLSVM in combination with the cluster distance d_{medH_B} is used. It is possible to select more than one bag to be queried at each iteration step. The algorithm has simply to select the corresponding count of bags at the selection of the most valuable ones at step 4

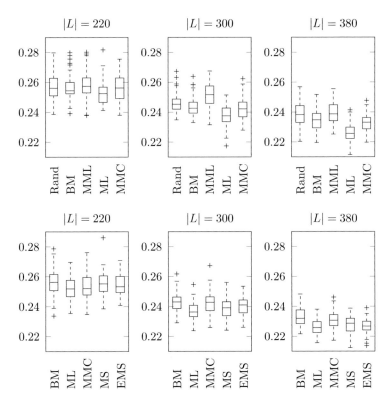

Fig. 3 The Hamming loss values (*down arrow*) of different multi-label active learning strategies achieved on MIMLImage (*left*) and the comparison of the MidSelect approaches (*right*) with the best performing multi-label active learning strategies on the data set MIMLImage

The random bag selection strategy (Rand) is used as a reference. Expect MML all strategies achieve better classification results than the random selection. Especially the ML strategy can break away from all other approaches, followed by MMC and BinMin. The experiment shows that the SIML strategies are able to perform well on an MIML data set in combination with MIMLSVM. Furthermore it shows that every other SIML active learning strategy can be useful executed on MIML data with the help of MIMLSVM. The right plot shows the performance of the MidSelect approaches compared with BinMin, ML, and MMC. Here it can be observed that ML and either ML MidSelect strategies can improve their classification performance during the active learning procedure. In addition the extended version of MidSelect seems to achieve more stable results than the base version. The experiment showed that the simple MidSelect strategies are able to perform well compared to the other three strategies. Only the ML strategy showed slightly better results than the two new strategies.

4   Conclusion

In this paper we investigated MIML and active MIMIL learning utilizing the MIMLSVM approach. We propose and investigate several variants of the Hausdorff distance measure and found that the median-based distance measure outperforms (in terms of Hamming loss) average-based and traditional Hausdorff distance on two MIML benchmark data. Furthermore, we introduce active learning for the MIML scenario. To the best of our knowledge it is the first time that active learning has been applied to MIML here we introduced two selection criteria to select the most informative data from the unlabeled data set. In addition we showed that SIML active learning strategies can achieve good results on an MIML scenario if MIMLSVM is used as their base classifier. Our two selection criteria were able to achieve a performance comparable with the best one of the applied SIML strategies.

References

Boutell, M. R., Luo, J., Shen, X., & Christopher, M. B. (2004). Learning multi-label scene classification. *Pattern Recognition, 37,* 1757–1771.

Brinker, K. (2005). On active learning in multi-label classification. In *Data and information analysis to knowledge engineering.* Proceedings of the 29th Annual Conference of the Gesellschaft für Klassifikation e.V. (pp. 206–213).

Dietterich, T. G., Lathrop, R. H., & Lozano-Perez, T. (1997). Solving the multiple-instance problem with axis-parallel rectangles. *Artificial Intelligence, 89,* 31–71.

Fu, J., & Yin, J. (2011). Bag-level active multi-instance learning. In *Eighth International Conference on Fuzzy Systems and Knowledge Discovery* (pp. 1307–1311).

Li, X., Wang, L., & Sung, E. (2004). Multi-label SVM active learning for image classification. In *IEEE 2004 International Conference on Image Processing* (pp. 2207–2210).

Mccallum, A. K. (1999). Multi-label text classification with a mixture model trained by EM. In *AAAI 99 Workshop on Text Learning.*

Schwenker, F., & Trentin, E. (2014). Pattern classification and clustering: A review of partially supervised learning approaches. *Pattern Recognition Letters, 37,* 3–14.

Settles, B. (2009). Active learning literature survey. *Computer Sciences Technical Report, 1648.*

Settles, B., Craven, M., & Soumya, R. (2008). Multiple-instance active learning. In *Neural Information Processing Systems* (pp. 1289–1296).

Wang, J., & Zucker, J. D. (2000). Solving the multiple-instance problem: A lazy learning approach. In *Proceedings of the 17th ICML* (pp. 1119–1126).

Yang, B., Sun, J. T., Wang, T., & Chen, Z. (2009). Effective multi-label active learning for text classification. In *15th ACM SIGKDD International Conference on KDDM* (pp. 917–926).

Zhang, M. L., & Zhou, Z. H. (2009). Multi-instance clustering with applications to multi-instance prediction. *Applied Intelligence, 31,* 47–68.

Zhou, Z., & Zhang, M. (2007). Multi-instance multi-label learning with application to scene classification. In *NIPS 19* (pp. 1609–1616).

Using Annotated Suffix Tree Similarity Measure for Text Summarisation

Maxim Yakovlev and Ekaterina Chernyak

Abstract The paper describes an attempt to improve the TextRank algorithm. TextRank is an algorithm for unsupervised text summarisation. It has two main stages: first stage is representing a text as a weighted directed graph, where nodes stand for single sentences, and edges are weighted with sentence similarity and connect consequent sentences. The second stage is applying the PageRank algorithm as is to the graph. The nodes that get the highest ranks form the summary of the text. We focus on the first stage, especially on measuring the sentence similarity. Mihalcea and Tarau suggest to employ the common scheme: use the vector space model (VSM), so that every text is a vector in space of words or stems, and compute cosine similarity between these vectors. Our idea is to replace this scheme by using the annotated suffix trees (AST) model for sentence representation. The AST overcomes several limitations of the VSM model, such as being dependent on the size of vocabulary, the length of sentences and demanding stemming or lemmatisation. This is achieved by taking all fuzzy matches between sentences into account and computing probabilities of matched concurrencies. For testing the method on Russian texts we made our own collection based on newspapers articles with some sentences highlighted as being more important. Using the AST similarity measure on this collection allows to achieve a slight improvement in comparison with using the cosine similarity measure.

1 Introduction

Automatic text summarisation is one of the key tasks in natural language processing. There are two main approaches to text summarisation, called abstractive and extractive approaches (Hahn and Mani 2000).

According to the abstractive approach, the summary of a text is another text, but much shorter, generated automatically to make the semantic representation of the text. This approach requires semantic analysis and usage of external vocabularies,

M. Yakovlev (✉) • E. Chernyak
National Research University Higher School of Economics, Moscow, Russia
e-mail: mail4mayase@gmail.com; echernyak@hse.ru; ek.chernyak@gmail.com

© Springer International Publishing Switzerland 2016
A.F.X. Wilhelm, H.A. Kestler (eds.), *Analysis of Large and Complex Data*, Studies in Classification, Data Analysis, and Knowledge Organization,
DOI 10.1007/978-3-319-25226-1_9

103

what may be sometimes rather complicated. According to extractive approach, the summary of a text is nothing else, but some important parts of the given text. If we compare both approaches to human activities, the abstractive approach is a sort of retelling, and the extractive approach is highlighting main words or sentences in the text. Obviously, the extractive approach is more simple and effective than the abstractive one, since it does not require any additional sources and basically may be reduced key phrase or key sentence extraction. We are going to focus on the extractive approach and follow the common simplification that the summary of a text is the set of important sentences.

The extractive summarisation problem can be formulated in the following way. Given a text T that is a consequence of sentences S that consists of words V, select a subset of the sentences S^* that are important in T. Therefore we need to define:

- what importance of a sentence is;
- how to measure importance of the sentence; Hence we need to introduce a function, *importance*(s), which measures the importance of a sentence. The higher *importance* is, the better. Next step is to build the summary. Let us rank all the sentences according to the values of *importance*. Suppose we look for the summary that consists of five sentences. Hence we take the five sentences with the highest values of *importance* and call them top-5 sentences according to *importance*. Generally, the summary of the text are the top-N sentences according to *importance* and N is set manually.

Surprisingly the best results for this statement of the problem are achieved in the paper by Mihalcea and Tarau (2004), where *importance*(s) is introduced as PageRank type function (Brin and Page 1998) without any kind of additional grammar, syntax or semantic information. The main idea of the suggested TextRank algorithm is to represent a text as a directed graph, where nodes stand for sentences and edges connect consequent sentences. The edges are weighted with sentence similarity. When PageRank is applied to this graph, every node receives its rank that is to be interpreted as the importance of the sentence, so that importance(s) = PageRank(s_{node}), where s_{node} is the node corresponding to sentence s.

Following Mihalcea and Tarau (2004) PageRank is "a way of deciding the importance of a node within a graph, based on global information recursively drawn from the entire graph". They see PageRank as a model of "voting" or "recommendation". When node A is connected by a directed edge to node B, it can be seen as node A voting for node B. Moreover, the importance of the node A itself is also taking into account. Hence, the resulting importance of the mode is defined by the votes for it and by the importance of the nodes that vote for it. When applied to the text graph, PageRank allows to rank the nodes (i.e. the sentences) according to the same principle.

To measure similarity of the sentences the authors of TextRank algorithm suggest to use the basic vector space model (VSM) scheme. First every sentence is represented as a vector in space of words or stems. Next cosine similarity between those vectors is computed.

There are several attempts to improve the TextRank algorithm. We can distinguish between three directions for TextRank improvement.

- Extension to the supervised learning paradigm (Cruz et al. 2006).
- Using cluster of sentences instead of single sentences as nodes. Usually, such clusters are referred as topics (Garg et al. 2009; Bougouin et al. 2013). After the topics are ranked according to the PageRank algorithm, the summary sentences are extracted from the top-ranked topics.
- Using different edge weighting scheme. In Erkan and Radev (2004) the LexRank weighting scheme is introduced. This scheme is based on computing eigenvectors of sentence connectivity matrix.

In this paper we also deal with a weighting scheme. Our main concern is that using any kind of VSM text representation requires using stemming procedure to map words to the stems. But because of the complexity of the Russian language and Russian derivation stemming procedures like Porter stemmer (Porter 1980) sometimes fail to map cognates to the same stem. For this reason we try to use annotated suffix tree (AST) similarity measure to estimate similarity between sentences (Pampapathi et al. 2008). This measure takes all fuzzy matches between sentences into account. It seems to be more adequate for the Russian language. Russian is an agglutinative language and the words are formed by stringing affixes (prefixes and suffixes). Porter stemmer does not cope with prefixes at all. It removes some of Russian suffixes from a complex word, but not all of them. However if we use fuzzy matching technique, we can easily get rid of all affixes and match two similar words.

The remainder is organised as follows. Section 2 briefly presents the TextRank algorithm and motivates the usage of the AST sentence similarity that is formally introduced in Sect. 3. Section 4 presents the experimental setup. The results are presented in Sect. 5. Section 6 concludes.

The article was prepared within the framework of the Academic Fund Program at the National Research University Higher School of Economics (HSE) in 2014-2015 (grant No 15-05-0041) and supported within the framework of a subsidy granted to the HSE by the Government of the Russian Federation for the implementation of the Global Competitiveness Program.

2 TextRank

To enable the application of PageRank to a text (Mihalcea and Tarau 2004) suggest the idea of a text graph. The text graph is a weighted directed graph $G = (V, E)$ with the set of nodes V and the set of edges E that is a subset of $V \times V$. Every node $v \in V$ stands for a single sentence from the given text. Two nodes are connected if they stand for consequent sentences. The node of preceding sentence links to node of succeeding sentence. The edge is weighted by the sentence similarity value. The sentence similarity can be computed in many different ways, but the most common one is to use the VSM for sentence representation and the cosine similarity measure.

According to VSM, every sentence is represented as a vector of word frequencies in the space of words. Given two sentences S_1 and S_2 and two corresponding vectors A and B, the cosine between them is $\frac{\sum_{i=1}^{n} A_i * B_i}{\sum_{i=1}^{n} A_i^2 * \sum_{i=1}^{n} B_i^2}$, where n is the total number of words. The VSM and the cosine similarity measure have two strong disadvantages. First, the similarity value is affected by the length of the sentence, since long sentences get poor representation. Second, the cosine similarity measure takes only exact matches between words or stems (if stemming procedure is exploited).

The so-called AST similarity measure overcomes both of this disadvantages. It does not depend on the size of the sentence and can cope with fuzzy matches between two sentences. We suggest to use the AST sentence similarity measure to put weights on the edges of the text graph. This measure exploits ASTs (see Fig. 1 for an example). This data structure stores not only all fragments off the strings, but also their frequencies. To find similarity between two sentences we have to construct an AS for every sentence, find the common subtree and score it. The scoring procedure is described later in detail. Its general idea is to estimate the average frequency of the common fragments of the sentences.

With this little exception we follow the TextRank algorithm. After the text graph is constructed, we can apply the PageRank algorithm introduced by Brin and Page (1998) to it. For every given node $v_i \in V$ we define $\text{In}(v_i)$ as a set of nodes, which link to it, and $\text{Out}(v_i)$ as a set of nodes v_i links to. Then the score of v_i is

$$s(V_i) = (1 - d) + d \times \sum_{j \in \text{In}(v_i)} \frac{1}{\text{Out}(v_j)} s(V_j), \qquad (1)$$

where d is damping factor and is usually set at 0.85. Starting from random values of $s(v_i)$ the computation iterates until it converges.

Fig. 1 An AST for string "mining"

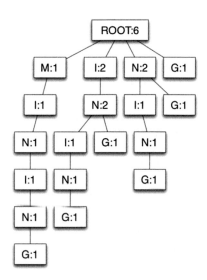

3 AST Similarity

3.1 Definition

According to the AST model introduced by Pampapathi et al. (2008), a text document is not a set of words or terms, but a set of the so-called fragments, the sequences of characters arranged in the same order as they occur in the text. Each fragment is characterised by a float number that shows the conditional probability of the fragment to occur in the text. The greater the number is, the more important the fragment is for the text. An AST (see Fig. 1) is a data structure used for computing and storing all fragments of the text and their frequencies. It is a rooted tree in which:

- Every node corresponds to one character.
- Every node is labeled by the frequency of the text fragment encoded by the path from the root to the node.

3.2 AST Construction

To construct an AST we use a naive algorithm following Gusfield (1997). First of all we need to split the whole article in sentences. After this splitting is done, we can construct an AST for every single sentence. To do this, we split every sentence into three-grams. A three-gram is the contiguous sequence of three words from a given sequence. Given the sentence "The rain in Spain stays mainly in the plain" we would generate the following three-grams: "The rain in", "rain in Spain", "in Spain stays", "Spain stays mainly" and so on.

These three-grams are used for the AST construction. We find all suffixes of the first three-gram and start with the first suffix. Here the i-th suffix of the string s is the substring, starting from the i-th position to the end of the string $s[i :]$, but not the grammatical suffix. For example, the fourth suffix of the string "rain in Spain" is the string "n in Spain". In the same fashion we define the i-th prefix of the string as the substring that starts from the first symbol and ends at the i-th symbol of the string.

So the first suffix forms the chain of nodes with frequencies equal to unity in the AST. Next we add the second suffix. We search for a match that is a path from the root of the tree that coincides with the prefix of the second suffix. If there is no match, we add the second suffix as the chain of nodes with frequencies equal to unity to the AST. If there is a match, we increase the frequencies of matched nodes by one, and add the unmatched symbols to the last node of the match. The same iterative process is repeated with all the left suffixes and three-grams.

To compute similarity between two sentences we:

- Find a common subtree for two ASTs
- Annotate it with average frequencies
- Score the whole subtree.

Fig. 2 An AST for string
"dinner"

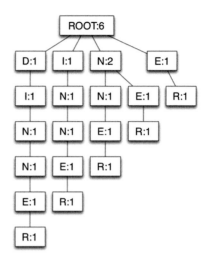

3.3 Constructing Common Subtree

To estimate the similarity between two sentences we find the common subtree of
the corresponding ASTs. We do the depth-first search for the common chains of
nodes that start from the root of the both ASTs. After the common subtree is
constructed we need to annotate and score it. We annotate every node of the common
subtree with the averaged frequency of the corresponding nodes in initial ASTs.
Consider, for example, two ASTs for strings "mining" and "dinner" (see Figs. 1 and
2, correspondingly). There are two common chains: "I N" and "N", the first one
consists of two nodes, the second one consists of a single node. Both this chains
form the common subtree. Let us annotate it. The frequency of the node "I" is equal
to 2 in the first AST and to 1 in the second. Hence, the frequency of this node in the
common subtree equals to $\frac{2+1}{2} = 1.5$. In the same way we annotate the node "N"
that follows after the node "I" with $\frac{2+1}{2} = 1.5$ and the node "N" on the first level
with $\frac{2+2}{2} = 2$. The root is annotated with the sum of the frequencies of the first level
nodes that is $1.5 + 2 = 3.5$.

3.4 Scoring Common Subtree

The score of the subtree is the sum of scores of every chain of nodes. According
to Pampapathi et al. (2008), the score of the path is the averaged sum of the
conditional probabilities of the nodes, where conditional probability of the node
is the frequency of the node divided by the frequency of its parent. For example,
the conditional probability of the node "G:1" on the third level of the AST on
Fig. 1 is 1/2. Let us continue with the example of "mining" and "dinner". There

Fig. 3 Common subtree of
ASTs for strings "mining"
and "dinner"

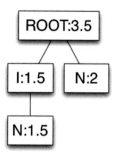

are two chains in their common subtree: "I N" and "N". The score of "I N" chain
is $(1.5/1.5 + 1.5/3.5)/2 = 0.71$, since there are two nodes in the chain. The
score of one node chain "N" is $1.5/3.5 = 0.42$. The score of the whole subtree
is $(0.71 + 0.42) = 1.13$ or $(0.71 + 0.42)/2 = 0.56$, if averaged by the number of
chains in the tree. Let us denote the first scoring technique as *AST* and the second as
AST.averaged. Note that the *AST.averaged* scoring technique results in values in
[0, 1], thus having properties of similarity measure. The similarity of two sentences
is scored in the same way, but for the sake of space we would not provide examples
of the AST for the whole sentence (Fig. 3).

4 Experimental Setup

The collection for experiments is made of 400 articles from Russian news portal
called Gazeta.ru. The articles are marked up in a special way, so that some of
sentences are highlighted because of being more important. This highlighting is
done either by the author of the article or by the editor on the basis of their own ideas.
In our experiments we considered those sentences as the summary of the article. We
tried to reproduce these summaries using TextRank with cosine and AST sentence
similarity measures. In total there were three similarity measures: *cosine*, *AST*,
AST.averaged. According to different similarity measures we get different ranking
of the sentences for every article. However the TextRank *importance* measure is
always the same. We take 10 sentences with the highest values of *importance* and
consider them as the summary of every article. To compare the similarity measures
we used two widely used characteristics: recall at 10 (*recall@10*) averaged and
precision at 10 (*precision@10*) averaged. Both of these measures are based on
the same idea: we check how many highlighted sentences appear among the 10
sentences with the highest values of *importance* for every article. Let us denote by
n_a the total number of sentences highlighted in the paper a and n_a^{10} be the number
of highlighted sentences among the top 10 sentences. To find *precision@10* we
divide n_a^{10} by 10 and to *recall@10* we divide n_a^{10} by n_a. Since both *precision@10*

Table 1 Results of using
three similarity measures in
TextRank algorithms

Similarity measure	recall@10	precision@10
All sentences		
Cosine	0.3646	0.1607
AST	0.4902	0.2166
AST.averaged	0.4374	0.1928
50 % edges		
Cosine	0.3663	0.1610
AST	0.3979	0.1763
AST.averaged	0.4060	0.1807

and *recall@*10 are calculated for a single article, the final output is *precision@*10 averaged over all 400 articles and *recall@*10 averaged over all 400 articles.

We also experimented with some parameters while constructing the graph (Table 1):

- *All sentences*: we kept all nodes in the graph so that they had at least one edge.
- *50 % edges*: we removed 50 % of all edges of lower weight from the graph, so that only those of higher weight remained.

5   Results

The highest recall and precision at 10 values are achieved by the AST similarity measure and *All sentences* graph construction technique. We should notice that no matter what parameters were used the AST similarity measure outperforms the cosine similarity measure. The low values of recall and precision are disappointing, but it may happen when unsupervised methods are applied to natural text processing tasks. They may be also a consequence of the random construction of the collection that we did not expect manually.

6   Conclusion

In this paper we presented an attempt to improve the TextRank algorithm for text summarisation, we focused on developing a new sentence similarity measure. These sentences were to be used as nodes in a graph. The nodes were connected with weighted edges, where the weights corresponded to sentence similarity.

We used the AST method to measure sentence similarity. More specifically we developed an algorithm for common subtree construction and annotation. The common subtrees were used to score the similarity between two sentences. Using this algorithm allows us to achieve some improvements according to cosine baseline on our own collection of Russian newspaper texts. The AST measure gained around 0.05 points of precision more than the cosine measure. This is a great figure for natural language processing task, taking into account how low the baseline precision of the cosine measure is. The fact that the precision is so low can be explained by some lack of consistency in the constructed collection: the authors of the articles use different strategies to highlight the important sentences. The text collection is heterogeneous: in some articles there are ten or more sentences highlighted, in some only the first one. Unfortunately, there is no other test collection for text summarisation in Russian. For further experiments we might need to exclude some articles, so that the size of summary would be more stable. Another issue of our test collection is the selection of sentences that form summaries. When the test collections are constructed manually, summaries are chosen to common principles. But we cannot be sure that the sentences are not highlighted randomly.

Although the AST technique is rather slow, it is not a big issue for the text summarisation problem. The summarisation problem is not that kind of problems where on-line algorithms are required. Hence the precision plays more significant part than time characteristics.

There are several directions of future work. First of all, we have to conduct experiments on the standard document understanding conference (DUC 2014) collections in English. Second, we are going to develop different methods for construction and scoring of common subtrees and compare it to each other. Finally, we may use some external and more efficient implementation of the AST method, such as EAST Python library by Mikhail Dubov, which uses annotated suffix arrays (EAST 2016).

Acknowledgements The article was prepared within the framework of the Academic Fund Program at the National Research University Higher School of Economics (HSE) in 2014–2015 (grant No 15-05-0041) and supported within the framework of a subsidy granted to the HSE by the Government of the Russian Federation for the implementation of the Global Competitiveness Program.

References

Bougouin, A., Boudin, F., & Daille, B. (2013). TopicRank: Graph-based topic tanking for keyphrase extraction. In *Proceedings of International Joint Conference on Natural Language Processing* (pp. 543–551).

Brin, S., & Page, L. (1998). The anatomy of a large-scale hypertextual Web search engine. In *Proceedings of the Seventh International Conference on World Wide Web 7* (pp. 107–117).

Cruz, F., Troyano, J. A., & Enruquez, F. (2006). Supervised TextRank. In *Advances in natural language processing* (pp. 632–639). Berlin/Heidelberg: Springer.

Document Understanding Conference. Retrieved October 20, 2014, http://www-nlpir.nist.gov/ (Web source)

Enhanced Annotated Suffix Tree . Retrieved January 15, 2015, https://pypi.python.org/pypi/EAST/ 0.2.2/ (Web source)

Erkan, G., & Radev, D. R. (2004). LexRank: graph-based lexical centrality as salience in text summarization. *Journal of Artificial Intelligence Research, 22*(1), 457–479.

Garg, N., Favre, B., Reidhammer, K., & Hakkani-Tur, D. (2009). ClusterRank: a graph based method for meeting summarization. In *Interspeech, ISCA* (pp. 1499–1502).

Gusfield, D. (1997). *Algorithms on strings, trees and sequences: Computer science and computational biology*. Cambridge: Cambridge University Press.

Hahn, U., & Mani, I. (2000). The challenges of automatic summarization. *Computer, 33*(11), 29–36.

Mihalcea, R., & Tarau P. (2004). TextRank: bringing order into text. In *Proceedings of the Conference on Empirical Methods in Natural Language Processing* (pp. 404–411).

Pampapathi, R., Mirkin, B., & Levene, M. (2008). A suffix tree approach to anti-spam email filtering. *Machine Learning, 65*(1), 309–338.

Porter, M. F. (1980). An algorithm for suffix stripping. *Program, 14*(3), 130–137.

Big Data Classification: Aspects on Many Features and Many Observations

Claus Weihs, Daniel Horn, and Bernd Bischl

Abstract In this paper we discuss the performance of classical classification methods on Big Data. We distinguish the cases many features and many observations. For the many features case we look at projection methods, distance-based methods, and feature selection. For the many observations case we mainly consider subsampling. The examples in this paper show that standard classification methods should not be blindly applied to Big Data.

1 Introduction

This paper is on Big Data Analytics (BDA). But what is Big Data? Unfortunately, the answer depends on whom you have asked when. In machine learning (ML) benchmarks in the 1990s (e.g., in the UCI repository) maximum 100s to 1000s of data points were available. In modern benchmarks we often have more than 10^6 data points. When you ask, e.g., Google, the answer might be "Big Data means that data are much too big for your computer storage, only streaming is possible from a cloud, only distributed analytics, ..." Another possibility is to define a "Big Data problem" by the impossibility to exactly solve the learning problem by computer time reasons.[1] Therefore, information in the data is not optimally utilizable. This definition is used in the very last example of this paper, where the question is: Which information brings us as fast as possible as near as possible to the solution and what

---

[1]Thanks to T. Glasmachers for suggesting this definition.

C. Weihs (✉)
Chair of Computational Statistics, Faculty of Statistics, TU Dortmund, Dortmund, Germany
e-mail: claus.weihs@tu-dortmund.de

D. Horn • B. Bischl
Department of Statistics, TU Dortmund University, Dortmund, Germany
e-mail: daniel.horn@tu-dortmund.de; bernd.bischl@tu-dortmund.de

© Springer International Publishing Switzerland 2016
A.F.X. Wilhelm, H.A. Kestler (eds.), *Analysis of Large and Complex Data*, Studies
in Classification, Data Analysis, and Knowledge Organization,
DOI 10.1007/978-3-319-25226-1_10

113

is a "perfect" approximation algorithm?[2] Note that those who think that the data in our paper is not big enough for being Big Data might also call our topic of interest "Large Scale Data Analysis."

In this paper we will discuss typical classification methods in the context of BDA. The message of this paper is that for BDA not all classical methods are adequate in all Big Data situations and that Big Data might even long for special methods. In order to demonstrate the extremes, we will particularly discuss the cases of many features (and small no. of observations) in Sect. 2, and the case of many observations (and small no. of features) in Sect. 3.

2 Many Features

With the advent of high throughput biotechnology data acquisition platforms such as micro arrays, SNP chips, and mass spectrometers, data sets with many more variables than observations are now routinely being collected (see, e.g., Kiiveri 2008). Most often, however, only a small part of these p features or a small number of directions in p-space are important for classification. Therefore, one might be tempted to thoughtlessly apply standard methods which are known to be only adequate for $p < n$ (not too big), but problematic in high dimensions (curse of dimensionality) and for very large n. In this paper, we will discuss some of the many available classification methods in this context. Let us start with projection-based methods.

2.1 Projection-Based Methods

One of the best known and most used projection-based classification methods in statistics is Fisher discrimination. The performance of this method in the case of more features than observations is discussed by Bickel and Levina (2004) showing the following property:

Consider two classes with Gauss distributions: $\mathcal{N}(\mu_1, \Sigma)$, $\mathcal{N}(\mu_2, \Sigma)$. Let the corresponding a priori probabilities be equal, i.e., $\pi_1 = \pi_2 = 0.5$. Then, for Fisher discrimination the classification function has the form $\delta_F(x) = (x - \mu)^T \Sigma^{-1}(\mu_1 - \mu_2)$ with $\mu = (\mu_1 + \mu_2)/2$. Let the corresponding samples be observed with equal sample sizes, i.e., $n_1 = n_2$. Then, the sample version of the classification rule is: Assign class 1 iff $\hat{\delta}_F(x) = (x - \bar{x})^T S^{-1}(\bar{x}_1 - \bar{x}_2) > \log(\pi_2/\pi_1) = 0$. If $p > n$, then the inverse of the estimated pooled covariance matrix S does not exist and the Moore–Penrose generalized inverse is used instead. For this situation, the following result is true under some regularity conditions which particularly state that the norm

[2]This part of the paper was supported by the Mercator Research Center Ruhr, grant Pr-2013-0015, see http://www.largescalesvm.de/.

of the mean vector should be limited. If $p \to \infty, n \to \infty$, and $p/n \to \infty$, then error $\to 0.5$, i.e., the class assignment is no better than random guessing.

This result states a strong warning concerning the application of Fisher discrimination in the case of many more features than observations. As have been motivated by Bickel and Levina (2004), the bad performance of Fisher discriminant analysis is due to the fact that the condition number of the estimated covariance matrix goes to infinity as dimensionality diverges even though the true covariance matrix is not ill-conditioned.

The regularity conditions mentioned above mainly state that the true covariance matrix should not be ill-conditioned and that the mean vectors of the classes should stay in a compact set. In the following simulations we thus consider two distinct cases, one where the mean vectors drift away from each other the higher the dimension p is, and one where the distance of the mean vectors of the classes stay the same for different p. Obviously, if the class distance is increasing with increasing p by taking the distances in the individual coordinates the same, the classification problem gets simpler. On the other hand, one can show that if the class distance stays the same for different p by means of shrinking the distance in the individual coordinates by $md/\sqrt{p}$, then the Bayes error stays the same if the covariance matrix is diagonal, i.e., $\Sigma = d \cdot I$. Indeed, in the second case the Bayes error is $1 - \Phi((md/2)/d^{0.5})$. This means that the difficulty of the classification problems stays mainly the same in the different dimensions p.

Noise accumulation can also be reduced by ignoring the covariance structure, i.e., by using a diagonal matrix as an estimate of the covariance matrix. In this context, Bickel and Levina (2004) derived the following asymptotic result for the so-called *independence rule (ir)*, i.e., linear discriminant analysis *(lda)* with diagonal covariance matrix:

Let Γ be a "regular" space of possible means and covariance matrices of the two classes, Σ the full covariance matrix in the two classes, Σ_0 the corresponding correlation matrix, $\lambda(\Sigma_0)$ an eigenvalue of Σ_0, and Φ the distribution function of the standard normal. Then, the following result is true:

If $\log(p)/n \to 0$, then $\lim \sup_{n \to \infty}$(maximal error in Γ) $= 1 - \Phi(\frac{\sqrt{K_0}}{1+K_0}c)$, where $K_0 = \max_\Gamma(\frac{\lambda_{max}(\Sigma_0)}{\lambda_{min}(\Sigma_0)})$ and $c^2 = \min_\Gamma((\mu_2 - \mu_1)^T \Sigma^{-1}(\mu_2 - \mu_1))$.

Therefore, if p is going slower to infinity than e^n, then for Big Data sets there is a bound for the maximal error in the space of possible data situations. In practice, this property may lead to a superiority of *ir* over the full *lda*.

Here, shrinking the distance in the individual coordinates by $md/\sqrt{p}$ and taking $\Sigma = d \cdot I$ leads to $K_0 = 1$ and to a limit for the maximal error of $1 - \Phi(c/2) = 1 - \Phi((md/d^{0.5})/2)$, which again is the Bayes error above.

Finally note that for normal distributions the independence rule is equivalent to the Naive Bayes method. In practice, however, the Naive Bayes method *(NB)* is typically implemented in a non-parametric way and not by assuming a certain type of distribution like the normal distribution. This generally leads to implementations different from the independence rule. For normal distributions as in our examples, *NB* is thus expected to be inferior to *ir*. Additionally, the linear support vector machine (*svm*), also looking for linear separations, will be discussed as an alternative to *lda* which can be adapted to the actual data by tuning the cost parameter.

Generic Data Generation (GDG). Let us demonstrate the above theoretical results by means of data examples. Let us start with a GDG step. We will always consider the ideal situation for the *lda*, i.e., two classes where the influential features are multivariate normally distributed with different mean vectors and the same covariance matrix. In the case where *pr* features influence class separation we choose the class means $m_1(i) = -md/2$, $m_2(i) = md/2$, where *md* = difference between the two class means, $i = 1, \ldots, pr$. The covariance matrices are built so that $\Sigma = \Sigma_R + d \cdot I$, where Σ_R is built of independent uniform random numbers between 0.1 and 1 and the multiple *d* of the identity is added in order to generate positive definiteness. Note that if *d* is large, then Σ is nearly diagonal, making our above discussion on Bayes errors for diagonal covariance matrices relevant. By choosing different distances *md* between the mean vectors or different *d* the Bayes error, interpreted as the difficulty of the classification problem, can be varied. Sometimes we add noise by means of features which do not have any influence on class separation by adding $(p - pr)$ normally distributed features with mean 0 and variance *d*. Overall, we assume that we have *p* features. Note that possibly $p = pr$. We typically use $n = 2 \cdot nel << p$ observations, *nel* observations for each class. Thus, *p* tends to be much bigger than *n*, the case we discuss in this section. The generation of *n* data points from the above normal distributions in *p* dimensions is repeated $rp = 200$ times using different random covariance matrices Σ. For the estimation of error rates, corresponding test samples with $nelt = 1000$ observations per class are generated from training distributions.

Example 1. Let us first assume that all involved features in fact influence the class choice, i.e., $p = pr = 12, \ldots, 2040$, and let $d = 25, nel = 6, md \in \{2.5, 20/\sqrt{pr}\}$ representing the above first and second case of class distance choice. By means of this variation of *p* with constant $n = 2 \cdot 6 = 12$ we vary the ratio p/n from 1 to 170. For $md = 2.5$ the classification problem tends to become easier for increasing *p* than the problem with $md = 20/\sqrt{pr}$. On the accordingly generated data (see GDG) different classification methods are compared. Let us start the discussion of the mean error rates in Table 1a with Bayes rules and approximate Bayes rules like (Fisher's) *lda*, the independence rule (*ir*), the naive Bayes method (*NB*), and the 1 nearest neighbor rule (*1NN*), as well as another standard linear separator, the linear *svm*.[3] Obviously, all methods benefit from higher dimensions in the case $md = 2.5$ as expected. In the case $md = 20/\sqrt{pr}$ all methods are suffering from higher dimensions. This was expected for *lda*, but appears also be true for the other methods. Notice that *svm* needs by far the most training time and is not distinctly better than the other methods (cp. column sec). Therefore, the choice of *svm* cannot be justified for the studied problems. Also note that runtime is near zero for 1NN because the training data set only consists of $n = 12$ observations.

---

[3]This simulation was carried out using the R-packages *BatchJobs* (Bischl et al. 2015) and *mlr* on the *SLURM* cluster of the Statistics Department of TU Dortmund University.

Table 1 Comparison of mean error rates (%): (a) all, (b) only $p/6$ features influence

p	12	120	240	360	480	600	1080	2040	12	120	240	360	480	600	1080	2040	sec
(a) All	$md = 2.5$								$md = 20/\sqrt{p}$								
lda	41	23	16	14	12	11	8	7	24	32	34	37	38	38	42	44	1.0
ir	32	14	10	9	8	8	7	6	8	25	31	34	37	39	41	45	0.1
NB	38	24	19	16	14	13	10	8	12	34	39	41	42	43	45	47	1.5
1NN	38	23	18	16	15	14	12	11	11	34	39	41	43	44	46	47	0.0
svm	35	15	11	9	8	8	7	6	9	26	32	35	37	39	42	44	160
(b) $p/6$	$md = 2.5$								$md = 20/\sqrt{p/6}$								
lda	48	45	45	44	42	41	38	32	22	37	42	43	44	44	45	46	1.0
lda+fs/6	43	39	36	34	31	31	25	18	5	25	30	33	34	37	38	40	10
ir	46	39	34	31	29	27	22	17	6	21	27	30	32	34	38	41	0.05
NB	48	44	42	40	39	38	35	32	13	32	38	40	41	42	44	46	1.5
1NN	48	42	40	38	36	35	31	27	9	28	34	37	39	40	43	45	0.0
svm	46	39	34	31	29	27	22	16	15	23	27	30	33	34	38	41	150

p no. of dimensions,
sec mean training time over both md in seconds for $p = 2040$,
md mean difference of classes in each dimension,
lda linear discriminant analysis (lda, package MASS, software R (R CORE TEAM 2014)),
fs/6 feature selection (best p/6 features, mutual information (symmetrical.uncertainty) criterion, package FSelector in R),
ir independence rule = lda with diagonal covariance matrix (sda, package sda in R, no shrinkage, diagonal = TRUE),
NB naive Bayes rule (naiveBayes, package e1071 in R),
1NN 1 nearest neighbor rule (knn, package class in R),
svm linear support vector machine (svm in R, package e1071, cost parameter tuned on grid $2^{-4}, \ldots, 2^4$ by leave-one-out)

Let us now compare this behavior with the case where only $pr = p/6$ features influence the classes. Looking at the results in Table 1b, the benefit for higher dimensions is much slower in case $md = 2.5$ because there is a much smaller class distance increase. Notice however that the methods *ir* and *svm* distinctly benefit the most, *ir* with much less training time than *svm*. In the case $md = 20/\sqrt{pr}$ the behavior is similar as for $pr = p$.

2.2 Distance Dependence

In the previous chapter we saw a distance dependency of classification quality. Let us now consider this dependency more exactly for a general class of distance-based classifiers.

For a plausible distance-based classifier g we only assume the following two properties: (a) g assigns X to class 1 if it is closer to each of the X_is in class 1 than it is to any of the X_js in class 2. (b) If g assigns X to class 1, then X is closer to at least one of the X_is in class 1 than to the most distant X_j in class 2.

Table 2 Mean error rates (%) of 1NN (left) and mean class distances MD (right)

$ip \mid p$	12	120	240	360	480	600	1080	2040	12	120	240	360	480	600	1080	2040
1	45	0	0	0	0	0	0	0	1.5	4.7	6.7	8.2	9.5	10.6	14.2	19.6
2	45	33	28	24	20	17	10	4	1.5	1.5	1.5	1.5	1.5	1.5	1.5	1.5
4	45	44	45	44	44	44	44	44	1.5	0.84	0.71	0.64	0.60	0.56	0.49	0.42

For such a method the following property is true (Fan et al. 2011):
Consider the model $X_{ij} = \mu_{kj} + \epsilon_{ij}$, $i \in G_k$, $k = 1, 2$, where X_{ij} is the jth component of X_i, μ_{kj} the j-th component of mean vector μ_k, and the ϵ_{ij} are independent identically distributed with mean 0 and finite fourth moment. Then, the probability that a distance-based classifier of the above kind classifies a new observation correctly converges to 1 iff $p = o(||\mu_2 - \mu_1||^4)$ for $p \to \infty$.

This property shows that with distance-based classifiers perfect class prediction is possible, but only if the distance of class means grows with the number of influential features so that $p^{1/4}/||\mu_2 - \mu_1|| \to 0$, i.e., that $MD = ||\mu_2 - \mu_1||$ grows faster than $p^{1/4}$. Note that this result is independent of sample size n. Let us now illustrate this property in more detail by means of an example.

Example 2. Consider the kNN method with $k = 1$ based on the Euclidean distance. Let the mean distance MD between the two classes increase with dimension p so that $MD = p^{1/ip-0.5} \cdot 1.5/12^{1/ip-0.5}$ guaranteeing a start distance of 1.5, $ip = 1, 2, 4$. Note that the mean distance automatically increases with $p^{0.5}$ if the distances between the classes are identical in all dimensions. In GDG we additionally choose $nel = 6$, $nelt = 1000$, $rp = 200$ and $\Sigma_R = 0 \cdot I$, $d = 25$, meaning that we sample from independent normal distributions with mean distance MD and standard deviation 5 in each dimension (cp. the above theoretical property). Table 2(left) shows that the start distance of 1.5 leads to a high mean test error rate of 45 %. However, the error rate benefits from more features if $ip < 4$, confirming the theoretical result. Also note that the distances in individual dimensions are shrunken for $ip > 2$ [see Table 2(right)] because of automatic increase of p-dimensional class distances by $p^{0.5}$.

2.3 Feature Selection

Let us now have a look on feature selection methods in high dimensions. Simple filters are the fastest feature selection methods. In filter methods, numerical scores s_i are constructed for the characterization of the influence of feature i on the dependent class variable. Filters are generally independent of classification models. Easy example filters are the χ^2-statistic for the evaluation of independence between (discretized) feature i and the class variable, the p-value of a t-test indicating whether the mean of feature i is different for the two classes, the correlation between feature i and the class variable, and the mutual information in feature i and the class variable.

Filters can be easily combined with a classification method. First calculate filter values (scores). Then sort features according to scores and choose the best k features. Finally, train the classification method on these k features. Let us demonstrate the possible effect of a filter by reconsidering Example 1.

Example 1 (cont). When only $p/6$ features influence the classes, the correct number of features is selected by feature selection (*lda+fs/6*). The corresponding error rates are then much lower than without feature selection but at the price of higher computation times (see column "sec") caused by the usage of a mutual information criterion for feature selection (see Table 1b).

The most important problem with feature selection is the adequate choice of k. Another idea is to apply dimension reduction, e.g., by principal components analysis (*pca*), before application of classification methods (see, e.g., Bair et al. 2006) since there is hope that projection dimensions put much more weight on features having large classification power. Unfortunately, the above result for *lda* can be generalized to the application of *lda* to any general projection on linear combinations. This is because such projection directions are constructed with probability 1 using essentially all features, so that the misclassification error tends to be big because of noise accumulation when not all features are relevant for class separation (Fan et al. 2011). This affects, e.g., *lda* applied to principal components, but also combinations with other projection methods like partial least squares (as proposed, e.g., by Boulesteix 2004).

Let us discuss whether a, at least nearly, correct finding of the real number of influential dimensions is helpful for *lda* and look at an example combining the two ideas, feature selection and *pca*.

Example 3. [4]Consider two classes in $p = 1000$ dimensions, where only $pr = 100$ dimensions really influence class membership. The idea is, first, to identify those m features with the highest effect on class separation, $m \in \{2, \ldots, p\}$, by means of feature selection on training data. Here, we use the linear correlation criterion, which is much faster than mutual information but only approximate for binary outputs (again from FSelector in R). In the above GDG we use $nel = 50, md \in \{0.5, 1.5, 2.5\}, d = 10, rp = 200, nelt = 1000$. Second, class separation is tried by means of *lda* on the first two principal components (*pcs*). Note that *pcs* are only determined up to sign. The sign might even differ for training and test sets resulting in an interchange of class labels in the test data. Therefore, $\min(mcr, 1 - mcr)$, mcr = estimated misclassification error, is used as the error rate. Table 3 shows mean *lda* error rates on the first two principal components of the same m dimensions of the test data identified on the training data. Obviously, choosing m near the correct $pr = 100$ is only acceptable for the easier problems ($md = 1.5, 2.5$). For the hardest problem with $md = 0.5$, higher m gave more acceptable results. This may be caused by a nearly inevitable imperfect feature selection. In any case, higher m, in our example near half the number of involved features (i.e., $m = 500$), appear to be on the save side in all cases.

---

[4]This example is inspired by Fan et al. (2011).

Table 3 Mean error rates (%) of *lda* on the first 2 principal components based on the best *m* features

md | m	2	10	50	90	100	110	200	500	800	900	1000
0.5	48	49	49	49	49	49	47	41	42	42	43
1.5	39	34	21	20	19	19	19	18	19	20	22
2.5	30	16	8	6	6	6	6	6	6	6	6

3 Many Observations

Let us also briefly look at cases with many more observations than features. In such cases, the standard idea is to split the data into smaller blocks, analyze these blocks, and recombine the results to an overall result. Let us concentrate here on the case where we actively split a too Big Data set and analyze the corresponding blocks. This idea is obviously adapted from cross validation and bagging. Then, we try to find a recombination method that gives a reasonable, as optimal as possible, approximation to that result which we would have seen if we would have looked at all observations at the same time. This leads us back to the definition of Big Data based on approximations in the introduction. Please note that we skip the streaming case here because of space restrictions. In such a case, the data is arriving already in blocks with the possibility of structural breaks in new blocks.

One example for splitting the data actively into subsamples and try to estimate the overall error rate from estimated error rates in the subsamples is the so-called cascade-svm (Graf et al. 2005), a version of which was realized in Meyer et al. (2013) in the following way:

1. Partition the data into *k* subsets of possibly the same size.
2. Parallelly train *svm* independently on each subset.
3. Generate new data sets by combination of the support vectors (*svs*) of pairs (or triplets,...) of such analyses.
4. Repeat steps 2 and 3 for some time.
5. Train an *svm* on all *svs* in step 4 leading to an *svm* model.

Here, the main idea for the estimation of the overall error rate is that for *svm* already the *svs* contain all information necessary for model building. For other classification methods, analogues might be constructed by identifying the important observations by means of the distance to the decision border.

Meyer et al. (2013) tested the method on examples with 67,000–581,000 observations comparing the full linear *svm*, pure bagging with majority voting, and different versions of the cascade-*svm*. The latter sometimes resulted in much better results than pure bagging and was much faster with only a little worse results than full *svm*.

For this paper we took a closer look on the approximation of the result of the full kernel *svm* by means of subsampling in the following way:

1. Optimize the cost parameter, the kernel-width, and the duality gap of the kernel *svm* method as well as the subsampling rate k with respect to two targets, namely the misclassification error and required training time, by means of sequential model-based multi-criteria optimization.
2. Randomly split the subsample once into 50 % training, 25 % test, and 25 % validation samples. For this, the order of the observations was permuted.
3. During the optimization, generate a training subsample by using the first $100 \cdot k$ percent of the training data. This way, the training sample for $k = 0.1$ is a subset of the sample for $k = 0.2$.
4. Train the kernel svm[5] on the training sample, calculate the training time, and estimate the misclassification rate on the validation sample.
5. Analyze the trade-off between the two targets by means of the Pareto-front.

With this method, we received very promising results for several large data sets[6] in that the subsample results built very promising alternatives to the full data result with only slightly higher errors produced in much less training time. In Fig. 1 on the left we see an expected Pareto-front. Using 93 % of the data, a small speed up is possible without a significant error loss, whereas using only 21 % we observe a speed up by factor 10, but twice the validation error. On the right we see a very promising result where we were able to reduce the training time by factor 100 nearly without any loss in accuracy.

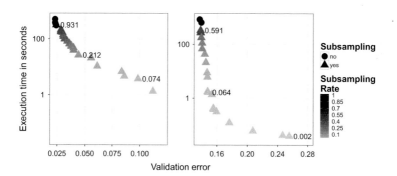

Fig. 1 Pareto-front for LibSVM with and without subsampling on binary versions of the data sets *mnist* (*left*, 70,000 samples, 780 features) and *vehicle* (*right*, 98,528 samples, 100 features). Selected subsampling rates are explicitly noted on the right of the mark

---

[5] We used the R-library *libSVM*, see http://www.csie.ntu.edu.tw/~cjlin/libsvm/.

[6] Data sets taken from http://www.csie.ntu.edu.tw/~cjlin/libsvmtools/datasets/.

4   Summary and Conclusion

In this paper we discussed the performance of standard classification methods on Big Data. We distinguished the cases many features and many observations. For the many features case we looked at projection methods, distance-based methods, and feature selection. If the class distance increases for higher dimensions, then error rates are decreasing, whereas for constant Bayes error the estimated errors are increasing up to nearly 0.5 for higher dimensions. Also, feature selection might help with finding better models in high dimensions. *ir* and *svm* performed best in high dimensions, *ir* in much less time than *svm*. For the many observations case, subsampling generated promising alternatives to the full data result by producing only slightly higher errors in much less training time.

References

Bair, E., Hastie, T., Paul, D., & Tibshirani, R. (2006). Prediction by supervised principal components. *Journal of the American Statistical Association, 101*, 119–137.

Bickel, P. J., & Levina, E. (2004). Some theory for Fisher's linear discriminant function, "naive Bayes", and some alternatives when there are many more variables than observations. *Bernoulli, 10*, 989–1010.

Bischl, B., Lang, M., Mersmann, O., Rahnenfuehrer, J., & Weihs, C. (2015). BatchJobs and BatchExperiments: Abstraction mechanisms for using R in batch environments. *Journal of Statistical Software, 64*(11), doi:10.18637/jss.v064.i11.

Boulesteix, A. L. (2004). PLS dimension reduction for classification with microarray data. *Statistical Applications in Genetics and Molecular Biology, 3*, 1–33.

Fan, J., Fan, Y., & Wu, Y. (2011). High-dimensional classification. In T. T. Cai, & X. Shen (Eds.), *High-dimensional data analysis* (pp. 3–37). New Jersey: World Scientific.

Graf, H.P., Cosatto, E., Bottou, L., Durdanovic, I., & Vapnik, V. (2005). Parallel support vector machines: The cascade SVM. *Advances in Neural Information Processing Systems, 17*, 521–528.

Kiiveri, H.T. (2008). A general approach to simultaneous model fitting and variable elimination in response models for biological data with many more variables than observations. *BMC Bioinformatics, 9*, 195. doi:10.1186/1471-2105-9-195

Meyer, O., Bischl, B., & Weihs, C. (2013). Support vector machines on large data sets: Simple parallel approaches. In M. Spiliopoulou, L. Schmidt-Thieme, & R. Jannings (Eds.), *Data analysis, machine learning, and knowledge discovery* (pp. 87–95). Berlin: Springer.

R Core Team (2014). R: A language and environment for statistical computing. Vienna, Austria: R Foundation for Statistical Computing. http://www.R-project.org/

Part III
Clustering

Bottom-Up Variable Selection in Cluster Analysis Using Bootstrapping: A Proposal

Hans-Joachim Mucha and Hans-Georg Bartel

Abstract Variable selection is a problem of increasing interest in many areas of multivariate statistics such as classification, clustering and regression. In contradiction to supervised classification, variable selection in cluster analysis is a much more difficult problem because usually nothing is known about the true class structure. In addition, in clustering, variable selection is highly related to the main problem of the determination of the number of clusters K to be inherent in the data. Here we present a very general bottom-up approach to variable selection in clustering starting with univariate investigations of stability. The hope is that the structure of interest may be contained in only a small subset of variables. Very general means, we make only use of non-parametric resampling techniques for purposes of validation, where we are looking for clusters that can be reproduced to a high degree under resampling schemes. So, our proposed technique can be applied to almost any cluster analysis method.

1 Introduction and Motivation

Cluster analysis aims at finding sub-populations (clusters). Usually, it considers several variables simultaneously. In this context, the observations within a cluster should be similar to each other, whereas objects from different clusters should be as dissimilar as possible to each other. However, taking all variables into account means that the discovery of clusters in the data is often impossible because of several masking and noisy variables. There are many papers on variable selection in clustering, mainly based on special cluster separation measures such as the Davies and Bouldin (1979) criterion: ratio of within-cluster dispersions and between-cluster separation. Meinshausen and Bühlmann (2010) introduced a much more general

H.-J. Mucha (✉)
Weierstrass Institute for Applied Analysis and Stochastics (WIAS), Mohrenstraße 39, 10117 Berlin, Germany
e-mail: mucha@wias-berlin.de

H.-G. Bartel
Department of Chemistry, Humboldt University, Brook-Taylor-Straße 2, 12489 Berlin, Germany
e-mail: hg.bartel@yahoo.de

© Springer International Publishing Switzerland 2016   125
A.F.X. Wilhelm, H.A. Kestler (eds.), *Analysis of Large and Complex Data*, Studies in Classification, Data Analysis, and Knowledge Organization,
DOI 10.1007/978-3-319-25226-1_11

"Stability selection", i.e., variable selection for high-dimensional problems, based also on subsampling (a special resampling technique) in combination with selection algorithms with a wide range of applicability.

In practical view of the occurrence of several masking and noisy variables, we propose a bottom-up variable selection starting with univariate cluster analysis, and going on to multivariate cluster analysis via bivariate clustering. The hope is that the structure of interest may be contained in only a small subset of variables. To our knowledge, the first (special) forward variable selection in clustering was proposed by Fowlkes et al. (1988). Gnanadesikan et al. (1995) showed that weighting and selection of variables can dramatically facilitate cluster recovery. Carmone et al. (1999) proposed a variable selection procedure based only on the J univariate clustering results. In our extremely general proposal, the assessment of stability of cluster analysis results plays a key role. It is based on resampling techniques such as bootstrapping and stability measures such as the adjusted Rand's measure (Hubert and Arabie 1985). Here we prefer bootstrapping as the favourite resampling technique because of its good performance in finding the number of clusters K (Mucha and Bartel 2014).

Let us start with a motivating real data application to archaeometry. Figure 1 displays both the original data values at the abscissa and the corresponding result of Ward's univariate hierarchical clustering based on the squared euclidean distances, for statistical details see below in Sect. 3. Univariate clustering simply means that the set of objects is reordered based on a single variable followed by dividing the total order of objects into homogeneous regions. Here archaeological objects (tiles) are clustered based on iron oxide. Iron oxide is one out of a set of 19 variables. The tiles were produced by two different Roman military units in the former Roman province *Germania Superior*. Obviously, the two archaeological groups can be reproduced only to a moderate degree by univariate clustering.

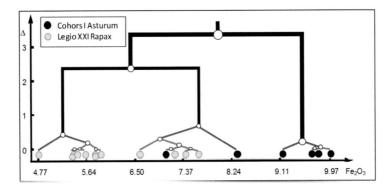

Fig. 1 From univariate data to dendrograms via squared euclidean distances: Ward's hierarchical cluster analysis based on measurements of the content of iron oxide (in mass-%). The ordinate reflects the increment Δ of the sum of within-cluster variances when merging two clusters (logarithmic scale)

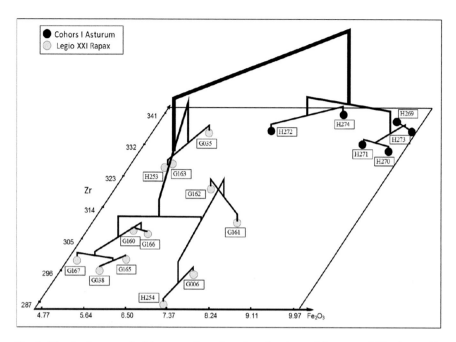

Fig. 2 Plot-dendrogram: the binary tree is projected on the plane of the two variables iron oxide and zirconium (Zr, in ppm). The dendrogram represents the result of Ward's bivariate hierarchical cluster analysis based on the squared euclidean distances of the standardized variables (Z-scores)

When taking into account an additional chemical element, here for instance Zr, Ward's cluster analysis finds the two archaeological groups perfectly. Figure 2 shows both the corresponding dendrogram and the two groups of tiles that are distinguished by the stamp of their military unit (marked by different symbols of the terminal nodes of the tree). Here in addition, each observation is marked by its identification number (for details on this dataset, see Mucha and Ritter 2009). These two groups of tiles cannot be uncovered by Ward's method (or by K-means clustering) when using all 19 variables of the dataset (Mucha et al. 2015). To be more precisely, at least one observation (i.e., "H272") is clearly misclassified.

2   Bottom-Up Variable Selection: A Proposal

In the case of high-dimensional data, it seems very realistic that class structures may be contained in a smaller set of variables. Looking for clusters in almost arbitrary subspaces is definitely intractable due to the computational complexity. Therefore here we propose a practicable idea. The basic bottom-up variable selection looks as follows:

1. The starting point is an assessment of the evidence of univariate clustering results based on bootstrapping. Concretely, we are looking for the most stable univariate clustering (i.e., the best variable) with respect to measures of the correspondence between two partitions such as the adjusted Rand's index (ARI) R_K or Jaccard's measure of correspondence between pairs of clusters (Hennig 2007).
2. Subsequently, we are looking for the best partner of the variable found in step 1. The hope is to find the most stable bivariate clustering in that way.
3. We are going ahead to find a third partner of the two variables found in step 2. Furthermore, we proceed the search for next variables as long as an "essential" improvement of the stability of the clustering can be realized.

In clustering, usually nothing is known about the true class structure, especially about the number of clusters K. Therefore, the performance or the stability of clustering cannot be assessed by counting the rate of misclassifications based on a confusion matrix. However, with the help of the non-parametric bootstrapping we are able to operate also on a confusion matrix. It comes from crossing two partitions: the original one and one coming from clustering a "bootstrap" sample. Then the ARI or other measures of stability can operate on such an "artificial" confusion matrix. Usually, hundreds of bootstrap samples are needed, see for details Mucha and Bartel (2014). Here we work with $B = 250$ bootstrap samples and we take the average (or median) of 250 ARI values to come to a final $R_K, K = 2, 3, \ldots$. The maximum R_K gives us an idea about the number of clusters K we are looking for.

To quantify what "essential" means, a stop criterion of increment of stability ΔR_K such as $\Delta R_K = 0.01$ can be used. The computational complexity decreases with the number of steps: J univariate (original) clustering results have to be assessed, $J - 1$ bivariate ones, $J - 2$ trivariate ones, and so on. Hierarchical clustering looks most fit and proper for our resampling proposal because of the (usual) unique and parallel clustering of the I observations into partitions of $K = 2, K = 3, \ldots$ clusters. Therefore, without loss of generality, here the hierarchical Ward's method is used. In addition, pairwise distances such as (3) (see next section), the usual starting point of hierarchical cluster analysis, are not changed by bootstrapping/subsampling techniques.

In contradiction, the results of partitional (iterative) clustering methods such as the K-means method are dependent on the initial partition into a fixed number of clusters K. Usually, 50 different initial partitions are used to get up to 50 different locally optimal solutions. The best solution is taken for the investigation of stability. Moreover, one has to do this for different K ($K = 2, K = 3, K = 4, \ldots, K = K_{\max}$). Finally, one has to do all things outlined above also for each bootstrap sample (or subsample). That means step 1 of our proposal needs altogether $50 * K_{\max} * (B+1) * J$ univariate partitional cluster analyses.

The proposed general variable selection procedure can be modified in several ways by

- switching to a top-down step to drop some variables in between the bottom-up selection direction,
- starting with $J * (J - 1)/2$ bivariate cluster analyses as a special variant, i.e., starting with step 2 of the general proposal. For instance, this makes sense in the case of rank data.

3 Application: Clustering of Swiss Bank Notes

The dataset contains six measurements made on 100 genuine and 100 forged counterfeit old-Swiss 1000-franc bank notes (Flury and Riedwyl 1988) with the following variables:

- *Length*: Length of bill (mm),
- *Left*: Width of left edge (mm),
- *Right*: Width of right edge (mm),
- *Bottom*: Bottom margin width (mm),
- *Top*: Top margin width (mm), and
- *Diagonal*: Length of diagonal (mm).

Without loss of generality we consider here the simplest model-based Gaussian clustering method. It seems to be an appropriate model for clustering this dataset. Let $\mathbf{X} = (x_{ij})$ be a data matrix consisting of I rows (observations) and J columns (variables). In particular, $\mathscr{C} = \{\mathbf{x}_1, \ldots, \mathbf{x}_i, \ldots, \mathbf{x}_I\}$ denotes the finite set of observations. Further, let $\mathscr{P} = \{\mathscr{C}_1, \ldots, \mathscr{C}_K\}$ of $\mathscr{C}$ be the partition we are looking for. The simplest Gaussian clustering model means the minimization of the sum of squares (SS) criterion

$$V_K(\mathscr{P}) = \sum_{k=1}^{K} \mathbf{W}_k, \tag{1}$$

which is equivalent to the minimization of

$$V_K(\mathscr{P}) = \sum_{k=1}^{K} \frac{1}{|\mathscr{C}_k|} \sum_{i \in \mathscr{C}_k} \sum_{h \in \mathscr{C}_k, h > i} d_{ih}, \tag{2}$$

where

$$d_{ih} = d(\mathbf{x}_i, \mathbf{x}_h) = (\mathbf{x}_i - \mathbf{x}_h)^T (\mathbf{x}_i - \mathbf{x}_h) = \|\mathbf{x}_i - \mathbf{x}_h\|^2 \tag{3}$$

is the squared euclidean distance between two observations $\mathbf{x}_i$ and $\mathbf{x}_h$, and $\mathbf{W}_k$ is the usual estimate of the within-cluster covariance matrix $\boldsymbol{\Sigma}_k$ of cluster $\mathscr{C}_k$. For details, see Mucha (2009).

Furthermore, also without loss of generality, we consider here the hierarchical Ward's method that minimizes the criterion (1) [or (2)]. Figure 3 summarizes the investigation of stability of the univariate Ward's clustering based on the ARI R_K coming from comparisons with cluster analyses of 250 bootstrap samples. The results of clustering based on variable "Diagonal" are most stable for $K = 2$, and also stable for $K = 3$. By the way, two errors are counted only by this two-cluster solution. Now we are looking for the best partner of "Diagonal". Figure 4 shows the results of the investigation of stability of the bivariate Ward's clustering in a similar manner as presented in Fig. 3. Obviously, the result of clustering based on the two variables "Diagonal" and "Top" is most stable for $K = 2$ with a significant decrease when splitting one of these clusters ($K = 3$). One error is counted here only. This is exactly the same result of Ward's method when using all six variables of the dataset.

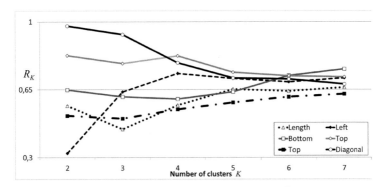

Fig. 3 The ARI R_K versus number of clusters in univariate hierarchical clustering of Swiss bank notes. The stability looks quite different for the six variables

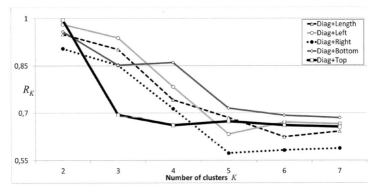

Fig. 4 The ARI R_K versus number of clusters in bivariate clustering. The stability of the clustering based on Diagonal + Top is extremely high for two clusters

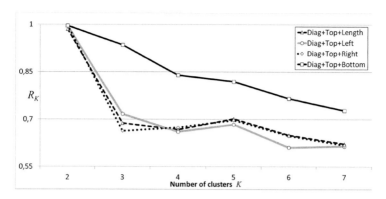

Fig. 5 The ARI R_K versus number of clusters in trivariate hierarchical clustering

Usually, the bottom-up procedure stops in the case of $\Delta R_K = 0.01$ because the ARI is very close to its maximum value 1. But, let us go on to trivariate clustering to see what happens. Figure 5 summarizes the investigation of stability of trivariate Ward's clustering. Now the four trivariate clustering vote for $K = 2$, but only three of them show a significant decrease when splitting one of these clusters ($K = 3$). The result of clustering based on the three variables "Diagonal", "Top" and "Bottom" is also stable for $K = 3$. That's interesting because, in fact, the class of forged bank notes is much more heterogeneous than the class of genuine bank notes (Mucha 1996). Maybe, the reason for this is that the forged banknotes stem from several different workshops.

4   Example: Variable Selection in Clustering of Synthetic Data

Figure 6 shows the bivariate density surface of the first two variables "$V1$" and "$V2$" of a randomly generated four-dimensional three class data. The other two variables "$R1$" and "$R2$" are masking variables without any class structure. Concretely, they are uniformly distributed in $(-5, 5)$. The three Gaussian sub-populations were generated with the following different parameters: cardinalities 80, 130, and 90; mean values $(-3, 3)$, $(0, 0)$, and $(3, 3)$, and standard deviations $(1, 1)$, $(0.7, 0.7)$, and $(1.2, 1.2)$.

Figure 7 summarizes the investigation of stability of the univariate Ward's clustering based on the ARI R_K coming from comparisons with cluster analyses of 250 bootstrap samples. Clearly, the clustering based on variable "$V1$" is most stable for $K = 3$ clusters with an additional most steep rise from $K = 2$ to $K = 3$. Here 32 errors were counted.

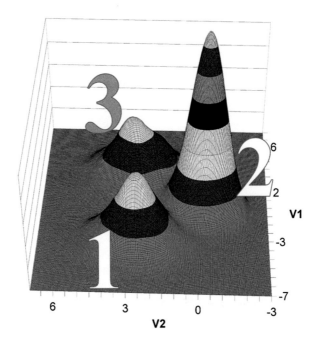

Fig. 6 Non-parametric density estimation of the first two variables of the synthetic three class dataset which carry the class structure

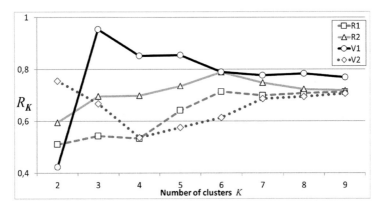

Fig. 7 Statistics of the ARI R_K versus number of clusters in univariate hierarchical clustering. The stability looks quite different for each of the four variables

However, the appropriate criterion for this data ought be the logarithmic sum-of-squares criterion (Mucha 2009)

$$V_K^*(\mathscr{P}) = \sum_{k=1}^{K} |\mathscr{C}_k| \log \operatorname{tr} \frac{\mathbf{W}_k}{|\mathscr{C}_k|} , \tag{4}$$

or, equivalently it holds

$$V_K^*(\mathscr{P}) = \sum_{k=1}^{K} |\mathscr{C}_k| \log \left(\sum_{i \in \mathscr{C}_k} \sum_{h \in \mathscr{C}_k, h > i} \frac{1}{|\mathscr{C}_k|^2} d_{ih} \right). \tag{5}$$

But, the most simple model (2) applied here did also a very good job in finding the true three classes.

Figure 8 shows the results of the investigation of stability of the bivariate Ward's clustering in a similar manner as Fig. 7. Obviously, the result of clustering based on the two variables "$V1$" and "$V2$" is most stable for $K = 3$ with a significant decrease in stability when merging two of these clusters ($K = 2$). Four errors were counted here only.

Figure 9 summarizes the investigation of stability of trivariate Ward's clustering. Now the stability of the two trivariate clustering are very low compared to bivariate clustering based on "$V1$" and "$V2$". Moreover, the stability of results of clustering based on all four variables obtain the lowest values. Here, obviously, the masking variables "$R1$" and "$R2$" make the discovery of the given class structure impossible.

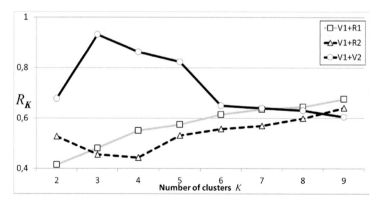

Fig. 8 The ARI R_K versus number of clusters in bivariate hierarchical clustering. The stability of the "$V1 + V2$" clustering is extremely high for $K = 3$ clusters

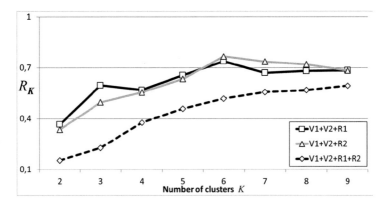

Fig. 9 The ARI R_K versus number of clusters in trivariate hierarchical clustering. In addition, the stability of clustering based on all variables is drawn

5   Conclusion

Our quite simple proposal of variable selection in clustering works without using special clustering criteria such as within-cluster or between-cluster variances. It is based on assessment of stability by non-parametric resampling, and it figures out criteria of stability such as the ARI R_K using confusion tables. As a stop criterion, a threshold for the increment of stability ΔR_K was used. Here further investigations are necessary. The proposed very general variable selection procedure can be modified in several ways such as by starting with $J * (J - 1)/2$ bivariate cluster analyses. Moreover, the statistical results of the investigations of stability of all J univariate (and/or $J - 1$ bivariate) cluster analyses can be useful for the development of other variable selection procedures (see, for example, Carmone et al. 1999).

References

Carmone, F. J., Kara, A., & Maxwell, S. (1999). HINoV: A new model to improve market segment definition by identifying noisy variables. *Journal of Marketing Research, 36*, 501–509.

Davies, D. L., & Bouldin, D. W. (1979). A cluster separation measure. *IEEE Transactions on Pattern Analysis and Machine Intelligence, 1*(2), 224–227.

Flury, B., & Riedwyl, H. (1988). *Multivariate statistics: A practical approach*. London: Chapman and Hall.

Fowlkes, E. B., Gnanadesikan, R., & Kettenring, J. R. (1988). Variable selection in clustering. *Journal of Classification, 5*, 205–228.

Gnanadesikan, R., Kettenring, J. R., & Tsao, S. L. (1995). Weighting and selection of variables for cluster analysis. *Journal of Classification, 12*, 113–136.

Hennig, C. (2007). Cluster-wise assessment of cluster stability. *Computational Statistics and Data Analysis, 52*, 258–271.

Hubert, L. J., & Arabie, P. (1985). Comparing partitions. *Journal of Classification, 2*, 193–218.

Meinshausen, N., & Bühlmann, P. (2010). Stability selection. *Journal of the Royal Statistical Society: Series B, 72*(4), 417–473.

Mucha, H.-J. (1996). ClusCorr: Cluster analysis and multivariate graphics under MS Excel. In H.-J. Mucha & H.-H. Bock (Eds.), *Classification and clustering: Models, software and applications*, Report 10 (pp. 97–106). Berlin: WIAS.

Mucha, H.-J. (2009). ClusCorr98 for Excel 2007: Clustering, multivariate visualization, and validation. In H.-J. Mucha & G. Ritter (Eds.), *Classification and clustering: Models, software and applications*, Report 26 (pp. 14–40). Berlin: WIAS.

Mucha, H.-J., & Bartel, H.-G. (2014). Soft bootstrapping in cluster analysis and its comparison with other resampling methods. In M. Spiliopoulou, L. Schmidt-Thieme, & R. Janning (Eds.), *Data analysis, machine learning and knowledge discovery* (pp. 97–104). Berlin: Springer.

Mucha, H.-J., Bartel, H.-G., Dolata, J., & Morales-Merino, C. (2015). An introduction to clustering with applications to archaeometry. In J. A. Barcelo & I. Bogdanovic (Eds.), *Mathematics and archaeology* (Chap. 9). Boca Raton: CRC Press.

Mucha, H.-J., & Ritter, G. (2009). *Classification and clustering: Models, software and applications*, Report 26 (pp. 114–125). Berlin: WIAS.

A Comparison Study for Spectral, Ensemble and Spectral-Mean Shift Clustering Approaches for Interval-Valued Symbolic Data

Marcin Pełka

Abstract Interval-valued data arise in practical situations such as recording monthly interval temperatures at meteorological stations, daily interval stock prices, etc. This paper presents a comparison study for clustering efficiency (according to adjusted Rand index) for spectral, ensemble, and spectral-mean shifted clustering methods for symbolic data. Evaluation studies with application of artificial data with known cluster structure (obtained from `mlbench` and `clusterSim` packages of R) show the usefulness and stable results of the ensemble clustering compared to spectral and spectral-mean shift method.

1 Introduction

Ensemble techniques based on aggregating information (results) from different (diverse) models have been applied with a success in context of supervised learning (discrimination and regression). The ensemble techniques are applied in order to improve the accuracy and stability of classification algorithms (Breiman 1996). In general ensemble clustering means combining (aggregating) results of N base clustering results (models) $P_1, \ldots, P_n$ into one final clustering (model) P^* with k^* clusters (Fred and Jain 2005). Many papers show the usefulness of ensemble learning in context of classical data (e.g., Ghaemi et al. 2009; Fred and Jain 2005; Stehl and Gosh 2002). However the idea of ensemble approach, that is combining (aggregating) the results of many base models, can be applied for cluster analysis of symbolic data.

The paper presents and compares the results obtained by applying two ensemble clustering approaches—co-association (co-occurrence) matrix (Fred and Jain 2005) and Leisch's adaptation of bagging (Leisch 1999) with the clustering results obtained by applying spectral clustering and spectral-mean shift clustering for symbolic data.

M. Pełka (✉)

Department of Econometrics and Computer Science, Faculty of Economics, Management and Tourism, Wrocław University of Economics, ul. Nowowiejska 3, 58-500 Jelenia Góra, Poland

e-mail: marcin.pelka@ue.wroc.pl

© Springer International Publishing Switzerland 2016   137

A.F.X. Wilhelm, H.A. Kestler (eds.), *Analysis of Large and Complex Data*, Studies in Classification, Data Analysis, and Knowledge Organization,

DOI 10.1007/978-3-319-25226-1_12

2 Symbolic Data

Each symbolic object can be described by following variables (Table 1 presents some examples of symbolic variables) (Bock and Diday 2000, p. 2–4; Billard and Diday 2006):

1. Quantitative (numerical) variables:

   - numerical single-valued variables,
   - numerical multi-valued variables,
   - interval variables,
   - histogram variables.

2. Qualitative (categorical) variables:

   - categorical single-valued variables,
   - categorical multi-valued variables,
   - categorical modal variables.

Regardless of their type symbolic variables also can be (Bock and Diday 2000) taxonomic variables with hierarchically structured categories, hierarchically dependent—rules which decide if a variable is applicable or not have been defined, logically dependent—logical or functional rules that affect variable's values have been defined.

Table 1 Examples of symbolic variables

Symbolic variable	Realizations	Variable type
Price of a car	(27,000, 42,000); (35,000, 50,000)	Interval-valued
(in PLN)	(20,000, 30,000); (25,000, 37,000)	(non-disjoint)
Engine's capacity	(1000, 1200); (1300, 1400)	Interval-valued
(in ccm)	(1500, 1800); (1900, 2200)	(disjoint)
Chosen color	{red, black, blue, green}	Categorical
	{orange, white, gray, magenta}	multi-valued
Preferred car	{Toyota (0.3); Volvo (0.7)}	Categorical
	{Audi (0.6), Skoda(0.35), VW (0.05)}	modal
	{BMW (1.0)}	
Distance	<10, 20> (0.65); <21, 30> (0.35)	Histogram
traveled	<10, 20> (0.25); <21, 30> (0.75)	

3 Ensemble, Spectral, and Spectral-Mean Shift Clustering for Symbolic Data

3.1 Spectral Clustering for Symbolic Data

Spectral approach, is not in fact a new clustering algorithm, but it is rather a new way of the data set preparation for some clustering algorithm (like k-means, pam, ward, etc.) (Ng et al. 2002; Von Luxburg 2006). Finite-sample properties of spectral clustering have been studied from a theoretical point of view by many scientists (Ng et al. 2002; Von Luxburg 2006; Shi and Malik 2000). Spectral clustering has the advantage of performing well in the presence of the non-Gaussian clusters. What is more, it also does not present the drawback of presence of the local minima. Furthermore, the convergence of the normalized spectral clustering is less difficult to handle than the unnormalized case. The results obtained by spectral clustering very often outperform the traditional approaches (see, for example, Von Luxburg 2006). The source of such success is that spectral clustering is based on the fact that it makes no assumptions on the form of the clusters—it can solve very general clustering problems (Von Luxburg 2006, p. 22).

The spectral clustering has some disadvantages. The choice of a good similarity graph is not trivial (usually the fully connected graph is applied). The spectral clustering can be quite unstable under different choices of the parameters for the neighborhood graphs. Many different kernels can be used, each can lead to different results (usually the Gaussian kernel is used at most cases)—Karatzoglou (2006) presents applications of different kernels in spectral clustering. Another important task is to choose a good σ parameter, that should minimize the inter-cluster distances for a given number of clusters. Karatzoglou (2006) has proposed a quite efficient way to estimate an appropriate σ parameter.

Spectral decomposition algorithm can be started in the following way:

1. Let **V** be a symbolic data table with n rows and m columns. Let u be the number of desired clusters.
2. Let $\mathbf{A} = [A_{ik}]$ be an affinity matrix of objects from **V**. A matrix can be calculated in many different ways. Most often its elements are defined as follows:

$$A_{ik} = \exp(-\sigma \cdot d_{ik}) \text{ for } i \neq k, \tag{1}$$

where: σ—scaling parameter that should minimize the sum of inter-cluster distances for a given number of clusters u; d_{ik}—distance measure between ith and kth object.

3. Calculation of the Laplacian $\mathbf{L} = \mathbf{D}^{\frac{1}{2}}\mathbf{A}\mathbf{D}^{\frac{1}{2}}$ (**D**—a diagonal weight matrix with sums of each row from **A** matrix on the diagonal).
4. Calculation of eigenvectors and eigenvalues of **L**. First u eigenvectors create a **E** matrix. Each eigenvector is treated as a column of the **E** (thus **E** has got dimensions $n \times u$).

5. Normalization of the **E** according to $y_{ij} = e_{ij}/\sqrt{\sum_{j=1}^{u} e_{ij}^2}$.
6. The **Y** matrix is clustered with one of well-known clustering algorithms (i.e., k-means, pam).

The main difference between spectral approach for classical and symbolic data is the distance measure applied in the Eq. (1). For details concerning distance measures for symbolic data, their properties, advantages, and disadvantages, e.g. Bock and Diday (2000), Billard and Diday (2006).

3.2 Spectral-Mean Shift Clustering for Symbolic Data

The mean-shift algorithm is a non-parametric clustering technique which does not require any prior knowledge of the number of clusters, and it does not constrain the shape of the clusters. This method was first proposed by Fukunaga and Hostetler (1975), later adapted by Cheng (1995) for image analysis purposes. Later on this method was extended by Comaniciu et al. (2001, 2003) to low-level vision problems, including segmentation and tracking.

The mean shift vector is defined (in general form) as (Cheng 1995; Comaniciu et al. 2003):

$$ms_h(y) = \frac{\sum_{i=1}^{n} -y_i k_y' \left(\left(\frac{y - y_i}{h} \right)^2 \right)}{\sum_{i=1}^{n} -k_y' \left(\left(\frac{y - y_i}{h} \right)^2 \right)} - y, \tag{2}$$

where: k' is a derivate of a kernel (Gaussian, Bessel, Epanechnikov, etc.) function.

This method iteratively calculates mean shift vector and translates coordinates of points until convergence is archived. The mean shift vector always points towards the direction of the maximum increase in the density. The spectral-mean shift algorithm can be described as follows:

1. Construct the symbolic data matrix **V**.
2. Calculate the affinity matrix **A**.
3. Calculate the elements of the Laplacian matrix **L**.
4. Calculate eigenvectors and eigenvalues of **L**. Calculate the elements of the **E** matrix.
5. Calculate the elements of the **Y** matrix.
6. Iteratively move points of the **Y** matrix towards the direction pointed by the mean shift until stationary points are reached.
7. Clustering: if the distance of an object from a prototype is lower than critical value (prior set) it is a member of the cluster; if not—this object is a new cluster prototype.

3.3 Ensemble Learning for Symbolic Data

Ensemble methods train multiple learners to solve the same problem. Ensemble learning is also known as committee-based learning, learning multiple classifier (model) systems (Zhi-Hua 2012, p. 15). The main idea of ensemble clustering is to aggregate (combine, join) results obtained from different models. There are several reasons of using ensembles, in case of supervised and unsupervised learning (Zhi-Hua 2012, pp. 16–17, 67–68; Polikar 2006, 2007):

1. There is a formal mathematical proof showing that in case of ensemble learning in supervised tasks, error reached by ensemble is lower than any of error of base models that form the ensemble.
2. Another important issue is the model selection. As in many cases many different methods can be applied (and each of them has some pros and cons). In such case ensemble learning allows to use different methods ("different points of view") and combine their results to obtain one final solution. However it is important to notice that there is no guarantee that the combination of multiple models will always perform better than the best single (individual) classifier (model). But even in such case ensemble reduces the overall risk of making a particularly poor selection.
3. Ensemble learning can be also useful when dealing the problem of too much or too little data. In case of big data sets ensemble learning allows to subsets of a data set. If we deal the problem of too little data then bootstrapping technique can be used to train different classifiers.
4. Sometimes we deal with data sets that are too complex, too difficult, to solve with a single classifier (model). In such case ensemble learning allows to "cut" a data set into smaller, easier to learn partitions. Such approach is sometimes called "divide and conquer" technique. Each model (classifier) learns only one of simpler partitions. The underlying complex decision boundary can then be approximated by an appropriate combination of different classifiers.
5. Also in many automated decision making problems, it is not unusual to receive data obtained from different sources that provide complementary information. Such approach is known as data (information) fusion. This kind of approach is usually applied in medicine.

In case of symbolic data there are three possible paths of ensemble learning:

1. Clustering algorithm for multiple relational matrices proposed by De Carvalho et al. (2012). This approach is based on different distance matrices. Those distance matrices can be obtained by applying different distance measures, or subsets of variables or subsets of objects. Distance matrices are used to calculate relevance weight vectors. Relevance weight vectors and distance matrices are then applied to cluster a set of objects into final clusters.
2. Applying well-known boosting algorithm for clustering ensembles. Boosting, like in supervised learning, in clustering means producing subsets from initial data set. Then each subset is clustered with some clustering method. Finally

results of clustering of each subset are combined. There are three proposals how to combine results of subsets clustering:

- Proposal made by Leisch (1999). In general the main idea of this approach is to cluster centers of clusters obtained from each of the subsets.
- Adaptation proposed by Dudoit and Fridlyand (2003)—the main idea is to permute cluster labels to get best agreement with clustering results for initial data set.
- Hornik's (2005) proposal—the main idea is to minimize the distance between the set of all possible consensus clusterings and the elements of the ensemble clustering.

3. Apply well-known consensus functions in clustering ensembles. There are five main types of consensus functions each with different assumptions (Ghaemi et al. 2009; Fred and Jain 2005; Pełka 2012):

- *Hypergraph partitioning* assumes that clusters can be represented as hyperedges on a graph. Their vertices correspond to the objects to be clusters. Each hyperedge describes a set of objects belonging to the same cluster. The problem of consensus clustering is reduced to finding the minimum-cut of a hypergraph.
- *Voting approach.* In this approach we permute cluster labels in such way that best agreement between the labels of two partitions is obtained. All the partitions from the cluster ensemble must be relabeled according to a fixed reference partition
- *Mutual information* assumes that the objective function of a clustering ensemble can be formulated as the mutual information between the empirical probability distribution of labels in the consensus partition and the labels in the ensemble. In this approach usually a generalized definition of mutual information is applied.
- *Co-association based functions.* The main assumption is that the objects belonging to the same cluster ("natural cluster") are more likely to be co-located in the same clusters in different data partitions. The elements of co-association (co-occurrence) matrix are defined as follows:

$$C(i,j) = \frac{n_{ij}}{N} \tag{3}$$

where: i, j—objects (pattern) numbers, n_{ij}—number of times patterns (i,j) are assigned to the same clusters among N partitions, N—total number of partitions

So the algorithm of building ensemble with application of co-association matrix is as follows (Fred and Jain 2005, p. 848). First we have to obtain different partitions (for example, by using different clustering methods, by using the same algorithm but with different parameters, etc.). Then upon of each of the partitions elements of co-occurrence matrix are calculated according to the Eq. (3). Finally the co-association matrix is used as the initial data set for some clustering algorithm to find final partition for a data set.

- *Finite mixture models.* The main assumption is that the output labels are modeled as random variables drawn from probability distribution described as a mixture of multinomial component densities. The objective of consensus clustering is formulated as a maximum likehood estimation.

4   Simulation Studies

In the empirical part of the paper spectral and spectral-mean shift clustering will be compared with clustering ensembles were Leisch's adaptation of bagging and co-association matrix are applied. In order to compare these different clustering approaches five different artificial data sets with known cluster structure were prepared with application of `cluster.Gen` function from `clusterSim` (Walesiak and Dudek 2014) package of R software:

1. *Model I.* Three elongated clusters in two dimensions. The observations are independently drawn from bivariate normal distribution with means (0, 0), (1.5, 7), (3, 14) and covariance matrix $\sum$ ($\sigma_{jj} = 1, \sigma_{jl} = -0.9$).
2. *Model II.* Five clusters in two dimensions. The observations are independently drawn from bivariate normal distribution with means (0, 0), (0, 10), (5, 5), (10, 0), (10, 10) and identity covariance matrix $\sum$ ($\sigma_{jj} = 1, \sigma_{jl} = 0$).
3. *Model III.* Five clusters in three dimensions. The observations are independently drawn from multivariate normal distribution with means (−4, 5, −4), (5, 14, 5), (14, 5, 14), (5, −4, 5) and identity covariance matrix $\sum$, where $\sigma_{jj} = 1$ ($1 \leq j \leq 3$) and $\sigma_{jl} = 0$ ($1 \leq j \neq l \leq 3$).
4. *Model IV.* Four clusters in three dimensions. The observations are independently drawn from multivariate normal distribution with means (−4, 5, −4), (5, 14, 5), (14, 5, 14). (5, −4, 5) and covariance matrices $\sum_1 = \begin{bmatrix} 1 & 0 & 0 \\ 0 & 1 & 0 \\ 0 & 0 & 1 \end{bmatrix}, \sum_2 = \begin{bmatrix} 1 & -0.9 & -0.9 \\ -0.9 & 1 & 0.9 \\ -0.9 & 0.9 & 1 \end{bmatrix}, \sum_3 = \begin{bmatrix} 1 & 0.9 & 0.9 \\ 0.9 & 1 & 0.9 \\ 0.9 & 0.9 & 1 \end{bmatrix}, \sum_4 = \begin{bmatrix} 3 & 2 & 2 \\ 2 & 3 & 2 \\ 2 & 2 & 3 \end{bmatrix}.$
5. *Model V.* Two elongated clusters in two dimensions. The observations in each of two clusters are independent bivariate normal random variables with means (0, 0), (1, 5), and covariance matrices $\sum_1 = \begin{bmatrix} 1 & -0.9 \\ -0.9 & 1 \end{bmatrix}, \sum_2 = \begin{bmatrix} 1 & 0.5 \\ 0.5 & 1 \end{bmatrix}.$

To obtain symbolic interval data the data were generated for each model twice into sets A and B and minimal (maximal) value of $\{x_{ij}^A, x_{ij}^B\}$ is treated as the beginning (the end) of an interval.

For each model and each clustering technique to be compared three different runs were made. First one for the data set without any noisy variables, second one for the data set with one noisy variable and third one for the data set with two noisy variables. The noisy variables are simulated independently from the uniform distribution. In clusterSim package of R it is required that the variations of noisy variables in the generated data are similar to non-noisy variables (Milligan and Cooper 1985; Qiu and Joe 2006, p. 322).

To compare different clustering techniques and models values of adjusted Rand index were calculated for each model and run, and at the end the average values of adjusted Rand index were compared (see Table 2). The bolded values in the Table 2 show the highest adjusted R and values for each model and method.

In case of ensemble clustering in each model 30 different results were merged. In case of Leisch's adaptation of bagging $\frac{2}{3}$ of a data set were drawn with replacement for each subset.

Besides artificial models, also three real data sets were used—Ichino's oil data, car data set, and Chinese meteorological stations data. These data sets are often applied to validate non-supervised and supervised methods, e.g. Bock and Diday (2000), Billard and Diday (2006). The same approach as in case of artificial data sets were used. For each of them all compared methods were used and average adjusted Rand index was calculated. As in the case of artificial data sets, Leisch's adaptation reached usually the best results.

Table 2 Results of simulation studies—values of adjusted Rand index

Model no.	Noisy variables	SMS	Spectral	Leisch	Co-association matrix
I	0	0.9560	0.9532	**0.9648**	0.9633
	1	0.3484	0.3482	0.3701	**0.3712**
	2	0.3425	0.3427	0.3598	**0.3609**
II	0	0.8528	0.8511	**0.8601**	0.8537
	1	0.2363	0.2370	**0.2531**	0.2467
	2	0.2356	0.2342	**0.2477**	0.2381
III	0	0.8710	0.8693	0.8801	**0.8812**
	1	0.2323	0.2311	0.4215	**0.4321**
	2	0.2328	0.2332	**0.4247**	0.4243
IV	0	0.8632	0.8649	**0.8788**	0.8691
	1	0.2360	0.2371	**0.2520**	0.2511
	2	0.2378	0.2377	0.2528	**0.2530**
V	0	0.8602	0.8792	**0.8908**	0.8868
	1	0.2345	0.2333	**0.2550**	0.2446
	2	0.2374	0.2381	**0.2559**	0.2451
ave.	0	0.8806	0.8792	**0.8908**	0.8868
	1	0.2575	0.2573	**0.3103**	0.3073
	2	0.2372	0.2572	**0.3082**	0.3043

Where: *ave.* average value of adjusted Rand index

5 Final Remarks

All presented clustering methods—spectral clustering, spectral clustering combined with mean-shift clustering (SMS), ensemble clustering that uses co-association matrix, and finally ensemble clustering based on Leisch's adaptation of bagging can be applied when dealing any kind of symbolic data. But a suitable distance measure for symbolic data is required.

The main advantage of SMS clustering is the kernel decomposition of a data matrix, via distance matrix, (*spectral approach*) combined with the movement towards the direction pointed by the mean shift (*mean shift approach*).

The main advantage of ensemble approach for symbolic data, compared to spectral and SMS approaches, is the utility of different clustering results ("different points of view") and then combining the results to find one final partition. There are many different ways of merging different clustering results into one final result that can be used in case of symbolic data. In most of them the only thing we need is just a suitable distance measure.

The simulation studies with five different data sets, and different number of noisy variables, show that Leisch's adaptation of bagging usually reaches the best results (in terms of adjusted Rand index) when compared with other clustering techniques. Co-association matrix is usually the second and the SMS is the third one.

An open issue is a comparison different ensemble clustering approaches for symbolic data when dealing non-spherical ("non typical") cluster shapes, noisy variables, and outliers.

References

Breiman, L. (1996). Bagging predictors. *Machine Learning, 24*(2), 123–140.

Bock, H.-H., & Diday, E. (Eds.) (2000). *Analysis of symbolic data. Explanatory methods for extracting statistical information from complex data*. Berlin/Heidelberg: Springer.

Billard, L., & Diday, E. (2006). *Symbolic data analysis: Conceptual statistics and data mining*. Chichester: Wiley.

Cheng, Y. (1995). Mean shift, mode seeking, and clustering. *IEEE Transactions on Pattern Analysis and Machine Intelligence, 17*(8), 790–799.

Comaniciu, D., Ramesh, V., & Meer, P. (2001). The variable bandwidth mean shift and data-driven scale selection. In *International Conference on Computer Vision* (Vol. I, pp. 438–445).

Comaniciu, D., Ramesh, V., & Meer, P. (2003). Kernel-based object tracking. *IEEE Transactions on Pattern Analysis and Machine Intelligence, 25*(5), 564–577.

de Carvalho, F. A. T., Lechevallier, Y., & de Melo, F. M. (2012). Partitioning hard clustering algorithms based on multiple dissimilarity matrices. *Pattern Recognition, 45*(1), 447–464.

Dudoit, S., & Fridlyand, J. (2003). Bagging to improve the accuracy of a clustering procedure. *Bioinformatics, 19*(9), 1090–1099.

Fred, A. L. N., & Jain, A. K. (2005). Combining multiple clustering using evidence accumulation. *IEEE Transactions on Pattern Analysis and Machine Intelligence, 27*, 835–850.

Fukunaga, K., & Hostetler, L. (1975) The estimation of the gradient of a destiny function, with applications in pattern recognition. *IEEE Transactions on Information Theory, 21*(1), 32–40.

Ghaemi, R., Sulaiman, N., Ibrahim, H., & Mustapha, N. (2009). A survey: Clustering ensemble techniques. In *Proceedings of World Academy of Science, Engineering and Technology* (Vol. 38, pp. 636–645).

Hornik, K. (2005). A CLUE for CLUster ensembles. *Journal of Statistical Software, 14*, 65–72.

Karatzoglou, A. (2006). *Kernel Methods. Software, Algorithms and Applications*. Doctoral thesis, Vienna University of Technology.

Leisch, F. (1999). Bagged clustering. *Adaptive Information Systems and Modeling in Economics and Management Science*, Working Papers, SFB, 51.

Milligan, G. W, & Cooper, M. C. (1985). An examination of procedures for determining the number of clusters in a data set. *Psychometrika, 2*, 159–179.

Ng, A., Jordan, M., & Weiss, Y. (2002). On spectral clustering: Analysis and an algorithm. In T. Dietterich, S. Becker, & Z. Ghahramani (Eds.), *Advances in Neural Information Processing Systems 14* (pp. 849–856). Cambridge: MIT Press.

Pełka, M. (2012). Ensemble approach for clustering of interval-valued symbolic data. *Statistics in Transition, 13*(2), 335–342.

Polikar, R. (2006). Ensemble based systems in decision making. *IEEE Circuits and Systems Magazine, 6*(3), 21–45.

Polikar, R. (2007). Bootstrap inspired techniques in computational intelligence: Ensemble of classifiers, incremental learning, data fusion and missing features. *IEEE Signal Processing Magazine, 24*(4), 59–72.

Qiu, W., & Joe, H. (2006). Generation of random clusters with specified degree of separation. *Journal of Classification, 23*, 315–334.

Shi, J., & Malik, J. (2000). Normalized cuts and image segmentation. *IEEE Transactions on Pattern Analysis and Machine Intelligence, 22*(8), 888–905.

Stehl, A., & Gosh, J. (2002). Cluster ensembles – A knowledge reuse framework for combining multiple partitions. *Journal of Machine Learning Research, 3*, 583–618.

von Luxburg, U. (2006). *A tutorial on spectral clustering*. Max Planck Institute for Biological Cybernetics, Technical Report TR-149.

Walesiak, M., & Dudek, A. (2014). *The clusterSim package*, http://www.R-project.org

Zhi-Hua, Z. (2012). *Ensemble methods. Foundations and algorithms*. Boca Raton: CRC Press.

Supervised Pre-processings Are Useful for Supervised Clustering

Oumaima Alaoui Ismaili, Vincent Lemaire, and Antoine Cornuéjols

Abstract Over the last years, researchers have focused their attention on a new approach, *supervised clustering*, that combines the main characteristics of both traditional clustering and supervised classification tasks. Motivated by the importance of pre-processing approaches in the traditional clustering context, this paper explores to what extent supervised pre-processing steps could help traditional clustering to obtain better performance on supervised clustering tasks. This paper reports experiments which show that indeed standard clustering algorithms are competitive compared to existing supervised clustering algorithms when supervised pre-processing steps are carried out.

1 Introduction

Over the last decade, the world has seen a real explosion of data due mainly to the web, social networks, etc. To exploit these high-dimensional sets of data, clustering and classification algorithms are efficient.

Clustering is an unsupervised learning approach that allows one to discover global structures in the data (i.e., clusters). Given a dataset, it identifies different data subsets which are hopefully meaningful (see Fig. 1a). The discovered clusters are deemed interesting if they are heterogeneous (i.e., their inter-similarity is low) while instances within each cluster share similar features (high intra-similarity). This clustering problem has motivated a huge body of work and has resulted in

O.A. Ismaili (✉)
Orange Labs, AV. Pierre Marzin, 22307 Lannion Cedex, France

AgroParisTech 16, rue Claude Bernard, 75005 Paris, France
e-mail: oumaima.alaouiismaili@orange.com

V. Lemaire
Orange Labs, AV. Pierre Marzin, 22307 Lannion Cedex, France
e-mail: vincent.lemaire@orange.com

A. Cornuéjols
AgroParisTech 16, rue Claude Bernard, 75005 Paris, France
e-mail: antoine.cornuejols@agroparistech.fr

© Springer International Publishing Switzerland 2016
A.F.X. Wilhelm, H.A. Kestler (eds.), *Analysis of Large and Complex Data*, Studies in Classification, Data Analysis, and Knowledge Organization,
DOI 10.1007/978-3-319-25226-1_13

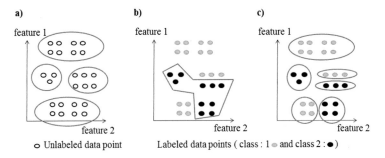

Fig. 1 Classification processes. (**a**) Unsupervised clustering. (**b**) Supervised classification. (**c**) Supervised clustering

a large number of algorithms (see, e.g., Jain et al. 1999). Clustering has thus been used in numerous real-life application domains [e.g., marketing (Berry and Linoff 1997), CRM (Berson et al. 2000)].

In contrast, classification is a supervised learning approach that consists to learn the link between a set of input variables and an output variable (*target class*). The main goal of this approach is to construct a learning model which is able to predict class membership for new instances (see Fig. 1b).

Recently, researchers have focused their attention on the combination of characteristics of both clustering and classification tasks with the goal to discover the internal structure of the target classes. This research domain is called *Supervised clustering* (for instance, see Al-Harbi and Rayward-Smith 2006 and Eick et al. 2004). The main idea is to construct or modify clustering algorithms in order to find clusters where instances are very likely to belong to the same class. Formally, *Supervised clustering* seeks clusters where instances in each cluster share characteristics (homogeneity) and class label. The generated clusters are labeled with the majority class of their instances. Figure 1 illustrates the difference between clustering, classification , and supervised clustering.

Generally, clustering tasks require an unsupervised pre-processing step [for example, see Milligan and Cooper (1988) or Celebi et al. (2013) for the k-means algorithm] in order to yield interesting clusters. For instance, this step might be aimed at preventing features with large ranges from dominating the distance calculations. Now, given the importance of pre-processing for the traditional clustering algorithms, it is natural to ask: could *supervised pre-processing* help standard clustering algorithms to reach good performance in a supervised clustering context? In other words, does a combination of a supervised pre-processing step and a standard clustering algorithm produce a good supervised clustering algorithm, meaning exhibiting high prediction accuracy (supervised criterion) while at the same time uncovering interesting clusters in the dataset.

The remainder of this paper is organized as follows. Section 2 briefly describes related work about supervised clustering. Section 3 presents classical unsupervised pre-processing methods and two supervised pre-processing approaches. Section 4 first compares the performance, in terms of prediction accuracy, when using a clustering technique combined with an *unsupervised pre-processing* step and a clustering technique combined with a *supervised pre-processing* step. A comparison between traditional clustering using a supervised pre-processing step with the techniques of supervised clustering algorithms is then carried out. Finally, a conclusion with future work is presented in the last section.

2 Related Work

In the last decade, many researchers focused their attention to build or modify standard clustering algorithms to identify class-uniform clusters where instances within each cluster are homogeneous. Several algorithms are developed to achieve that objective (e.g., Aguilar-Ruiz et al. 2001; Sinkkonen et al. 2002; Qu and Xu 2004; Finley and Joachims 2005; Bungkomkhun 2012).

In this section, we present two methods proposed by Al-Harbi and Rayward-Smith (2006) and Eick et al. (2004) which modify the K-means algorithm. The experimental results of these algorithms will be compared in Sect. 4.2.2 to the results obtained by using a standard K-means algorithm preceded by a supervised pre-processing step.

Al-Harbi and Rayward-Smith (2006) developed a K-means algorithm in such a way to use it as a classifier algorithm. First of all, they replaced the Euclidean metric used in a standard K-means by a weighted Euclidean metric. This modification is carried out in order to be able to estimate the distance between any two instances that have the same class label. The vector of weights is chosen in such a way to maximize the confidence of the partitions generated by the k-means algorithm. This confidence is determined by calculating the percentage of correctly classified objects with respect to the total number of objects in the dataset. To solve this problem of optimization, they used simulated annealing (a generic probabilistic metaheuristic for the global optimization problem). This iterative process is repeated until an optimal confidence is obtained. In this algorithm, the number of clusters is an input.

Eick et al. (2004) introduced four representative-based algorithms for supervised clustering: *SRIDHCR*, SPAM, TDS, and *SCEC*. In their experimentation, they used the first one (i.e., *SRIDHCR*). The greedy algorithm *SRIDHCR* (or Single Representative Insertion/Deletion Steepest Decent Hill Climbing with Randomized Start) is mainly based on three phases. The first one is the initialization of a set of representatives that is randomly selected from the dataset. The second is the primary cluster creation phase, where instances are assigned to the cluster of their closest representative. The third one is the iteration phase where the algorithm is run r times: In each time "r," the algorithm tries to improve the quality of clustering, for instance, by adding a non-representative instance or by deleting a representative instance. To measure this quality, they use a supervised criterion. It takes into account two points: (1) The impurity of the clustering which defined as a percentage of misclassified observations in the different clusters and (2) a penalty condition which used in a manner to keep a lowest number of clusters. In this greedy algorithm, the number of clusters is an output.

3 Pre-processing

The following notation is used below:

Let $D = \{(X_i, Y_i)\}_1^N$ denote a training dataset of size N, where $X_i = \{X_{i1}, \dots, X_{id}\}$ is a vector of d features and $Y_{i \in \{1, \dots, N\}} \in \{C_1, \dots, C_J\}$ is the target class of size J. Let K denote the number of clusters.

3.1 Unsupervised Pre-processing

A pre-processing step is a common requirement for clustering tasks. Several unsupervised pre-processing approaches have been developed depending on the nature of features: continuous or categorical. In this paper, we have used the most common unsupervised pre-processing approach, that is normalization (see, e.g., Milligan and Cooper 1988).

For continuous features, to the best of our knowledge, data normalization is the most frequently used. It acts to weight the contribution of different features with the aim of making the distance between instances unbiased. Formally, normalization scales each continuous feature into a specific range such that one feature cannot dominate the others. The common data normalization approaches are: Min–Max, *statistical*, and *rank* normalization.

– *Min–Max Normalization* **(NORM):** If the minimum and maximum values are given for each continuous feature, it can be then transformed to fit in the range $[0, 1]$ using the following formula: $X'_{iu} = \frac{X_{iu} - \min_{i=1,\dots,N} X_{iu}}{\max_{i=1,\dots,N} X_{iu} - \min_{i=1,\dots,N} X_{iu}}$. Where X_{iu} is the original value of feature u. If minimum and maximum values are equal, then X'_{iu} is set to zero.

– *Statistical Normalization* **(SN):** This approach transforms data derived from any normal distribution into a standard normal distribution $N(0, 1)$. The formula that allows this transformation is: $X'_{iu} = \frac{X_{iu} - \mu}{\sigma}$ where μ is the mean of the feature u, σ is its standard deviation.

– *Rank Normalization* **(RN):** The purpose of rank normalization is to rank continuous feature values and then scale the feature into $[0, 1]$. The different steps of this approach are: (1) Rank feature values u from lowest to highest values and and then divide the resulting vector into H intervals, where H is the number of intervals. (2) Assign for each interval a label $r \in \{1, \dots, H\}$ in increasing order, (3) If X_{iu} belong to the interval r, then $X'_{iu} = \frac{r}{H}$.

For categorical features, among the existing approaches of unsupervised pre-processing, we use in this study the *Basical Grouping Approach (BGB)*. It aims at transforming feature values into a vector of Boolean values. The different steps of this approach are: (1) group feature values into g groups with "at best" equal frequencies, where g is a parameter given by the user, (2) assign for each group a label $r \in \{1, \ldots, g\}$, (3) use a full disjunctive coding.

3.2 Supervised Pre-processing

In this paper, we suggest that one way to help a standard algorithm to reach a good performance in terms of prediction accuracy is to incorporate information given by the target class in a pre-processing step. To prove this, we proposed two supervised pre-processing approaches called *Conditional Info* and *Binarization*. These approaches are based on two steps: (1) supervised representation and (2) recoding. The first one is a common step for the two approaches. It aims at giving information about variables distribution conditionally to a target class. There are several methods that could achieve the above objective. In this study, we have used the *MODL* (a Bayes optimal pre-processing method for continuous and categorical features) approach. It seeks to estimate the univariate conditional density $(P(X|C))$. To obtain this estimation a supervised discretization method is used for continuous features (Boullé 2006) and a supervised grouping method is used for categorical ones (Boullé 2005).

To exploit the information given by the first step, a recoding phase is then used as second (common) step. In this paper, we present two ways of recoding (i.e., C.I and BIN). The following methods are compared in Sect. 4.

- *Conditional Info (C.I)*: Each feature from the instance X_i is recoded in a qualitative attribute containing I_J recoding values. The resulting vector for this instance is $X_i = X_{i1_1}, \ldots, X_{i1_J}, \ldots, X_{id_1}, \ldots X_{id_J}$. Where $X_{id_1}, \ldots X_{id_J}$ represent the recoding values for the feature d with respect to the number of a class label $(X_{id_J} = \log(P(X_{id}|C_J))$. As a result, the initial vector containing d features (continuous and categorical) becomes a vector containing $d \times J$ real components: $\log(P(X_{im}|C_j)), j \in \{1, \ldots, J\}, m \in \{1, \ldots, d\}$.

  The most remarkable point in this pre-processing process is that if two instances are close in term of distance, they are close also in term of their class membership. A detailed description of this process exists in Lemaire et al. (2012). Besides, the recoding step provides, for each feature, an amount of information related to the target class. That is by calculating $\log(P(X_{im}|C_j))$. This recoding allows one to obtain a new feature space of apriori-fixed size which corresponds to the total number of class labels in the dataset. The similarity between instances is interpreted as a Bayesian distance: $\mathrm{Dist}(X_i, X_j) = \sum_{m=1}^{d} \sum_{l=1}^{J} \left[\log(P(X_{im_l}|C_l)) - \log(P(X_{jm_l}|C_l)) \right]^2$.

  However, it does not allow keeping the notion of instances: two different instances belonging to different intervals (or groups of modalities) can have equal values of $\log(P(X_{im}|C_j))$.

- **_Binarization (BIN)_**: In this process, each feature is described on t Boolean features. Where t is a number of intervals or groups of modalities generated by MODL or an other supervised approach. The synthetic feature takes 1 as a value if the real value of the original feature belongs to the corresponding interval or group of modalities and is zero otherwise.

    The recoding step of this approach is based on the full disjunctive coding. It transforms each feature into a vector of Boolean features. The size of the vector depends on the number of intervals or groups of modalities associated with each feature. Hence, the size of the new feature space mainly depends on the number of intervals or groups of modalities for all features. Besides, the similarity between instances is determined such that similar instances belong to the same interval or group of modalities.

4 Experimentation

In this section, we present and compare first the average performance of both supervised and unsupervised pre-processing approaches using the k-means algorithm. Then, we compare and discuss the average performance of both supervised pre-processing and other supervised clustering algorithms. These experiments are intended to assess the ability of supervised pre-processing to provide better results than unsupervised pre-processing and also to evaluate the competitiveness of a traditional clustering algorithm (k-means) preceded by a supervised pre-processing step compared to some supervised algorithms in a supervised clustering context.

4.1 Protocol

To test the validity of our assumption, we choose to use the standard K-means algorithm (Macqueen 1967) which is traditionally viewed as the most popular algorithm in unsupervised clustering. To reduce at best the problem that the K-means algorithm does not guarantee to reach a global minimum: (1) the k-means++ algorithm (Arthur and Vassilvitskii 2007) is used to initialize centers, (2) the algorithm is realized 100 times.

   At this stage, it is important to define what the best partition is. To be consistent with the definition of *supervised clustering*, we search a criterion that allows us to choose the closest partition to the one given by the target class. In fact, the main aim is to get a compromise between intra-similarity and prediction. The intra-similarity criterion is guaranteed by the K-means algorithm (trade-off between inertia inter- and intra-cluster) and the class membership of instances inside each cluster is verified by the chosen criterion; knowing that a supervised/unsupervised pre-processing step is used. For this, we use the Adjusted Rand Index (ARI) (Hubert and Arabie 1985) criterion to select the best partition. It is computed by comparing

Table 1 The used pre-processing approaches

Unsupervised pre-processing			Supervised pre-processing		
Name	Num features	Cat features	Name	Num features	Cat features
RN-BGB	RN	BGB	BIN-BIN	BIN	BIN
CR-BGB	CR	BGB	C.I-C.I	C.I	C.I
NORM-BGB	NORM	BGB			

Table 2 Datasets from UCI used in experiment (*Var* Variable, *Cat* Categorical, and *Num* Numerical)

Dataset	N	# Var	# Cat	# Num	Dataset	N	# Var	# Cat	# Num
Auto-import	205	26	11	15	Heart-stat-log	270	13	3	10
Breast cancer	699	9	0	9	Iris	150	4	0	4
Contraceptive	1473	9	7	2	Pima	768	8	0	8
Glass	214	10	0	10	Vehicle	846	18	0	18

the partition of the target class labels with the partition of the k-means algorithm. For pre-processing approaches, we use those presented above in Sect. 3. Table 1 presents a list of these approaches.

To evaluate and compare the behavior of different pre-processing approaches in term of their capacity to help traditional clustering in a supervised context, some tests are performed on different databases of the UCI repository (Lichman 2013). Table 2 presents the databases used in this study.

In order to compare the obtained results with some supervised clustering algorithms, we do: (1) 10×5 fold cross classification (like in Al-Harbi and Rayward-Smith (2006) experiment) for Auto-import, Breast cancer, Contraceptive, and Pima datasets. These datasets are also modified in the same way as in Al-Harbi and Rayward-Smith (2006), (2) 5×10 fold cross classification (like in Eick et al. (2004) experiment) for Glass, Heart-stat-log, Vehicle, and Iris datasets.

4.2 Results

4.2.1 Part 1: Comparing Supervised and Unsupervised Pre-processing

Table 3 presents the average performance of the K-means algorithm in term of predictions [Accuracy (ACC) criterion], using each pre-processing approach (Sect. 3.2) for six datasets. In this case, the number of clusters is selected following the next procedure. First, the value of K is varied from 1 to 64. Then, for each value of K, a x-fold (see Sect. 4.1) cross validation is performed and the mean value of the ARI is calculated. Finally, the optimal value of K corresponds to the closest partition to the one given by the target class (higher value of ARI versus the value of K in train dataset). Based on this value of K, the ACC is calculated from the corresponding

Table 3 Average performance of k-means algorithm in term of predictions using several pre-processing approaches

			ARI	ACC				ARI	ACC
		K	train	test			K	train	test
H	**RN-BGB**	**2**	**0.422**	**0.815 ± 0.071**	I	RN-BGB	3	0.675	0.851 ± 0.087
	SN-BGB	2	0.365	0.796 ± 0.074		SN-BGB	3	0.641	0.833 ± 0.099
	NORM-BGB	3	0.241	0.754 ± 0.077		NORM-BGB	3	0.726	0.879 ± 0.080
	BIN-BIN	2	0.452	0.813 ± 0.069		**BIN-BIN**	**3**	**0.872**	**0.929 ± 0.069**
	C.I-C.I	2	0.451	0.807 ± 0.079		C.I-C.I	3	0.836	0.899 ± 0.092
C	RN-BGB	2	0.069	0.627 ± 0.025	V	RN-BGB	7	0.196	0.546 ± 0.036
	SN-BGB	2	0.052	0.604 ± 0.025		SN-BGB	8	0.157	0.507 ± 0.049
	NORM-BGB	3	0.067	0.616 ± 0.030		NORM-BGB	8	0.159	0.510 ± 0.044
	BIN-BIN	**3**	**0.093**	**0.630 ± 0.027**		BIN-BIN	5	0.256	0.558 ± 0.039
	C.I-C.I	3	0.075	0.621 ± 0.026		**C.I-C.I**	**5**	**0.283**	**0.589 ± 0.033**
P	RN-BGB	2	0.132	0.671 ± 0.038	B	RN-BGB	2	0.898	0.973 ± 0.012
	SN-BGB	2	0.177	0.705 ± 0.034		SN-BGB	2	0.850	0.959 ± 0.016
	NORM-BGB	5	0.135	0.673 ± 0.041		NORM-BGB	2	0.854	0.962 ± 0.015
	BIN-BIN	3	0.148	0.694 ± 0.039		**BIN-BIN**	**2**	**0.904**	**0.974 ± 0.011**
	C.I-C.I	**2**	**0.244**	**0.736 ± 0.034**		C.I-C.I	2	0.870	0.961 ± 0.036

H Heart, C Contraceptive, P Pima, I Iris, V Vehicle, B Breast
Bold values show highest accuracy among methods for the particular data set

Fig. 2 Auto-import: average performance of k-means (k is an output) using supervised pre-processing (the two first boxplots) and unsupervised pre-processing (the three last boxplots)

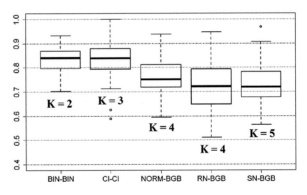

partition in a test dataset. The results in this table show that: (1) supervised pre-processing approaches have most of the time a better performance than unsupervised pre-processing approaches, (2) Binarization (BIN) and Conditional Info (C.I) are close with a small preference for BIN.

In the case where K is given (K is equal to the cardinality of the target class), we obtain also the same result. For example, Figs. 2 and 3 present respectively the case where K is an output and where K is an input for Auto-import and the Glass dataset. This result shows clearly the influence of supervised pre-processing steps (the two first boxplots) on the K-means performance [using the accuracy (ACC) criterion].

Fig. 3 Glass: average performance of the k-means (k is an input) using supervised pre-processing (the two first boxplots) and unsupervised pre-processing (the three last boxplots)

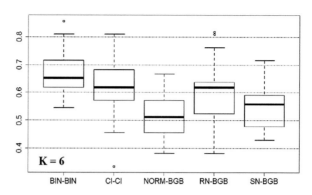

Table 4 Comparing with Eick and Al-Harbi algorithms

Comparing with Eick algorithm: (K is an output)

	Glass dataset		Heart dataset		Iris dataset	
	K	ACC test	K	ACC test	K	ACC test
Eick algorithm	34	0.636	2	0.745	3	0.973
K-means with BIN	6	0.664 ± 0.070	2	0.813 ± 0.069	3	0.929 ± 0.068
K-means with C.I	5	0.627 ± 0.080	2	0.808 ± 0.079	3	0.898 ± 0.091

Comparing with Al-Harbi algorithm: (K is an input)

	Auto-import dataset		Breast dataset		Pima dataset	
	K	ACC test	K	ACC test	K	ACC test
Al-Harbi algorithm	2	0.925	2	0.976	2	0.746
K-means with BIN	2	0.830 ± 0.051	2	0.974 ± 0.012	2	0.672 ± 0.041
K-means with C.I	2	0.809 ± 0.102	2	0.961 ± 0.035	2	0.735 ± 0.033

4.2.2 Part 2: Comparing Supervised Pre-processing to Other Supervised Clustering Algorithms

We compare the obtained results using the standard k-means algorithm preceded by a supervised pre-processing step (BIN or C.I) to a supervised k-means algorithm proposed by Eick or Al-Harbi. The results for the later algorithms are available in Eick et al. (2004) and Al-Harbi and Rayward-Smith (2006), respectively. Table 4 presents a summary of the average performance of the used methods in term of predictions in the case where K is estimated (Eick) and where K is given (Al-Harbi). The results obtained in the experiments using a standard k-means preceded by a supervised pre-processing are competitive with the mean results of Eick or Al-Harbi (who performed a single x-fold cross validation). We also observe that a standard k-means with a supervised pre-processing step tends to conserve a lower number of clusters (in Glass dataset, $k = 34, 7$, and 6 for respectively Eick, Binarization and Conditional Info approaches).

5 Conclusion

This paper has presented the influence of a supervised pre-processing step on the performance of a traditional clustering (especially K-means) in term of predictions. The experimental results showed the competitiveness of a traditional clustering using a supervised pre-processing step comparing to unsupervised pre-processing approaches and other methods of supervised clustering from the literature (especially Eick and Al-Harbi algorithms). Future works will be done (1) to compare supervised pre-processing approaches to others supervised clustering algorithms from the state of the art, (2) to combine supervised pre-processing presented in this paper with supervised K-means and (3) to define a better supervised pre-processing approach to combine the advantages of BIN and C.I without their drawbacks.

References

Aguilar-Ruiz, J. S., Ruiz, R., Santos, J. C. R., & Girldez, R. (2001). SNN: A supervised clustering algorithm. In L. Monostori, J. Vncza, & M. Ali (Eds.), *IEA/AIE*. Lecture Notes in Computer Science (Vol. 2070, pp. 207–216). Heidelberg: Springer.

al-Harbi, S. H., & Rayward-Smith, V. J. (2006). Adapting k-means for supervised clustering. *Journal of Applied Intelligence, 24*(3), 219–226.

Arthur, D., & Vassilvitskii, S. (2007). K-means++: The advantages of careful seeding. In *Proceedings of the Eighteenth Annual ACM-SIAM Symposium on Discrete Algorithms. SODA '07* (pp. 1027–1035).

Berry, M., & Linoff, G. (1997). *Data mining techniques for marketing, sales, and customer support*. New York: Wiley.

Berson, A., Smith, S., & Thearling, K. (2000). *Building data mining applications for CRM*. New York: McGraw-Hill.

Boullé, M. (2005). A Bayes optimal approach for partitioning the values of categorical attributes. *Journal of Machine Learning Research, 6*, 1431–1452.

Boullé, M. (2006). MODL: A Bayes optimal discretization method for continuous attributes. *Journal of Machine Learning, 65*(1), 131–165.

Bungkomkhun, P. (2012). Grid-based supervised clustering algorithm using greedy and gradient descent methods to build clusters. In *National Institute of Development Administration*. http://libdcms.nida.ac.th/thesis6/2012/b175320.pdf.

Celebi, E. M., Kingravi, H. A., & Vela, P. A. (2013). A comparative study of efficient initialization methods for the k-means clustering algorithm. *Journal of Expert Systems with Applications, 40*(1), 200–210.

Eick, C. F., Zeidat, N., & Zhao, Z. (2004). Supervised clustering algorithms and benefits. In *16th IEEE International Conference on Tools with Artificial Intelligence, 2004. ICTAI 2004, Boca Raton* (pp. 774–776).

Finley, T., & Joachims, T. (2005). Supervised clustering with support vector machines. In *Proceedings of the 22nd International Conference on Machine Learning. ICML '05* (pp. 217–224). New York, NY: ACM.

Hubert, L., & Arabie, P. (1985). Comparing partitions. *Journal of Classification, 2*(1), 193–218.

Jain, A. K., Murty, M. N., & Flynn, P. J. (1999). Data clustering: A review. *ACM Computing Surveys, 31*(3), 264–323.

Lemaire, V., Clérot, F., & Creff, N. (2012). K-means clustering on a classifier-induced representation space: Application to customer contact personalization. *Real-world data mining applications*. Annals of Information Systems (pp. 139–153). Cham: Springer.

Lichman, M. (2013). UCI Machine Learning Repository [http://archive.ics.uci.edu/ml/]. Irvine, CA: University of California, School of Information and Computer Science.

Macqueen, J. B. (1967). Some methods for classification and analysis of multivariate observations. In L. M. L. Cam & J. Neyman (Eds.), *Proceedings of the Fifth Berkeley Symposium on Mathematical Statistics and Probability* (pp. 281–297). Berkeley, CA: University of California Press.

Milligan, G., & Cooper, M. (1988). A study of standardization of variables in cluster analysis. *Journal of Classification, 5*(2), 181–204.

Qu, Y., & Xu, S. (2004). Supervised cluster analysis for microarray data based on multivariate gaussian mixture. *Journal of Bioinformatics, 20*(12), 1905–1913.

Sinkkonen, J., Kaski, S., & Nikkil, J. (2002). Discriminative clustering: Optimal contingency tables by learning metrics. *Machine learning: ECML 2002* (Vol. 2430, pp. 418–430). Heidelberg: Springer.

Similarity Measures on Concept Lattices

Florent Domenach and George Portides

Abstract This paper falls within the framework of Formal Concept Analysis which provides classes (the extents) of objects sharing similar characters (the intents), a description by attributes being associated to each class. In a recent paper by the first author, a new similarity measure between two concepts in a concept lattice was introduced, allowing for a normalization depending on the size of the lattice.

In this paper, we compare this similarity measure with existing measures, either based on cardinality of sets or originating from ontology design and based on the graph structure of the lattice. A statistical comparison with existing methods is carried out, and the output of the measure is tested for consistency.

1 Introduction

Measures of similarities have been widely used, particularly in biomedical domain (Nguyen and Al-Mubaid 2006) or in semantic web for natural language processing with the use of WordNet (Seco et al. 2004). However most of these applications are relying on an ontology tree-like structure to quantify the degree to which two concepts are similar. The purpose of this paper is to extend such similarity measures to a more general framework provided by lattices and Formal Concept Analysis, and to statistically evaluate and compare the new measure introduced in (Domenach 2015).

This paper is organized as follows: after recalling the main definitions of Formal Concept Analysis framework in Sect. 2, we describe briefly in Sect. 3 the new similarity measure as well as existing similarities. We also explain the theoretical rational for the need of this new measure. In Sect. 4, we describe an experimental simulation that was carried out in order to validate the efficacy of the new measure, followed by a discussion and perspectives in Sect. 5.

F. Domenach (✉) • G. Portides
Computer Science Department, University of Nicosia, 46 Makedonitissas Av., P.O. Box 24005, 1700 Nicosia, Cyprus
e-mail: domenach.f@unic.ac.cy; portides.g@unic.ac.cy

© Springer International Publishing Switzerland 2016
A.F.X. Wilhelm, H.A. Kestler (eds.), *Analysis of Large and Complex Data*, Studies in Classification, Data Analysis, and Knowledge Organization,
DOI 10.1007/978-3-319-25226-1_14

2 Background Definitions

2.1 Formal Concept Analysis

We recall here the standard Formal Concept Analysis (FCA) notations and we refer
readers to Ganter and Wille (1999) for details and proofs. A *formal context* (G, M, I)
is defined as a set G of objects, a set M of attributes, and a binary relation $I \subseteq G \times M$.
$(g, m) \in I$ means that "the object g is related with the attribute m through the relation
I". Two derivation operators can be defined on sets of objects and sets of attributes
as follows, $\forall A \subseteq G, B \subseteq M$:

$$A' = \{m \in M : \forall g \in A, (g, m) \in I\}$$

$$B' = \{g \in G : \forall m \in B, (g, m) \in I\}$$

The two operators $(\cdot)'$ define a Galois connection between the power set of
objects $\mathscr{P}(G)$ and the power set of attributes $\mathscr{P}(M)$. A pair $(A, B), A \subseteq G, B \subseteq M$,
is a *formal concept* iff $A' = B$ and $B' = A$. A is called the *extent* and B the *intent* of
the concept.

The set of all formal concepts, ordered by inclusion of extents (or dually by
inclusion of intents), i.e., $(A_1, B_1) \leq (A_2, B_2)$ iff $A_1 \subseteq A_2$ (or dually $B_2 \subseteq B_1$),
forms a complete lattice (Barbut and Monjardet 1970), called *concept lattice*. A
Hasse diagram can be associated to the concept lattice as the graph of the cover
relation $((A_1, B_1) \prec (A_2, B_2)$ when there is no concept (A_3, B_3) such that $(A_1, B_1) <
(A_3, B_3) < (A_2, B_2))$ where each concept of the lattice is represented as a vertex in
the plane and edges that goes upward from (A_1, B_1) to (A_2, B_2) whenever $(A_1, B_1) \prec
(A_2, B_2)$. The concept lattice associated with our toy example is shown in Fig. 1,

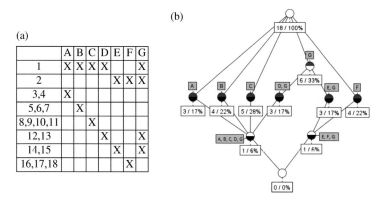

Fig. 1 Toy data set and its Galois lattice. (**a**) Toy data set. (**b**) Galois lattice associated with
table (**a**)

omitting the objects label and displaying only the object count associated with each concept.

2.2 Overhanging Relation

An interesting aspect of concept lattices is the many existing equivalent cryptomorphisms (for a survey see Caspard and Monjardet 2003, for a thorough introduction Caspard et al. 2012). One of them, called overhanging relation (Domenach and Leclerc 2004), is defined on the power set of the set of objects as follows:

A binary relation $\mathscr{O}$ on $\mathscr{P}(M)$ is an *overhanging relation* if it satisfies, for any $X, Y, Z \subseteq M$:

- $(X, Y) \in \mathscr{O}$ implies $X \subseteq Y$
- $X \subset Y \subset Z$ imply $(X, Z) \in \mathscr{O} \iff ((X, Y) \in \mathscr{O} \text{ or } (Y, Z) \in \mathscr{O})$
- $(X, X \cup Y) \in \mathscr{O}$ implies $(X \cap Y, Y) \in \mathscr{O}$

An alternative definition which links overhanging relation and concept lattice is the following: the overhanging relation associated with the concept lattice is a binary relation $\mathscr{O}$ on $\mathscr{P}(M)$ such that:

$$(X, Y) \in \mathscr{O} \iff X \subset Y \text{ and } X'' \subset Y''$$

In other words, two sets are overhanged if one is a subset of the other and they have a different closure. This binary relation finds its roots in consensus theory, where it appeared under the term of nesting in Adams (1986). Applying this for our example of table (a) in Fig. 1, we have $(\{D\}, \{C, D\}) \in \mathscr{O}$.

3 Existing Similarity Measures Between Concepts

Although, to the best of the authors' knowledge, there is no literature where similarity measures are defined on lattices, although many existing similarity measures can be adapted to concept lattices. They can be roughly divided in three main categories: the first one, based on Tversky (1977) model, only considers concepts as sets, here of attributes, in order to calculate the similarity between two concepts. The second one, taking its origin in studies of ontologies, uses the Hasse diagram associated with the concept lattice to evaluate distances between concepts. Lastly, the third category is concerned with semantic similarity measures using information content.

3.1 Set-Based Similarities

Set-based similarity measures can be expressed using Tversky similarity model. It is defined as follows: given two concepts $C_1 = (A_1, B_1)$ and $C_2 = (A_2, B_2)$, with $\alpha, \beta \geq 0$,

$$S(C_1, C_2) = \frac{|A_1 \cap A_2|}{|A_1 \cap A_2| + \alpha|A_1 - A_2| + \beta|A_2 - A_2|}$$

Depending on the values of α and β, the Tversky index can be seen as a generalization of Jaccard index and Dice's coefficient.

When taking $\alpha = 1$ and $\beta = 1$, we have the Jaccard similarity:

$$S_{\text{jaccard}}(C_1, C_2) = \frac{|A_1 \cap A_2|}{|A_1 \cup A_2|}$$

Using $\alpha = \beta = 1/2$ in the Tversky similarity model, we obtain the Dice's coefficient:

$$S_{\text{dice}}(C_1, C_2) = \frac{2|A_1 \cap A_2|}{|A_1| + |A_2|}$$

With $\alpha = 0, \beta = 1$, we have the inclusion measure:

$$S_{\text{inclusion}}(C_1, C_2) = \frac{|A_1 \cap A_2|}{|A_1|}$$

3.2 Ontology-Based Similarities

The following measures of similarities are inspired by the large corpus of work existing for the studies of Description Logic (DL) ontologies, i.e., ontologies based on a set of concepts, relations and individuals represented within a DL (Baader et al. 2003). The $\mathcal{EL}$ DL allows for conjunction ($\wedge$) and existential restriction in definitions of concepts. We simply assume that the least common subsumer (lcs) of two concepts in $\mathcal{EL}$ always exists, provided that there is no cycle in concept definitions (Baader et al. 1999).

The ordered structure of the lattice is used to calculate the similarity between two concepts C_1 and C_2, by considering only the taxonomic links of the ontology and the lattice $\mathbb{L}$ as a generalization of a tree. In order to define those similarities in this framework, we need to define the least common subsumer as lcs = $\text{lcs}(C_1, C_2) = C_1 \wedge C_2$, the length $\text{length}(C_1, C_2)$ as topological distance in the covering graph of the lattice between C_1 and C_2, and $\text{depth}(C_1) = \text{length}(C_1, 0_L)$ as

Table 1 Ontology-based similarities

Rada et al. (1989)	$S_{\text{rada}} = \frac{1}{\text{length}(C_1.C_2)+1}$
Wu and Palmer (1994)	$S_{\text{wup}} = \frac{2*\text{depth}(\text{lcs})}{\text{depth}(C_1)+\text{depth}(C_2)}$
Leacock and Chodorow (1998)	$S_{\text{lc}} = -\log\left(\frac{\text{length}(C_1,C_2)+1}{2\times\text{depth}(L)}\right)$
Pekar and Staab (2002)	$S_{\text{ps}} = \frac{\text{depth}(\text{lcs})}{\text{length}(C_1,\text{lcs})+\text{length}(C_2,\text{lcs})+\text{depth}(\text{lcs})}$
Zhong et al. (2002)	$S_{\text{Zho}} = 1-\left(\frac{1}{2^{\text{depth}(C_1)+1}} + \frac{1}{2^{\text{depth}(C_2)+1}} - \frac{1}{2^{\text{depth}(\text{lcs})}}\right)$
Nguyen and Al-Mubaid (2006)	$S_{\text{NguAl}} = \log(2+(\text{length}(C_1,C_2)-1)\times(\text{depth}(L)-\text{depth}(\text{lcs})))$

Table 2 IC functions

Resnik (1995)	$\text{IC}_{\text{res}} = \frac{|O_1|}{|O|}$
Seco et al. (2004)	$\text{IC}_{\text{seco}} = 1 - \frac{\log(\text{hypo}(C))}{\log(|L|-1)}$
Zhou et al. (2008)	$\text{IC}_{\text{zhou}} = k\times\left(1-\frac{\log(\text{hypo}(C))}{\log(|L|-1)}\right) + (1-k)\times\frac{\log(\text{depth}(C)+1)}{\log(\text{depth}(L)+1)}$
Sanchez et al. (2011)	$\text{IC}_{\text{san}} = -\log\left(\frac{\frac{\text{leaves}(C)}{\text{hypo}(C)}+1}{\text{number of leaves}+1}\right)$

the depth of concept C_1, i.e. distance between C_1 and bottom concept of L. The depth of the lattice is $\text{depth}(L) = \max_{x\in L}(\text{depth}(x))$. The ontology-based similarities implemented can be found in Table 1.

3.3 Information Content-Based Similarities

Another approach, particularly used in the study of semantic similarity between words in WordNet, improves on previous measures by augmenting concepts with Information Content (IC) derived from sense-tagged corpora or from raw unannotated corpora (Resnik 1995). We can apply it in our concept lattice framework by first defining the notion of IC in FCA in the following way: the IC of a concept provides an estimation of its degree of generality/concreteness, and is an increasing function: i.e., a is hypernym of $b \Rightarrow \text{IC}(a) < \text{IC}(b)$. IC is a measure of specificity for a concept. Higher values are associated with more specific concepts, while those with lower values are more general.

The different IC functions in Table 2 capture different aspects of information content. For Resnik (1995), it is the probability of appearance of concept in corpora, i.e. infrequent terms are more informative. For Seco et al. (2004), it is based on the number of concept hyponyms, where $\text{hypo}(C) = $ number of concepts above C. Zhou et al. (2008) complement hyponym-based IC computation with the relative depth of the concept, and Sanchez et al. (2011) is based on the number of leaf hyponyms, i.e. the number of co-atoms above the concept. Each similarity of Table 3 was implemented using each of the IC functions of Table 2.

Table 3 IC-based similarities

Resnik (1995)	$S_{\text{res}} = \text{IC(lcs)}$
Jiang and Conrath (1997)	$S_{\text{JC}} = \frac{1}{\text{IC}(C_1)+\text{IC}(C_2)-2\times\text{IC(lcs)}}$
Lin (1998)	$S_{\text{lin}} = \frac{2\times\text{IC(lcs)}}{\text{IC}(C_1)+\text{IC}(C_2)}$

3.4 Overhanging-Based Similarity

Let C be a concept of the concept lattice $\mathbb{L}$ and define $o(C)$ as the set of attributes that C is overhanged with:

$$o(C) = \{k \in M : (C, C \cup \{k\}) \in \mathcal{O}\}$$

$o(C)$ is the set of attributes that when added to C, generates a concept different than concept C. Continuing with our recurring example, we have $o(\{D,G\}) = \{A,B,C,E,F\}$.

From this mapping we can define a similarity measure based on the previous definition as follows: for any concepts C_1 and C_2 of $\mathbb{L}$,

$$S_o(C_1, C_2) = \frac{|o(C_1 \wedge C_2)|}{|o(C_1) \cup o(C_2)|}$$

The idea behind this measure is to take into account the width of the lattice. Two concepts will be more similar if they are close in the lattice and are not sharing attributes with other concepts. In other words, it indicates the similarity of two concepts in relation to all the other concepts.

As shown in Domenach (2015), we have the following properties for S_o, for any $C_1, C_2 \in \mathbb{L}$:

- S_o is normalized, i.e. takes its values in $[0, 1]$;
- $S_o(C_1, C_2) = 0$ if and only if $C_1 \wedge C_2 = 0_{\mathbb{L}}$;
- $S_o(C_1, C_2) = 1$ if and only if $C_1 = C_2$;
- if $C_1 \leq C_2$, $S_o(C_1, C_2) = \frac{|o(C_2)|}{|o(C_1)|}$.

Consider, in order to illustrate the interest of this similarity measure, the lattice of Fig. 1. We have $S_o(\{F\}, \{E,G\}) = \frac{4}{7}$ and $S_o(\{A\}, \{B\}) = \frac{2}{7}$. In this example, we have the Tversky similarities between $\{F\}$ and $\{E,G\}$ on the one hand, and $\{A\}$ and $\{B\}$ on the other hand, all equal to zero as the concepts don't share elements. Similarly none of those existing similarity measures allows us to discriminate between these two concepts—and one can argue that $\{F\}$ and $\{E,G\}$ are closer together, since they describe all the information of $\{E,F,G\}$, than $\{A\}$ and $\{B\}$, containing only partial information of their parent concept $\{A,B,C,D,G\}$.

4 Experimental Evaluation

4.1 Experiment on Similarity Methods

The proposed measure has been investigated through a simulation implemented[1] in C♯ and ran on a computer with Intel Core 2 Duo 3.16 GHz with 2 GB of memory. More precisely, we have randomly generated three 20×20 Boolean tables. We were limited on the size of the table due to the exponential time needed to calculate concept lattices and similarity values. For each table the entries had a density of 20, 30 and 40 %, respectively. Furthermore, for each Boolean table we generated the associated concept lattice using the CbO algorithm. It is well known that the number of concepts in the lattice increases with the density. Indicatively, for the table with a 20 % density there are approximately 65 concepts, with a 30 % density there are approximately 130 concepts and with a 40 % density there are approximately 260 concepts. We expect that an increase in the size of the lattice will lead to similarity values that are more spread.

For the simulation, we randomly selected two different concepts and calculated 22 similarities between them. The similarities used were: the set-based similarities (Jaccard, Dice, Inclusion), the ontology-based ones (Rada, Leacock, Wu-Palmer, Pekar-Staab, Ngueyn, Zhong) and the combination of IC-based similarities (Resnik, Lin, Jiang-Conrath) versus the IC functions (Resnik, Seco, Zhou, Sanchez). We also included a random number taken between 0 and 1 to use as a baseline. This simulation was repeated 1000 times for each Boolean table.

4.2 Results and Discussion

The simulation data were statistically investigated. The data obtained are presented in the histogram of Fig. 2 (left). It is evident that the histogram depicts two populations of data. In the first population, we have the results that produce the value of zero, i.e. indicating that there is no similarity between concepts, their least common ancestor being the bottom element of the lattice. Due to the large number of zeros in the Boolean tables such results are to be expected as concept lattices are more structurally simple and less dense. More precisely, the results produced 715, 580, and 489 zeros out of 1000 concepts for the Boolean tables of 20 %, 30 % and 40 %, respectively. The second population concerns the non-zero results, see Fig. 2 (right). These data depict a symmetric-like distribution for which the normal probability distribution perhaps fits adequately.

---

[1] Source code available on demand.

a
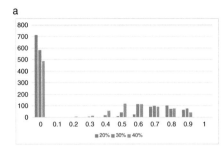
b
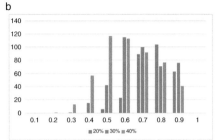

Fig. 2 Frequency distribution of overhanging similarity, including 0 (*left*) or excluding 0 (*right*)

Table 4 Descriptive of overhanging similarity

	20 %	30 %	40 %		20 %	30 %	40 %
N-valid	1000	1000	1000	*N*-excluding 0	285	420	511
Mean	0.217	0.289	0.311	Mean	0.763	0.688	0.608
Std. err. of mean	0.0103	0.0112	0.0110	Std. err. of mean	0.0061	0.0071	0.0070
Median	0.000	0.000	0.308	Median	0.750	0.684	0.600
Minimum	0.0	0.0	0.0	Minimum	0.5	0.3	0.2
Maximum	0.9	0.9	0.9	Maximum	0.9	0.9	0.9

This population classification leads to a separate statistical analysis according to prior information:

- *Case A: No information*, i.e. there is no information as to whether there are similarities within concepts (Table 4, left). Here we observe that the estimated mean obtained has an increasing trend against the density of the Boolean tables. The large number of zeros obtained, as mentioned above, is evident from the values of the median being equal to zero. The standard error of the mean remains at a similar level for all Boolean tables.
- *Case B: Similarity exists*, i.e. there is information that similarity exists between concepts (Table 4, right). In effect, here we ignore zeros obtained. It is interesting to observe that the estimated mean and median, now have a decreasing trend against the density of the Boolean tables. The standard error of the mean remains at a similar level for all Boolean tables as well.

A clustering with average linkage was ran on Pearson's correlations between all similarities (dendrogram in Fig. 3). There are two obvious clusters depicted, but no clear characteristics stand out. However, similarities are to an extent clustered together following the classification of Sect. 3, see, for example, the set-based similarities.

The Pearson's correlation of the overhanging similarity measure with the measures used in the simulation, with its significance, is presented in Table 5. There is clear evidence that the measure is significantly different to the baseline. The remaining correlation estimates should, however, be taken with caution as the

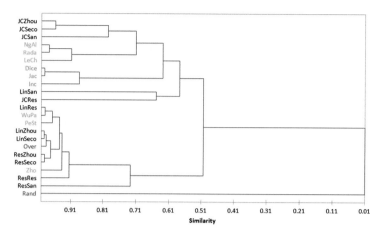

Fig. 3 Average linkage of correlation matrix of all similarities

Table 5 Pearson's correlation and significance of the overhanging similarity measure with the 22 other measures

Jac	Dice	Inc	LeCh	PeSt	Rada	WuPa	LinRes	ResRes	JCRes	LinSec
0.444	0.422	0.389	0.588	0.917	0.575	0.950	0.934	0.893	0.258	0.982
0.000	0.000	0.000	0.000	0.000	0.000	0.000	0.000	0.000	0.000	0.000
ResSec	JCSec	LinZho	ResZho	JCZho	NgAl	Zho	LinSan	ResSan	JCSan	Rand
0.967	0.613	0.988	0.976	0.680	0.680	0.949	0.242	0.707	0.296	0.014
0.000	0.000	0.000	0.000	0.000	0.000	0.000	0.000	0.000	0.000	0.650

linearity between the aforementioned measures is still to be investigated. Albeit, one can see that there are several very strong correlations, both positive and negative, in particular with the Lin and Resnik IC-based similarities. This may be interpreted as capturing similar information as those functions using only the structure of the lattice.

5   Conclusion and Perspectives

The simulation and the resulting analysis have allowed for a comparison of the overhanging similarity with other similarity measures. Further analysis will be carried out on the computational complexities of the different measures and on possible relationships, either linear or nonlinear, between these measures. In turn, heteroscedasticity and homoscedasticity will need to be investigated. Lastly, robust statistics, in order to minimize/eliminate, the effect of influential values, may be needed to be employed. We are also planning to do such a comparison on real data sets selected from UCI machine learning repository.

References

Adams III, E. N. (1986). N-trees as nestings: Complexity, similarity, and consensus. *Journal of Classification, 3*, 299–317.

Baader, F., Calvanese, D., Mcguiness, D. L., NARDI, D., & Patel-Schneider, P. F. (Ed.) (2003). *The description logic handbook: Theory, implementation, and applications.* Cambridge: Cambridge University Press.

Baader, F., Kusters, R., & Molitor, R. (1999). Computing least common subsumers in description logics with existential restrictions. In *IJCAI* (pp. 96–103).

Barbut, M., & Monjardet, B. (1970). *Ordres et classification: Algèbre et combinatoire (tome II).* Paris: Hachette.

Caspard, N., & Monjardet, B. (2003). The lattices of Moore families and closure operators on a finite set: A survey. *Discrete Applied Mathematics, 127*, 241–269.

Caspard, N., Leclerc, B., & Monjardet, B. (2012). *Finite ordered sets: Concepts, results and uses.* Cambridge: Cambridge University Press.

Domenach, F. (2015). Similarity measures of concept lattices. In B. Lausen, S. Krolak-Schwerdt, & M. Böhmer (Eds.), *Data science, learning by latent structures, and knowledge discovery. Studies in classification, data analysis, and knowledge organization* (pp. 89–99). Berlin/Heidelberg: Springer. doi:10.1007/978-3-662-44983-7_8. http://dx.doi.org/10.1007/978-3-662-44983-7_8. ISBN: 978-3-662-44982-0.

Domenach, F., & Leclerc, B. (2004). Closure systems, implicational systems, overhanging relations and the case of hierarchical classification. *Mathematical Social Sciences, 47*(3), 349–366.

Ganter, B., & Wille, R. (1999). *Formal concept analysis: Mathematical foundations.* New York: Springer.

Jiang, J. J., & Conrath, D. W. (1997). Semantic similarity based on corpus statistics and lexical taxonomy. In *International Conference on Research on Computational Linguistics (ROCLING X)*

Leacock, C., & Chodorow, M. (1998). Combining local context and wordnet similarity for word sense identification. In C. Fellbaum (Ed.), *WordNet: An electronic lexical database* (pp. 265–283). Cambridge: MIT Press.

Lin, D. (1998). An information-theoretic definition of similarity. In J. W. Shavlik (Ed.), *Proceedings of the Fifteenth International Conference on Machine Learning (ICML '98)* (pp. 296–304). San Francisco, CA: Morgan Kaufmann.

Nguyen, H. A., & Al-Mubaid, H. (2006). New ontology-based semantic similarity measure for the biomedical domain. In *Proceedings of the IEEE International Conference on Granular Computing* (pp. 623–628).

Pekar, V., & Staab, S. (2002). Taxonomy learning: Factoring the structure of a taxonomy into a semantic classification decision. In *COLING '02 Proceedings of the 19th International Conference on Computational Linguistics* (Vol. 1, pp. 1–7).

Rada, R., Mili, H., Bicknell, E., & Blettner, M. (1989). Development and application of a metric on semantic nets. *IEEE Transactions on Systems, Man, and Cybernetics, 19*, 17–30.

Resnik, P. (1995). Using information content to evaluate semantic similarity in a taxonomy. In C. S. Mellish (Ed.), *Proceedings of the 14th International Joint Conference on Artificial intelligence (IJCAI'95)* (pp. 448–453).

Sanchez, D., Batet, M., & Isern, D. (2011). Ontology-based information content computation. *Knowledge-Based Systems, 24*(2), 297–303.

Seco, N., Veale, T., & Hayes, J. (2004). An intrinsic information content metric for semantic similarity in WordNet. In *Proceedings of 16th European Conference on Artificial Intelligence, ECAI 2004* (pp. 1089–1090).

Tversky, A. (1977). Features of similarity. *Psychological Reviews, 84*(4), 327–352.

Wu, Z., & Palmer, M. (1994). Verb semantics and lexical selection. In *Proceedings of 32nd Annual Meeting of the Associations for Comp. Linguistics* (pp. 133–138).

Zhong, J., Zhu, H., Li, J., & Yu, Y. (2002). Conceptual graph matching for semantic search. In *Proceedings of the 10th International Conference on Conceptual Structures* (pp. 92–106).

Zhou, Z., Wang, Y., & Gu, J. (2008). A new model of information content for semantic similarity in WordNet. In *Proceedings of Second International Conference on Future Generation Communication and Networking Symposia, FGCNS 2008* (pp. 85–89). Sanya: IEEE Computer Society.

Part IV
Classification

Multivariate Functional Regression Analysis with Application to Classification Problems

Tomasz Górecki, Mirosław Krzyśko, and Waldemar Wołyński

Abstract Multivariate functional data analysis is an effective approach to dealing with multivariate and complex data. These data are treated as realizations of multivariate random processes; the objects are represented by functions. In this paper we discuss different types of regression model: linear and logistic. Various methods of representing functional data are also examined. The approaches discussed are illustrated with an application to two real data sets.

1 Introduction

Much attention has been paid in recent years to methods for representing data as functions or curves. Such data are known in the literature as functional data (Ramsay and Silverman 2005). Applications of functional data can be found in various fields, including medicine, economics, meteorology and many others. In many applications there is a need to use statistical methods for objects characterized by multiple features observed at many time points (doubly multivariate data). Such data are called multivariate functional data. The pioneering theoretical work was that of Besse (1979), in which random variables take values in a general Hilbert space. Saporta (1981) presents an analysis of multivariate functional data from the point of view of factorial methods (principal components and canonical analysis). In this paper we focus on the problem of classification via regression for multivariate functional data. Functional regression models have been extensively studied; see, for example, James (2002), Müller and Stadmüller (2005), Reiss and Ogden (2007), Matsui et al. (2008) and Li et al. (2010). Various basic classification methods have also been adapted to functional data, such as linear discriminant analysis (Hastie et al. 1995), logistic regression (Rossi et al. 2002), penalized optimal scoring (Ando 2009), $k$nn (Ferraty and Vieu 2003), SVM (Rossi and Villa 2006), and neural

T. Górecki (✉) • M. Krzyśko • W. Wołyński
Faculty of Mathematics and Computer Science, Adam Mickiewicz University, Umultowska 87, 61-614 Poznań, Poland
e-mail: tomasz.gorecki@amu.edu.pl; mkrzysko@amu.edu.pl; wolynski@amu.edu.pl

© Springer International Publishing Switzerland 2016
A.F.X. Wilhelm, H.A. Kestler (eds.), *Analysis of Large and Complex Data*, Studies in Classification, Data Analysis, and Knowledge Organization,
DOI 10.1007/978-3-319-25226-1_15

networks (Rossi et al. 2005). Moreover, the combining of classifiers has been extended to functional data (Ferraty and Vieu 2009).

In the present work we adapt multivariate regression models to the classification of multivariate functional data. Apart from well-known classical regression models like the linear and logistic models, we also use models based on functional principal components. We focus on the binary classification problem. There exist several techniques for extending the binary problem to multi-class classification problems. A brief overview can be found in Krzyśko and Wołyński (2009). The accuracy of the proposed methods is demonstrated using biometrical examples. Promising results were obtained for future research.

The remainder of the paper is organized as follows. In Sect. 2 we describe the classification model for multivariate functional data. In Sect. 3 we introduce a multivariate regression model with scalar response and functional predictors. Section 4 contains examples and brief discussion of the results. Concluding remarks and topics of future research are presented in Sect. 5.

2 Classification via Regression

The classical classification problem involves determining a procedure by which a given object can be assigned to one of K populations based on observation of p features of that object.

The object being classified can be described by a random pair $(\boldsymbol{X}, Y)$, where $\boldsymbol{X} = (X_1, X_2, \ldots, X_p)' \in \mathbf{R}^p$ and $Y \in \{0, 1, \ldots, K-1\}$.

The optimum Bayesian classifier then takes the form (Anderson 1984):

$$d(\boldsymbol{x}) = \arg \max_{k \in \{0,1,\ldots,K-1\}} P(Y = k | \boldsymbol{X} = \boldsymbol{x}).$$

We shall further consider only the case $K = 2$. Here

$$d(\boldsymbol{x}) = \begin{cases} 1, & P(Y = 1 | \boldsymbol{X} = \boldsymbol{x}) \geq P(Y = 0 | \boldsymbol{X} = \boldsymbol{x}); \\ 0, & P(Y = 1 | \boldsymbol{X} = \boldsymbol{x}) < P(Y = 0 | \boldsymbol{X} = \boldsymbol{x}). \end{cases}$$

We note that

$$P(Y = 1 | \boldsymbol{X} = \boldsymbol{x}) = E(Y | \boldsymbol{X} = \boldsymbol{x}) = r(\boldsymbol{x}),$$

where $r(\boldsymbol{x})$ is the regression function of the random variable Y with respect to the random vector $\boldsymbol{X}$.

Hence

$$d(\boldsymbol{x}) = \begin{cases} 1, & r(\boldsymbol{x}) \geq 1/2; \\ 0, & r(\boldsymbol{x}) < 1/2. \end{cases}$$

We now assume that the object being classified is described by a p-dimensional random process $\boldsymbol{X}(t) = (X_1(t), X_2(t), \ldots, X_p(t))'$, with continuous parameter $t \in I$ and that $\boldsymbol{X}(t) \in L_2^p(I)$, where $L_2^p(I)$ is the Hilbert space of p-dimensional square-integrable functions.

Typically data are recorded at discrete moments in time. The process of transformation of discrete data to functional data is performed for each variable separately.

Let x_{kj} denote an observed value of the feature X_k, $k = 1, 2, \ldots p$ at the jth time point t_j, where $j = 1, 2, \ldots, J$. Then our data consist of the pJ pairs (t_j, x_{kj}). These discrete data can be smoothed by continuous functions $x_k(t)$, where $t \in I$ (Ramsay and Silverman 2005). Here I is a compact set such that $t_j \in I$, for $j = 1, \ldots, J$. Let us assume that the function $x_k(t)$ has the following representation:

$$x_k(t) = \sum_{b=0}^{B_k} c_{kb}\varphi_b(t), \ t \in I, \ k = 1, \ldots, p, \tag{1}$$

where $\{\varphi_b\}$ are orthonormal basis functions, and $c_{k0}, c_{k1}, \ldots, c_{kB_k}$ are the coefficients.

Let $\boldsymbol{x}_k = (x_{k1}, x_{k2}, \ldots, x_{kJ})'$, $\boldsymbol{c}_k = (c_{k0}, c_{k1}, \ldots, c_{kB_k})'$ and $\boldsymbol{\Phi}_k$ be a matrix of dimension $J \times (B_k + 1)$ containing the values $\varphi_b(t_j), b = 0, 1, \ldots, B_k, j = 1, 2, \ldots, J$, $k = 1, \ldots, p$. The coefficient $\boldsymbol{c}_k$ in (1) is estimated by the least squares method:

$$\hat{\boldsymbol{c}}_k = \left(\boldsymbol{\Phi}_k'\boldsymbol{\Phi}_k\right)^{-1}\boldsymbol{\Phi}_k'\boldsymbol{x}_k, \ k = 1, \ldots, p.$$

The degree of smoothness of the function $x_k(t)$ depends on the value B_k (a small value of B_k causes more smoothing of the curves). The optimum value for B_k may be selected using the Bayesian information criterion (BIC) (Shmueli 2010).

Let

$$\hat{\boldsymbol{c}} = (\hat{\boldsymbol{c}}_1', \ldots, \hat{\boldsymbol{c}}_p')', \tag{2}$$

$$\boldsymbol{\Phi}(t) = \begin{bmatrix} \boldsymbol{\varphi}_1'(t) & \boldsymbol{0} & \ldots & \boldsymbol{0} \\ \boldsymbol{0} & \boldsymbol{\varphi}_2'(t) & \ldots & \boldsymbol{0} \\ \ldots & \ldots & \ldots & \ldots \\ \boldsymbol{0} & \boldsymbol{0} & \ldots & \boldsymbol{\varphi}_p'(t) \end{bmatrix}, \tag{3}$$

where $\boldsymbol{\varphi}_k(t) = (\varphi_0(t), \ldots, \varphi_{B_k}(t))', k = 1, \ldots, p$.

Then $\boldsymbol{x}(t)$ can be represented as

$$\boldsymbol{x}(t) = \boldsymbol{\Phi}(t)\hat{\boldsymbol{c}}. \tag{4}$$

3   Multivariate Functional Regression Analysis

We now consider the problem of the estimation of the regression function $r(x)$, or more precisely the regression function $r(x(t))$.

Let us assume that we have an n-element training sample

$$\mathscr{L}_n = \{(x_1(t), y_1), (x_2(t), y_2), \ldots, (x_n(t), y_n)\},$$

where $x_i(t) \in L_2^p(I)$ and $y_i \in \{0, 1\}$.

Analogously as in Sect. 2, we assume that the functions $x_i(t)$ are obtained as the result of a process of smoothing n independent discrete data pairs (t_j, x_{kij}), $k = 1, \ldots, p, j = 1, \ldots, J, i = 1, \ldots, n$.

Thus the functions $x_i(t)$ have the following representations:

$$x_i(t) = \Phi(t)\hat{c}_i, \quad i = 1, 2, \ldots, n. \tag{5}$$

3.1   Multivariate Linear Regression

We take the following model for the regression function:

$$r(x) = \beta_0 + \int_I \beta'(t)x(t)dt. \tag{6}$$

We seek the unknown parameters in the regression function by minimizing the sum of squares

$$S(\beta_0, \beta) = \sum_{i=1}^n \left(y_i - \beta_0 - \int_I \beta'(t)x_i(t)dt \right)^2. \tag{7}$$

We assume that the functions $x_i(t)$, $i = 1, 2, \ldots, n$ have the representation (5). We adopt an analogous representation for the p-dimensional weighting function $\beta(t)$, namely

$$\beta(t) = \Phi(t)d, \tag{8}$$

where $\beta(t) - (\beta_1(t), \ldots, \beta_p(t))'$, $d = (d_1', \ldots, d_k')'$ and $d_k = (d_{k0}, d_{k1}, \ldots, d_{kB_k})'$.

Then

$$\int_I \beta'(t)x_i(t)dt = \int_I d'\Phi'(t)\Phi(t)\hat{c}_i dt = d'\hat{c}_i, \quad i = 1, 2, \ldots, n.$$

Hence

$$S(\beta_0, \boldsymbol{\beta}) = S(\beta_0, \boldsymbol{d}) = \sum_{i=1}^{n}(y_i - \beta_0 - \boldsymbol{d}'\hat{\boldsymbol{c}}_i)^2.$$

We define $\boldsymbol{y} = (y_1, y_2, \ldots, y_n)'$ and

$$\boldsymbol{Z} = \begin{bmatrix} 1 & \hat{\boldsymbol{c}}_1' \\ 1 & \hat{\boldsymbol{c}}_2' \\ \vdots & \vdots \\ 1 & \hat{\boldsymbol{c}}_n' \end{bmatrix}, \quad \boldsymbol{\gamma} = \begin{bmatrix} \beta_0 \\ \boldsymbol{d} \end{bmatrix}.$$

Then

$$S(\beta_0, \boldsymbol{\beta}) = S(\boldsymbol{\gamma}) = (\boldsymbol{y} - \boldsymbol{Z}\boldsymbol{\gamma})' \, (\boldsymbol{y} - \boldsymbol{Z}\boldsymbol{\gamma}) \, .$$

Minimizing the above sum of squares leads to the choice of a vector $\boldsymbol{\gamma}$ satisfying

$$\boldsymbol{Z}'\boldsymbol{Z}\boldsymbol{\gamma} = \boldsymbol{Z}'\boldsymbol{y}. \tag{9}$$

Provided the matrix $\boldsymbol{Z}'\boldsymbol{Z}$ is non-singular, Eq. (9) has the unique solution

$$\hat{\boldsymbol{\gamma}} = (\boldsymbol{Z}'\boldsymbol{Z})^{-1}\boldsymbol{Z}'\boldsymbol{y}. \tag{10}$$

We obtain the following form for the estimator of the regression function for the multivariate functional data:

$$\hat{r}(\boldsymbol{x}) = \hat{\beta}_0 + \hat{\boldsymbol{d}}'\hat{\boldsymbol{c}},$$

where $\hat{\boldsymbol{\gamma}} = (\hat{\beta}_0, \hat{\boldsymbol{d}})'$ is given by the formula (10) and $\hat{\boldsymbol{c}}$ by the formula (2).

In the case of functional data we can also use the smoothed least squares method discussed by Ramsay and Silverman (2005).

3.2 Regression Model Based on MFPCA

An alternative strategy for fitting the model (6) is to carry out a preliminary multivariate functional principal components Analysis (MFPCA) of the covariate functions, and effectively reduce the dimensionality of the covariate space (Ferraty and Vieu 2006; Górecki and Krzyśko 2012; Jacques and Preda 2014). For the random process $\boldsymbol{X}(t)$, the lth multivariate functional principal component (MFPC) has the form

$$U_l = \, < \boldsymbol{u}_l(t), \boldsymbol{X}(t) >$$

and satisfy the conditions

$$\text{Var}(< \boldsymbol{u}_l(t), \boldsymbol{X}(t) >) = \sup_{\boldsymbol{u}(t)\in L_2^p(I)} \text{Var}(< \boldsymbol{u}(t), \boldsymbol{X}(t) >),$$

$$< \boldsymbol{u}_{\kappa_1}(t), \boldsymbol{u}_{\kappa_2}(t) >= \delta_{\kappa_1\kappa_2}, \qquad \kappa_1, \kappa_2 = 1, \dots, l.$$

It may be assumed that the vectors of the weighting function $\boldsymbol{u}_l(t)$ and realizations of the process $\boldsymbol{X}(t)$ are in the same space, i.e. the functions $\boldsymbol{u}_l(t)$ and $\boldsymbol{X}(t)$ can be written in the form:

$$\boldsymbol{u}_l(t) = \boldsymbol{\Phi}(t)\boldsymbol{u}_l, \quad \boldsymbol{X}(t) = \boldsymbol{\Phi}(t)\boldsymbol{C},$$

where $\boldsymbol{u}_l, \boldsymbol{C} = (\boldsymbol{C}_1, \boldsymbol{C}_2, \dots, \boldsymbol{C}_p)' \in \mathbf{R}^{B+p}, B = B_1 + B_2 + \dots + B_p$. Then

$$< \boldsymbol{u}_l(t), \boldsymbol{X}(t) >=< \boldsymbol{\Phi}(t)\boldsymbol{u}_l, \boldsymbol{\Phi}(t)\boldsymbol{C} >= \boldsymbol{u}_l' < \boldsymbol{\Phi}(t), \boldsymbol{\Phi}(t) > \boldsymbol{C} = \boldsymbol{u}_l'\boldsymbol{C}$$

and $\text{Var}(< \boldsymbol{u}_l(t), \boldsymbol{X}(t) >) = \boldsymbol{u}_l' \text{Var}(\boldsymbol{C})\boldsymbol{u}_l = \boldsymbol{u}_l'\boldsymbol{\Sigma}\boldsymbol{u}_l$.

The unknown matrix $\boldsymbol{\Sigma}$ is estimated using the training sample $\mathscr{L}_n$. We have

$$\boldsymbol{x}_i(t) = \boldsymbol{\Phi}(t)\hat{\boldsymbol{c}}_i, \quad i = 1, 2, \dots, n.$$

Let

$$\tilde{\boldsymbol{C}} = (\tilde{\boldsymbol{c}}_1, \tilde{\boldsymbol{c}}_2, \dots, \tilde{\boldsymbol{c}}_n), \quad \tilde{\boldsymbol{c}}_i = \hat{\boldsymbol{c}}_i - \frac{1}{n}\sum_{j=1}^{n}\hat{\boldsymbol{c}}_j.$$

Then

$$\hat{\boldsymbol{\Sigma}} = \frac{1}{n-1}\tilde{\boldsymbol{C}}\tilde{\boldsymbol{C}}'.$$

Let $\hat{\gamma}_1 \geq \hat{\gamma}_2 \geq \dots \geq \hat{\gamma}_s$ be non-zero eigenvalues of matrix $\hat{\boldsymbol{\Sigma}}$, and $\hat{\boldsymbol{u}}_1, \hat{\boldsymbol{u}}_2, \dots, \hat{\boldsymbol{u}}_s$ the corresponding eigenvectors, where $s = \text{rank}(\hat{\boldsymbol{\Sigma}})$.

Then for the ith realization $\boldsymbol{x}_i(t)$ of the process $\boldsymbol{X}(t)$, we have

$$\boldsymbol{x}_i(t) = \sum_{l=1}^{s} \hat{u}_{il}\hat{\boldsymbol{u}}_l(t) \tag{11}$$

where

$$\hat{\boldsymbol{u}}_l(t) = \boldsymbol{\Phi}(t)\hat{\boldsymbol{u}}_l, \quad l = 1, \dots, s$$

and

$$\hat{u}_{il} =< \hat{\boldsymbol{u}}_l(t), \boldsymbol{x}_i(t) >=< \boldsymbol{\Phi}(t)\hat{\boldsymbol{u}}_l, \boldsymbol{\Phi}(t)\hat{\boldsymbol{c}}_i >= \hat{\boldsymbol{u}}_l' < \boldsymbol{\Phi}(t), \boldsymbol{\Phi}(t) > \hat{\boldsymbol{c}}_i = \hat{\boldsymbol{u}}_l'\hat{\boldsymbol{c}}_i,$$

for $i = 1, 2, \dots, n, l = 1, 2, \dots, s$.

Let us now consider again the regression model given by (6). To solve the problem of minimizing the sum of squares (7), we use the representation of the function $x_i(t)$ in the form (11). Moreover, we assume that

$$\boldsymbol{\beta}(t) = \sum_{l=1}^{s} b_l \hat{\boldsymbol{u}}_l(t). \tag{12}$$

Then

$$\int_I \boldsymbol{\beta}'(t) \boldsymbol{x}_i(t) dt = \boldsymbol{b}' \hat{\boldsymbol{u}}_i,$$

where $\boldsymbol{b} = (b_1, b_2, \ldots, b_s)'$ and $\hat{\boldsymbol{u}}_i = (\hat{u}_{i1}, \hat{u}_{i2}, \ldots, \hat{u}_{is})'$.
Therefore

$$S(\beta_0, \boldsymbol{\beta}) = S(\beta_0, \boldsymbol{b}) = \sum_{i=1}^{n} (y_i - \beta_0 - \boldsymbol{b}' \hat{\boldsymbol{u}}_i)^2.$$

Following transformations analogous to those in Sect. 3.1, we obtain an estimate of the vector $\boldsymbol{\delta} = (\beta_0, \boldsymbol{b}')'$ in the form

$$\hat{\boldsymbol{\delta}} = (\tilde{\boldsymbol{Z}}' \tilde{\boldsymbol{Z}})^{-1} \tilde{\boldsymbol{Z}}' \boldsymbol{y}, \qquad \text{where} \qquad \tilde{\boldsymbol{Z}} = \begin{bmatrix} 1 & \hat{\boldsymbol{u}}_1' \\ 1 & \hat{\boldsymbol{u}}_2' \\ \vdots & \vdots \\ 1 & \hat{\boldsymbol{u}}_n' \end{bmatrix}.$$

Therefore

$$\hat{r}(\boldsymbol{x}) = \hat{\beta}_0 + \hat{\boldsymbol{b}}' \hat{\boldsymbol{u}},$$

where $\hat{\boldsymbol{u}} = (\hat{u}_1, \ldots, \hat{u}_s)'$ and $\hat{u}_l = \langle \hat{\boldsymbol{u}}_l(t), \boldsymbol{x}(t) \rangle = \hat{\boldsymbol{u}}_l' \hat{\boldsymbol{c}}$.

3.3 Multivariate Functional Logistic Regression

We adopt the following logistic regression model for functional data:

$$r(\boldsymbol{x}) = \frac{\exp(\beta_0 + \int_I \boldsymbol{\beta}'(t) \boldsymbol{x}(t) dt)}{1 + \exp(\beta_0 + \int_I \boldsymbol{\beta}'(t) \boldsymbol{x}(t) dt)}. \tag{13}$$

Using the representation of the function $x(t)$ given by (4) and the weighting function $\boldsymbol{\beta}(t)$ given by (8) we reduce (13) to a standard logistic regression model in the form

$$r(x) = \frac{\exp(\beta_0 + \boldsymbol{d}'\hat{\boldsymbol{c}})}{1 + \exp(\beta_0 + \boldsymbol{d}'\hat{\boldsymbol{c}})}. \tag{14}$$

To estimate the unknown parameters of the model, we use the training sample $\mathscr{L}_n$ and the analogous representation for the functions $\boldsymbol{x}_i(t)$, $i = 1, 2, \ldots, n$ given by (5).

The functional logistic regression model (13) can be reduced to a standard model by taking representations of the function $x(t)$ in the basis defined by the multivariate functional principal components (MFPCA).

Assuming that

$$x(t) = \sum_{l=1}^{s} \hat{u}_l \hat{\boldsymbol{u}}_l(t)$$

and that the representation of the weighting function $\boldsymbol{\beta}(t)$ is given by (12), we obtain

$$r(x) = \frac{\exp(\beta_0 + \boldsymbol{b}'\hat{\boldsymbol{u}})}{1 + \exp(\beta_0 + \boldsymbol{b}'\hat{\boldsymbol{u}})}. \tag{15}$$

To estimate the unknown parameters of the model, we again use the training sample $\mathscr{L}_n$ and the representations of the functions $\boldsymbol{x}_i(t)$, $i = 1, 2, \ldots, n$ given by (11).

From now on we shall use the following notation for classifiers: MFLM for the multivariate functional linear model, MFPCR for the multivariate functional linear model based on MFPCA, MFLG for the multivariate functional logistic model, and MFLGR for the multivariate functional logistic model based on MFPCA.

4   Examples

Experiments were carried out on two data sets, these being labelled data sets whose labels are given. The data sets originate from Olszewski (2001).

The *ECG* data set uses two electrodes (Fig. 1) to collect data during one heartbeat. Each heartbeat is described by a multivariate time series (MTS) sample with two variables and an assigned classification of normal or abnormal. Abnormal heartbeats are representative of a cardiac pathology known as supraventricular premature beat. The ECG data set contains 200 MTS samples, of which 133 are normal and 67 are abnormal. The length of an MTS sample is between 39 and 152.

The *Wafer* data set uses six vacuum-chamber sensors (Fig. 2) to collect data while monitoring an operational semiconductor fabrication plant. Each wafer is described by an MTS sample with six variables and an assigned classification of normal or

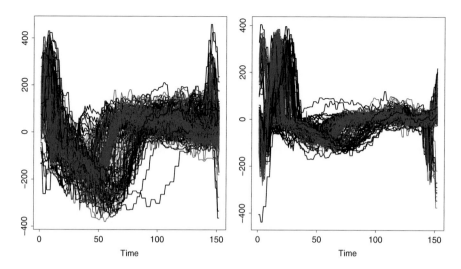

Fig. 1 Extended *ECG* data set (*red*—normal, *black*—abnormal)

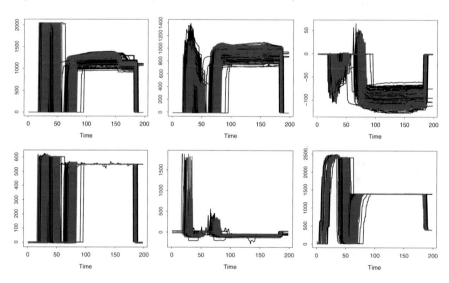

Fig. 2 Extended *Wafer* data set (*red*—normal, *black*—abnormal)

abnormal. The data set used here contains 327 MTS samples, of which 200 are normal and 127 are abnormal. The length of an MTS sample is between 104 and 198.

The multivariate samples in the data sets are of different lengths. For each data set, the multivariate samples are extended to the length of the longest multivariate sample in the set (Rodriguez et al. 2005). We extend all variables to the same length. For a short univariate instance x with length J, we extend it to a long instance x_{ex}

with length $J_{\max}$ by setting

$$x_{\mathrm{ex}}(t_j) = x(t_i), \quad \text{for} \quad i = \left\lceil \frac{j-1}{J_{\max}-1}(J-1) + 0.5 \right\rceil \quad (j = 1, 2, \ldots, J_{\max}).$$

Some of the values in a data sample are duplicated in order to extend the sample. For instance, if we wanted to extend a data sample of length 75 to a length of 100, one out of every three values would be duplicated. In this way, all of the values in the original data sample are contained in the extended data sample.

For the classification process, we used the classifiers described above. For each data set we calculated the classification error rate using the leave-one-out cross-validation method (LOO CV). We obtained error rates (in percent): 11.50, 11.50, 20.00 and 22.50 for the data set *ECG*, respectively, for the methods MFLM, MFLG, MFPCR and MFLGR. Analogously for the *Wafer* data set we obtained the error rates (in percent): 0.59, 0.17, 2.01 and 0.92. As we can see, moving to the space of principal components appears not to improve the results, and in fact makes them significantly worse. In this situation the best classification method would appear to be MFLG.

5 Conclusion

This paper develops and analyzes methods for constructing and using regression methods of classification for multivariate functional data. These methods were applied to two biometrical multivariate time series. In the case of these examples it was shown that the use of multivariate functional regression methods for classification gives good results. Of course, the performance of the algorithm needs to be further evaluated on additional real and artificial data sets. In a similar way, we can extend other regression methods, such as partial least squares regression—PLS, least absolute shrinkage and selection operator—LASSO, or least-angle regression—LARS, to the multivariate functional case. This will be the direction of our future research.

References

Anderson, T. W. (1984). *An introduction to multivariate statistical analysis*. New York: Wiley.
Ando, T. (2009). Penalized optimal scoring for the classification of multi-dimensional functional data. *Statistcal Methodology, 6*, 565–576.
Besse, P. (1979). *Etude descriptive d'un processus*. Ph.D. thesis, Universit'e Paul Sabatier.
Ferraty, F., & Vieu, P. (2003). Curve discrimination. A nonparametric functional approach. *Computational Statistics & Data Analysis, 44*, 161–173.
Ferraty, F., & Vieu, P. (2006). *Nonparametric functional data analysis: Theory and practice*. New York: Springer.

Ferraty, F., & Vieu, P. (2009). Additive prediction and boosting for functional data. *Computational Statistics & Data Analysis, 53*(4), 1400–1413.

Górecki, T., & Krzyśko, M. (2012). Functional Principal components analysis. In J. Pociecha & R. Decker (Eds.), *Data analysis methods and its applications* (pp. 71–87). Warszawa: C.H. Beck.

Hastie, T. J., Tibshirani, R. J., & Buja, A. (1995). Penalized discriminant analysis. *Annals of Statistics, 23*, 73–102.

Jacques, J., & Preda, C. (2014). Model-based clustering for multivariate functional data. *Computational Statistics & Data Analysis, 71*, 92–106.

James, G. M. (2002). Generalized linear models with functional predictors. *Journal of the Royal Statistical Society, 64*(3), 411–432.

Krzyśko, M., & Wołyński, W. (2009). New variants of pairwise classification. *European Journal of Operational Research, 199*(2), 512–519.

Li, Y., Wang, N., & Carroll, R. J. (2010). Generalized functional linear models with semi parametric single-index interactions. *Journal of the American Statistical Association, 105*(490), 621–633.

Matsui, H., Araki, Y., & Konishi, S. (2008). Multivariate regression modeling for functional data. *Journal of Data Science, 6*, 313–331.

Müller, H. G., Stadmüller, U. (2005). Generalized functional linear models. *Annals of Statistics, 33*, 774–805.

Olszewski, R. T. (2001). *Generalized Feature Extraction for Structural Pattern Recognition in Time-Series Data.* Ph.D. Thesis, Carnegie Mellon University, Pittsburgh, PA.

Ramsay, J. O., & Silverman, B. W. (2005). *Functional data analysis.* New York: Springer.

Reiss, P. T., & Ogden, R. T. (2007). Functional principal component regression and functional partial least squares. *Journal of the American Statistcal Assosiation, 102*(479), 984–996.

Rodriguez, J. J., Alonso, C. J., & Maestro, J. A. (2005). Support vector machines of interval based features for time series classification. *Knowledge-Based Systems, 18*, 171–178.

Rossi, F., Delannayc, N., Conan-Gueza, B., & Verleysenc, M. (2005). Representation of functional data in neural networks. *Neurocomputing, 64*, 183–210.

Rossi, F., & Villa, N. (2006). Support vector machines for functional data classification. *Neural Computing, 69*, 730–742.

Rossi, N., Wang, X., Ramsay, J. O. (2002). Nonparametric item response function estimates with EM algorithm. *Journal of Educational and Behavioral Statistics, 27*, 291–317.

Saporta, G. (1981). *Methodes exploratoires d'analyse de donn'ees temporelles.* Ph.D. thesis, Cahiers du Buro.

Shmueli, G. (2010). To explain or to predict? *Statistical Science, 25*(3), 289–310.

Incremental Generalized Canonical Correlation Analysis

Angelos Markos and Alfonso Iodice D'Enza

Abstract Generalized canonical correlation analysis (GCANO) is a versatile technique that allows the joint analysis of several sets of data matrices through data reduction. The method embraces a number of representative techniques of multivariate data analysis as special cases. The GCANO solution can be obtained noniteratively through an eigenequation and distributional assumptions are not required. The high computational and memory requirements of ordinary eigendecomposition makes its application impractical on massive or sequential data sets. The aim of the present contribution is twofold: (a) to extend the family of GCANO techniques to a split-apply-combine framework, that leads to an exact implementation; (b) to allow for incremental updates of existing solutions, which lead to approximate yet highly accurate solutions. For this purpose, an incremental SVD approach with desirable properties is revised and embedded in the context of GCANO, and extends its applicability to modern big data problems and data streams.

1 Introduction

Numerous procedures for relating multiple sets of variables have been described in the literature (Van der Burg 1988; Gifi 1990; Kroonenberg 2008). In this paper, we consider a generalized version of canonical correlation analysis (GCANO) developed by Carroll (1968). The central problem of GCANO is to construct a series of components, or canonical variates, aiming to maximize the association or homogeneity among the multiple variable sets. This version is most attractive because the solution can be obtained through an eigenequation, strict distributional

A. Markos (✉)
Department of Primary Education, Democritus University of Thrace, Xanthi, Greece
e-mail: amarkos@eled.duth.gr; amarkos@gmail.com

A.I. D'Enza
Department of Economics and Law, Università di Cassino e del Lazio Meridionale, Cassino FR, Italy
e-mail: iodicede@unicas.it; iodicede@gmail.com

© Springer International Publishing Switzerland 2016
A.F.X. Wilhelm, H.A. Kestler (eds.), *Analysis of Large and Complex Data*, Studies in Classification, Data Analysis, and Knowledge Organization,
DOI 10.1007/978-3-319-25226-1_16

185

assumptions are not required and, most importantly, the method subsumes a number of representative techniques of multivariate data analysis as special cases (Takane et al. 2008; Van de Velden and Takane 2012). In the case of two data sets with continuous variables, GCANO reduces to canonical correlation analysis. When one of the two sets of variables consists of indicator variables, the method specializes into canonical discriminant analysis and into correspondence analysis when both sets consist of indicator variables. In the case of more than two data sets with indicator variables, GCANO specializes into multiple correspondence analysis, and into principal component analysis when each of the data sets consists of a single continuous variable. Therefore, a useful modification of the GCANO algorithm has far reaching implications beyond what is normally referred to as GCANO (Takane et al. 2008).

GCANO has been profitably applied as a tool for integrating data obtained from different sources (e.g., subjects, stimuli, locations), in fields ranging from marketing (Bijmolt and Van de Velden 2012) to neuroimaging (Correa et al. 2010). In the last few years, new application frameworks emerged that usually involve large/massive amounts of data. Some of the examples that can be given in this context are the continuous monitoring of consumer preferences plotted on perceptual maps, fusing data concurrently acquired from different imaging modalities, and monitoring of word associations that are present in data pulled on-the-fly from social networking sites. In all these examples there is a high rate of data accumulation coupled with constant changes in data characteristics. Such type of data is often referred to as data streams or data flows and require fast response time and efficient memory use.

In fact, the capability of GCANO is challenged by such data and requires a different approach. The problem is that finding the (generalized) canonical correlations requires a computationally expensive eigendecomposition. More specifically, the application of ordinary eigenvalue decomposition (EVD) or singular value decomposition (SVD) to large and high-dimensional data becomes infeasible because of high computational and memory requirements. This subsequently makes the application of GCANO impractical on massive or sequential data, i.e., when new data arrive, one needs to re-run the method with the original data augmented by the new data and the whole data structures being decomposed have to be kept in memory.

Literature offers several proposals aiming to overcome the EVD and SVD-related limitations via efficient eigensolvers (e.g., Baglama and Reichel 2007) or via the update (or downdate) of existing EVD/SVD solutions according to new data (see Baker et al. 2012 for an overview). Another strategy for tackling large data problems is the so-called split-apply-combine (Wickam 2011): the full data set is split into blocks, each block is analyzed separately and the results are combined to obtain the global solution. In that case, the solution corresponding to the decomposition of the starting data block has to be incrementally updated each time new data comes in.

The aim of the present contribution is twofold: (a) to extend the family of GCANO techniques to a split-apply-combine framework, that leads to an exact solution, i.e. to the exact calculation of canonical correlations; (b) to allow for incremental updates of existing solutions, which lead to approximate yet highly accurate solutions. For this purpose, an incremental SVD approach with desirable properties is revised and embedded in the context of GCANO, and extends its applicability to modern big data problems and data streams.

The paper is organized as follows: Sect. 2 presents GCANO as a matrix decomposition technique. Section 3 focuses on an incremental SVD approach with desirable properties in the context of GCANO. Based on this approach, two algorithmic modifications for incrementally computing the GCANO solution are proposed in Sect. 4. In Sect. 5 we illustrate a real-world application on data gathered from a social networking site. The paper concludes in Sect. 6.

2 Generalized Canonical Correlation Analysis

Let $\mathbf{Z}_k$ denote an n by p_k matrix of variables of the kth data set ($k = 1, \cdots K$), where n is the number of cases. We assume that $\mathbf{Z}_k$ is column-wise standardized. Let $\mathbf{Z}$ denote an n by $p = (\Sigma_k p_k)$ row block matrix, $\mathbf{Z} = [\mathbf{Z}_1, \ldots, \mathbf{Z}_K]$. Let $\mathbf{W}_k$ denote a p_k by d matrix of weights assigned to each variable in $\mathbf{Z}_k$, where d is the number of dimensions. Let $\mathbf{F}$ denote an n by d matrix of low dimensional data representations, known as object scores, which characterize the association or homogeneity among all $\mathbf{Z}_k$'s. The aim of GCANO is to obtain $\mathbf{W}$ which maximizes

$$\phi(\mathbf{W}) = \text{tr}(\mathbf{W}^\mathsf{T}\mathbf{Z}^\mathsf{T}\mathbf{Z}\mathbf{W}) \tag{1}$$

subject to the restriction that $\mathbf{W}^\mathsf{T}\mathbf{D}\mathbf{W} = \mathbf{I}_t$, where $\mathbf{D}$ is a p by p block diagonal matrix formed from $\mathbf{D}_k = \mathbf{Z}_k^\mathsf{T}\mathbf{Z}$ as the kth diagonal block.

The solution can be achieved through different algorithmic approaches (see Takane et al. 2008), but the most relevant one to our purpose is via the generalized singular value decomposition (GSVD) of matrix $\mathbf{Z}\mathbf{D}^-$ with column metric $\mathbf{D}$, where $\mathbf{D}^-$ is a generalized inverse of $\mathbf{D}$. In other words, GCANO can be performed as a principal component analysis applied to the global table $\mathbf{Z}$, with metric corresponding to the block diagonal matrix composed of the inverses of the variance-covariance matrices internal to every group $\mathbf{Z}_k$. This is equivalent to obtaining the ordinary SVD of:

$$\mathbf{Z}\mathbf{D}^{-1/2} = \mathbf{U}\boldsymbol{\Sigma}\mathbf{V}^\mathsf{T}. \tag{2}$$

Then canonical weights are given by $\mathbf{W} = \mathbf{D}^{-1/2}\mathbf{V}$ and canonical variates by $\mathbf{F} = \mathbf{Z}\mathbf{W}\boldsymbol{\Sigma}^{1/2}$ (Takane et al. 2008).

A convenient choice of $\mathbf{D}^{-1/2}$ is $\mathbf{D}^-\mathbf{D}^{1/2}$, where $\mathbf{D}^-$ is an arbitrary g-inverse and $\mathbf{D}^{1/2}$ is the symmetric square root factor of $\mathbf{D}$. Different choices of $\mathbf{D}^{-1/2}$ lead to the different methods which lie under the GCANO framework. For instance, in the case of correspondence analysis $\mathbf{D}^{-1/2}$ is replaced by $(\mathbf{D}^+)^{1/2}$, where $\mathbf{D}^+$ is the Moore–Penrose inverse of $\mathbf{D}$ and each $\mathbf{Z}_k$ is a set of column-centered indicator matrices. In the case of principal component analysis, the solution is simply given by the SVD of $\mathbf{Z}$; thus $\mathbf{D}^{-1/2} = \mathbf{I}$.

3   Enhancing the SVD Computation

A downside of SVD and related EVD is their high computational cost. The SVD has a computational complexity of $\mathcal{O}(n^2 p)$, assuming $n \geq p$ (Golub and Van Loan 1996). Therefore, both methods become computationally infeasible for large data sets. In the literature, we find two broad classes of methods which lead to efficient eigendecompositions.

The first class, known as batch methods, requires that all the data is available in advance to perform the decomposition. This class includes methods such as iterative EVD (Golub and Van Loan 1996) and bilinear diagonalization Lanczos (Baglama and Reichel 2007). Although these methods enable very efficient computations, their application is not possible in cases where the data size is too large to fit in memory, or if the full data set is not available in advance, as in the case of data streams. In the latter case, it may be advantageous to perform the computations as the data become available.

The so-called incremental methods can operate on streaming data and aim to update (or downdate) an existing SVD or EVD solution when new data is processed. This class of methods includes some expectation-maximization (e.g., Tipping and Bishop 1999), stochastic approximation (e.g., Herbster and Warmuth 2001), and sequential decomposition approaches (see Baker et al. 2012; Iodice D' Enza and Markos 2015, for an overview). These methods have an advantage over batch methods as they can be applied to sequential data blocks without the need to store past data in memory.

In this paper, we utilize the desirable properties of a sequential decomposition algorithm, (herein referred to as Incremental SVD) described by Ross et al. (2008), which is based on a series of efficient block updates instead of a full and expensive SVD. The procedure allows to keep track of the data mean, so as to simultaneously update the center of the low dimensional space of the solution. This property is important when the data sets consist of indicator variables, as in the case of correspondence analysis (Iodice D' Enza and Markos 2015). Second, the method offers a computational advantage over alternatives in that the decomposition can be computed in constant time regardless of data size. This property makes it more appealing for an incremental GCANO implementation in the case of data streams.

3.1   Incremental SVD

In this section, we present an algorithm which exploits the fact that a low-rank update to the eigenbasis is decomposable into efficient block operations. In the original description of the method (Ross et al. 2008), the data is updated column-wise, but here we derive a row-wise update formulation of the method. This is subsequently utilized in the following section to provide an incremental GCANO approach, suitable for analyzing data streams.

Before describing the core of the method, we introduce some necessary definitions. An *eigenspace* is a collection of the quantities needed to define the result of a matrix eigendecomposition as it involves eigenvalues (singular values), eigenvectors (singular vectors), data mean, and size. In particular, with respect to the SVD, for a $n^{(x)} \times p$ matrix $\mathbf{X}$ and a $n^{(y)} \times p$ matrix $\mathbf{Y}$, we define two eigenspaces as

$$\Omega^{(x)} = \left(n^{(x)}, \mu^{(x)}, \mathbf{U}^{(x)}, \boldsymbol{\Sigma}^{(x)}, \mathbf{V}^{(x)} \right) \quad \text{and} \quad \Omega^{(y)} = \left(n^{(y)}, \mu^{(y)}, \mathbf{U}^{(y)}, \boldsymbol{\Sigma}^{(y)}, \mathbf{V}^{(y)} \right).$$

The aim of incremental decomposition is to obtain an eigenspace $\Omega^{(xy)}$ for the matrix $\begin{bmatrix} \mathbf{X} \\ \mathbf{Y} \end{bmatrix}$, using uniquely the information in $\Omega^{(x)}$ and $\Omega^{(y)}$.

The total number of statistical units and the global data mean are easily updated: $n^{(xy)} = n^{(x)} + n^{(y)}$ and $\mu^{(xy)} = \frac{n^{(x)}\mu^{(x)} + n^{(y)}\mu^{(y)}}{n^{(xy)}}$. Adding the eigenspaces acts to rotate the eigenvectors (or singular vectors) and to scale the eigenvalues (or singular values) relating to data spread; furthermore, the new eigenvectors must be a linear combination of the old.

Let $\tilde{\mathbf{X}}$ and $\tilde{\mathbf{Y}}$ be the centered versions of $\mathbf{X}$ and $\mathbf{Y}$, respectively. In order to take into account the varying mean, the row vector $\sqrt{\frac{n^{(x)}n^{(y)}}{n^{(x)}+n^{(y)}}} \left(\mu^{(y)} - \mu^{(x)} \right)$ is added to the $\tilde{\mathbf{Y}}$ matrix. Given the SVD of $\tilde{\mathbf{X}} = \mathbf{U}^{(x)}\boldsymbol{\Sigma}^{(x)}\left(\mathbf{V}^{(x)}\right)^{\mathsf{T}}$, the projection $\mathbf{L}$ of $\tilde{\mathbf{Y}}$ onto the orthogonal basis $\left(\mathbf{V}^{(x)}\right)^{\mathsf{T}}$ is described by:

$$\mathbf{L} = \tilde{\mathbf{Y}} \left(\mathbf{V}^{(x)} \right)^{\mathsf{T}}.$$

Let $\mathbf{H}$ be the component of $\tilde{\mathbf{Y}}$ orthogonal to the subspace spanned by $\left(\mathbf{V}^{(x)}\right)^{\mathsf{T}}$:

$$\mathbf{H} = \tilde{\mathbf{Y}} \left(\mathbf{I} - \left(\mathbf{V}^{(x)}\right)^{\mathsf{T}} \mathbf{V}^{(x)} \right) = \tilde{\mathbf{Y}} - \mathbf{L}\mathbf{V}^{(x)}.$$

$\mathbf{H}$ is decomposed such that an orthogonal matrix $\mathbf{V}^{(h)}$ is obtained, as follows:

$$\mathbf{V}^{(h)} = \text{orth} \left(\tilde{\mathbf{Y}} - \mathbf{L}\mathbf{V}^{(x)} \right).$$

Then, $\begin{bmatrix} \tilde{\mathbf{X}} \\ \tilde{\mathbf{Y}} \end{bmatrix}$ is given by

$$\begin{bmatrix} \tilde{\mathbf{X}} \\ \tilde{\mathbf{Y}} \end{bmatrix} = \left(\begin{bmatrix} \mathbf{U}^{(x)} & \mathbf{0} \\ \mathbf{0} & \mathbf{I} \end{bmatrix} \mathbf{U}^{(r)} \right) \boldsymbol{\Sigma}^{(r)} \left(\mathbf{V}^{(r)} \begin{bmatrix} \mathbf{V}^{(x)} \\ \mathbf{V}^{(h)} \end{bmatrix} \right), \tag{3}$$

where $\mathbf{U}^{(r)}\boldsymbol{\Sigma}^{(r)}\left(\mathbf{V}^{(r)}\right)^{\mathsf{T}}$ is the SVD of $\mathbf{R} = \begin{bmatrix} \boldsymbol{\Sigma}^{(x)} & \mathbf{0} \\ \mathbf{L} & \mathbf{H}\left(\mathbf{V}^{(h)}\right)^{\mathsf{T}} \end{bmatrix}$.

Finally, $\mathbf{U}^{(xy)} = \begin{bmatrix} \mathbf{U}^{(x)} & \mathbf{0} \\ \mathbf{0} & \mathbf{I} \end{bmatrix} \mathbf{U}^{(r)}$, $\boldsymbol{\Sigma}^{(xy)} = \boldsymbol{\Sigma}^{(r)}$ and $\mathbf{V}^{(xy)} = \mathbf{V}^{(r)} \begin{bmatrix} \mathbf{V}^{(x)} \\ \mathbf{V}^{(h)} \end{bmatrix}$.

It should be noted that in many practical situations, it suffices to obtain only $\mathbf{V}^{(xy)}$, the basis for the right singular subspace.

4   Dynamic Modifications of GCANO Solutions

This Section describes two dynamic modifications of the GCANO algorithm, referred to as "Exact" and "Live." The Exact approach is an incremental GCANO which leads to the same exact solution as the one obtained using ordinary GCANO. This approach requires all the data to be available from the start. On the other hand, the Live approach is suitable for analyzing data blocks incrementally as they arrive, but leads to an approximate GCANO solution. The main difference between the two approaches lies in the calculation of the block diagonal matrix $\mathbf{D}$. With regard to the Exact approach, its calculation is based on the "global" variance-covariance matrices internal to every data block, that is, the matrices which correspond to the whole data matrix, which is available in advance. For the Live approach, the whole matrix is unknown and the variance-covariance matrices internal to every data block are approximated by the "local" matrices, that is, the average variance-covariance matrices of the data analyzed insofar. A detailed description of the two implementations follows and the corresponding pseudo-code is provided in Algorithms 1 and 2.

4.1   Exact GCANO

Algorithm 1 summarizes the Exact approach. The procedure is iterated k times, where k is the total number of incoming blocks leading to k updates. The superscript "(x)" indicates a quantity referred to the current block, whereas the superscript "(y)" refers to the incoming block. The updated quantities are then indicated by the superscript "(xy)". The merging of eigenspaces is achieved using the SVD-based method described in Sect. 3.1.

   In terms of time complexity, the Exact GCANO algorithm is data dependent and is expected to yield much lower space complexity than ordinary GCANO. More specifically, computing an eigenspace model of size $n \times p$ using the SVD, usually incurs a computational cost of $O(n^2p)$. Therefore, using the batch approach, the merging of two consecutive eigenspaces requires approximately $O((n^{(x)} + n^{(y)})^2p)$ operations. Now we focus on the parts which dominate the computational complexity of the Exact algorithm. The starting and the incoming eigenspaces (Steps 4 and 8) need a total of $O(((n^{(x)})^2 + (n^{(y)})^2)p)$ operations and eigenspace merging (Step 11) requires at most $O((d + n^{(y)} + 1)^2p)$. Note that d is the number of the largest singular values retained in each step.

Algorithm 1: Exact GCANO

Require: $\mathbf{Z} = [\mathbf{Z}_1, \cdots, \mathbf{Z}_K]$ $\{n \times p, K \text{ sets}\}$
1:   $\mathbf{D}[k, k] = \mathbf{Z}_k^\top \mathbf{Z}_k$ $\{\text{form block-diagonal } \mathbf{D}\}$
2:   $\mathbf{X} = \mathbf{Z}\mathbf{D}^{-1/2} = [\mathbf{X}_1; \mathbf{X}_2; \cdots ; \mathbf{X}_s]$ $\{\text{calculate } \mathbf{X} \text{ and split into } s \text{ blocks}\}$
3:   $\mathbf{X}^{(x)} = \mathbf{X}_1$
4:   $\mathbf{\Omega}^{(x)} = (n^{(x)}, \mu^{(x)}, \mathbf{U}^{(x)}, \mathbf{\Sigma}^{(x)}, \mathbf{V}^{(x)})$ $\{\text{starting eigenspace}\}$
5:   **while** (incoming data block) **do**
6:      $\mathbf{X}^{(y)} = \mathbf{X}_{m+1}$ $\{\text{incoming block}\}$
7:      $\mu^{(y)} = \left(n^{(y)}\right)^{-1} \left(\mathbf{X}^{(y)}\right)^\top \mathbf{1}_{n^{(y)}}$ $\{\text{mean vector}\}$
8:      $\mathbf{\Omega}^{(y)} = (n^{(y)}, \mu^{(y)}, \mathbf{U}^{(y)}, \mathbf{\Sigma}^{(y)}, \mathbf{V}^{(y)})$ $\{\text{new eigenspace}\}$
9:      $n^{(xy)} = n^{(x)} + n^{(y)}$ $\{\text{data size update}\}$
10:     $\mu^{(xy)} = \left[\mu^{(x)}n^{(x)} + \mu^{(y)}n^{(y)}\right]\left(n^{(x)}\right)^{-1}$ $\{\text{mean vector update}\}$
11:     $\mathbf{\Omega}^{(xy)} = \mathbf{\Omega}^{(x)} \oplus \mathbf{\Omega}^{(y)} = (n^{(xy)}, \mu^{(xy)}, \mathbf{U}^{(xy)}, \mathbf{\Sigma}^{(xy)}, \mathbf{V}^{(xy)})$ $\{\text{eigenspace update}\}$
12:     $\mathbf{W}^{(xy)} = \mathbf{D}^{-1/2}\mathbf{V}^{(xy)}$ $\{\text{canonical weights}\}$
13:     $\mathbf{F} = \mathbf{Z}\mathbf{W}^{(xy)}\left(\mathbf{\Sigma}^{(xy)}\right)^{-1/2}$ $\{\text{canonical variates}\}$
14:     $n^{(x)} = n^{(xy)}, \mu^{(x)} = \mu^{(xy)}, \mathbf{U}^{(x)} = \mathbf{U}^{(xy)}, \mathbf{\Sigma}^{(x)} = \mathbf{\Sigma}^{(xy)}, \mathbf{V}^{(x)} = \mathbf{V}^{(xy)}$ $\{\text{update}\}$
15:   **end while**

4.2 Live GCANO

This section introduces the Live GCANO algorithm and discusses some of its basic properties, in comparison with the Exact case of the previous section. The pseudo-code for the Live case is presented in Algorithm 2. Since the data is not available from the beginning, k, the total number of blocks being analyzed, is not defined in advance. The most crucial difference between the Exact and Live approach lies in the computation of the block diagonal matrix $\mathbf{D}$ of the incoming block. In fact, an additional step (Step 7) is required within the loop of Algorithm 2, where the average of the "local" variance-covariance matrices is obtained.

The complexity of the Live algorithm can be summarized in a similar way to that of the Exact case. Each update (Step 13) requires approximately $O((d + n^{(y)} + 1)^2 p)$ flops, versus $O((n^{(x)} + n^{(y)})^2 p)$ with the naive approach. Another important feature is that the decomposition can be computed in constant time by constraining the update block size (see Iodice D' Enza and Markos 2015 for details). The storage required for the $n^{(y)}$ new rows reduces significantly the space complexity to approximately $O((d + n^{(y)} + 1)p)$, down from $O((n^{(x)} + n^{(y)})^2 p)$ required for the naive SVD.

Algorithm 2: Live GCANO

Require: $\mathbf{Z} = [\mathbf{Z}_1; \mathbf{Z}_2; \cdots]$ {incoming data stream}

1: $\mathbf{Z}^{(x)} = \mathbf{Z}_1$

2: $\mathbf{D}^{(x)}[k,k] = \left(\mathbf{Z}_k^{(x)}\right)^{\mathsf{T}} \mathbf{Z}_k^{(x)}$ {form block-diagonal $\mathbf{D}_x$}

3: $\mathbf{X}^{(x)} = \mathbf{Z}^{(x)} \left(\mathbf{D}^{(x)}\right)^{-1/2}$ {calculate $\mathbf{X}^{(x)}$}

4: $\boldsymbol{\Omega}^{(x)} = (n^{(x)}, \mu^{(x)}, \mathbf{U}^{(x)}, \boldsymbol{\Sigma}^{(x)}, \mathbf{V}^{(x)})$ {starting eigenspace}

5: **while** (incoming data block) **do**

6: $\quad \mathbf{Z}^{(y)} = \mathbf{Z}_{m+1}$ {incoming block}

7: $\quad \mathbf{D}^{(xy)}[k,k] = \left(\left(\mathbf{Z}^{(x)}\right)^{\mathsf{T}} \mathbf{Z}^{(x)} + \left(\mathbf{Z}^{(y)}\right)^{\mathsf{T}} \mathbf{Z}^{(y)}\right)/n^{(x)}$ {update block-diagonal $\mathbf{D}$}

8: $\quad \mathbf{X}^{(y)} = \mathbf{Z}^{(y)} \left(\mathbf{D}^{(xy)}\right)^{-1/2}$ {calculate $\mathbf{X}^{(y)}$}

9: $\quad \mu^{(y)} = \left(n^{(y)}\right)^{-1} \left(\mathbf{X}^{(y)}\right)^{\mathsf{T}} \mathbf{1}_{n^{(y)}}$ {mean vector}

10: $\quad \boldsymbol{\Omega}^{(y)} = (n^{(y)}, \mu^{(y)}, \mathbf{U}^{(y)}, \boldsymbol{\Sigma}^{(y)}, \mathbf{V}^{(y)})$ {new eigenspace}

11: $\quad n^{(xy)} = n^{(x)} + n^{(y)}$ {data size update}

12: $\quad \mu^{(xy)} = \left[\mu^{(x)} n^{(x)} + \mu^{(y)} n^{(y)}\right] \left(n^{(x)}\right)^{-1}$ {mean vector update}

13: $\quad \boldsymbol{\Omega}^{(xy)} = \boldsymbol{\Omega}^{(x)} \oplus \boldsymbol{\Omega}^{(y)} = (n^{(xy)}, \mu^{(xy)}, \mathbf{U}^{(xy)}, \boldsymbol{\Sigma}^{(xy)}, \mathbf{V}^{(xy)})$ {eigenspace update}

14: $\quad \mathbf{W}^{(xy)} = \left(\mathbf{D}^{(xy)}\right)^{-1/2} \mathbf{V}^{(xy)}$ {canonical weights}

15: $\quad \mathbf{F}^{(xy)} = \mathbf{Z}^{(xy)} \mathbf{W}^{(xy)} \left(\boldsymbol{\Sigma}^{(xy)}\right)^{-1/2}$ {canonical variates}

16: $\quad n^{(x)} = n^{(xy)}, \mu^{(x)} = \mu^{(xy)}, \mathbf{U}_{(x)} = \mathbf{U}^{(xy)}, \boldsymbol{\Sigma}^{(x)} = \boldsymbol{\Sigma}^{(xy)}, \mathbf{v}^{(x)} = \mathbf{V}^{(xy)}, \mathbf{Z}^{(x)} = \mathbf{Z}^{(y)}$ {update}

17: **end while**

---

5 Application

The two proposed approaches were applied to a real-world data set. For convenience, we consider only sets of categorical variables, thus GCANO results coincide with those of multiple correspondence analysis (MCA). A detailed description of Algorithms 1 and 2 in the case of MCA can be found in Iodice D' Enza and Markos (2015). The data refers to a small corpus of messages or tweets mentioning seven major hotel brands. It was gathered by continuously querying and archiving the Twitter Streaming API service, which provides a proportion of the most recent publicly available tweets, along with information about the user. The data was collected using the twitteR package in R (Gentry 2011). A total of about 10,000 tweets in English were extracted within a time period of 6 consecutive days.

Apart from brand name, two additional variables were considered. A sentiment score was assigned to each tweet by first counting the number of occurrences of "positive" and "negative" words according to a sentiment dictionary and then subtracting the number of occurrences of negative words from the number of positive. Larger negative scores correspond to more negative expressions of sentiment, neutral (or balanced) tweets net to zero, and very positive tweets score larger, positive numbers. A sentiment polarity variable of either "positive +", "neutral +/−" or "negative −" sentiment was finally obtained by simply taking the sign of the sentiment score. Another variable, user visibility or popularity, as measured

(a)   (b)

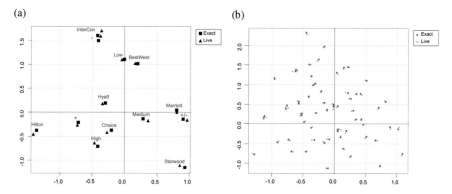

Fig. 1 Exact and Live MCA map of attributes and tweets. (**a**) Attributes. (**b**) Tweets

by the number of followers each user had, was also included in the data set. The variable was categorized into three groups, "low," "medium," and "high."

The potential of the proposed methodology lies in monitoring the competitive position of each brand relative to rival brands over time, as new data blocks are processed, according to standard MCA interpretation. The first block consisted of 500 rows (tweets), and five equally sized blocks were consequently added to update the original solution. In Fig. 1, we plot the final solutions of both approaches (Exact and Live) on the same map for the attributes and tweets, respectively. The similarity between the two configurations was measured by the R index, that equals $\sqrt{1 - m^2}$, where m^2 is the symmetric orthogonal Procrustes statistic. The index ranges from 0 to 1 and can be interpreted as a correlation coefficient. The similarity was found very high ($R = 0.997$ for both tweets and attributes), which indicates that the "Live" approach in this case is highly accurate. An implementation in the R language available on http://www.amarkos.gr/research/dynMCA/ allows the reader to directly experiment with the proposed methods.

6   Conclusions

The applicability of GCANO and related SVD-based methods has been extended to big data settings and data streams. The so-called Exact implementation leads to an exact solution and follows the split-apply-combine paradigm; this is a desirable feature for the incremental processing of previously available data or for parallel execution. The "Live" implementation extends the applicability of GCANO in cases when the whole data set is not available from the beginning, as in the case of data streams. The discrepancy between Live and ordinary approaches, as well as the accuracy of the Live approach are thoroughly discussed in Iodice D' Enza and Markos (2015), albeit only in the case of indicator variables. These properties are left to be studied in the case of continuous variables. Since the proposed incremental approach was based on sequential decomposition, an interesting perspective would be to investigate its relationship with a probabilistic approach, i.e. based on stochastic approximation. We defer consideration of these possibilities to future work.

References

Baglama, J., & Reichel, L. (2007). Augmented implicitly restarted Lanczos bidiagonalization methods. *SIAM Journal on Scientific Computing, 27*, 19–42.

Baker, C., Gallivan, K., & Van Dooren, P. (2012). Low-rank incremental methods for computing dominant singular subspaces. *Linear Algebra and its Applications 436*(8), 2866–2888.

Bijmolt, T. H., & Van de Velden, M. (2012). Multiattribute perceptual mapping with idiosyncratic brand and attribute sets. *Marketing Letters, 23*(3), 585–601.

Carroll, J. D. (1968). A generalization of canonical correlation analysis to three or more sets of variables. In *Proceedings of the 76th Annual Convention of the American Psychological Association* (pp. 227–228).

Correa, N. M., Eichele, T., Adali, T., Li, Y., & Calhoun, V. D. (2010). Multi-set canonical correlation analysis for the fusion of concurrent single trial ERP and functional MRI. *Neuroimage, 50*, 1438–1445.

Gentry, J. (2011). *twitteR: R based twitter client*, http://cran.r-project.org/web/packages/twitteR/

Gifi, A. (1990). *Nonlinear multivariate analysis*. New York: Wiley.

Golub, G., & Van Loan, A. (1996). *Matrix computations*. Baltimore: John Hopkins University Press.

Herbster, M., & Warmuth, M. K. (2001). Tracking the best linear predictor. *Journal of Machine Learning Research, 1*, 281–309.

Iodice D' enza, A., & Markos, A. (2015). Low-dimensional tracking of association structures in categorical data. *Statistics and Computing, 25*(5), 1009–1022.

Kroonenberg, P. M. (2008). *Applied multiway data analysis*. New York: Wiley.

Ross, D., Lim, J., Lin, R. S., & Yang, M. H. (2008). Incremental learning for robust visual tracking. *International Journal of Computer Vision, 77*, 125–141.

Takane, Y., Hwang, H., & Abdi, H. (2008). Regularized multiple-set canonical correlation analysis. *Psychometrika, 73*(4), 753–775.

Tipping, M. E., & Bishop, C. M. (1999). Probabilistic principal component analysis. *Journal of the Royal Statistical Society: Series B (Statistical Methodology), 61*(3), 611–622.

Van de Velden, M., & Takane, Y. (2012). Generalized canonical correlation analysis with missing values. *Computational Statistics, 27*(3), 551–571.

Van der Burg, E. (1988). *Nonlinear canonical correlation and some related techniques*. Leiden: DSWO Press.

Wickam, H. (2011). A split-apply-combine strategy for data analysis. *Journal of Statistical Software 11*(1), 1–29.

Evaluating the Necessity of a Triadic Distance Model

Atsuho Nakayama

Abstract Various studies have examined multi-way proximity generalizations of multidimensional scaling (MDS). Some of these have proposed one-mode three-way proximity data analyses to investigate triadic relationships among three objects. However, the results of a triadic distance model are generally similar to those of a one-mode two-way MDS. Moreover, no technique for judging whether a triadic distance model or one-mode two-way MDS is more appropriate has been developed. Thus, it would be valuable to establish a technique for examining the need for a one-mode three-way MDS analysis. Here, we propose a technique to evaluate the need for a triadic distance model using a log-linear model. When the analysis of the log-linear model shows that three objects, i, j, and k, are not independent, the one-mode three-way proximity data should be analyzed with a triadic distance model. However, one-mode three-way proximity data should not be analyzed with a triadic distance model when the analysis of the log-linear model shows that the three objects i, j, and k are independent.

1 Introduction

Multidimensional scaling (MDS) can be classified according to the number of directions and modes used (Carroll and Arabie 1980). A single-symmetric proximity matrix has I rows and I columns, where I indexes the same ordered set of I objects for both rows and columns. Carroll and Arabie (1980) referred to this kind of single matrix as two-way, because it has both rows and columns; this corresponds to an MDS involving only one input matrix as a two-way analysis. Because both directions correspond to the same set of objects, the model only includes one mode. However, the two modes have two different directions, such as objects and sources. Thus, proximity matrices for objects i and i, according to the k-th source, are considered two-mode three-way matrices. One-mode three-way proximity data consist of numerical values assigned to triplets of objects. Researchers have often

A. Nakayama (✉)
Tokyo Metropolitan University, 1-1 Minami-Ohsawa, Hachioji-shi, 192-0397 Tokyo, Japan
e-mail: atsuho@tmu.ac.jp

© Springer International Publishing Switzerland 2016   195
A.F.X. Wilhelm, H.A. Kestler (eds.), *Analysis of Large and Complex Data*, Studies
in Classification, Data Analysis, and Knowledge Organization,
DOI 10.1007/978-3-319-25226-1_17

used one-mode two-way MDS (Kruskal 1964a,b). One configuration, $\mathbf{X}$ of n points $x_i = (x_{i1}, \ldots, x_{ip})$, is assumed, for $i = 1, \ldots, n$, in p-dimensional Euclidean space, where the x_i-coordinate corresponds to the point for object i. Dyadic distances d_{ij} between two points, representing objects i and j in the configuration, are given by

$$d_{ij} = \left(\sum_{t=1}^{p} \left(x_{it} - x_{jt} \right)^2 \right)^{1/2}. \tag{1}$$

The dyadic distances d_{ij} are determined by finding the $\hat{d}_{ij}$ that satisfies the following conditions:

$$\delta_{ij} < \delta_{rs} \Rightarrow \hat{d}_{ij} > \hat{d}_{rs} \text{ for all } i < j, \ r < s, \tag{2}$$

where δ_{ij} represents the one-mode two-way proximity data. The badness-of-fit measure of d_{ij} to δ_{ij} is called the stress S and is based on the stress formula, defined below (Kruskal and Carroll 1969):

$$S = \sqrt{\sum_{i<j}^{n} (d_{ij} - \hat{d}_{ij})^2 \Big/ \sum_{i<j}^{n} (d_{ij} - \bar{d}_{ij})^2}. \tag{3}$$

However, a model capable of analyzing proximities data that differ from one-mode two-way proximities data is needed. A new model is required to explain high-level phenomena among objects. For example, one-mode three-way proximity data among three objects are used in various areas of sociology. Applications include joint purchases of items, friendships among persons, and trade among countries. We can reveal hidden aspects of such data by fitting certain models to the data. Various models have been proposed for analyzing one-mode three-way proximity data among three-objects, including the triadic distance model (e.g., Cox et al. 1991; De Rooij 2002; De Rooij and Gower 2003; Gower and De Rooij 2003; Joly and Le Calvé 1995; Heiser and BennaniI 1997). De Rooij and Gower (2003) used symmetric functions of triadic distances, including the perimeter distance, generalized Euclidean distance, generalized dominance distance, variance function, area of the triangle, and the product model. The triadic distance among points i, j, and k is denoted by d_{ijk}. As is the case in one-mode two-way MDS, one configuration, $\mathbf{X}$ of n points $x_i = (x_{i1}, \ldots, x_{ip})$, is assumed for $i = 1, \ldots, n$, in p-dimensional Euclidean space, where an x_i-coordinate corresponds to the point for object i. For example, the generalized Euclidean distance model is given by

$$d_{ijk} = (d_{ij}^2 + d_{jk}^2 + d_{ik}^2)^{1/2}, \tag{4}$$

where d_{ij} is the Euclidean dyadic distance between the points i and j. A monotonic regression can be used to find the $\hat{d}_{ijk}$ that satisfies

$$\delta_{ijk} < \delta_{rst} \Rightarrow \hat{d}_{ijk} \geq \hat{d}_{rst} \text{ for all } i < j < k, \ r < s < t, \tag{5}$$

where δ_{ijk} represents one-mode three-way proximity data. The badness-of-fit measure of d_{ijk} to δ_{ijk} is called the stress S and is obtained from the stress formula, defined below (Kruskal and Carroll 1969):

$$S = \sqrt{\sum_{i<j}^{n}(d_{ij} - \hat{d}_{ij})^2 \Big/ \sum_{i<j}^{n}(d_{ij} - \bar{d}_{ij})^2}. \tag{6}$$

The triadic distance d_{ijk} satisfies the following five properties:

$$d_{ijk} \geq 0, \tag{7}$$

$$d_{ijk} = d_{ikj} = \cdots = d_{(\text{every permutation of } i,j,k)}, \tag{8}$$

$$d_{ijk} = 0 \text{ only if } i = j = k, \tag{9}$$

$$d_{iji} = d_{ijj}, \text{ and} \tag{10}$$

$$2d_{ijk} \leq d_{ikl} + d_{jkl} + d_{ijl}. \tag{11}$$

Thus, the three-way distances d_{ijk} must first satisfy non-negativity and three-way symmetry for all i, j, and k. The third condition requires that three-way self-dissimilarities should not differ from zero, and the fourth specifies that, when one object is identical to one of the others, the lack of resemblance between the two non-identical objects should remain invariant regardless of which two are identical. By symmetry, this condition must also be satisfied for $d_{iij} = d_{jji}$, $d_{iji} = d_{ijj}$, $d_{ijj} = d_{jji}$, and so on. The last condition plays a role similar to that of the triangle inequality in the context of two-way distances. The details were formulated by Heiser and BennaniI (1997). In summary, Eq. (4) will be called a three-way distance if and only if Eq. (4) satisfies (7)–(11).

Triadic distance models extend one-mode two-way non-metric MDS and allow visualization of relationships among objects as a way to better understand them. However, Gower and De Rooij (2003) stated that the results obtained from triadic distance models were likely to resemble those obtained from a one-mode two-way MDS. If the results of the triadic distance model are almost consistent with those of one-mode two-way MDS, the reasonable choice for the analysis would be the one-mode two-way MDS because it has simpler restrictions than a triadic distance model. However, no technique for judging whether triadic distance model analysis or one-mode two-way MDS analysis is more appropriate has been available. Here, we propose a technique that examines the appropriateness of these analyses using a log-linear model.

2   The Technique

We discuss the relationships between the triadic distances and a three-way contingency table. A three-way contingency table is a cross-classification of observations by the levels of three categorical variables. Thus, one-mode three-way proximity data are a special case in which the three categorical variables are the same in a three-way contingency table. A two-way contingency table is also a cross-classification of observations by the levels of two categorical variables, and one-mode two-way proximity data are a special case in which the two categorical variables in a two-way contingency table are the same. One-mode three-way proximity data represent the frequencies of co-occurrences of three objects of the first, second, and third ways when the one-mode three-way proximity data are calculated from *objects* × *sources* binary data that display co-occurrences among triadic objects, where the value "1" indicates the presence, and the value "0" indicates the absence of an object. If the one-mode two-way proximity data are calculated from the same binary data, the one-mode two-way proximity data represents the frequencies of co-occurrences between two objects of the first and second ways without objects of the third way. In this context, we regard the one-mode two-way proximity data as the marginal data collapsed by the objects of the third way in the one-mode three-way proximity data.

A three-way $I \times J \times K$ cross-classification of response variables F, S, T has several potential types of interdependence. The interdependence of the variables can be estimated by analyzing the differences given by

$$\pi_{ijk} = \mu \alpha_i^F \alpha_j^S \alpha_k^T \eta_{ij}^{FS} \eta_{jk}^{ST} \eta_{ki}^{FT} \eta_{ijk}^{FST} \tag{12}$$

in a log-linear model. Here, π_{ijk} represents the elements of the three-way contingency tables, μ is a general constant, α denotes the main effects of the variables, and η is the association between variables. Additionally, the log relationship is given by:

$$\log(\pi_{ijk}) = \lambda + \lambda_i^F + \lambda_j^S + \lambda_k^T + \lambda_{ij}^{FS} + \lambda_{jk}^{ST} + \lambda_{ik}^{FT} + \lambda_{ijk}^{FST}, \tag{13}$$

where $\log(\mu) = \lambda$, $\log(\alpha_i^F) = \lambda_i^F$, ... If no restrictions are imposed on the parameters, Eq. (13) specifies a saturated model. A three-way association model (the saturated model) has a three-factor association and is denoted by (*FST*). De Rooij (2002) rewrote Eq. (13) as

$$\log(\pi_{ijk}) = \lambda_{ijk}^I + \lambda_{ijk}^{II}, \tag{14}$$

where $\lambda_{ijk}^I = \lambda + \lambda_i^F + \lambda_j^S + \lambda_k^T$ are the terms that will not be transformed to distances but will be kept in the model, and λ_{ijk}^{II} represents the two-way and three-way association terms that will be transformed to distances. This can be modeled by a monotonic decreasing function of the multiplicative association parameters. The parameters for both the log-linear model and the multiplicative model are not always easily interpretable. Thus, transformation of the parameters into a distance model may enhance interpretability. De Rooij (2002) used a monotone decreasing function of the multiplicative association parameters. A family of transformations is given by the exponential $-p$ similarity function as follows:

$$\eta = \exp(-d^p). \tag{15}$$

For $p \geq 1$, d is a distance satisfying the metric axioms. A small distance corresponds to a large association, and thus to a large number than can be expected based on the marginal parameters (i.e., the set λ_{ijk}^I). A large distance corresponds to a low association, and thus to a smaller number than can be expected from the marginal parameters. We do not use subscripts here because we will apply this transformation to both two-and three-way association parameters. Taking the natural logarithm on both sides, the transformation can be written as $\lambda = d^p$. De Rooij (2002) generalized that observation by including the exponential $-p$ similarity function and other Minkowski metrics as

$$-\lambda_{ijk}^{II} = d_{ijk}^p = d_{ij}^p + d_{jk}^p + d_{ik}^p. \tag{16}$$

De Rooij (2002) transformed all of the two-way and three-way association terms of the model shown in Eq. (13) to a triadic distance model, where $\lambda_{ijk}^{II} = \lambda_{ij}^{FS} + \lambda_{jk}^{ST} + \lambda_{ki}^{TF} + \lambda_{ijk}^{FST}$. The model then becomes

$$\log(\pi_{ijk}) = \lambda_{ijk}^I - d_{ijk}^p = \lambda_{ijk}^I - d_{ij}^p + d_{jk}^p + d_{ik}^p. \tag{17}$$

De Rooij (2002) explained the three two-way associations as triadic distances. They were not modeling any three-way association with the triadic distance models, but only two-way marginal association. Triadic distance models are useful, but they do not model three-way associations. Triadic distance models represent a homogeneous association model.

However, Gower and De Rooij (2003) stated that the results obtained from triadic distance models were likely to resemble those obtained from a one-mode two-way MDS. De Rooij (2002) explained the three two-way associations as triadic distances and assumed a homogeneous association. If the results of the triadic distance model are almost consistent with those of one-mode two-way MDS, it would not be appropriate to represent three two-way associations as triadic distances, and the assumption of a homogeneous association is not appropriate in the three-way contingency table. We should examine other association models that exclude one or more two-way associations or set some associations equal to specified values. The independence and association patterns of the three-way contingency table are equivalent to that of a log-linear model. Here, we use a log-linear model to examine the appropriateness of triadic distance model analysis. The result is that the relationships among the objects may or may not be interdependent. We decide whether the three-way contingency table should or should not be analyzed by a triadic distance model based on the result.

If no restrictions are imposed on the parameters, Eq. (13) specifies a saturated model. The models of interest are constructed to restrict sets of parameters. Accordingly, we compare the saturated model with the restricted models. The restrictions are generally designed to (a) exclude three-way associations and one or more two-way associations and (b) set some associations equal to specified values. A complete independence model, a joint independence model, and a conditional independence model have three, two, and one pair of conditionally independent variables, respectively. In the latter two models, the doubly subscripted terms (such as λ_{ij}^{FS}) pertain to conditionally dependent variables. A homogeneous association model permits all three pairs to be conditionally dependent. If a homogeneous association is appropriate, then one-mode three-way proximity data should be analyzed by a triadic distance model. Effects may change after collapsing over any variable when the model contains all two-factor effects (Agresti 2002). The triadic distance model is preferable to the one-mode two-way MDS. If a homogeneous association is not appropriate, we need a careful examination of the collapsibility of the one-mode three-way proximity data. According to the selected association model, we have to examine which dyadic distances should be excluded in Eq. (17).

3   An Application

We applied the proposed technique to Japanese beer brand-image survey data. A brand-image survey of college students who had taken a course was conducted to assess consumer impressions of various brands on the Japanese beer market. The students were asked to select similar brands from a list of ten brands of beer sold in Japan (Table 1) after watching TV commercials for each of the beers. Ten brands from four companies were used (Table 1). Brands 1, 2, 3, and 4 are "ordinary"

Table 1 The ten beer brands from the four companies

	Taste	Malt	Price	History
Brand 1 (company A)	Mild	Not-all malt	Middle price	Traditional brand
Brand 2 (company A)	Rich	All malt	Middle price	New brand
Brand 3 (company B)	Mild	Not-all malt	Middle price	New brand
Brand 4 (company C)	Mild	Not-all malt	Middle price	Traditional brand
Brand 5 (company D)	Rich	All and pure malt	Premium price	New brands
Brand 6 (company C)	Rich	All and pure malt	Premium price	New brands (resale)
Brand 7 (company A)	Mild	Other than malt	Low price	New brand
Brand 8 (company B)	Mild	Other than malt	Low price	New brand
Brand 9 (company D)	Mild	Other than malt	Low price	New brand
Brand 10 (company C)	Mild	Other than malt	Low price	New brand

malt beers brewed in Japan. Brands 5 and 6 are "premium" beers made from rich, pure malt, using carefully selected ingredients and original brewing methods. The premium beers are more expensive than the others because of their higher-quality ingredients. Brands 7, 8, 9, and 10 constitute a third beer category. The third-category beer is a new kind of alcoholic beverage. It tastes like beer, but is actually brewed from ingredients other than malt. The third-category beer uses ingredients such as corn, soybeans, and peas instead of malt to reduce the cost. Overall, 79 questionnaires were returned and were used as the sample. Based on these results, we selected binary data that displayed the co-occurrence of triplets of beer brands chosen from the ten brands because of similar brand images. One-mode three-and two-way similarity data were calculated from these binary data. The $10 \times 10 \times 10$ one-mode three-way symmetric similarity data indicated the frequencies with which each triplet of brands was chosen based on perceived similarity. The 10×10 one-mode two-way symmetric similarity data indicated the frequencies with which each pair of brands was chosen based on perceived similarity and was not dependent on third objects, so they were considered the collapsed marginal data of the $10 \times 10 \times 10$ one-mode three-way symmetric similarity data.

The one-mode three-way proximity data were analyzed via the log-linear model. We start with the saturated model. The models of interest are constructed to restrict sets of parameters. The restrictions are designed to exclude three-way associations and one or more two-way associations and to set some associations equal to specified values. We compared the saturated model with these restricted models. The results obtained from the log-linear model (Table 2) show that the three objects i, j, and k were not independent. (In Table 2, F denotes the first way, S denotes the second way, and T denotes the third way.) Thus, the triadic distance model is preferable to the one-mode two-way MDS.

To check the validity of the proposed technique, the results obtained from triadic distance model and one-mode two-way MDS were compared. First, the one-mode three-way similarity data were analyzed via a triadic distance model, based on a generalized Euclidian distance model (De Rooij and Gower 2003). Next, the one-mode two-way similarity data were analyzed via one-mode two-way MDS (Kruskal

Table 2 Results of the log-linear model

	Likelihood ratio	Degrees of freedom	P-value	AIC
Complete independence (F, S, T)	4463.6	972	0.000	4519.6
Joint independence (T, FS)	2862.1	891	0.000	3080.1
Joint independence (F, ST)	2862.1	891	0.000	3080.1
Joint independence (S, FT)	2862.1	891	0.000	3080.1
Conditional independence (FS, ST)	1260.5	810	0.000	1640.5
Conditional independence (FS, FT)	1260.5	810	0.000	1640.5
Conditional independence (ST, FT)	1260.5	810	0.000	1640.5
Homogeneous association (FS, ST, FT)	696.1	729	0.805	1238.1
Three-way association (FST)	0.0	1	1.000	2000.0

1964a,b). These analyses were performed using maximum dimensionalities of eight through four and a minimum dimensionality of one. The resulting minimized stress values for the one-mode three-way similarity data in five- to one-dimensional spaces were 0.303, 0.303, 0.325, 0.354, and 0.467, respectively, and the corresponding values for the one-mode two-way similarity data in five- to one-dimensional spaces were 0.000, 0.000, 0.031, 0.144, and 0.280, respectively. We were not able to make a simple comparison between the stress values obtained from the triadic distance model and the one-mode two-way MDS. Thus, the stress values of the one-mode two-way model were calculated using Eq. (6). The converted stress values in five-through one-dimensional spaces from the one-mode two-way MDS were 0.352, 0.355, 0.358, 0.438, and 0.498, respectively. The triadic distance model provided a better fit for the relationships among objects than did the one-mode two-way MDS.

Next, we checked the differences visually between the results of triadic distance model and one-mode two-way MDS. As an example, we compared the results for two dimensions and three dimensions. We were unable to make a simple comparison between the results of the triadic distance model and the one-mode two-way MDS. To compare the results of the triadic distance model MDS with those of the one-mode two-way analysis, the configuration of the one-mode two-way analysis was matched to the configuration of the triadic distance model using Procrustes analysis. Figure 1 shows the two-dimensional configuration of the results for two dimensions and jointly represents the configurations obtained from the triadic distance model and the one-mode two-way MDS. This configuration represents the similarities among the ten brands. The configuration of the triadic distance model reveals the triadic relationships among members of each set of three brands that shared similar impressions. The configuration of the one-mode two-way MDS reveals the dyadic relationships between members of each set of two brands that shared similar impressions. The triadic distance model configuration had almost the same tendencies as the one-mode two-way MDS.

The three-dimensional configuration of the results for three dimensions is presented separately. It is divided into configurations for dimensions 1 and 2 and dimensions 1 and 3. Figure 2a shows a two-dimensional configuration for dimensions 1 and 2 from the results for three dimensions. As in the two-dimensional configuration of the results for two dimensions, there is little difference between the triadic distance model and one-mode two-way MDS configurations. The two configurations show almost the same tendencies in terms of the similarities among the ten brands. Figure 2b shows the two-dimensional configuration for dimensions 1 and 3 from the results for three dimensions. There are some differences between the configurations of the triadic distance model and the one-mode two-way MDS in the analysis of three dimensions. In vertical dimension 3 of the configuration of the triadic distance model, brands generating an impression of high quality, such as Brands 3, 4, 6, 7, and 8, have a negative value, whereas those generating an impression of good taste (Brands 1, 2, 5, 9, and 10) have a positive value. Vertical dimension 3 of the configuration of the triadic distance model represents brands considered to be of high quality versus brands considered to taste good. However, the tendency of vertical dimension 3 for one-mode two-way MDS is less clear than that for the triadic distance model. The positive and negative coordinate values of Brands 1 and 6 are reversed in vertical dimension 3 (see Fig. 2b). The configuration of the one-mode three-way MDS contains information that cannot be expressed by one-mode two-way MDS. For two dimensions (low dimensionality), the triadic relationships found using the triadic distance model were almost consistent with the dyadic relationships obtained from the results of the one-mode two-way MDS. However, the triadic relationships differed from the dyadic relationships for three dimensions (higher dimensionality). Triadic relationships in one-mode three-way similarity data cannot be explained in terms of dyadic relationships based on one-mode two-way MDS. They can only be represented by a triadic distance model. The visual comparisons of the results obtained from the triadic distance model and the one-mode two-way MDS are consistent with the results of the log-linear model.

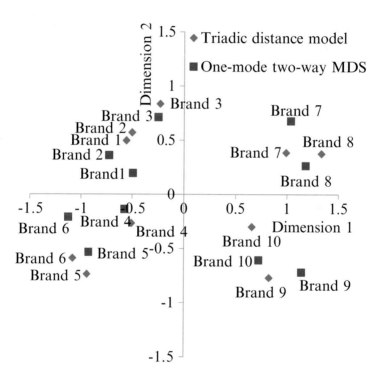

Fig. 1 Two-dimensional configuration obtained from the two-dimensional solution and joint representation of the configuration obtained from triadic distance model and one-mode two-way MDS

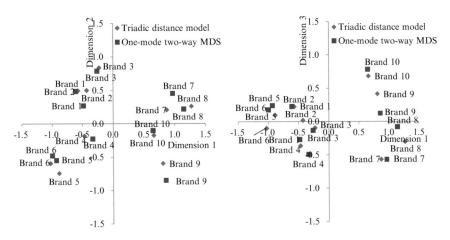

Fig. 2 Two-dimensional view of the three-dimensional configuration. The *left-hand figure* (**a**) defines a plane using dimensions 1 and 2 and presents a joint representation of the configurations obtained from triadic distance model and one-mode two-way MDS. The *right-hand figure* (**b**) defines a plane using dimensions 1 and 3 and presents a joint representation of the configurations obtained from triadic distance model and one-mode two-way MDS

4 Conclusions

We propose a technique for examining the appropriateness of a triadic distance model using a log-linear model. The proposed technique seems to have provided accurate results in the present analysis and can be used successfully to evaluate the need for a triadic distance model. For future study, we are interested in establishing the validity of the proposed technique for various data and would like to consider the possibility of combining the log-linear and distance models in a single model. Application of the log-linear model would be influenced by sample size. However, we have not discussed the relationship between the need for a triadic distance model and the influence of the sample size in the present study. In future research, we would like to investigate the influence of sample size in more detail.

Acknowledgements We express our gratitude to the anonymous referees for their valuable reviews. This work was supported by a Grant-in-Aid for Young Scientists (B) (No. 25730019) from the Japan Society for the Promotion of Science.

References

Agresti, A. (2002). *Categorical data analysis*. New York: Wiley.
Carroll, J. D., & Arabie, P. (1980). Multidimensional scaling. In M. R. Rosenzweig & L. W. Porter (Eds.), *Annual review of psychology* (Vol. 31, pp. 607–649). Palo Alto: Annual Reviews.
Cox, T. F., Cox, M. A. A., & Branco, J. A. (1991). Multidimensional scaling for *n*-tuples. *British Journal of Mathematical and Statistical Psychology, 44*, 195–206.

De Rooij, M. (2002). Distance models for three-way tables and three-way association. *Journal of Classification, 19*, 161–178.

De Rooij, M., & Gower, J. C. (2003). The geometry of triadic distances. *Journal of Classification, 20*, 181–220.

Gower, J. C., & De Rooij, M. (2003). A comparison of the multidimensional scaling of triadic and dyadic distances. *Journal of Classification, 20*, 115–136.

Heiser, W. J., & Bennanii, M. (1997). Triadic distance models: axiomatization and least squares representation. *Journal of Mathematical Psychology, 41*, 189–206.

Joly, S., & Le CalvÉ, G. (1995). Three-way distances. *Journal of Classification, 12*, 191–205.

Kruskal, J. B. (1964a). Multidimensional scaling by optimizing goodness of fit to a nonmetric hypothesis. *Psychometrika, 29*, 1–27.

Kruskal, J. B. (1964b). Nonmetric multidimensional scaling: a numerical method. *Psychometrika, 29*, 115–129.

Kruskal, J. B., & Carroll, J. D. (1969). Geometrical models and badness-of-fit functions. In: P. R. Krishnaiah (Ed.), *Multivariate analysis* (Vol. 2, pp. 639–671). New York: Academic.

Assessing the Reliability of a Multi-Class Classifier

Luca Frigau, Claudio Conversano, and Francesco Mola

Abstract Multi-class learning requires a classifier to discriminate among a large set of L classes in order to define a classification rule able to identify the correct class for new observations. The resulting classification rule could not always be robust, particularly when imbalanced classes are observed or the data size is not large.

In this paper a new approach is presented aimed at evaluating the reliability of a classification rule. It uses a standard classifier but it evaluates the reliability of the obtained classification rule by re-training the classifier on resampled versions of the original data. User-defined misclassification costs are assigned to the obtained confusion matrices and then used as inputs in a Beta regression model which provides a cost-sensitive weighted classification index. The latter is used jointly with another index measuring dissimilarity in distribution between observed classes and predicted ones. Both indices are defined in $[0, 1]$ so that their values can be graphically represented in a $[0, 1]^2$ space. The visual inspection of the points for each classifier allows us to evaluate its reliability on the basis of the relationship between the values of both indices obtained on the original data and on resampled versions of it.

1 Introduction

In a classification problem it is common practice testing a wide variety of learning algorithms by varying threshold values and by using different tuning parameters. In that way different classifiers are obtained which can be compared in order to evaluate their predictive ability, which is usually evaluated starting from the confusion matrix. This is a contingency table in which each column represents the observations in a predicted class, while each row represents those in an actual class. Notationally, given a classification problem on L classes observed on n cases, let

L. Frigau (✉) • F. Mola
Department of Business and Economics, University of Cagliari, Cagliari, Italy
e-mail: frigau@unica.it; mola@unica.it

C. Conversano
Department of Mathematics and Informatics, University of Cagliari, Cagliari, Italy
e-mail: conversa@unica.it

© Springer International Publishing Switzerland 2016
A.F.X. Wilhelm, H.A. Kestler (eds.), *Analysis of Large and Complex Data*, Studies in Classification, Data Analysis, and Knowledge Organization,
DOI 10.1007/978-3-319-25226-1_18

Q be a confusion matrix resulting from a classifier k. In this framework rows of Q refer to the true classes, and columns of Q to the predicted ones. By checking rows, the elements $q_{\ell j}$ indicate how many cases have been classified in each predicted class $\hat{\ell}_j$ ($j = 1, \ldots, L$). By checking columns, the elements $q_{i\ell}$ indicate how many cases of each predicted class have been classified as ℓ_i ($i = 1, \ldots, L$). Starting from the confusion matrix Q several measures and approaches have been proposed to evaluate classifier performance (accuracy, sensitivity, specificity, etc.). Likewise, the confusion entropy index (Wei et al. 2010), the global performance index (Freitas et al. 2007), the entropy of a confusion matrix (Van Son 1995), the transmitted information of the classifier (Abramson 1963), and the relative classifier information (Sindhwani et al. 2001) are all measures that have been defined in order to compare classifiers performance on the basis of the misclassification cells obtained from confusion matrices. Among all these measures, accuracy is the most known. It refers to the proportion of true results (both true positives and true negatives) among the total number of cases examined. This measure is very plain, overlooking a lot of information about the costs of different elements of misclassification (Hand and Till 2001).

The goal of this paper is to propose a new approach that enables us to compare performances of several classifiers in the framework of multi-class learning (i.e., when a new observation has to be classified into one, and only one, of L non-overlapping classes). The output is a bivariate classifier performance index obtained from two different measures. The first one refers to a cost-sensitive weighted classification accuracy index. The second one refers to an index measuring the similarity in distribution between the n observations which have been classified in one of the L classes by a classifier and the original distribution of the n cases among the L classes. Both indices are defined in $[0, 1] \in \mathbb{R}$, so that a comparison of different classifier performance can be represented in a $[0, 1]^2$ space. Additionally, introducing a measure which is not one-dimensional allows us to study the reliability of each classifier by re-training the classifier on resampled versions of the original data and computing the convex hull of the area obtained in the 2 dimensions in which values of the bivariate classifier performance index are projected.

The rest of the paper is organized as follows. Section 2 presents the main features of the proposed bivariate classifier performance index and describes the three steps characterizing it, while Sect. 3 concentrates on reliability. Section 4 presents the results of the performance of the proposed approach on real data and Sect. 5 ends the paper with some concluding remarks.

2   The Bivariate Classifier Performance Index

The bivariate classifier performance index derives from a three steps procedure to be carried out for each candidate classifier. They can be briefly identified with: (1) the model-based measurement of classification accuracy; (2) the measurement of the similarity in distribution between observed classes and predicted ones; (3) the visualization of the results of the previous steps in order to assess global classifier performance.

2.1 Model-Based Measurement of Classification Accuracy

In this section, we present a model-based and cost-sensitive index for measuring accuracy of a multi-class classifier. The basic idea is to use the cells of the observed confusion matrix, i.e., the confusion matrix obtained from training a classifier on the original data, within a regression model in order to derive the estimated cost-sensitive classification accuracy. The regression model is firstly estimated using data obtained from simulated confusion matrices which present the same marginal frequencies of the observed confusion matrix but they refer to situations in which a perfect or random classification is observed. Next, cells of the observed confusion matrix are used together with the estimated regression parameters to derive the value of the index. Let $\pi \in [0, 1]$ be a misclassification level, so that $1 - \pi$ is the classification accuracy level. If K different classifiers are considered, K values of π can be observed and those values, defined in $[0, 1]$, can be modeled on the basis of other information related to each classifier. The model specified for π allows us to assess classifier performance through a model-based classification accuracy index.

In a regression modeling framework characterized by a continuous response variable Y defined in $[0, 1]$, data are usually transformed in order to map the domain of Y in the real line and then a standard linear regression analysis is applied. This approach has some shortcomings (see Cribari-Neto and Zeileis 2010), such as heteroskedasticity and difficulties in the interpretation of estimated parameters, which are expressed in terms of the transformed variable instead of the original one. Ferrari and Cribari-Neto (2004) proposed a regression model for continuous variables that assumes values in $[0, 1]$, called *Beta Regression Model*. The assumption of this model is that the response variable is beta-distributed, $Y \sim \text{Beta}(a, b)$ with $a, b > 0$. The authors proposed a particular parameterization of the beta density in order to obtain a regression structure for the mean of the response along with a precision parameter. They showed that, through setting $\mu = a/(a + b)$ and $\phi = a + b$, it is possible to express expectation and variance of Y as $E(Y) = \mu$ and $\text{VAR}(Y) = \mu(1 - \mu)/(1 + \phi)$, respectively. The parameter ϕ conveys a rate of precision because for larger ϕ $\text{VAR}(Y)$ decreases.

The Beta regression model introduced in Ferrari and Cribari-Neto (2004) is applied in the framework of the present study in order to estimate π and, indirectly, $1 - \pi$. Specifically, the goal is to estimate a Beta regression model using a large number of simulated confusion matrices weighted by some proximity measures and misclassification costs, in order to obtain estimated regression parameters and associated π values. Weighting is very important in this framework, because it conveys essential information to the model about the different importance attributed to possible different misclassification levels. Once the model is estimated, it is applied to the confusion matrix resulting from each classifier in order to estimate a *cost-sensitive (model-based) weighted classification index*. For a classifier k ($k =$

$1, \ldots, K$) and assuming $\pi_k \sim \text{Beta}(\mu_k, \phi)$, the Beta regression model is defined as

$$g(\mu_k) = \sum_{i=1}^{L} \sum_{j=1}^{L} \beta_{ij} q_{ij}^k d(\ell_i, \ell_j) = \eta_k \tag{1}$$

where $d(\ell_i, \ell_j)$ is a cost-weighted proximity measure as defined in Eq. (2), q_{ij}^k is the frequency of the cell of the i-th row and j-th column of the confusion matrix resulting from the classifier k, and β_{ij} is the model coefficient that expresses the contribution of q_{ij}^k to global misclassification of classifier k. Finally, $g(\cdot)$ is a link function. In Eq. (1) the probit distribution is chosen for specifying the link function $g(\cdot)$, so that the expectation of π_k can be defined as $\mu_k = g^{-1}(\eta_k) = \Phi(\eta_k)$, where $\Phi(\cdot)$ is the cumulative distribution function of a standard normal distribution. As already mentioned, for estimating the β_{ij} in Eq. (1) a large number B of confusion matrices are simulated. A proportion α with $\pi = 0$ and non-zero elements in the diagonal only, and the other proportion $1 - \alpha$ with random assigned elements in order to simulate random classifications, so that $\pi = 1$. A random classified confusion matrix is quite simple to obtain. All confusion matrices stemmed by classifiers have the same marginal row frequencies. In fact, since they come from the same dataset the number of true classes is fixed for all matrices. Hence, it is sufficient to simulate matrices with uniformly distributed rows by setting their marginal row frequencies equal to those of the confusion matrices resulting from the classifiers. Next step consists in excluding diagonal cells from simulated matrices, leaving just cells that convey misclassification information. Additionally, the cells of the simulated confusion matrices are weighted by some proximity measures, which are defined, for all entries q_{ij} (with $i \neq j$) corresponding to off-diagonal elements of confusion matrix, as

$$d(\ell_i, \ell_j) = \begin{cases} \dfrac{\ell_L - \ell_1}{|\ell_i - \ell_j|} w_{ij} & \text{if } x \text{ is numerical} \\[2ex] \dfrac{L-1}{|i-j|} w_{ij} & \text{if } x \text{ is ordinal} \\[2ex] w_{ij} & \text{if } x \text{ is nominal} \end{cases} \tag{2}$$

where w_{ij} is a weight, fixed by the researcher, that specifies the importance in terms of misclassification cost attributed to the proximity level between ℓ_i and ℓ_j. As such, weighting is motivated by the idea of adding information deriving from expert knowledge. Once the simulated matrices are weighted, the model could be fitted through them in order to derive the estimated value $\hat{\mu}_k$ of π_k for the k-th classifier as

$$\hat{\mu}_k = \Phi \left(\sum_{i=1}^{L} \sum_{j=1}^{L} \hat{\beta}_{ij} q_{ij}^k d(\ell_i, \ell_j) \right) \tag{3}$$

$\hat{\mu}_k$ is the model-based classification accuracy index used in the rest of the paper.

2.2 Similarity in Distribution Index

One of the main problem in the framework of classifier performance measurement is the choice of the best classifier once that two (or more) classifiers present the same value of the classification accuracy $1 - \pi$ but the latter derives from different confusion matrices. To define a classifier performance measure that also considers information about the difference in distribution among classifier confusion matrices, a *normalized similarity in distribution index* is considered. It derives from a dissimilarity index introduced by Gini and used, among others, in Rachev (1985). In general, for a L-class classification problem D, the Gini index of dissimilarity in distribution, is defined as

$$D = \sqrt{\frac{1}{L^2 - 1} \sum_{h=1}^{L^2-1} |F_h^{v_1} - F_h^{v_2}|^2} \tag{4}$$

where $F_h^{v_1}$ and $F_h^{v_2}$ are the cumulative frequencies in h of the vectors v_1 and v_2, whereas $\sqrt{L^2 - 1}$ is equal to the maximum value of this index, and it is used to normalize it. D is defined in $[0, 1]$ and is susceptible to change in values as long as one or more observations are assigned to the class j instead of the true class i ($i \neq j$ and $i, j \in \{1, \ldots, L\}$).

In the framework of the bivariate classifier performance index described so far, the dissimilarity in distribution index introduced in Eq. (4) is reformulated in terms of a similarity in distribution index. To this aim, let us consider two confusion matrices, Q_{k_1} and Q_{k_2}, corresponding to classifiers k_1 and k_2, respectively. They refer to a situation in which the value of classification accuracy is the same for both classifiers, even if the two confusion matrices are clearly different. Measuring similarity between Q_{k_1} and Q_{k_2} requires the comparison of each element of the two matrices with those of a common reference matrix $Q_{\max}$. The latter is the matrix which refers to the situation of maximum accuracy so that all predicted values correspond to observed ones. To make such a comparison, the matrices $Q_{\max}$, Q_{k_1} and Q_{k_2} are transformed into vectors $v_{\max}$, v_{k_1}, and v_{k_2} by writing the matrix elements in row-major order. To compute the similarity in distribution for Q_{k_1} and Q_{k_2}, it is necessary to compare the distribution of v_{k_1} and v_{k_2} with that of $v_{\max}$. Considering the difference $1 - D$, where D has been defined in Eq. (4), we define a similarity in distribution index for Q_{k_1} and Q_{k_2} whose values are in $[0, 1]$ as

$$S_{Q_{k_i}} = 1 - \sqrt{\frac{\sum_{h=1}^{L^2-1} |F_h^{v_{k_i}} - F_h^{v_{\max}}|^2}{L^2 - 1}}, \qquad \forall i = 1, 2 \tag{5}$$

2.3   Visualization

Once both values of the *cost-sensitive (model-based) weighted classification index*
introduced in Sect. 2.1 and the *normalized similarity in distribution index* introduced
in Sect. 2.2 are available for each classifier, their values can be projected in a
$[0, 1]^2$ space in order to evaluate their performance from the perspective of both
classification accuracy and similarity in distribution. The possibility of analyzing
classifier performance in a two-dimensional space is very useful since it facilitates
the comparison among different classifiers and allows the user to understand which
of the two considered items (weighted classification and similarity in distribution)
mostly influences classifier performance. Of course, the two-dimensional representation is particularly helpful when the number of considered classifiers is very large.

3   Assessing Reliability

Besides measuring the performance of a classifier on the basis of classification
accuracy and similarity in distribution, it is very important to define its reliability.
The *cost-sensitive (model-based) weighted classification index* can be used to
accomplish this goal also. In fact, the measurement of the performance of a classifier
can be used as a tool in order to define a measure of its reliability. To this purpose,
the basic idea is that applying the same classifier to slightly modified versions of the
original data, we expect that its results are rather similar, so that the closer they are
to each other the more reliable the classifier can be considered. Thus, the proximity
of the results obtained from the same classifier by resampling and measured by
the bivariate classifier performance index of Sect. 2 is considered as a measure
of classifier performance reliability. Formally, if we have p different measures of
classification accuracy of a classifier k (including $\hat{\mu}_k$ and $S_{Q_{k_i}}$) we can measure such
a proximity as the convex hull of a set of points $\mathscr{P}$ in p dimensions. The convex hull
is computed by measuring the intersection of all convex sets containing $\mathscr{P}$. For N
points $p_1, \ldots, p_N$, the convex hull $\mathscr{C}$ is then given by:

$$\mathscr{C} = \left\{ \sum_{j=1}^{N} \lambda_j p_j : \lambda_j \geq 0 \quad \forall j \quad \text{and} \quad \sum_{j=1}^{N} \lambda_j = 1 \right\} \tag{6}$$

In the case of a bivariate index, like the one introduced in Sects. 2.1 and 2.2,
this proximity is measured by the convex hull of a set of points defined in the
Euclidean space obtained with respect to the two dimensions of the bivariate
classifier performance index. In order to obtain this measure of reliability three steps
are necessary:

1. Re-train the classifier B times on resampled versions of the original data;

2. Use the resulting B confusion matrices as inputs for the two indices measuring cost-weighted accuracy and similarity in distribution;
3. Measure the classifier reliability as the area of the convex hull $\mathscr{C}$ of the set of points $\mathscr{P}$ defined by the values of two indices obtained over the B runs.

In our computations, the area of $\mathscr{C}$ is measured with the function `convhulln` implemented in the R package `geometry` (Habel et al. 2014).

4   Real Data Example: Classification of Botany Seeds

During the last decades, one of the most important target for botanists is to call a halt to the loss of plant diversity. To achieve that, two strategies are possible: *In situ* and *ex situ* plant conservation. *In situ* conservation consists in protecting threatened plant species in their natural habitat, whereas *ex situ* conservation consists in protecting them outside their natural habitat. Although the *in situ* conservation strategy is considered the best one for preserving plant diversity, its measures are more expensive than *ex situ* ones. For this reason, in the last two decades, the latter conservation approach has been used more often. Among all *ex situ* methods, the most effective is storage of plant seeds in seed banks. It allows us to save large amounts of genetic material in a small space and with minimum risk of genetic damage. Therefore, several seed banks and other structures have been established. Due to the increasing number of seeds gathered, more attention has been focused on classification of accessions in entry. Manual classification of seeds is still a common practice. It is labor-intensive, subjective, and suffers from inconsistencies and errors. It is also a time-consuming task even for highly specialized botanists, and the increasing number of seeds to classify is making the time spent for classification unbearable. For those reasons, application of statistical classifiers for seeds classification is ever more useful and common. Hence botanists require a tool that helps them to evaluate performance and reliability of classifiers, in order to be able to choose among them.

In this study a dataset containing seven variables and $n = 5712$ cases is considered. The response variable is plant family and has five classes (Cyperaceae, Dipsacaceae, Fabaceae, Iridaceae, Lamiaceae). The other six variables are used as predictors and consist in measurements of colorimetric characteristics of seeds. These are the mean of hue, the saturation, the luminance as well as the red channel, green channel, and blue channel intensity.

To measure classification accuracy and reliability the original data were randomly split into two subsets: a proportion of $0.5 \cdot n$ defines the training set and the remaining observations the test set. The experiment involves three different classifiers: CART-like recursive partitioning (CART), Random Forests (RF), and Support Vector Machines (SVM). The choice of these classifiers is based on the consideration that CART is notably known as unstable in terms of reliability of the classification outcome whereas the other two methods are presumably more reliable and able to provide more accurate classification. The bivariate classification accuracy index and the classifier reliability measured and visualized through the convex hull are used to verify that the approach presented in Sects. 2 and 3 provides new insights for the analyzed dataset.

4.1   Results

When classifying botany seeds the goal was to measure the performance and reliability of three classifiers using the approach discussed above. It is worth to remember that the *cost-sensitive (model-based) weighted classification index* is made up of two measures: (1) the model-based measurement of classification accuracy and (2) the measurement of the similarity in distribution between observed classes and predicted ones.

To obtain the cost-sensitive weighted classification accuracy index as defined in Eq. (3) it is necessary to define a proximity measure between each pair of classes of the response variable. To this purpose, observations of the training set are standardized and the proximity is measured as the normalized Euclidean distance between the centroids related to pairs of response classes. Furthermore, for estimating the coefficients of the Beta regression model introduced in Eq. (1), $B = 1000$ confusion matrices were simulated, with a proportion $\alpha = 0.5$ of cases of perfect classification ($\pi = 0$) and the same proportion of cases of random classification ($\pi = 1$). The classifier (CART, SVM , or RF) was trained on the training set observations and predicted classes for the test set observations were used to obtain the confusion matrices, which are the input of the Beta regression model estimated according to the specification introduced in Eq. (1). As for the measurement of the similarity in distribution between observed classes and predicted ones, the Eq. (4) was applied to the three confusion matrices obtained by predicting the response classes of the test set observations for the classifiers CART, RF, and SVM , respectively.

Results are summarized in Table 1, where the two above-mentioned measures are compared with other measures which are frequently used to evaluate the accuracy of a classifier, namely: the proportion of data points in the main diagonal of the confusion matrix; the Rand index and the confusion entropy index (Wei et al. 2010). In order to assess reliability of the three classifiers we used the approach explained in Sect. 3. Firstly, we re-trained each classifier on 100 resampled versions of the training set. Next, we used the 100 confusion matrices obtained from each sample as inputs for the two considered accuracy indexes. Finally, we computed the convex hull $\mathscr{C}$ of the area defined by the values of two indexes obtained over the 100 runs as a measure of reliability.

Table 1 Accuracy and reliability results for the Random Forest (RF), Support Vector Machine (SVM), and CART-like recursive partitioning classifiers

Classifier	Diag	Rand	Cen	$(1 - \hat{\pi}_k)$	$\hat{S}_Q$	$\mathscr{C}$
RF	0.687	0.697	0.403	0.833	0.952	0.193
	(0.667)	(0.686)	(0.423)	(0.778)	(0.950)	
SVM	0.677	0.654	0.346	0.810	0.948	0.155
	(0.674)	(0.653)	(0.350)	(0.802)	(0.947)	
CART	0.623	0.618	0.408	0.602	0.943	0.409
	(0.616)	(0.618)	(0.411)	(0.588)	(0.940)	

Notes: diag is the proportion of data points in the main diagonal of the confusion matrix; rand is the Rand index; cen is the confusion entropy index; $(1 - \hat{\pi}_k)$ is the accuracy measure defined in Eq. (3); $\hat{S}_Q$ is the similarity in distribution as defined in Eq. (4); $\mathscr{C}$ is the reliability of a classifier as defined in Eq. (6). Each cell reports the value of the index obtained for test set observations and, in parentheses, the same value obtained as an average from 100 resampled versions of the original data

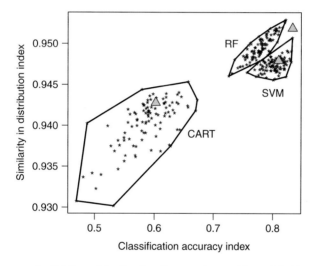

Fig. 1 Accuracy and reliability of the Random Forests, Support Vector Machines, and CART-like recursive partitioning classifiers. The *triangles* correspond to the cost-sensitive (model-based) weighted classification index and the similarity in distribution index obtained from the original data, whereas the *stars* are values of the same indices obtained on resampled versions of the original data. Reliability is measured through the convex hull of the area defined by each set of points

As it is possible to note from both Table 1 and Fig. 1, Random Forest is the best classifier in this example with respect to accuracy. In fact, it has both the highest classification accuracy (0.833) and the highest similarity in distribution (0.952). In contrast, the most reliable classifier is SVM as it provides the smallest convex hull area ($\mathscr{C} = 0.155$). As expected, CART has to be considered as the worst one for both accuracy and reliability.

5   Concluding Remarks

Cost-sensitive classification is one of the mainstream research topics in data mining and machine learning that induces models from data with an unbalanced class distribution and impacts by quantifying and tackling the unbalance. In this paper a bivariate index based on a model-based accuracy measure and a similarity in distribution measure has been introduced. In addition, classifier performance reliability is also considered by computing the convex hull of the set of points in the two-dimensional space defined by the values of the above-mentioned bivariate index computed on resampled versions of the original data. Results obtained for a real data classification problem involving botanic seeds provide evidence about the effectiveness of the proposed approach, since they confirm the expectation that less accurate and less reliable classifiers (CART-like recursive partitioning) do not outperform more robust and accurate ones (SVM and Random Forest). Future research efforts will be directed to the identification and computation of other possible dimensions of accuracy and reliability (like those mentioned in Sect. 1). In addition, following the approach proposed in Müssell et al. (2012), the proposed measures will be framed within the context of Pareto dominance through the visualization of the relative Pareto fronts. Next, our method for measuring cost-sensitive classification accuracy and reliability will be tested on several datasets, with particular attention to multi-class learning problems characterized by an unbalanced distribution of the response classes and/or a reduced data size.

References

Abramson, N. (1963). *Information theory and coding*. New York: McGraw-Hill.

Cribari-Neto, F., & Zeileis, A. (2010). Beta regression in R. *Journal of Statistical Software, 34*(2), 1–24.

Ferrari, S., & Cribari-Neto, F. (2004). Beta regression for modelling rates and proportions. *Journal of Applied Statistics, 31*(7), 799–815.

Freitas, C. O., De Carvalho, J. M., Oliveira, J. R., Aires, S. B., & Sabourin, R. (2007). Confusion matrix disagreement for multiple classifiers. In *Progress in pattern recognition, image analysis and applications* (pp. 387–396). Berlin: Springer.

Habel, K., Grasman, R., Stahel, A., & Sterrat, D. C. (2014). Geometry: Mesh generation and surface tesselation. R package version 0.3-5, http://CRAN.R-project.org/package=geometry

Hand, D. J., & Till, R. J. (2001). A simple generalisation of the area under the ROC curve for multiple class classification problems. *Machine Learning, 45*(2),171–186.

Müssell, C., Lausser, L., Maucher, M., & Kester, H. A. (2012). Multi-objective parameter selection for classifiers. *Journal of Statistical Software, 46*(5), 1–27.

Rachev, S. T. (1985). The Monge-Kantorovich mass transference problem and its stochastic applications. *Theory of Probability and Its Applications, 29*(4), 641–676.

Sindhwani, V., Bhattacharya, P., & Rakshit, S. (2001). Information theoretic feature crediting in multiclass support vector machines. In *Proceedings of the First SIAM International Conference on Data Mining* (pp. 5–7). Philadelphia, PA. SIAM.

Van Son, R. (1995). A method to quantify the error distribution in confusion matrices. In *Proceedings of Eurospeech 95*, Madrid, 22772280.

Wei, J.-M., Yuan, X.-J., Hu, Q.-H., & Wang, S.-Q. (2010). A novel measure for evaluating classifiers. *Expert Systems with Applications, 37(5)*,3799–3809.

Part V
Regression and Other Statistical Techniques

Reviewing Graphical Modelling of Multivariate Temporal Processes

Matthias Eckardt

Abstract Graphical models provide a suitable approach of dealing with uncertainty and complexity by using conditional independence statements and factorizations of joint densities. Static undirected as well as directed graphical models have been applied frequently to pattern analysis, decision modelling, machine learning or image filtering. Several temporal extensions have been published including dynamic Bayesian networks or temporal Markov random fields. Although, graphical models are most commonly used within computer science there has been a growing interest in adjacent disciplines. Recently, a few temporal extensions have been applied to multivariate time series data and event histories.

1 Introduction

Statistical models for the analysis of highly complex data and processes have gained strong attraction within the last decade. Graphical models have been proven to be a sufficient approach of dealing with high dimensionality and uncertainty. A reduction of complexity is achieved by using factorizations of joint densities. Plotting the graph offers a simple and intuitive visualization of the underlying dependence structure. Besides, several statistical models can be seen as special cases of a general graphical model formalism (e.g. mixture models, factor analysis).

Most research on graphical models have focussed on cross-sectional data, henceforth referred to as *static* graphical models. As most prominent cases, this class includes *directed acyclic graphs* (also known as *Bayesian networks*) introduced by Pearl (1988) as well as *undirected graphical models*, which are also labelled as *conditional independence graphs* or *Markov random fields* (cf. Lauritzen 1996). Applications of static models can be found in different scientific fields including machine learning, decision modelling, artificial intelligence or image analysis. Markov random fields have played a prominent role in spatial econometrics and spatial statistics with regard to lattice data. Currently, a strong increase of interest

M. Eckardt (✉)
Institute of Computer Science, Humboldt-Universität zu Berlin, Berlin, Germany
e-mail: eckardtm@cms.hu-berlin.de

© Springer International Publishing Switzerland 2016
A.F.X. Wilhelm, H.A. Kestler (eds.), *Analysis of Large and Complex Data*, Studies in Classification, Data Analysis, and Knowledge Organization,
DOI 10.1007/978-3-319-25226-1_19

221

in resp. usage of (social) network analysis emerged in economics as well as risk assessment or natural sciences. A profound treatment of static graphical models is presented in Lauritzen (1996), Cowell et al. (1999), Edwards (2000), Cox and Wermuth (1996), Pearl (1988), Spirtes (2000), Whittacker (2008) and Koller and Friedman (2010).

Recently, several extensions have been introduced aiming to model temporal dynamics and changing dependence structures including time series data as well as event histories. The objective of this paper is to review these extensions.

2   Graph Theoretic Preliminaries

Let $\mathscr{G} = (\mathscr{V}, \mathscr{E})$ denote a graph with $\mathscr{V} = \{v_1, \ldots, v_k\}$ as finite set of *vertices* and $\mathscr{E} \subseteq \mathscr{V} \times \mathscr{V}$ as set of *edges*—joining the vertices where $\mathscr{E}(\mathscr{G}) \cap \mathscr{V}(\mathscr{G}) = \emptyset$. Two vertices v_i, v_j are called *adjacent* if and only if $(v_i, v_j) \in \mathscr{E}(\mathscr{G})$. Otherwise, v_i, v_j are *non-adjacent*. If an edge e_i consists of a pair of identical nodes (v_i, v_i) we refer to e_i as a loop of $\mathscr{G}$. Furthermore, if a distinct pair (v_i, v_j) is joined by more than one edge, we label this *multiple edges* or *parallel edges*. In order to emphasize graphs containing multiple edges we term any such graph *multigraph*. Similarly, a graph is *simple* if it is not a multigraph. An *undirected* or *unoriented* edge exists if the pairs (v_i, v_j) and (v_j, v_i) are both in the edge set $\mathscr{E}(\mathscr{G})$ given $v_i \neq v_j$. We use $v_j \sim v_i$ to indicate undirected edges. To specify adjacency in case of undirected edges we define the *neighbourhood* as ne$(v_j) = \{v_i : v_j \sim v_i\}$. If all edges in $\mathscr{E}(\mathscr{G})$ are undirected $\mathscr{G}$ is an *undirected graph*. In contrast, an edge is called *directed* or *oriented* if the ordered pair $(v_i, v_j) \in \mathscr{E}(\mathscr{G})$, $v_i \neq v_j$. Thus, only $(v_i, v_j) \in \mathscr{E}(\mathscr{G})$ while $(v_j, v_i) \notin \mathscr{E}(\mathscr{G})$. Directed edges are also referred to as *arcs* and we write $v_i \longrightarrow v_j$ if there is a directed edge from v_i to v_j. Formally, for $v_i \neq v_j$ we define pa$(v_j) = \{v_i : v_i \longrightarrow v_j\}$ as the *parents*. In addition, if $v_i \longrightarrow v_j$ we call $\{v_j\}$ *children* of v_i and define ch$(v_i) = \{v_j : v_i \longrightarrow v_j\}$ (see Kolaczyk 2009; Bondy and Murty 2008). Additionally, let co-pa$(v_i) = \{v_j : \text{ch}(v_j) \cap \text{ch}(v_i) \neq \emptyset\}$ be the *co-parents* of v_i. Hence, if co-pa$(v_i) = v_j$ it follows that v_i and v_j share a common child whereas v_i and v_j have not necessarily to be joined by an edge. Consequently, a *directed* graph or *digraph* is a graph exclusively build on directed edges. A detailed discussion on digraphs is given in Bang-Jensen (2001). In addition, a *multiple digraph* is a multiple graph exclusively build on directed edges. Following Kolaczyk (2009), *mutual* edges are directed parallel edges with diametrically direction (e. g. if $v_i \longrightarrow v_j$ and $v_j \longrightarrow v_i$ are in $\mathscr{E}(\mathscr{G})$) whereas *multi-arcs* are multiple edges with identical orientation.

Let $(v_0, e_1, v_1, e_2, \ldots, v_{k-1}, e_k, v_k)$ be a sequence of vertices and edges of $\mathscr{G}$ with endpoints v_0 and v_k such that $\forall\ e_i, 1 \leq i \leq k$ the pair v_{i-1} and v_i is joined by e_i. We call this sequence w_{ik} of potentially repeating pairs of vertices a *walk* of *length* k in $\mathscr{G}$. If a walk passes through every node of a sequence exactly once we label this as a *path*. A path with identical endpoints is a *cycle* and a cycle of length one is a loop. Thus, a *directed acyclic graph* is a digraph without any directed cycles which is linked to undirected graphs by an operation termed *moralization*. In a first step an undirected edge is inserted joining every co-parents in $\mathscr{G}$. Hereafter, every directed edge in $\mathscr{G}$ is substituted by an undirected edge.

A graph build on directed as well as undirected edges is called a *mixed graph*. As a special case hereof, a *chain graph* is a simple mixed graph without any partially directed cycle where $\mathscr{V}(\mathscr{G})$ is partitioned into k *blocks* $\mathscr{B}_k$ such that $\mathscr{V}(\mathscr{G}) = \mathscr{B}_1 \cup \mathscr{B}_2 \cup \ldots \cup \mathscr{B}_{k-1} \cup \mathscr{B}_k$ and

1. $v_i \longrightarrow v_j$ if and only if $v_i \in \mathscr{B}_i$ and $v_j \in \mathscr{B}_j, i < j$
2. $v_i \sim v_j$ if and only if $v_i \in \mathscr{B}_i$ and $v_j \in \mathscr{B}_i$.

Hence, undirected and directed acyclic graphs are special cases of chain graphs.

3 Causality in Graphical Models

Graphical models relate conditional independence statements among random variables of a multivariate probability distribution to graphs such that the nodes represent random variables and edges encode the dependence structure. Recently, these models have been extended to the time domain based on different definitions of causality. Eichler (2013), Didelez (2011) as well as Eichler and Didelez (2010) have discussed alternative formulations including *Granger causality*, *Sims causality*, *local dependence* and causality in terms of *interventions*. Hereof, Granger causality and local dependence have been most prominent with regard to graphical models. As shown in Florens and Fougere (1996) local dependence can be seen as a continuous time version of Granger causality.

Definition 1 (Granger Causality). Let $\{X(t)\}$ and $\{Y(t)\}$ be stochastic processes on $(\Omega, \mathscr{F}, \mathbb{P})$ where $t \in T \subset \mathbb{Z}$. Given $\{\Omega(t)\}$ as all information in the universe $\{X(t)\} \subset \{\Omega(t)\}$ is causal with respect to $\{Y(t)\}$ if the prediction is less precise based on $\{\Omega(t)\}\backslash\{X(t)\}$ (ct. Granger 1969; Lütkepohl 2005).

Different levels of Granger causal relations have been discussed in Florens and Fougere (1996). Obviously, conditional on all information in the universe seems impracticable and might be replaced by all available information with regard to a vector valued process.

Local dependence has been introduced by Schweder (1970) who focussed on transition intensities in discrete state-space Markov processes. As an extension hereof, Aalen (1987) considered non-causality in continuous time restricted to processes satisfying the Doob–Meyer decomposition.

Definition 2 (Local Independence). Let $\mathscr{F}_t$ denote the information which is available at time $t \in T \subset \mathbb{R}_+ = [0, \infty)$. Then, $\{Y(t)\}$ is locally independent of $\{X(t)\}$ given $\{Z(t)\}$ if the compensator $\Lambda^Y(t)$ of $\{Y(t)\}$ remains unchanged whether conditional on $\mathscr{F}_{t-}^{XYZ}$ or conditional on $\mathscr{F}_{t-}^{YZ}$ (ct. Schweder 1970; Aalen 1987; Florens and Fougere 1996).

4   Graphical Modelling of Temporal Processes

Several graphical models have been developed to approach temporal stochastic processes either in the time as well as in the frequency domain. Misleadingly, different models have been named identical. Generally, two different classes can be differentiated with regard to the underlying definition of the nodes. Firstly, the vertex set can encode random variables at different times. Secondly, the components of a vector valued process can be represented by individual nodes which lead to a coarser modelling of the graphical structure.

4.1   Time Series Data

Most of the research regarding graphical modelling of time series data focussed on vector valued stationary processes in discrete time. Different approaches of graphical modelling towards autoregressive processes are summarized in Songsiri et al. (2010).

Static graphical models have been applied to time series data manifold. Chain graph models in which time slices are represented by blocks have been discussed in Lynggarrd and Walther (1993) and Dahlhaus and Eichler (2003). Furthermore, Queen and Smith (1992) and Anacleto and Queen (2013) introduced a *dynamic chain graph* model based on multivariate Bayesian dynamic models.

Extensions of directed acyclic graphs (so-called *dynamic Bayesian networks*) have extensively been treated by Murphy (2002). Defined as a sequence of directed acyclic graphs such that time is displayed in form of stacked time slices these models include hidden Markov models and Kalman filters besides other latent state-space models as special cases. An overview of different models belonging to this class of graphs is given in Barber and Cemgil (2010). Anacleto et al. (2013a,b) extended the multiregression dynamic model as introduced in Queen and Smith (1993) and used directed acyclic graphs to analyse multivariate time series from traffic flows. These graphs have been called *Bayesian dynamic graphical models*.

Several papers have focussed on structural vector autoregressive processes and discussed the sufficiency of learning structural constraints from static graphical models. Moneta (2008) proposed the usage of a structural learning algorithm with regard to directed acyclic graphs in order to obtain the contemporaneous dependence structure. Alternatively, Oxley et al. (2008), Meurk et al. (2007) as well as Penny and Reale (2004) presented a two-step estimation procedure based on a novel graph linkage called *demoralization*. Thereby, the optimal directed acyclic graph is chosen from a list of suitable graphs obtained from an undirected graphical model by the inverse moralization operation.

Additionally, sequences of undirected Gaussian graphical models arranged in independent and identically distributed blocks have been used by Talih and Hengartner (2005), Talih (2003) and Cai and Li (2012) to capture changing dependence structures of multivariate time series. Thus, a new block only emerges if a new edge is included or excluded into $\mathscr{E}(\mathscr{G})$ such that consecutive blocks contain different graphs. Xuan and Murphy (2007) presented a similar approach. Gao and Tian (2010) proposed a mixed graphical model called *latent ancestral graph* to model latent variables in case of structural vector autoregressive processes.

A first approach focussing on graphical models related to components of vector valued stationary time series in discrete time has been presented by Brillinger (1996) in frequency domain. Dahlhaus (2000) introduced a refined version hereof called *partial correlation graph* based on partial spectral coherence between components of a multivariate time series. The resulting graph is a simple undirected graph in which the nodes correspond to the components of a vector valued process. Partial spectral coherence measures the dependence of two components after removing linear time invariant effect of the remaining series (cf. Brillinger 1981). Thus, two components are conditional orthogonal if the partial error processes are uncorrelated after filtration. This is equivalent to conditional independence only in cases of Gaussian time series. Efficiently, instead of computing the correlation between partial error processes conditional orthogonality can equivalently be achieved from the partial spectra coherence and similarly been read of zero entries of the inverse spectral matrix. Partial correlation graphs have been applied in various fields by Gather et al. (2002), Fried et al. (2004), Feiler et al. (2005) and Allali et al. (2008). Avventi et al. (2013) discussed the usage of these graphs in case of autoregressive moving average processes. Additionally, structural learning has been treated in Bach and Jordan (2004).

Eichler (1999, 2012) has introduced a mixed graph called *Granger causality graph* since it encodes Granger causal as well as contemporaneous relationships between time series components. This graph has also been called *dynamic chain graph* by Murphy (2002). Corander and Villani (2003, 2006) discussed Granger causal graphs from a Bayesian perspective. Additionally, Marttinen and Corander (2009) dealt with the task of Bayesian learning of such graphs. Application of Granger causal graphs are presented in Wild et al. (2010), Allali et al. (2011) and Arnold et al. (2007). Focussing on latent variables Eichler (2010) recently introduced extended Granger causality graphs related to autoregressive moving average processes which he called *dynamic maximal anchestral graphs*. In difference to Granger causality graphs these graphs consist of one additional edge type.

4.2 Event History Data

Based on the concept of local dependence as described in Sect. 3 Didelez (2000) introduced extensions of directed acyclic graphs with regard to counting processes which she termed *local dependence graphs*. These models are defined in case of marked point processes (Didelez 2008) as well as composable finite Markov processes (Didelez 2007). Similarly, Gottard (2007) presented so-called *graphical duration models* as extensions of chain graph models displaying marked point processes. Dreassi and Gottard (2007) dealt with Bayesian estimation of this duration models. This graphs might also allow to model frailty terms besides the hierarchical structures discussed in Gottard and Rampichini (2007). Additionally, Fosen et al. (2006) derived *dynamic path analysis* models based upon Aalens additive hazard model. A further discussion of these models is given in Aalen et al. (2008), whereas the large sample properties are derived in Martinussen (2010). Alternatively, dynamic Bayesian networks build on two time slices have been used to model duration data. Confusingly, Donat et al. (2008, 2010) named these models also *graphical duration models*.

Graphical modelling of counting processes in the frequency domain has been described in Dahlhaus et al. (1997) and Eichler et al. (2003). In both papers partial correlation graphs are used to model intensity functions.

4.3 Beyond Discrete Time

Besides discrete time a limited number of papers have been published with regard to stochastic processes that evolve in continuous time. Shelton et al. (2010) and Nodelman et al. (2002, 2003) introduced continuous time Bayesian networks (CTBN). Alternatively, El-Hay et al. (2006) presented an extended version of undirected graphical models which they called continuous time Markov network (CTMN).

5 Discussion

This review has shown the state of the art and the great variety of graphical models in the temporal domain. Mostly, they are closely related to classical graphical models and satisfy the traditional Markov properties. The main difference between the approaches arises with regard to the inherent definition of the nodes. Relating the vertices to the components of a multivariate process leads to a coarser modelling of the dependence structure. Oppositely, the dimension of the graphical model strongly increases in case of temporally separated nodes as $dim(\mathcal{G}) = N \times T$. This might negatively impact the costs of estimation and computational efficiency, especially in case of high dimensional data structures evolving in time. Contrary, parameter as well as structural estimation can be built on already existing algorithms when applying static models to time series data.

Additionally, Granger causality graphs are strongly effected by the choice of the correct time intervals taking into account. Hence, larger intervals correspond to marginalization over time and create additional correlation. Nonetheless, the graphs are suitable models for structural learning in high dimensional time series and might easily be extended to more complex dimensions. Furthermore, extensions to non-linear relationships can be achieved conditioning on σ-algebras.

References

Aalen, O. O. (1987). Dynamic modelling and causality. *Scandinavian Actuarial Journal, 1987,* 177–190.

Aalen, O. O., Borgan, Ø., & Gjessing, H. K. (2008). *Survival and event history analysis: A process point of view.* Berlin: Springer.

Allali, A., Oueslati, A., & Trabelsi, A. (2008). The analysis of the interdependence structure in international financial markets by graphical models. *International Research Journal of Finance and Economics*, (15), 283–298.

Allali, A., Oueslati, A., & Trabelsi, A. (2011). Detection of information flow in major international financial markets by interactivity network analysis. *Asia-Pacific Financial Markets, 18*, 319–344.

Anacleto, O., & Queen, C. (2013). *Dynamic chain graph models for high-dimensional time series.* Technical Report, The Open University, Milton Keynes.

Anacleto, O., Queen, C., & Albers, C. J. (2013a). Multivariate forecasting of road traffic flows in the presence of heteroscedasticity and measurement errors. *Journal of the Royal Statistical Society: Series C (Applied Statistics), 62*, 251–270.

Anacleto, O., Queen, C., & Albers, C. J. (2013b). Forecasting multivariate road traffic flows using Bayesian dynamic graphical models, splines and other traffic variables. *Australian and New Zealand Journal of Statistics, 55*, 69–86.

Arnold, A., Liu, Y., & Abe, N. (2007). Temporal causal modelling with graphical granger methods. In P. Berkhin, R. Caruana & X. Wu (Eds.), *Proceedings of the 13th ACM SIGKDD International Conference on Knowledge Discovery and Data Mining* (pp. 66–75). New York: ACM.

Avventi, E., Lindquist, A. G., & Wahlberg, B. (2013). ARMA identification of graphical models. *IEEE Transactions on Automatic Control, 58*, 1167–1178.

Bach, F. R., & Jordan, M. I. (2004). Learning graphical models for stationary time series. *IEEE Transactions on Signal Processing, 52*, 2189–2199.

Barber, D., & Cemgil, A.T. (2010). Graphical models for time series. *IEEE Signal Processing Magazine, Special Issue on Graphical Models, 27*, 18–28.

Bang-Jensen, J. (2001). *Digraphs: Theory, algorithms and applications.* Berlin: Springer.

Bondy, J. A., & Murty, U. S. R. (2008). *Graph theory.* Berlin: Springer.

Brillinger, D. R. (1981). *Time series: Data analysis and theory.* New York: Holt, Rinehart and Winston.

Brillinger, D. R. (1996). Remarks concerning graphical models for time series and point processes. *Revista de Econometrica, 16*, 1–23.

Cai, F., & Li, Y. (2012). Modelling changing graphical structure. *International Journal of Engineering and Manufacturing, 2*, 50–57.

Corander, J., & Villani, M. A. (2003). *Causality in vector autoregressions: A Bayesian graphical modelling approach.* Technical Report. Statistiska Institutionen, Stockholms Universitet.

Corander, J., & Villani, M. A. (2006). Bayesian approach to modelling graphical vector autoregressions. *Journal of Time Series Analysis, 27*, 141–156.

Cowell, R. G., Dawid, A. P., Lauritzen, S., & Spiegelhalter, D. J. (1999). *Probabilistic networks and expert systems.* Berlin: Springer.

Cox, D. R., & Wermuth, N. (1996). *Multivariate dependencies: Models, analysis and interpretation.* Boca Raton: Chapman & Hall CRC.

Dahlhaus, R. (2000). Graphical interaction models for multivariate time series. *Metrika, 51*, 157–172.

Dahlhaus, R., & Eichler, M. (2003). Causality and graphical models in time series analysis. In P. J. Green, N. L. Hjort & S. Richardson (Eds.), *Highly structured stochastic systems* (pp. 115–137). Oxford: Oxford University Press.

Dahlhaus, R., Eichler, M., & Sandkühler, J. (1997). *Identification of synaptic connections in neural ensembles by graphical models.* Maastricht: Maastricht University.

Didelez, V. (2000). *Graphical models for event history analysis based on local independence.* Universität Dortmund: Dortmund.

Didelez, V. (2007). Graphical models for composable finite Markov processes. *Scandinavian Journal of Statistics, 34,* 169–185.

Didelez, V. (2008). Graphical models for marked point processes based on local independence. *Journal of the Royal Statistical Society, Series B, 70,* 245–264.

Didelez, V. (2011). Graphical models for stochastic processes. In P. J. Green, N. L. Hjort & S. Richardson (Eds.), *Highly Structured Stochastic Systems* (pp. 138–141). Oxford: Oxford University Press.

Donat, R., Bouillaut, L., Aknin, P., & Leray, P. (2008). A dynamic graphical model to represent complex survival distributions. In T. Bedford, J. Quigley, L. Walls, B. Alkali, A. Daneshkhah, & G. Hardman (Eds.), *Advances in mathematical modelling for reliability* (pp. 17–24). Amsterdam: IOS Press.

Donat, R., Leray, P., Bouillaut, L., & Aknin, P. (2010). A dynamic Bayesian network to represent discrete duration models. *Neurocomputing, 73,* 570–577.

Dreassi, E., & Gottard, A. (2007). A Bayesian approach to model interdependent event histories by graphical models. *Statistical Methods and Applications, 16,* 39–49.

Edwards, D. I. (2000). *Introduction to graphical modelling.* Berlin: Springer.

Eichler, M. (1999). *Graphical models in time series analysis.* Heidelberg: Universität Heidelberg.

Eichler, M. (2010). Graphical Gaussian modelling of multivariate time series with latent variables. *Journal of Machine Learning Research - Proceedings Track, 9,* 193–200.

Eichler, M. (2012). Graphical modelling of multivariate time series. *Probability Theory and Related Fields, 153,* 233–268.

Eichler, M. (2013). Causal inference with multiple time series: Principles and problems. *Philosophical Transactions of the Royal Society A: Mathematical, Physical and Engineering Sciences, 371*(1997), 20110613.

Eichler, M., & Didelez, V. (2010). On Granger-causality and the effect of interventions in time series. *Lifetime Data Analysis, 16,* 3–32.

Eichler, M., Dahlhaus, R., & Sandkühler, J. (2003). Partial correlation analysis for the identification of synaptic connections. *Biological Cybernetics, 89,* 289–302.

El-hay, T., Friedman, N., Koller, D., & Kupferman, R. (2006). Continuous time Markov networks. In R. Dechter & T. Richardson (Eds.), *Proceedings of the Twenty-Second Conference Annual Conference on Uncertainty in Artificial Intelligence (UAI-06)* (pp. 155–164). AUAI Press.

Feiler, S., Müller, K., Müller, A., Dahlhaus, R., & Eich, W. (2005). Using interaction graphs for analysing the therapy process. *Psychotherapy and Psychosomatics, 74,* 93–99.

Florens, J.-P., & Fougere, D. (1996). Noncausality in continuous time. *Econometrica, 64,* 1195–1212.

Fosen, J., Borgan, Ø., Weedon-Fekjær, H., & Aalen, O. O. (2006). Dynamic analysis of recurrent event data using the additive hazard model. *Biometrical Journal, 48,* 381–398.

Fried, R., Didelez, V., & Lanius, V. (2004). Partial correlation graphs and dynamic latent variables for physiological time series. In D. Baier & K.-D. Wernecke (Eds.), *Innovations in classification, data science, and information systems* (pp. 259–266). Berlin: Springer.

Gao, W., & Tian, Z. (2010). Latent ancestral graph of structure vector autoregressive models. *Journal of Systems Engineering and Electronics, 21,* 233–238.

Gather, U., Imhoff, M., & Fried, R. (2002). Graphical models for multivariate time series from intensive care monitoring. *Statistics in Medicine, 21,* 2685–2701.

Gottard, A. (2007). On the inclusion of bivariate marked point processes in graphical models. *Metrika, 66,* 269–287.

Gottard, A., & Rampichini, C. (2007). Chain graphs for multilevel models. *Statistics and Probability Letters, 77,* 312–318.

Granger, C. (1969). Investigating causal relations by econometric models and cross-special methods. *Econometrica, 37,* 424–438.

Kolaczyk, E. D. (2009), *Statistical analysis of network data: methods and models.* Berlin: Springer.

Koller, D., & Friedman, N. (2010). *Probabilistic graphical models.* Cambridge, MA: MIT Press.

Lauritzen, S. (1996). *Graphical models*. Oxford: Oxford University Press.

Lütkepohl, H. (2005). *New introduction to multiple time series analysis*. Berlin: Springer.

Lynggarrd, H., & Walther, K. H. (1993). *Dynamic modelling with mixed graphical association models*. Aalborg: Aalborg University.

Martinussen, T. (2010). Dynamic path analysis for event time data: Large sample properties and inference. *Lifetime Data Analysis, 16*, 85–101.

Marttinen, P., & Corander, J. (2009). Bayesian learning of graphical vector autoregressions with unequal lag-lengths. *Machine Learning, 75*, 217–243.

Meurk, C. S., Brown, J. A., & Reale, M. (2007). Graphical modelling of ecological time series data. In L. Oxley & D. Kulasiri (Eds.), *MODSIM 2007 International Congress on Modelling and Simulation. Modelling and Simulation Society of Australia and New Zealand* (pp. 1393–1398).

Moneta, A. (2008). Graphical causal models and VARs: An empirical assessment of the real business cycles hypothesis. *Empirical Economics, 35*, 275–300.

Murphy, K. (2002). *Dynamic Bayesian networks: Representation, inference and learning*. Berkeley: University of California.

Nodelman, U., Shelton, C. R., & Koller, D. (2002). Continuous time Bayesian networks. In A. Darwiche & N. Friedman (Eds.), *Proceedings of the 18th Conference on Uncertainty in Artificial Intelligence*. San Francisco: Morgan Kaufmann.

Nodelman, U., Shelton, C. R., & Koller, D. (2003). Learning continuous time Bayesian networks. In C. Meek & U. Kjærulff (Eds.), *Proceedings of the 19th Conference on Uncertainty in Artificial Intelligence* (pp. 451–458). San Francisco: Morgan Kaufmann.

Oxley, L., Reale, M., & Wilson, G. T. (2008). *Constructing structural VAR models with conditional independence graphs*. Technical Report, University of Canterbury, Canterbury.

Pearl, J. (1988). *Probabilistic reasoning in intelligent systems: Networks of plausible inference*. San Francisco: Morgan Kaufmann Publishers Inc.

Penny, R. N., & Reale, M. (2004). Using graphical modelling in official statistics. *Quaderni di Statistica, 6*, 31–48.

Queen, C. M., & Smith, J. Q. (1992). Dynamic graphical models. In J. M. Bernado, J. O. Berger, A. P. Dawid & A. F. M. Smith (Eds.), *Bayesian Statistics 4. Proceedings of the 4th Valencia International Meeting* (pp. 741–751). Oxford: Oxford University Press.

Queen, C. M., & Smith, J. Q. (1993). Multiregression dynamic models. *Australian and New Zealand Journal of Statistics, 49*(3), 221–239.

Shelton, C. R., Fan, Y., Lam, W., Lee, J., & Xu, J. (2010). Continuous time Bayesian network reasoning and learning engine. *Journal of Machine Learning Research, 11*, 1137–1140.

Schweder, T. (1970). Composable Markov processes. *Journal of Applied Probability, 7*, 400–410.

Songsiri, J., Dahl, J., & Vandenberghe, L. (2010). Graphical models of autoregressive processes. In D. P. Paloma & Y. C. Eldar (Eds.), *Convex optimization in signal processing and communications* (pp. 89–116). Cambridge: Cambridge University Press.

Spirtes, P. (2000). *Causation, prediction, and search*. Cambridge, MA: MIT Press.

Talih, M. (2003). *Markov random fields on time-varying graphs, with an application to portfolio selection*. Yale University: Yale.

Talih, M., & Hengartner, N. (2005). Structural learning with time-varying components: Tracking the cross-section of financial time series. *Journal of the Royal Statistical Society: Series B (Statistical Methodology), 67*, 321–341.

Whittacker, J. C. (2008). *Grapical models in applied multivariate statistics*. New York: Wiley.

Wild, B., Eichler, M., Friederich, H. C., Hartmann, M., Zipfel, S., & Herzog, W. (2010). A graphical vector autoregressive modelling approach to the analysis of electronic diary data. *BMC Medical Research Methodology, 10*, 28.

Xuan, X., & Murphy, K. (2007). Modeling changing dependency structure in multivariate time series. In Z. Ghahramani (Ed.), *Proceedings of the 24th International Conference on Machine Learning* (pp. 1055–1062). New York: ACM.

The Weight of Penalty Optimization for Ridge Regression

Sri Utami Zuliana and Aris Perperoglou

Abstract A method of weight optimization is introduced when fitting penalized ridge regression models. A penalty term added to a likelihood may be viewed in the light of a hierarchical likelihood. Under this context a method to estimate the variance of a random effect in a mixed model can be employed to obtain an estimate of the penalization weight. We review the theory of ridge penalties from a Bayesian point of view and show how an algorithm for estimating the variance of a random effect can be combined with hierarchical likelihood. The method is compared with other commonly used methods to obtain a penalty weight, such as leave-one-out cross validation, generalized cross validation, penalized quasi-likelihood methods and principal components estimation. Simulation studies are performed to compare the different approaches. For each of the methods we use packages already publicly available in the statistical software R.

1 Introduction

Ridge regression (Hoerl and Kennard 1970; Hoerl et al. 1975) is used in many applications to shrink estimates of coefficients towards zero. It was introduced originally within the family of linear models but also implemented in generalized linear models (Le Cessie and Van Houwelingen 1992), survival analysis (Perperoglou 2014) as well as within the context of high-dimensional data and machine learning.

On all these approaches, a penalty term is added to the likelihood, controlled by a weight λ. It is up to the researcher to decide what should the penalty weight be. A common method used to optimize the penalty is to select a series of different lambdas, fit the model for each of the weights and choose a model that would maximize a criterion such as Akaike's Information Criterion (Akaike 1974), the corrected version (AICc) (Hurvich and Tsai 1989) or Bayesian Information criterion (BIC) (Schwarz 1978). In other cases generalized cross validation (GCV) may be used (Golub et al. 1979). Examples of the latter approach can be found in

S.U. Zuliana (✉) • A. Perperoglou
Department of Mathematical Sciences, University of Essex, Colchester CO4 3SQ, UK
e-mail: sutami@essex.ac.uk; aperpe@essex.ac.uk

© Springer International Publishing Switzerland 2016
A.F.X. Wilhelm, H.A. Kestler (eds.), *Analysis of Large and Complex Data*, Studies in Classification, Data Analysis, and Knowledge Organization,
DOI 10.1007/978-3-319-25226-1_20

Le Cessie and Van Houwelingen (1992) for logistic regression, or in simple linear regression one may use function `lm.ridge` available in package `MASS` (Venables and Ripley 2002) within R (R Core Team 2014) software. More recently, Goeman (2010) suggested leave-one-out cross validation which was implemented in package `penalized` (Goeman 2012).

All of these approaches can be computationally expensive. In more complicated models where estimation time may be an issue, penalty optimization through a grid search of weights is counter-productive. Xue et al. (2007) suggested simple remedies to address the problem, within the framework of survival analysis, which were shown however to be inferior in simulation studies (Perperoglou 2014). Recently, within the field of econometrics Kibria (2003) investigated penalty weights that are obtained by dividing the residual mean square estimate with the maximum, mean, median, etc of the coefficients and came up with suggestions in their follow up paper (Muniz and Kibria 2009). More recently Mansson and Shukur (2011) investigated the performance of these estimators for Poisson regression. Cule and De Iorio (2013) introduced a four-step algorithm to fit penalized models based on principal components of the eigenvectors of the regressors. This approach is implemented in package `ridge` (Cule 2014), for linear and logistic regression.

Here we present an approach that is based on mixed models methodology. We view the penalty as a random effect added to the model and then we employ mixed model machinery to estimate optimal weight. Under that umbrella λ becomes a parameter to be estimated from the model with a repeating algorithm. Our approach is similar to the one suggested by Rigby and Stasinopoulos (2014) for optimizing the penalty weights of smoothing parameters when fitting generalized additive models for scale shape and location. They have implemented their method in package `gamlss` (Rigby and Stasinopoulos 2005).

The paper is organized as follows: In Sect. 2, we present the background theory on penalized regression methods in generalized linear models. We present the general framework and show how to optimize the penalty weight using a mixed models approach. The emphasis is on a special case of a GLM, a simple linear model. We use this simple case to illustrate the Bayesian viewpoint of our suggested algorithm and present simulation studies that evaluate the performance of the suggested algorithm and also compare it with other methods. The paper closes with a discussion.

2   Ridge Regression in Generalized Linear Models

Consider the form of any generalized linear model as:

$$g(y) = \eta = X\beta$$

where y is a response variable coming from any of the exponential distributions, g() is the link function and $\eta = X\beta$ is the linear part of the model for X, an $n \times p$ matrix of p covariates on n observations and β is the vector of unknown coefficients. Let $l(\beta)$ denote the likelihood function of that general model and define the penalized likelihood function as:

$$l^*(\beta) = l(\beta) - \frac{1}{2}\lambda \sum_{i=1}^{p} \beta_i^2$$

To estimate the model an iterative weighted least squares (IWLS) algorithm can be used which takes the form:

$$\hat{\beta} = (X'WX + \lambda I)^{-1} X'Wz$$

where W is a matrix of appropriate weights, z is the intermediate variable given by $z = W^{-1}(y - \hat{\mu}) + X\beta$ and I is a $p \times p$ identity matrix.

The choice of penalty weight is crucial. In cases where λ tends to infinity coefficients become zero, while when λ approaches zero coefficients are allowed to vary freely.

2.1 Ridge Regression from Bayesian Perspective

Any penalized model may be seen as a mixed model. Let $p_\beta(x, y)$ be the joint density function of observed data x and unobserved data y when parameter β is known. We can then define the likelihood for β and y as:

$$L(\beta, y) = p_\beta(x|y)p_\beta(y). \tag{1}$$

Lee and Nelder (1996) defined Eq. (1) as an *h-likelihood* while Green and Silverman (1993) as *penalized likelihood*. h-likelihood can also be seen mathematically as a Bayesian posterior distribution. The first part of the (1) corresponds to the likelihood of the simple model multiplied by the likelihood that corresponds random part, in this case, the ridge penalty. Hierarchical likelihood has many similarities to Bayesian methods.

Consider a simple linear model

$$y = X\beta + \epsilon \tag{2}$$

with X an $n \times p$ matrix of covariates and β a $p \times 1$ vector of coefficients. For simplicity we assume that no constant term β_0 exists in the model. Then where $y \sim N(X\beta, \sigma^2 I)$ and let $\beta \sim N(0, \tau^2 I)$.

Then the likelihood can be written as:

$$L(\beta|y) \propto \exp\left(-\frac{1}{2\sigma^2}(y - X\beta)'(y - X\beta)\right) \exp\left(-\frac{1}{2\tau^2}\hat{\beta}'\hat{\beta}\right) \tag{3}$$

.U. Zuliana and A. Perperoglou

Taking the logarithm of (3) leads to:

$$-\mathrm{log}L(\beta|y) = \frac{1}{2\sigma^2}(y - X\beta)'(y - X\beta) + \frac{1}{2\tau^2}\hat{\beta}'\hat{\beta}$$
$$= \frac{1}{2\sigma^2}\left((y - X\beta)'(y - X\beta) + \lambda\hat{\beta}'\hat{\beta}\right)$$

with $\lambda = \frac{\sigma^2}{\tau^2}$.

Looking at model (2) from a mixed model perspective one needs to estimate, along with the coefficients, the variance of the random effects as well. Schall (1991) defined a two-step algorithm for fitting mixed models and estimating the variance of the random effect. In the first step, given estimates of $\hat{\tau}^2$ the least square estimates of $\hat{\beta}$ can be obtained. In the second step, given estimates of the coefficients, variance estimators are obtained from:

$$\sigma^2 = \frac{(y - X\beta)'(y - X\beta)}{n - \mathrm{ED}}$$

and

$$\tau^2 = \frac{\hat{\beta}'\hat{\beta}}{\mathrm{ED}}$$

where ED stands for effective dimensions and is the trace of the hat matrix of the mode (Hoaglin and Welsch 1978). An estimate of the penalty weight can be then given by:

$$\lambda = \frac{\mathrm{ED}}{\hat{\beta}'\hat{\beta}}$$

The algorithm can be initialized with any value for λ and usually converges within a small number of steps. For further applications see Perperoglou (2014) and Perperoglou and Eilers (2010). An implementation of the method is also part of the `coxRidge` package in R (Perperoglou 2013).

3   Simulation

A simulation study was designed to investigate the performance of different approaches to maximize penalty weight. The sample size of the full data was $n = 500$. The response variable y was simulated from

$$y = \beta z + 0.2\epsilon$$

where z comes from a standard normal distribution ($z \sim N(0, 1)$), and the true value of the coefficient is 1 ($\beta = 1$). Some noise is added in the form of a random vector $\epsilon \sim N(0, 1)$ which is independent of z.

In a second step, the simulated values of z were used to create a set of correlated regressors, given as:

$$x_1 = z + \epsilon_1$$
$$x_2 = z + \epsilon_2$$
$$x_3 = x_1 + x_2 + 0.05\epsilon_3$$

where the errors $\epsilon_1, \epsilon_2, \epsilon_3$ are once again random numbers generated from a normal distribution and assumed to be independent from z. The data set was then split into a training (labelled d_1) and testing data set (labelled d_2), of size $n_1 = 400$ and $n_2 = 100$, respectively, and a linear model of the form $y = \beta_1 x_1 + \beta_2 x_2 + \beta_3 x_3$ was fitted on the training set. A simple linear regression model was fitted to the training data along with five more penalized approaches based on different methods of penalty weight optimization. These approaches were: leave-one-out cross validation using package `penalized`, penalized quasi-likelihood optimization using package `gamlss`, optimization by principal components using package `ridge`, generalized cross validation using package `MASS` and optimization via random effects models suggested here using Schall's algorithm.

Once a model has been fitted, the prediction error on the testing data set was obtained based on the estimates of each approach as

$$\text{p.error} = \sum_{i \in d_2}(y_{i \in d_2} - \hat{\beta}X_{i \in d_2})^2.$$

The whole process was repeated 1000 times.

Figure 1 illustrates the distribution of lambdas as they were obtained by the different methods. As it should be expected, the mixed models approach suggested here is almost identical to the penalized quasi-likelihood optimization. On the other hand, leave-one-out-cross validation produces a wider range of λ values and a higher median value than all other approaches. On the other extreme of the spectrum, principal components optimization leads to very small weights and almost no penalization. Generalized cross validation also selects small penalty weights when compared with mixed models and leave-one-out cross validation.

Including a penalty term not only shrinks estimates towards zero, but in cases where collinearity is present, it reduces mean squared prediction error and corrects coefficient signs. Table 1 illustrates the average prediction error of all approaches. As expected the simple linear model has the largest prediction error. Although the differences amongst the models are small, using our proposed algorithm produces the smallest prediction error (the value is printed in italic). When no penalization is applied, estimates obtained from the ordinary least squares model have an opposite sign from the real one. Table 1 presents in the third column the percentage of

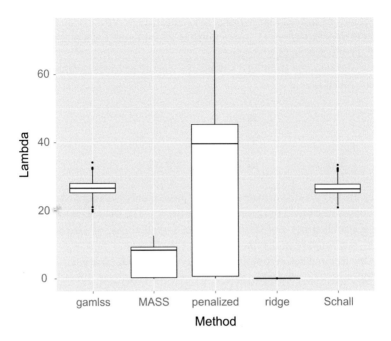

Fig. 1 Distribution of lambdas based on different methods of optimization

Table 1 MSE from correlated count data in different correlation coefficients and different sample sizes

Method	Prediction error	% of $\hat{\beta}_3 < 0$
OLS	37.64	49.7
penalized	37.57	20.4
gamlss	37.55	0
ridge	37.58	0
MASS	37.58	18.2
Schall	37.53	0

cases where $\hat{\beta}_3$ coefficient was mistakenly estimated as negative. Three out of five methods estimate a correct sign for the coefficient.

A second simulation study was also applied to investigate the performance of the methods. This time, the regressors had the same distributional assumptions, however, correlation amongst them was 0. The data were simulated this way to investigate how each method performs when in fact penalization is not necessary. Figure 2 illustrates the distribution of estimated lambdas. The graph reveals that both methods based on extended likelihoods (labelled as Schall and gamlss) overestimate the importance of the penalty. The median lambda weight was 4.8 in both while in the one obtained by cross validation, either leave-one-out or generalized, was 0.2.

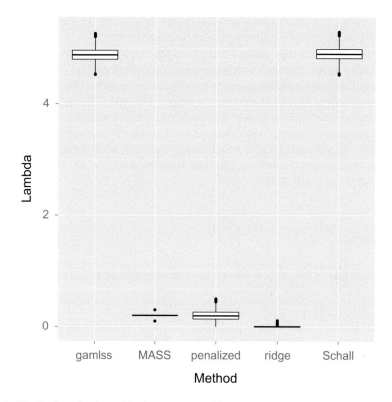

Fig. 2 Distribution of estimated lambdas based on different methods of optimization

4 Discussion

We have introduced a method for optimizing a penalty weight in ridge-type regression problems. The method is based on mixed models algorithms although in practice one does not need to regard the penalization as a random effect. We have shown the theory and illustrated application in two small simulation studies.

The suggested method can work in any type of regression model, regardless of the distribution assumption of the response or the link function. In this work we have shown the advantages of our approach within the context of linear regression. The second author has showed in other texts how the method can be used in survival analysis (Perperoglou 2014). In future work we aim to show how the method performs when fitting Poisson or binary data.

We presented two simulation studies. As discussed earlier, some caution is needed when applying penalized methods in data that do not require that complexity from the model. Cross validation methods were able to perform quite well in the absence of collinearity and showed that lambda has to be near zero, i.e. they ended up with no shrinkage of the coefficients. When mixed models methods were applied, some shrinkage was always present in the model. In any case, preliminary analysis of the data should reveal whether a penalty is needed or not.

It should be noted that using a mixed models approach as the one discussed here is similar to the approach within gamlss models. Both methods use a restricted maximum likelihood(REML) approach to estimate the variance of a random effect, and use that variance to obtain the penalty weight. The only difference is that Rigby and Stasinopoulos (2014) used their approach to optimize a roughness penalty when fitting regression splines for smoothing. An extension of these methods would be very useful in smoothing via a roughness penalty, or when modelling in more than one dimensions. A similar idea has been explored in the PRIDE models (Perperoglou and Eilers 2010).

References

Akaike, H. (1974). A new look at the statistical model identification. *IEEE Transactions on Automatic Control, 19*(6), 716–723.
Cule, E. (2014). ridge: Ridge regression with automatic selection of the penalty parameter. R package version 2.1-3. http://CRAN.R-project.org/package=ridge
Cule, E., & De Iorio, M. (2013). Ridge regression in prediction problems: Automatic choice of the ridge parameter. *Genetic Epidemiology, 37*(7), 704–714.
Goeman, J. (2010). L1 penalized estimation in the Cox proportional hazards model. *Biometrical Journal, 52*(1), 70–84.
Goeman, J. (2012). Penalized R package, version 0.9-42.
Golub, G. H., Heath, M., & Wahba, G. (1979). Generalized cross-validation as a method for choosing a good ridge parameter. *Technometrics, 21*(2), 215–223.
Green, P. J., & Silverman, B. W. (1993). *Nonparametric regression and generalized linear models: A roughness penalty approach.* Boca Raton, FL: CRC Press.
Hoaglin, D. C., & Welsch, R. E. (1978). The hat matrix in regression and ANOVA. *The American Statistician, 32*(1), 17–22.
Hoerl, A. E., & Kennard, R. W. (1970). Ridge regression: Biased estimation for nonorthogonal problems. *Technometrics, 12*(1), 55–67.
Hoerl, A. E., Kennard, R. W., & Baldwin, K. F. (1975). Ridge regression: Some simulations. *Communications in Statistics-Theory and Methods, 4*(2), 105–123.
Hurvich, C. M., & Tsai, C. L. (1989). Regression and time series model selection in small samples. *Biometrika, 76*(2), 297–307.
Kibria, B. M. G. (2003). Performance of some new ridge regression estimators. *Communications in Statistics: Simulation and Computation, 32*(2), 419–435.
Le Cessie, S., & Van Houwelingen, J. C. (1992). Ridge estimators in logistic regression. *Applied Statistics, 41*(1), 191–201.
Lee, Y., & Nelder, J. A. (1996). Hierarchical generalized linear models. *Journal of the Royal Statistical Society, Series B (Methodological), 58*, 619–678.
Mansson, K., & Shukur, G. (2011). A Poisson ridge regression estimator. *Economic Modelling, 28*, 1475–1481
Muniz, G., & Kibria, G. (2009). On some ridge regression estimators: An empirical comparisons. *Communications in Statistics: Simulation and Computation, 38*, 621–630.
Perperoglou, A. (2013). CoxRidge: Cox models with dynamic ridge penalties. R package version 0.9.1. http://CRAN.R-project.org/package=CoxRidge
Perperoglou, A. (2014). Cox models with dynamic ridge penalties on time-varying effects of the covariates. *Statistics in Medicine, 33*(1), 170–180.

Perperoglou, A., & Eilers, P. H. (2010). Penalized regression with individual deviance effects. *Computational Statistics, 25*(2), 341–361.

R Core Team (2014). *R: A language and environment for statistical computing.* Vienna, Austria: R Foundation for Statistical Computing. http://www.R-project.org/

Rigby, R. A., & Stasinopoulos, D. M. (2005). Generalized additive models for location, scale and shape (with discussion). *Journal of the Royal Statistical Society: Series C (Applied Statistics), 54*(3), 507–554.

Rigby, R. A., & Stasinopoulos, D. M. (2014). Automatic smoothing parameter selection in GAMLSS with an application to centile estimation. *Statistical Methods in Medical Research, 23*(4), 318–332.

Schall, R. (1991). Estimation in generalized linear models with random effects. *Biometrika, 78*(4), 719–727.

Schwarz, G. E. (1978). Estimating the dimension of a model. *Annals of Statistics, 6*(2), 461–464.

Venables, W. N., & Ripley, B. D. (2002). *Modern applied statistics with S* (4th ed.). New York: Springer.

Xue, X., Kim, M., & Shore, R. (2007) Cox regression analysis in presence of collinearity: An application to assessment of health risks associated with occupational radiation exposure. *Lifetime Data Analysis, 13*, 333–350.

Monitoring a Dynamic Weighted Majority Method Based on Datasets with Concept Drift

Dhouha Mejri, Mohamed Limam, and Claus Weihs

Abstract Monitoring changes during a learning process is an interesting area of research in several online applications. The most important problem is how to detect and explain these changes so that the performance of the learning model can be controlled and maintained. Ensemble methods have perfectly coped with concept drift. This paper presents an online classification ensemble method designed for concept drift entitled dynamic weighted majority (DWM) algorithm. It adds and removes experts based on their performance and adjusts learner's weights taking into account their age in the ensemble as well as their historical correct prediction. The idea behind this paper is to monitor the classification error rates of DWM based on a time adjusting control chart which adjusts the control limits each time an adjustment condition is satisfied. Moreover, this paper handles datasets with concept drift and analyzes the impact of the diversity of base classifiers, noises, permutations and number of batches. Experiments tested with ANOVA and confirmed by Tukey's test have shown that monitoring classification errors with DWM algorithm has a perfect reaction capacity to different types of concept drift.

1 Introduction

There is surely a lot of research in the classification domain. Most of the proposed techniques in machine learning are based on learning concepts from static data whereas in real life domain issues are changing over time and the target concept

D. Mejri (✉)
Technische Universität Dortmund, Dortmund, Germany

ISG, University of Tunis, Tunis, Tunisia
e-mail: dhouha.mejri@tu-dortmund.de; mejri_dhouha@yahoo.fr

M. Limam
ISG, University of Tunis, Tunis, Tunisia

Dhofar University, Salalah, Oman
e-mail: mohamed.limam@isg.rnu.tn

C. Weihs
Chair of Computational Statistics, Faculty of Statistics, TU Dortmund, Dortmund, Germany
e-mail: claus.weihs@t-online.de; claus.weihs@tu-dortmund.de

© Springer International Publishing Switzerland 2016

A.F.X. Wilhelm, H.A. Kestler (eds.), *Analysis of Large and Complex Data*, Studies in Classification, Data Analysis, and Knowledge Organization,
DOI 10.1007/978-3-319-25226-1_21

to be learned may change accordingly. Indeed, instead of training all the available data from the beginning, data arrives online in instances or batches.

A robust online classification method has been recently proposed by Asensio et al. (2014) based on a streaming ensemble algorithm (SEA) and hyperplane datasets with concept drift. This method called supervised neural constructivist system (SNCS) is analyzed with different characteristics of the dataset and it has been proven that it can function with different real applications of concept drift. Gama and Kosina (2014) proposed a new system monitoring the evolution of a meta-learning algorithm in detecting recurrence of context which self adapts the system according to the degradation process. Their method was evaluated on two real datasets.

Kuncheva (2009) proposed the use of control charts to detect concept drift in streaming data by examining the classification accuracy based on control charts. Her approach is a window resizing technique which uses a variable size adjustment of the window for more sensitivity on the changes. She applies her approach to the Shewhart chart and the sequential probability ratio test (SPRT).

In this paper, we are interested in classification methods of online streaming data. Our aim is to monitor classification error rates of an ensemble method and to detect concept drift based on classifier's performance during an online process. The best solution we propose is to adapt the combination of classifiers in an ensemble with each new batch arriving over time by removing some bad classifiers and adding new ones. There is a need to detect a change without forgetting previous knowledge about the age of the classifiers as well as the past correct prediction in the ensemble and also a need to distinguish between concept drift and out of control situations caused by the non-stationarity of the environment. This paper proposes to use dynamic control charts to monitor misclassification error rates and to detect concept drift when classifier's performance decreases over time. We denote the dynamic control chart time adjusting control limits (TACL) chart for monitoring the error rates of DWM-WIN of Mejri et al. (2013) which is an improved version of Kolter and Maloof (2007)'s algorithm. We discuss an application of detecting concept drifts in SEA datasets with concept drift. We study the impact of the classifier diversity, the noise level, the permutation of the sequences of concept drift, and the number of batches on the DWM-WIN capacity to react to the concept drift.

The paper is outlined as follows: Sect. 2 explains the proposed two-step learning model. Results and analyses on datasets with concept drift are detailed in Sect. 3. Moreover, we analyze the impact of classifier diversity on the DWM-WIN performance on SEA datasets. Section 4 contains our concluding remarks.

2 Two-Step Learning Model

In this section, we explain the main idea behind this paper. The proposed method is based on a two-step learning model. The first step, Step 1, is a learning step, where we learn the model based on an ensemble of classifiers in an incremental way.

The second step, Step 2, is a control step based on a time adjusting control chart which adapts the control limits (CLs) of the control chart after a drift detection. The objective of this step is to control the dynamic learning process of an online classification method in order to detect the concept drift in the data stream.

2.1 Dynamic Learning Process

The proposed method is based on a two-step model. First, the dataset is subdivided into many batches in order to solve the original decision problem. For each incoming example $(\vec{x}, y)$, the current ensemble of classifiers makes a prediction of the class label y. Each classifier has a weight equal to 1 at the beginning. Then, if the classifier makes a correct prediction, the classifier's weight increases based on a parameter $\gamma > 1$. If the classifier makes a wrong prediction, its weight is decreased using a parameter $\beta < 1$. This procedure is applied to all instances in each batch. After many instances, if the weight of one classifier is under a threshold θ, then the classifier is removed from the ensemble. If the global prediction of the ensemble of the classifiers is different from the class label, then a new classifier is installed.

This procedure is repeated for each new batch arriving over time, and each time a concept drift exists in the dataset, until having the best combination of classifiers which copes with the concept drifting data stream. This heuristic highlights the fact that the presence of concept drift in Step 1 perturbed the classification decision of the classifiers existing in the pool and in consequence it requires the assessment of new classifiers and the removal of non-performant ones.

In Step 2, the class label of the model is either 1 if the model is in control due to a correct classification of the data or 0 if the model is out of control due to a misclassification of the data. Accordingly, a data point is $(\vec{x}, 1)$ meaning that the ensemble of classifier made a correct prediction or $(\vec{x}, 0)$ meaning that in Step 1 the ensemble of classifiers misclassified many instances in the batch and made wrong predictions.

2.2 Time Adjusting Control Chart Limits (TACL)

In order to deal with a concept drifting data stream, we assume that the only information available is the misclassification error rates. Our aim is to monitor the error rate and to detect concept drift of the distribution through the behavior of the ensemble of classifiers. For this, a TACL chart based on DWM-WIN error rates is proposed to handle concept drifting data streams and improve phase I and II of the monitoring process. The TACL method can be described as follows.

2.2.1 First Step

We assume that $z_1, z_2, z_3, \ldots z_n$ are the misclassification error rates of each batch in the data. For example z_1 represents the misclassification error rate of the first batch in the data. We compute the mean and the variance of z_i for $i = 1, \ldots, n$ as follows:

$$\mu_n = \overline{z_n}, \quad \sigma_n^2 = \frac{1}{n}\sum_{i=1}^{i=n}(z_i - \overline{z_n})^2. \tag{1}$$

2.2.2 Second Step

Phase I: We consider $z_1, z_2, \ldots, z_n$ as the observations of phase I. The CLs set in Phase I are:

$$\text{UCL} = \overline{z_n} + L\sigma_n, \text{ LCL} = \overline{z_n} - L\sigma_n \tag{2}$$

where L is a constant equal to 3.
Phase II:

• **Check on the condition adjustment**
When a fixed number of instances denoted P is detected to be out of control based on the pth percentile of the batches $i = n + 1, \ldots, n + k$, then the adjustment condition is satisfied. For example, for a process with 800 batches, the fifth percentile is equal to 40. So, the CL is adjusted only after 40 batches were detected out of control.

• **If condition adjustment is satisfied, then adjust**
Decide whether $z_{n+1}, z_{n+2}, z_{n+3}, \ldots, z_{n+k-1}$ where $k \geq 1$ are in or out of control based on the fact that if z_i exceeds the CLs, then the instance is out of control. Then, if the adjustment condition is satisfied, CLs are adjusted by:

$$x_{(n+k)} = \frac{\text{UCL}_{n+k} + \text{LCL}_{n+k}}{2}, \tag{3}$$

$$\text{UCL}_{n+k+1} = \lambda x_{n+k} + (1 - \lambda)z_{n+k} + L\sigma_n, \tag{4}$$

$$\text{LCL}_{n+k+1} = \lambda x_{n+k} + (1 - \lambda)z_{n+k} - L\sigma_n. \tag{5}$$

2.2.3 Third Step

For monitoring Phase II observations, a batch is declared of control if z_t for $t = 1, \ldots, N$, where N is the total number of batches of the monitored process, exceeds a CL.

3   Experimental Results

The proposed method is applied on different variants of SEA dataset of Street and Kim (2001). We use the R package mlr of Bischl et al. (2014) for calling the R classifiers rpart, kknn, and naiveBayes. We use only ensembles based on decision trees (rpart), nearest neighbors models (kknn), and naive Bayes models (naiveBayes), respectively. In the following, we study the impact of the batch size, the noise level, permutations, and the capacity of DWM-WIN on detecting and adapting to the concept drift. All results are summarized in Table 1. We analyze the results by means of ANOVA using Friedman's test (see Tables 2 and 3) and by Tukey's test (see Table 4).

3.1   SEA Concepts

The SEA dataset was first used by Street and Kim (2001) to evaluate the SEA. It was then used by Kolter and Maloof (2005) to test the Add-Exp algorithms. It is downloaded from Stream Data Mining repository (http://www.cse.fau.edu/xqzhu/stream.html) of Zhu (2010).

SEA presents a binary classification problem with 60,000 observations. Features are independent and identically distributed based on a Uniform distribution $U[0, 10]$. The target concept to be learned is determined based on the function $x_1 + x_2 \leq b$, where $b \in \{7, 8, 9, 9.5\}$. Two classes are distinguished, one when this condition is satisfied and one where it is not. First the data is divided into 20, 50, and 100 batches. Four different concepts occur in the data by adaptation of the class labels in SEA dataset when changing the value of b. For the first 250 batches, the target concept is $b = 8$, e.g., for the second concept $b = 9$, the third target $b = 7$, and the fourth $b = 9.5$. We consider all permutations of the ordering of these four concepts.

3.2   Impact of the Permutation on the Concept Drift

Reaction to the Concept Drift   Two types of drift are distinguished: the gradual concept drift and the sudden concept drift. Gradual concept drift is represented by the sequences (7, 8, 9, 9.5) and (9.5, 9, 8, 7). Whereas the other sequences are considered as sudden drift since there is no special characteristic in the concept sequences.

Statistical Tests   According to one way ANOVA given in Table 2, there is a significant difference in the error means between the different permutations. This result is also confirmed by the two way ANOVA (see Table 3) where the interactions between permutation and learner and permutation and noise are significant. To understand the behavior of the permutations, we applied Tukey's test. Results are

Table 1 Mean error rates for different number of batches (20, 50, 100) versus different basic classifiers, N.perm=24, N.rep=100, prob.noise= (0.1, 0.2) for DWM-WIN algorithm

Classifier	rpart						kknn						naiveBayes					
	Noise = 10%			Noise = 20%			Noise = 10%			Noise = 20%			Noise = 10%			Noise = 20%		
1.(7, 8, 9, 9.5)	0.19	0.18	0.19	0.29	0.29	0.29	0.18	0.19	0.19	0.31	0.31	–	0.2	0.19	0.2	0.29	0.29	0.29
2. (7, 8, 9.5 9)	0.22	0.21	0.2	0.29	0.3	0.29	0.19	0.18	0.18	0.31	0.31	0.31	0.32	0.32	0.32	0.28	0.28	0.28
3. (7, 9.5 8, 9)	0.19	0.19	0.22	0.3	0.29	0.31	0.18	0.18	0.18	0.3	0.3	0.3	0.18	0.19	0.19	0.24	0.26	0.25
4. (9.5, 7, 8, 9)	0.195	0.19	0.19	0.28	0.27	0.28	0.17	0.18	0.18	0.27	0.27	–	0.17	0.145	0.15	0.25	0.25	0.25
5. (9.5, 7, 9, 8)	0.19	0.17	0.19	0.3	0.3	0.3	0.155	0.154	–	0.3	0.31	0.31	0.14	0.145	0.15	0.27	0.28	0.28
6. (7, 9.5, 9, 8)	0.17	0.18	0.19	0.32	0.29	0.31	0.17	0.17	0.17	0.3	0.31	–	0.17	0.18	0.17	0.27	0.28	0.27
7. (7, 9, 9.5, 8)	0.18	0.19	0.19	0.26	0.25	0.26	0.16	0.16	0.16	0.26	0.27	0.27	0.17	0.16	0.25	0.25	0.24	0.23
8. (7, 9, 8, 9.5)	0.16	0.18	0.19	0.27	0.28	0.29	0.15	0.15	–	0.27	0.27	0.27	0.17	0.17	0.24	0.25	0.24	0.24
9. (9, 7, 8, 9.5)	0.22	0.22	0.23	0.29	0.28	0.32	0.21	0.2	–	0.3	0.3	–	0.18	0.18	0.19	0.27	0.26	0.27
10. (9, 7, 9.5 8)	0.19	0.2	0.20	0.29	0.27	0.28	0.17	0.17	–	0.3	0.29	–	0.17	0.17	0.17	0.24	0.25	0.24
11. (9, 9.5 7, 8)	0.2	0.2	0.21	0.27	0.29	0.3	0.18	0.18	–	0.28	0.3	–	0.16	0.18	0.16	0.26	0.25	0.26
12. (9.5, 9, 7, 8)	0.19	0.2	0.19	0.27	0.29	0.28	0.18	0.16	–	0.29	0.28	–	0.16	0.15	0.16	0.25	0.25	0.25
13. (9.5, 9, 8, 7)	0.2	0.2	0.2	0.26	0.27	0.26	0.17	0.17	–	0.26	0.27	–	0.18	0.17	0.18	0.24	0.27	0.25
14. (9, 9.5 8, 7)	0.2	0.2	0.2	0.28	0.28	0.28	0.19	0.18	–	0.27	0.28	–	0.17	0.17	0.17	0.26	0.26	0.25
15. (9, 8, 9.5, 7)	0.19	0.2	0.19	0.27	0.27	0.27	0.17	0.17	–	0.27	0.26	–	0.17	0.17	0.17	0.27	0.26	0.26
16. (9, 8, 7, 9.5)	0.18	0.19	0.2	0.27	0.28	0.27	0.175	0.18	–	0.26	0.27	–	0.16	0.17	0.18	0.26	0.27	0.26
17. (8, 9, 7, 9.5)	0.2	0.19	0.2	0.25	0.28	0.25	0.174	0.17	–	0.25	0.24	–	0.17	0.18	0.19	0.22	0.23	0.23
18. (8, 9, 9.5, 7)	0.2	0.2	0.2	0.27	0.28	0.28	0.184	0.18	–	0.28	0.28	–	0.18	0.19	0.18	0.25	0.25	0.26
19. (8, 9.5, 9, 7)	0.19	0.21	0.2	0.3	0.31	0.3	0.17	0.17	0.2	0.31	0.29	–	0.17	0.18	0.19	0.3	0.31	0.3
20. (9.5, 8, 9, 7)	0.21	0.22	0.2	0.3	0.3	0.29	0.19	0.2	–	0.28	0.29	–	0.18	0.18	0.16	0.27	0.27	0.28
21. (9.5, 8, 7, 9)	0.18	0.2	0.21	0.27	0.29	0.3	0.17	0.17	–	0.29	0.27	–	0.16	0.18	0.18	0.26	0.26	0.26
22. (8, 9.5, 7, 9)	0.2	0.2	0.19	0.31	0.3	0.3	0.19	0.19	–	0.32	0.31	–	0.18	0.18	0.18	0.28	0.27	0.28
23. (8, 7, 9.5 9)	0.2	0.22	0.2	0.29	0.3	0.3	0.19	0.18	–	0.29	0.29	–	0.18	0.17	0.18	0.25	0.25	0.25
24. (8, 7, 9, 9.5)	0.2	0.2	0.19	0.29	0.3	0.29	0.16	0.16	–	0.29	0.3	–	0.18	0.17	0.18	0.26	0.27	0.26

Table 2 Friedman test for one way ANOVA

Measure	Df	Sum Sq	Mean Sq	F value	Pr (>F)
p	23	0.323	0.0014	9.1656	$2\,e^{-16}$***
Learner	2	0.02859	0.01429	93.2841	$2\,e^{-16}$***
Noise	1	0.63480	0.63480	4143.0868	$2\,e^{-16}$***
N.batches (20, 50)	1	0.00002	0.00002	0.1389	0.7097
N.batches(20, 50, 100)	2	0.00025	0.00012	1.0534	0.3503
Residuals	254	0.03892	0.00015		

* 0.05
** 0.01
*** 0.001

Table 3 Friedman test for two way ANOVA

Measure	Df	Sum Sq	Mean Sq	F value	Pr (>F)
p vs. learner	46	0.00492	0.00011	1.7223	0.007411**
p vs. noise	23	0.01491	0.00065	10.4332	$2.2\,e^{-16}$***
p vs. batch	23	0.00118	0.00005	0.8286	0.691683
Learner vs. noise	2	0.00778	0.00389	62.6127	$2.2\,e^{-16}$***
Learner vs. batch	2	0.00036	0.00018	2.91	0.057417
Noise vs. batch	1	0.000	0.000	0.0612	0.804962
Residuals	157	0.00975	0.00006		

* 0.05
** 0.01
*** 0.001

Table 4 Tukey's test

	Diff	lwr	upr	p_{adj}
Learner				
kknn vs. rpart	−0.0117	−0.016	−0.007	0
naiveBayes vs. rpart	−0.024	−0.028	−0.0203	0
kknn vs. naiveBayes	−0.0128	−0.017	−0.0085	0
Noise				
10 vs. 20 %	0.094	0.0915	0.0973	0
Batch				
20−50	0.00054	−0.0023	0.0034	0.709

not detailed here because of lack of space. We only analyze the most important cases.

According to Fig. 1, the error rates are relatively stable and the algorithm perfectly deals with the gradual drift by learning the drift and using stored information to adapt the algorithm after each drift detection. Concerning some sudden drifts,

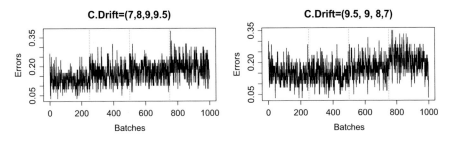

Fig. 1 Reaction capacity of DWM-WIN error rates on gradual drift

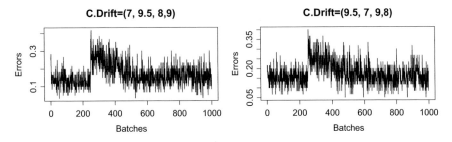

Fig. 2 Reaction capacity of DWM-WIN error rates on sudden drift

DWM-WIN shows a different behavior in the error rates. As shown in Fig. 2, DWM-WIN performs better after the first concept change detection. In fact both figures show that DWM-WIN error rates are stable after the first concept change. Thus, differences between sudden and gradual drifts might be bigger than between gradual changes themselves.

Conclusion DWM-WIN performs very well with concept drift for gradual as well as sudden drift.

3.3 Impact of Varying the Batch Size on the Error Rate

Reaction to the Concept Drift The performance of the algorithm is not affected by the number of batches. The DWM-WIN does not show a significant change in the error rate levels. This is due to the good and quick reaction capacity of DWM-WIN in detecting the concept drift and adapting the classifier's ensemble to this change.

Statistical Tests Based on Friedman's test on one way ANOVA, shown in Table 2, we do not reject the hypothesis that the algorithms have the same performance on average for different number of batches. For this test, we consider two situations. First when testing the difference between three different batch sizes (20, 50, 100) without considering kknn since it doesn't work in all cases when $N = 100$. In this case, we have an F value of 0.1399 and a p-value of 0.7079. In the second situation, we consider DWM-WIN based on rpart, kknn, and naiveBayes but we test the difference only between number of batches of 20 and 50. Results are an F-statistic of 1.053 and a p-value of 0.3505. This result is also confirmed by Tukey's test in Table 4 where lwr indicates the lower end point of the interval and upr indicates the upper end point of the interval. Based on the two way ANOVA the interactions of the number of batches with the other factors are not significant at the 5 % level. Respective p-values in Table 3 are 0.804, 0.69, and 0.0574.

Conclusion DWM-WIN has a noticeable robustness on concept drift for the different batch values. Indeed, it is quite interesting that DWM-WIN quickly adapt itself to the concept drift for small batches (10 instances per batch when the number of batches is 100) as well as for large batches (50 instances per batch when the number of batches is 20).

3.4 Impact of Varying the Noise Level to the Error Rate

Reaction to Concept Drift As expected, changing the level of the noise, shows high impact on the error rate. Changing the noise level from 10 to 20 % impacts the error which increases from 19 to 29 % during the first permutation (7, 8, 9, 9.5) for a number of batches of 20 using an ensemble of rpart in DWM-WIN. For naiveBayes, e.g., in the 19th permutation the rate increases by 14 % with number of batches of 50 (see Table 1).

Results of DWM-WIN based on naiveBayes are better than those of DWM-WIN based on rpart and kknn for the small noise level=10 %. However, when increasing the noise level, naiveBayes based DWM-WIN approximately achieves the same level of error rates as the other methods.

Statistical Tests Friedman's test rejected the null hypothesis that the algorithms have similar performance when changing the noise level. In fact, given F statistic, $F = 4143.0868$ and p-value$= 2 \cdot 10^{-16}$, shown in Table 2, the algorithm performs differently when the noise level differs.

These results were confirmed by the two way ANOVA in Table 3 where the Friedman test rejected the null hypothesis that the algorithms have similar performance on average when considering the two factors noise and permutation or noise and type of the learner with a p-value $= 2.2 \cdot 10^{-16}$. This result is also confirmed by Tukey's test as shown in Table 4.

Conclusion We conclude that introducing more noise in the data impacts on the general error level. Naive Bayes performs better than other classifiers when handling problems of concept drift.

4 Conclusion

In this paper mining online data streams where the data distribution may change over time and the concepts may drift is discussed. The suitability of control charts with online ensemble methods for detecting concept drifts in data streams is analyzed. A two-step model was proposed where step 1 consists of applying the DWM-WIN heuristic on datasets with concept drift and compute the error rates of the online classification methods. Step 2 uses the TACL chart to monitor the output of step 1 in order to detect changes in data distribution through the decrease of classifiers performance.

Based on the SEA dataset, we studied the impact of the performance capacity reaction of DWM-WIN in detecting the concept drift and adapting the algorithm to the drift. It has been shown that DWM-WIN has a robust capacity to adapt different situations of concept drift with several variants of the data. It quickly adjusts itself after a concept drift detection and maintains a high performance for different drift situations with different noise levels. Further work can be carried out in studying the learners stability in the ensemble and to introduce other situations of real concept drift using generators under MOA of Albert et al. (2010b). Also, we would like to investigate comparisons with other methods such as ADWIN algorithms of Albert et al. (2010a).

References

Albert, B., Eibe, F., Geoffrey, H., & Bernhard, P. (2010a). Accurate ensembles for data streams: Combining restricted Hoeffding trees using stacking, *JMLR: Workshop and Conference Proceedings* (Vol. 13, pp. 225–240).

Albert, B., Geoff, H., Richard, K., & Bernhard, P. (2010b). MOA: Massive online analysis. *Journal of Machine Learning Research, 11*, 1601–1604.

Asensio, S. A., Puig, O. A., & Golobardes, E. (2014). Robust on-line neural learning classifier system for data stream classification tasks. *Journal of Soft Computing, 18*(8), 1441–1461.

Bischl, B., Lang, M., & Richter, J. (2014). mlr: Machine learning in R. https://github.com/berndbischl/mlr

Gama, J., & Kosina, P. (2014). Recurrent concepts in data streams classification. *Knowledge and Information Systems Recurrent, 40*(3), 489–507.

Kolter, Z. J., & Maloof, M. A. (2005). Using additive expert ensembles to cope with concept drift. In *Proceedings of the Twenty Second International Conference on Machine Learning* (pp. 449–456). New York, NY: ACM Press.

Kolter, Z. J., & Maloof, M. A. (2007). Dynamic weighted majority: An ensemble method for drifting concepts. *Journal of Machine Learning Research, 8*(13), 2755–2790. JMLR.org.

Kuncheva, L. I. (2009). Using control charts for detecting concept change in streaming data Technical Report, BCS-TR-001-2009, School of Computer Science, Bangor University, UK.

Mejri, D., Khanchel, R., & Limam, M. (2013). Ensemble method for concept drift in nonstationary environment. *Journal of Statistical computation and Simulation, 83*, 1115–1128.

Street, W., & Kim, Y. (2001). A streaming ensemble algorithm (SEA) for large-scale classification. *Proceedings of the 7th SIGKDD Conference* (pp. 377–382). New York: ACM Press.

Zhu, X. (2010). Stream data mining repository. http://www.cse.fau.edu/~xqzhu/stream.html

Part VI
Applications

Specialization in Smart Growth Sectors vs. Effects of Change of Workforce Numbers in the European Union Regional Space

Elżbieta Sobczak and Marcin Pełka

Abstract The purpose of the study is to identify the relations between the level of specialization in smart growth sectors and the effects of change of workforce numbers in NUTS (*The Nomenclature of Territorial Units for Statistics*) 2 regions of the European Union countries. Multivariate data analysis methods, structural-geographic shift-share method and regional specialization indices were applied in the study. The structure of workforce in economic sectors, separated based on the intensity of research and development activities in NUTS 2 regions in the period 2009–2012, constituted the subject matter of the analysis. The application of shift-share analysis allowed for determining the effects of workforce structure, competitiveness and number changes in smart growth sectors against the reference area, i.e. the European Union regional area. Multivariate data analysis methods facilitated the typology of the analyzed regions against the level of specialization and the type of effects resulting from the change of workforce numbers in smart growth sector, as well as determining the relations between them.

1 Introduction

The analysis of regional specialization can be performed having considered the sector structure of economy. Currently the significance of economy sectors, based on the implementation of knowledge and innovation, is constantly increasing. In 2010 the European Union adopted Europe 2020 development strategy, which defined the goals to facilitate member states in ensuring, among others, smart growth consisting in the development of knowledge and innovation based economy (Europe 2020 2010). The concept of smart growth refers to numerous previously

E. Sobczak (✉)
Department of Regional Economics, Wrocław University of Economics, Nowowiejska 3, 58-500 Jelenia Góra, Poland
e-mail: elzbieta.sobczak@ue.wroc.pl

M. Pełka
Department of Econometrics and Computer Science, Faculty of Economics, Management and Tourism, Wrocław University of Economics, ul. Nowowiejska 3, 58-500 Jelenia Góra, Poland
e-mail: marcin.pelka@ue.wroc.pl

© Springer International Publishing Switzerland 2016
A.F.X. Wilhelm, H.A. Kestler (eds.), *Analysis of Large and Complex Data*, Studies in Classification, Data Analysis, and Knowledge Organization,
DOI 10.1007/978-3-319-25226-1_22

presented theoretical concepts and models of regional growth, among which the dominating role is played by the regional innovation systems (Cooke et al. 1997), innovation environments—*milieu innovateur* (analyzed by GREMI research group), learning regions (Florida 1995; Morgan 1997) and innovation clusters (Porter 1998). Smart specialization of workforce structure constitutes one of the instruments and components of smart development.

The study focuses on analyzing workforce structure in economy sectors separated in accordance with the intensity of research and development activities, also referred to as technological intensity and defined as the relation of expenditure on R&D (research and development) against added value of manufacturing sector (Hatzichronoglou 1996). The distribution of workforce in these sectors (called smart growth ones) represents the basic determinant of regional smart growth. The objective of the study is to:

- identify the intensity and diversification of regional specialization in the European Union,
- classify the European NUTS 2 level regions with regard to the allocation effects of structural, regional and the change of workforce numbers,
- identify the relations occurring between the level of regional specialization in smart growth sectors and the effects of the change of workforce numbers in NUTS 2 regions of the European Union countries.

2   The Information Basis and the Research Procedure Stages

The statistical information, indispensable for conducting empirical research, was obtained based on Eurostat database. Workforce structure constitutes the reference basis of performed analyses, in the cross-section of the following technological intensity sectors, prepared by Eurostat and OECD: HMH—high and medium high-technology manufacturing, LML—low and medium low-technology manufacturing, KIS—knowledge-intensive services, LKIS—less knowledge-intensive services, OTHER sectors (farming, hunting, forestry, fishing, mining, production and supply of electricity, gas, water, construction).

The study included 237 European Union regions selected following NUTS 2 classification. Due to the unavailability of statistical data the analysis did not cover 35 NUTS 2 regions. The research time range covered the period of 2009–2012. From 1 January 2008 the updated NACE classification (*Statistical Classification of Economic Activities in the European Community*) (NACE Rev. 2) and the definitions of high-tech manufacturing and knowledge-intensive services have changed. Therefore data comparisons, before and after 2008, have to be approached cautiously or the discussed changes have to be regarded as breaks in the continuity of data. Moreover, in 2008, due to the unavailability of statistical data the analysis could not include another 50 regions.

The following research procedure was applied:

1. The identification of regional specialization level by means of applying Krugman specialization index.
2. The quantification of structural, regional and allocation effects of the change of workforce numbers applying both classical and dynamic shift-share analysis.
3. The assessment of relationships occurring between regional specialization and structural and regional effects of the change of workforce numbers.

Shift-share analysis determines what portions of regional economic growth or decline can be attributed to national, economic industry and regional factors. The analysis helps identify industries where a regional economy has competitive advantages over the larger economy. In the classical form the shift-share analysis (also known as the comparative static model) examines changes in the economic variable between 2 years. Changes are calculated for each industry in the analysis. In the dynamic form shift components are determined annually for the entire study period with the base year continuously changing. The annual components are summed over the entire multiyear period.

Stage 1. The identification of regional specialization level by means of applying Krugman specialization index.

The study presents the assessment of regional specialization by applying Krugman's specialization index (Krugman 1991) for NUTS 2 regions with regard to reference area defined as the regional space of 28 European Union member states. This index was defined as the sum of absolute differences between the sector shares of workforce in a particular NUTS 2 region in the total workforce employed in this region against the total sector workforce share in the overall workforce number in the European Union. Krugman specialization index takes the following form:

$$K_r^* = \sum_{i=1}^{S} |u_i^r - u_i|, \tag{1}$$

$$u_i^r = \frac{x_{ri}}{x_{r.}}, \tag{2}$$

$$u_i = \frac{x_{.i}}{x_{..}}, \tag{3}$$

where: $r = 1, \ldots, R$—region number; $i = 1, \ldots, S$—sector number; x_{ri}—workforce number in r-th region and i-th sector; $x_{r.}$—total workforce number in r-th region, $x_{.i}$—workforce number in i-th sector of the reference area (EU 28); $x_{..}$—total workforce number in the reference area (EU 28).

Stage 2. The quantification of structural, regional and allocation effects of workforce structure changes by applying classical and dynamic shift-share analysis.

Structural and geographic analysis of workforce was performed considering the intensity of R&D (research and development) activities in the European NUTS 2 regions, having applied the classical Dunn shift-share analysis and the dynamic competitive model by Barff and Knight (1988). Structural and regional effects of changes were defined as recurring for each couple of years in the analyzed period, which was later aggregated in accordance with Barff–Knight concept. It is a dynamic model shift-share analysis, where the shift components are determined annually for the entire study period with the base year continuously changing. The annual components are summed over the entire multiyear period. The shift-share analysis of the change of workforce numbers in the EU regions allowed to: identify key sectors responsible for regional development, specify structural and regional effects of the change of workforce numbers in sectors distinguished based on R&D intensity, classify EU regions by positive and negative change effects values: structural and competitive ones, classify EU regions by the components of allocation effects: specialization and competitiveness.

The net total effect (surplus of regional average of employment over the EU) is decomposed into two effects: regional effect and structural effect. Structural effect is equal to the weighted average deviation between the average growth in the sectors and the growth rate of the EU (the average growth rate in the sector is the same in all analyzed regions). Regional effect is equal to the weighted average deviations in the various sectors and the EU average (it is the effect of internal changes in a particular region).

The classical equation of shift-share analysis indicates that the interregional diversification of the average of the change of workforce numbers rate can be represented as the consequence of two reasons: different regional workforce structures (the structural effect of changes), as well as the diversification of the change of workforce numbers dynamics in high-tech intensity sectors characteristic for these regions (the regional effect of changes). A positive structural effect indicates that the change of workforce numbers rate, in a given region, was more favourable than in other regions regarding sector oriented employment structure characteristic for this particular region. A positive regional effect informs that the change of workforce numbers, in a given region, was higher than in other regions since the sectors of this particular region were characterized by more favourable dynamics of the change of workforce numbers than in case of other regions. The analysis of allocation effects with reference to workforce structure results, among others, in the identification of regions featuring smart specialization and the occurrence of competition advantages (Suchecki 2010). A region is characterized by the specialized smart workforce structure if the share of workforce in HMH or KIS in this particular region is higher than in the respective EU sectors. If, in the analyzed region, the rate of workforce number changes in HMH or KIS sector is higher than in the respective EU sectors, then such a region is characterized by certain competitive advantages.

Stage 3. The assessment of relationships between regional specialization and structural and regional effects of the change of workforce numbers.

Regression analysis was applied at this stage of research procedure.

3   The Empirical Analysis Results

Figure 1 presents the distribution of Krugman index values of regional specialization in smart growth sectors in NUTS 2 regions in the period 2009–2012. The discussed values did not change significantly in terms of the median value referring to the studied period, however, they were characterized by large variations. The variation coefficient amounted to about 50 %.

Figure 2 shows NUTS 2 regions characterized by the highest and the lowest values of Krugman index. Seven out of eight Romanian regions, the Spanish

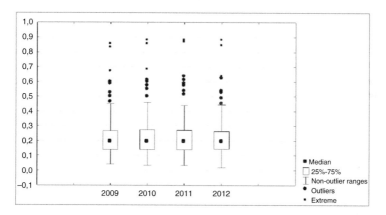

Fig. 1 Krugman index values of regional specialization in the sectors distinguished by R&D intensity in NUTS 2 regions in the period 2009–2012. *Source*: Authors' compilation using STATISTICA program

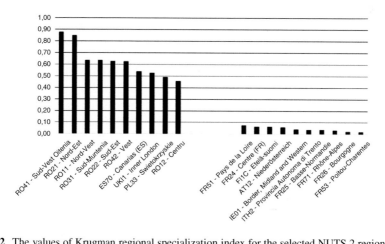

Fig. 2 The values of Krugman regional specialization index for the selected NUTS 2 regions (10 regions presenting the highest and 10 the lowest values) in 2012. *Source*: Authors' compilation based on Eurostat database

Table 1 Workforce structure in the regions featuring the highest level of regional specialization in 2012

Regions	Workforce structure (%)				
	HMH	LML	KIS	LKIS	OTHER
EU 28	5.7	10.0	39.1	30.7	14.5
RO41—South-West Oltenia	3.6	7.0	15.6	15.5	58.3
RO21—North-East	2.4	16.0	19.6	22.3	39.7
RO11—North-West	3.8	19.6	17.3	22.8	36.6
RO31—South-Mutentia	6.8	12.5	18.3	19.9	42.4
RO22—South-East	2.4	16.0	19.6	22.3	39.7
RO42—West	14.7	17.1	16.9	21.8	29.6
ES70—Canarias	0.4	3.1	30.2	57.7	8.7
UK11—Inner London	1.0	1.7	65.4	26.4	5.5
PL33—Swietokrzyskie	3.4	13.4	24.8	22.7	35.6
RO12—Centru	5.4	22.6	21.7	25.7	24.7

Source: Authors' compilation based on Eurostat database

region of Canarias, British capital region of Inner London and Polish region of Świętokrzyskie were characterized by the highest level of regional specialization. Among ten regions featuring the lowest specialization (workforce structure most similar to the EU structure) four French, two Italian, one Austrian and one Finnish region were listed.

Table 1 presents NUTS 2 regions characterized by workforce sector structure different, to the greatest extent, from workforce structure in the European Union regional space.

Based on Fig. 2 and Table 1 the economy sectors, characterized by workforce share definitely different from workforce share in the EU, can be identified. In case of Romanian regions and the Polish region, regional specialization consists in the significantly larger share of workforce in the so-called other sectors. In the Romanian region of South-West Oltenia, featuring the highest specialization indicator, the share of workforce in the so-called OTHER sectors is over 43 % points higher than in the EU. As one can infer from Fig. 3, the Spanish region of Canarias specializes in the sector of less knowledge-intensive services and the services related to tourist traffic (workforce share in LKIS sector is higher than in the EU by 27 % points). On the other hand, the British capital region of Inner London specializes in knowledge-intensive services (workforce share in KIS sector is higher than in the EU by 26 % points).

The next stage of research procedure consisted in quantifying structural, regional and allocation effects of the change of workforce numbers applying both classical and dynamic shift-share analysis. Table 2 presents the results of classical shift-share analysis covering 2009 and 2012 in relation to the previous year and referring to the overall results and structural effects of employment changes in the distinguished sectors.

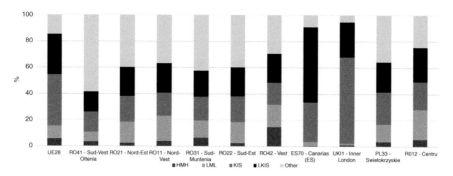

Fig. 3 Workforce structure in the regions featuring the highest level of regional specialization in 2012. *Source*: Authors' compilation based on Eurostat database

Table 2 The results of classic shift-share analysis with regard to the effects of employment changes in the sectors distinguished by R&D intensity

Effects of employment changes in NUTS 2 (%)		2012/2009
Total effect (employment growth rate in EU)		−0.87
Gross structural effect	HMH	−1.30
	LML	−5.44
	KIS	2.39
	LKIS	0.22
	OTHER	−5.83
Net structural effect	HMH	−0.43
	LML	−4.57
	KIS	3.26
	LKIS	1.09
	OTHER	−4.96

Source: Authors' compilation based on Eurostat database

The general tendency of decline in employment in the European Union is responsible for 0.87 % of workforce number drop rate in every region and economy sector in 2012. Net structural effects were defined by the decreasing gross effects against workforce growth rate in the EU. Changes in employment in the KIS sector in 2012 resulted in higher workforce number in all NUTS 2 regions, on average by 3.26 %. Employment growth rate in LKIS sector in 2012 also influenced the slight growth of workforce number in the EU countries (1.09 %). It is unclear how employment in the OTHER sector was related to the drop of employment. Table 3 presents the results of the studied regions' classification by positive and negative effects of the change of workforce numbers in the period 2009–2012.

Class I (see Table 3) includes the largest number of regions featuring positive influence of both structural and regional effects on the employment changes. Class III covers the smallest number of regions characterized by the positive impact of the regional factor only. Forty-five percent of the analyzed NUTS 2 regions are characterized by the positive influence of structural and regional factors on the employment structure changes. In 1 % of the studied regions both structural

Table 3 Classification of NUTS 2 regions by positive and negative effect values: structural and regional (DSSA 2012/2009)

Group	Division criterion	Country (number of regions)	Number of regions
I	Effects: Structural (+) Regional (+)	Belgium (11), Germany (25), France (12), Italy (6), Luxembourg (1), Netherland (5), Austria (6), Finland (3), Sweden (8), UK (25)	106
		Cyprus (1), Hungary (1), Malta (1), Romania (1)	
II	Effects : Structural (+) Regional (–)	Denmark (5), Germany (5), Ireland (2), Greece (1) Spain (12), France (5), Italy (5), Netherland (6) Portugal (1), Finland (1), UK (8)	57
		Bulgaria (1), Czech Rep. (1), Croatia (1), Poland (1), Slovenia (1), Slovakia (1)	
III	Effects: Structural (–) Regional (+)	Germany (1), Italy (5), Austria (3) Czech Rep. (4), Estonia (1), Hungary (1), Poland (6) Romania (4), Slovakia (1)	30
IV	Effects: Structural (–) Regional (–)	Greece (5), Spain (5), Italy (4), Portugal (3) Bulgaria (5), Czech Rep. (3), Lithuania (1), Latvia (1) Poland (9), Slovenia (1), Hungary (1), Croatia (1) Romania (3), Slovakia (2)	44

Source: Authors' compilation based on Eurostat database

Table 4 The effects of workforce number allocation

Smart specialization of a region and a competitive advantage	Countries (NUTS 2 regions)	Number of regions
HMH sector (EU share of workforce 5.6 % the rate of workforce number change –1.3 %)	Belgium (1), Czech Rep. (1), Germany (12) Ireland(1), Spain (3), France (3), Italy (5) Hungary (5), Netherland (1), Austria (4) Poland (3), Romania (1), Slovakia (4) Finland (1), United Kingdom (5)	55 EU15 36 EU13 19
KIS sector (EU share of workforce 39.1 % the rate of workforce number change 2.39 %)	Belgium (11), Germany (10), France (11) Luxembourg (10), Hungary (1), Malta (1) Netherland (1), Austria (1) Slovenia (1), Finland (3), Sweden (6) Romania (1), UK (16)	64 EU15 60 13 4

Source: Authors' compilation based on Eurostat database

and regional factors exerted negative impacts on employment structure changes. This class covered the largest number of the latest EU enlargement regions (27). Table 4 presents the composition of classes distinguished by the allocation effects of workforce structure changes. 55 NUTS 2 regions featuring smart specialization of workforce structure in 2012, as well as competitive advantages in the period 2012/2009, were identified in high and medium high-technology manufacturing. Smart specialization and competitive advantage in knowledge-intensive services

Table 5 NUTS 2 specialized regions characterized by competitive advantages in HMH and KIS sectors

Countries	NUTS 2 regions	Specialization		Competitive advantage		K_r^*
		HMH	KIS	HMH	KIS	
EU 28		5.6	39.1	−1.3	2.4	–
Belgium	BE28—Prov. Oost-Vlaanderen	6.0	46.0	25.8	2.8	0.15
Germany	DE71—Darmstadt	10.1	45.0	19.1	2.4	0.23
	DE93—Lüneburg	8.7	39.4	33.3	19.0	0.11
France	FR53—Poitou-Charentes	5.7	39.2	10.5	18.6	0.02
Finland	FI19—Länsi-Suomi	7.0	40.0	0.0	7.36	0.10
United Kingdom	UKD6—Cheshire	5.8	48.0	0.0	7.0	0.18
	UKF1—Derbyshire and Nottinghamshire	6.0	45.0	18.4	6.3	0.12
	UKJ3—Hampshire and Isle of Wight	5.8	51.0	15.2	6.4	0.24

Source: Authors' compilation based on Eurostat database

were characteristic for the group covering 64 NUTS 2 regions which included only four EU13 regions—regions of so-called new EU countries—Malta, Bucuresti–Ilfov, Zahodna Slovenija and Közép-Magyarország.

The subsequent research procedure stage was focused on evaluating the relations occurring between regional specialization and structural and regional effects of the change of workforce numbers. Table 5 lists the regions characterized by two-sector smart specialization and two-sector competitive advantages. Among 237 analyzed regions as few as 8, representing five of the so-called old EU15 countries, met the presented criteria. The discussed group covered 3 British regions, 2 German, 1 Belgian, 1 French and 1 Finnish region. However, in accordance with Krugman specialization index the listed regions were characterized by the low level of regional specialization.

Figure 4 presents the relations existing between the level of regional specialization in smart growth sectors in accordance with Krugman specialization index and the structural and regional effects of the change of workforce numbers in the studied NUTS 2 regions. The dependence occurs between structural effects of changes in the number of working population and the values of Krugman regional specialization index, however, no dependence was observed between regional effects of the change of workforce numbers and the values of Krugman regional specialization index.

In order to identify sectors responsible for the EU NUTS 2 development, Pearson's linear correlation coefficient of structural effects (distinguished in accordance with the dynamic shift-share analysis) and the share of workforce in particular economy sectors was calculated.

Definitely, the strongest positive relation was observed between structural effects and the share of workforce in knowledge-intensive services sector (0.93). A positive relation, however, statistically irrelevant for the accepted significance level $\alpha = 0.05$ was characteristic for structural effects and the share of workforce in less

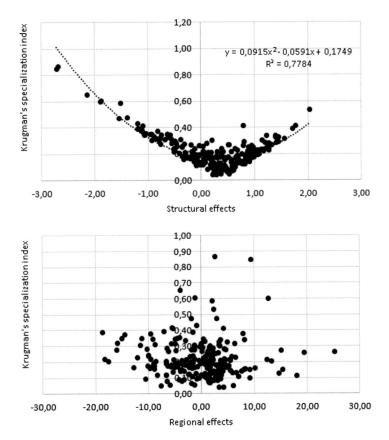

Fig. 4 The relations occurring between the level of regional specialization in smart growth sectors and the effects of the change of workforce numbers. *Source*: Authors' compilation based on Eurostat database

knowledge-intensive services sector. The other sectors, as well as low and mid low-tech sector, featured negative correlation on structural effects in the NUTS 2 regions. The strongest negative correlation was registered with reference to structural effects and the share of workforce in other sectors (−0.83).

4 Conclusions

In the period 2009–2012 the analyzed regions showed the significant diversification of regional specialization in the sectors distinguished by R&D intensity. The variability coefficient of regional specialization indices amounted to over 50 %.

Sector economic structures, and thus also workforce structures featuring R&D intensity, are usually characterized by slow, evolutionary transformations over time and therefore the analyzed values of regional specialization indices presented relative stability in the studied period. The Romanian region South-West Oltenia showed the highest specialization indicator, the share of workforce in the other sectors was over 43 % points higher than in the EU28.

Forty-five percent of the analyzed NUTS 2 regions were characterized by positive effect of structural and regional factors on employment structure changes. In case of 18 % studied regions both structural and regional factors had negative effect on employment structure changes. Among 237 analyzed regions as few as 8, representing five of the so-called old EU15 countries, were characterized by two-sector smart specialization and two-sector competitive advantages (in HMH and KIS sectors). The discussed regions were, however, characterized by the low level of regional specialization measured by Krugman index. The dependence was observed between structural effects of changes in the number of working population and the values of Krugman regional specialization index. Such dependence was, however, absent in case of regional effects.

Workforce share in KIS sector had strong positive impact on the structural effect value, whereas workforce share in OTHER sectors exerted strong negative impact on the structural effect value. Workforce share in HTM sector had no such impact. In the regions characterized by high values of Krugman specialization index and positive structural effects workforce share in KIS sector was higher than the EU 28 average. In the regions featuring high values of Krugman specialization index and negative structural effects workforce share in the so-called OTHER sectors was significantly higher than the EU 28 average. Shift-share analysis is used in regional specialization research. The analysis results can constitute the basis for determining the regional specialization specificity identified by means of Krugman index application.

References

Barff, R. A., & Knight III, P. L. (1988). Dynamic shift-share analysis. *Growth and Change, 19*(2), 1–10.

Cooke, P. et al. (1997). Regional innovation systems: Institutional and organizational dimensions. *Research Policy, 26*, 475–491.

Europe 2020. (2010). The strategy for smart and sustainable growth facilitating social inclusion. In *The Communication by the Commission*. Brussels: The EU Commission.

Florida, R. (1995). Towards the learning region. *Futures, 27*, 527–536.

Hatzichronoglou, T. (1996). *Revision of the high-technology sector and product classification*. Paris: OECD.

Krugman, P. R. (1991). *Increasing Returns and Economic Geography*. Working Paper, no 3275, National Bureau of Economic Research.

Morgan, K. (1997). The learning region: Institutions, innovations and regional renewal. *Regional Studies, 31*, 491–503.

Porter, M. E. (1998). *The competitive advantage of nations*. London: Macmillan.

Suchecki, B. (Ed.) (2010). *Ekonometria przestrzenna. Metody i modele analizy przestrzennej* [Spatial econometrics. Methods and models for spatial analysis]. Warsaw: C.H. Beck Publishers.

Evaluation of the Individually Perceived Quality from Head-Up Display Images Relating to Distortions

Sonja Köppl, Markus Hellmann, Klaus Jostschulte, and Christian Wöhler

Abstract The head-up display (HUD) projects a virtual image in the driver's field of vision. Here, the image quality plays an important role. However, assembly tolerances cause image distortions. The evaluation of these distortions is a current issue, because procedures for the assessment of optical aberrations cannot be applied. Therefore new features and methods are implemented, which evaluate the subjective impression of distortions. The overall objective is to investigate the correlation between subjective labels and objective features. A total of 13 features are required to describe the image quality. Subsequently, the relationship between the labels and the features is adapted to a regression equation. For it, representative images are needed, which are selected by cluster analytical methods.

1 Introduction

The HUD system projects important information such as navigation instructions and indications of different driver-assistance systems, see Fig. 1, directly in the field of vision. The system housing is mounted behind the steering wheel in the dashboard. The device consists of mirror optics, a full-coloured display module and the windscreen. Due to the optical path, the virtual image seems to hover just above the front end of the hood at nearly 2 m distance to the driver (Blume et al. 2013).

S. Köppl (✉) • K. Jostschulte
Daimler AG, Stuttgart, Germany

Daimler Research & Development, Ulm, Germany
e-mail: sonja.koeppl@tu-dortmund.de; klaus.jostschulte@daimler.com

M. Hellmann
Daimler AG, Stuttgart, Germany
e-mail: markus.hellmann@tu-dortmund.de

C. Wöhler
TU Dortmund, Dortmund Germany
e-mail: christian.woehler@tu-dortmund.de

© Springer International Publishing Switzerland 2016
A.F.X. Wilhelm, H.A. Kestler (eds.), *Analysis of Large and Complex Data*, Studies
in Classification, Data Analysis, and Knowledge Organization,
DOI 10.1007/978-3-319-25226-1_23

265

Fig. 1 The virtual image
shows driver relevant
information. (Own illustration
adapted from Jordan 2014)

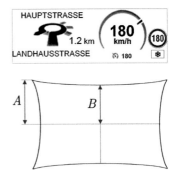

Fig. 2 SMIA TV-Distortion
(image taken from SMIA
2004)

The image quality is influenced by assembly tolerances. Observable distortions
are generated that affect the overall impression (Díaz 2005). Possible types of
distortions are rotation, trapezoid, aspect deviation, etc. (Eichhorn and Zink 2012).
This results in the demand for a method that is able to evaluate the individually
perceived quality from HUD images. Therefore, calculation rules are required to
assess distortions quantitatively. Additionally, it is evaluated which factors cause
distortions and how these defects are perceived by drivers.

2 Existing Methods to Handle Distortions

Perceived distortions can be corrected by complex image warping. Warping assumes
that an undistorted image is transformed to a distorted image by an optical system.
Conversely, a suitable pre-distorted image passed through the optical system is
converted into a straight undistorted image (Miličić 2009).

There exist many standards that specify occurring distortions. The descriptions
are mainly based on the aberrations of optical systems. A practical and common
method to measure distortions is named TV-Distortion, defined in the SMIA
specification (SMIA 2004, Sect. 5.20). It is the difference in height of the image
corners and the image centre as shown in Eq. (1). The reported value is the average
distortion of the four corners of the image (SMIA 2004).

$$\text{TV-Dist} = \frac{A - B}{B} \cdot 100\,\% \text{ (SMIA 2004).} \tag{1}$$

The values A and B, needed to calculate the TV-Distortion rate, are schematically
shown in Fig. 2.

With respect to the HUD, occurring distortions are not limited to the aberrations
of optical systems. The distortions are rarely symmetrically, and they occur irregu-
larly on all four sides. For these reasons, a geometric description of the distortion is
more useful.

Table 1 Distortion types considered annoying

Answers from participants		Evaluation features	Feature-no.
Skewing	75 %	Contour	v_6, v_7, v_8
Bending	59 %	Contour	v_{11}, v_{12}, v_{13}
Discomposure	50 %	Local deviations	v_4, v_5
Tapered	17 %	Contour	v_9, v_{10}
Stretching	8 %	Size difference	v_1, v_2, v_3

3   Objectification of Distortions

In order to make distortions visible, a test pattern is displayed as virtual image. The used pattern consists of 9×21 dots with the central dot highlighted. From a driver's perspective a camera image of the virtual HUD image is made. This camera image is analysed by existing image processing algorithms, which calculate the x–y centre coordinates of the dots (Eichhorn and Zink 2012).

According to Eichhorn and Zink (2012) and Neumann (2012), the 9×21 x–y coordinates of the dots are converted into 13 objective features. Each feature is supplied with a unique number $v_1 \ldots v_{13}$, which is referenced in the course of the document. The features are based on the geometric characterisation and represent 13 distortion types numerically. The first three values describe the size differences and the aspect deviation (v_1, v_2, v_3). The following two features are used to determine local magnifications (v_4, v_5) and the contour (rotation, trapezoid, misalignment, smile...) of the image is captured with the remaining eight values ($v_6, \ldots v_{13}$) (Eichhorn and Zink 2012; Neumann 2012).

To become familiar with the evaluation features, a ranking by relevance is created. Therefore, a volunteer study is carried out. During this enquiry, 39 images are shown to 12 test persons. The images are produced artificially and show 3 different occurrences for each of the 13 distortion types. Therefore, each kind of distortion type is considered separately. Subsequently, the participants are asked: "Which distortion type interferes most strongly?". The answers are listed in Table 1. In order to determine a subjective ranking, the used evaluation features are assigned to the statements of the test persons.

The analysis of the answers shows that skewing is mentioned first, followed by bending, discomposure, a tapered shape and a stretched image.

The generated ranking of the features is needed for the determination of the cluster condition, see Sect. 5.2.

4   Simulation of Assembly Tolerances

To check which kind of assembly tolerances cause image distortions, an existing hardware arrangement is used. Starting from the ideal position, different tolerances are simulated. From a driver's view, a camera image of the virtual image, which shows the test image, is generated. Subsequently, for each camera image the 13

Fig. 3 Distortions caused by assembly tolerances of the HUD housing (illustration adapted from Jordan 2014, distortion types according to Eichhorn and Zink 2012 and Neumann 2012)

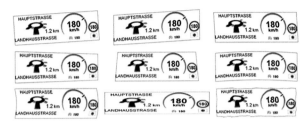

objective feature values $v_1 \ldots v_{13}$ are calculated and the subjective labels y are determined for some representative images.

4.1 Effect of Different Assembly Tolerances

Since the virtual image is generated by the HUD housing and the windscreen, assembly tolerances of these two components are simulated. The resulting virtual images are considered subjectively. The housing can be installed shifted or rotated in the dashboard. Several test series show that a displaced HUD does not cause visible distortions. The same applies to assembly tolerances of the windscreen. It can be positioned wrongly or with restraints, but no perceivable distortions are generated. In contrast, a rotated HUD housing generates visible distortions see Fig. 3.

The resulting image quality is a combination of all 13 single distortion types. Overall, rotational assembly tolerances of the housing change the optical performance substantially.

4.2 Generation of the Database

To examine the effect of distortions caused by assembly tolerances, a database is generated. Thereby, only impacts of a rotated HUD housing are examined. Unfortunately, the irregularities can occur in a very large variety of combinations. The investigation is thus limited on the main permutations. Overall, the measurement consists of 23,965 combinations. The resulting virtual images are captured from three different camera positions, for a small, tall and average driver. Altogether 71,895 camera images are generated and the objective evaluation features are determined for each image.

5 Determine the Subjective Impression

To examine the customer acceptance, the images of the database are evaluated subjectively. Unfortunately, the rating of nearly 72,000 images is very time consuming and cannot be obtained by asking test persons. Therefore the evaluation is limited to some representative images, which are detected by cluster analytical methods. During clustering, similar images are grouped together. The aim is that one group contains only images with the same subjective evaluation. Additionally, the rating of one single image should correspond to the evaluation of all other images in the same group. Consequently, the expenditure of a customer questioning can be reduced, because only one image from each group must be assessed.

5.1 Used Cluster Analytical Methods

K-means: According to Bortz and Schuster (2010), an initial predefined number of cluster centres are chosen randomly. For all data points v, the (usually Euclidean) distance to each cluster centre is computed, and the cluster membership of each data point is set according to the cluster centre with the smallest distance. In the next step the cluster centres are refined by computing the mean vectors of the data points assigned to them. This procedure is repeated until the assignment of the data points to the cluster centres does not change any more. The final clustering result depends on the initialisation (Bortz and Schuster 2010).

Ward: According to Bortz and Schuster (2010), each data point initially forms a cluster of its own. From all possible pairs of cluster centres, the cluster centres V_i and V_i' whose combination results in the minimal increase of the sum of squared differences *QSe* of the corresponding elements (number of images in one cluster, n_i and n_i') are merged, see Eq. (2).

$$QSe = \frac{n_i \cdot n_{i'}}{n_i + n_{i'}} \cdot \sum_{j=1}^{13}(V_{ij} - V_{i'j})^2 \text{ for merging cluster centres } Vi \text{ and } Vi'. \quad (2)$$

Merging of cluster pairs is stopped once a predefined threshold for the value of QSe is exceeded (Bortz and Schuster 2010).

Combine Ward and K-means: It is often favourable to utilise the result of the *Ward* algorithm as an initialisation to the *K-means* algorithm. This circumvents the property of the *Ward* algorithm that the assignment of a specific data point to a cluster remains unchanged, and the result of the *K-means* algorithm does not depend on an arbitrary random initialisation any more (Bortz and Schuster 2010).

Mean-shift: According to Cheng (1995), for an initial position in feature space a kernel based approximation to the gradient of the distribution of data points is computed in an iterative manner, and the cluster centre is moved in the direction of this approximated gradient. This procedure is repeated until convergence. Cluster centres thus correspond to local maxima of the distribution of the points in feature space. For a distribution of data points with several local maxima, the algorithm needs to be started from a (possibly large) number of different initial points in order to determine all local maxima of the distribution, where the number of detected local maxima may also depend on the form of the utilised kernel function (Cheng 1995).

5.2 Numeric Determination of Cluster Condition

The minimal number of clusters should be found that exhibits no subjective difference between the images in one group. The requirement of subjective equality in a cluster is examined and numerically quantified in a preliminary investigation. Therefore, the 71,895 images are analysed in different cluster solutions. Due to the large amount of data and the temporal expenditure, the examination is executed with only six test persons. For clustering the simple *K-means* algorithm is used, where the number of clusters is known in advance. Successively, the images for each cluster are shown on an external monitor. As soon as a difference between the images in one cluster can be perceived, the questioning is aborted and restarted with modified parameters. The resulting clusters depend on the selected features and the chosen cluster number. Due to the different subjective relevance, see Table 1, not all evaluation features are needed to describe the subjective difference. Until a suitable arrangement of the images is found, either a further feature is taken for clustering or the cluster number is increased. It must be ensured that the start allocation of the *K-means* algorithm remains the same. The cluster condition is reached, if the participants confirm that no subjective difference exist between the images in one cluster. Then, the required cluster condition can be obtained based on the involved features.

The study shows that only five evaluation features (v_6, v_7, v_8, v_{11}, v_{12}) are needed to describe the subjective perceptible difference. The images are properly sorted if the maximum difference of the feature values $\triangle_{\text{rate}}$ inside the clusters is smaller than 1 HUD pixel (0.58×0.58 mm), see Eq. (3). The distribution of the values of the remaining eight features has no effect on the subjective perception. It is shown that images with similar objective feature values (v_6, v_7, v_8, v_{11}, v_{12}) cause the same quality impression.

$$\triangle_{\text{rate}} < 1 \text{ HUD pixel} \begin{cases} \text{Contour - skewing. Feature-No. } v_6, v_7, v_8 \\ \text{Contour - bending: Feature-No. } v_{11}, v_{12} \end{cases} . \qquad (3)$$

5.3 Clustering Results

After the objective of clustering is known, the 71,895 images are divided into subjective groups. Therefore, all 13 features are used. The minimum number of clusters is determined where no subjective difference between the images could be perceived. The images in one cluster are all annotated with the same subjective label and have objective values of features v_6, v_7, v_8, v_{11} and v_{12} that vary by less than 1 HUD pixel, see Eq. (3). To obtain the best solution, the results of *Ward*, *K-means*, combined *Ward–K-means* and *Mean-shift* clustering are compared. The used cluster analytical methods are implemented in MATLAB.

The resulting cluster numbers are summarised in Table 2. Since the cluster solution of the *K-means* algorithm depends on the randomly selected starting condition, the results of 100 different initial conditions are considered.

The comparison between the solutions shows that the combination of *Ward and K-means* yields, with 1007 groups, the smallest number of clusters. Thus, it can be considered as the best solution. In contrast, the largest number of clusters is obtained from the *Mean-shift* algorithm. The results of the *Ward* method and the *K-means* algorithm are in between these values.

5.4 Subjective Evaluation

The previously found cluster solution is now used as basis for another customer survey, where only 1007 representative images are assessed subjectively. A total of 12 test persons are asked. The images that are next to the theoretical cluster centre are shown on an external monitor. The images are evaluated on a scale from 1 (very annoying) to 5 (imperceptible), according to the ITU-R 500 directive, see Table 3 (ITU-R 2002). This scale is initially used to assess the subjective perception

Table 2 Minimum numbers of clusters resulting from different cluster methods

Clustering method	Number of clusters
Ward	1239
K-means (average of 100 starting positions)	1146
Ward and K-means	1007
Mean-Shift	4288

Table 3 Evaluation of representative HUD images

	Rating scale (ITU-R 2002)	Number of images
5	Imperceptible	5
4	Perceptible, but not annoying	9
3	Slightly annoying	35
2	Annoying	551
1	Very annoying	407

of television images and is adopted for this study to quantify the perceived quality of HUD images.

It could be seen that only five clusters with the highest score exist. On the contrary, there exist 407 clusters with the lowest score. Besides, the occurring distortions are corrected by image warping before the vehicle is sold.

6 Correlation Objective: Subjective

To analyse the relationship between the feature values v and the subjective labels y, the correlation coefficients (r_{vy}) are determined. The correlation between each single feature and the perceived quality of the whole image is calculated, see Eq. (4). The correlation coefficient can take values between ± 1. A value of ± 1 indicates a complete positive or negative linear correlation. In contrast, 0 shows that the variables are linearly independent from each other (Kohn and Öztürk 2010).

$$r_{vy} = \frac{\sum_{j=1}^{806}(v_{ij} - \bar{v}_i) \cdot (y_j - \bar{y})}{\sqrt{\sum_{j=1}^{806}(v_{ij} - \bar{v}_i)^2 \cdot \sum(y_j - \bar{y})^2}} \quad \text{correlation between } v_i, y. \tag{4}$$

The resulting correlation coefficients are shown in Table 4. As it can be seen, the values are between -0.14 and 0.17. This indicates that there is almost no correlation between a single feature and the perceived quality of the image. The subjective label cannot be mapped to only one single feature.

The subjective labels correspond to a variety of combinatorics of all feature types. Therefore, the relationship between the subjective labels and all objective features is adapted to a regression equation ($d(v)$), which is determined by a full polynomial approach. If all polynomial terms up to a certain degree (G) are used, the polynomial length (M) is given by Eq. (5).

$$M = \frac{(q+G)!}{G! \cdot q!} \quad \text{polynomial length with } q \text{ factors (Schürmann 1996).} \tag{5}$$

According to Schürmann (1996), this general polynomial consists of one constant term a_0 followed by linear terms, quadratic terms, cubic terms and so on, up to an arbitrary degree G:

$$d(v) = a_0 + a_1 v_1 + \cdots + a_q v_q + a_{q+1} v_1^2 + a_{q+2} v_1 v_2 + \cdots + a_{M-1} v_q^G.$$

Table 4 Correlation between the subjective labels and the single feature types

Feature-No.	v_1	v_2	v_3	v_4	v_5	v_6	v_7	v_8	v_9	v_{10}	v_{11}	v_{12}	v_{13}
r		0.07	0.06	0.12	0.17	-0.03	0.01	-0.14	0.03	-0.14	-0.11	-0.06	0.06	0.13

Table 5 Success analysis of the predicted ratings

Degree of the polynomial	First	Second
RMSE	1.34	1.38

The correlation between the subjective labels and the combination of all objective features is mapped to the regression coefficients $a_0 \ldots a_{M-1}$ (Schürmann 1996). The equation is thus able to estimate the subjective labels (a value between 1 and 5) from given objective feature values. To develop meaningful results, the 1007 labelled images are divided into two groups. The training set includes 806 images (80 % of the labelled data) and is used to determine the regression coefficients. The remaining 201 images are used to test the detected equation. Because the original customer labels (y_{given}) are known the predicted labels (y_{model}) can be checked. According to Schürmann (1996), the root mean square error (RMSE) is determined, see Eq. (6).

$$\text{RMSE} = \sqrt{\frac{1}{n} \cdot \sum_{i=1}^{n} (y_{model_i} - y_{given_i})^2}. \qquad (6)$$

The RMSE is a measure of difference between the labels. If all prognoses exactly apply, the RMSE is 0. The larger the RMSE, the worse the quality of the prediction.

The regression analysis is performed two times, for a polynomial of the first and the second order. For this purpose, a special MATLAB script is developed. The resulting RMSE values are summarised in Table 5. The values show that the RMSE is lowest for the first order equation. It can be concluded that the correlation between the subjective labels and the combination of all objective features is best reproduced by a regression equation of the first order. Thus, an RMSE of 1.34 could be obtained.

7 Summary

To evaluate possible distortions in the virtual HUD image, 13 features are determined, which are based on the geometric characterisation. These features represent occurring distortions (rotation, trapezoid, aspect deviation, etc.; Eichhorn and Zink 2012) numerically. Because distortions are mainly caused by rotatory assembly tolerances of the HUD housing, an existing hardware arrangement is used to generate numerous distorted images. The result is a database of 71,895 images. To analyse the customer acceptance, the images are labelled subjectively. To keep the effort of the questioning low, the evaluation is limited to representative images, which are selected by cluster analytical methods. Therefore the large database is split into several subgroups. For each cluster, the image that is closest to the cluster centre is considered to be representative. Besides, the number of clusters results from the subjective impression. The best result, with 1007 groups, is achieved by the combination of the *Ward and K-means* clustering. It could be confirmed that images with equal objective values cause a very similar quality impression. After executing a volunteer study, the customer requirements and the achieved quality judgements of virtual HUD images could be inferred. Subsequently, the relationship between the subjective labels and the objective features is adapted to a regression equation, which is determined by a full polynomial approach. To develop meaningful results, the 1007 labelled images are divided into two groups. The training set includes 806 images and is used to determine the regression coefficients. The remaining 201 images are used to test the detected equation. The best result is achieved by a regression equation of the first order. Thus, an RMSE of 1.34 is obtained.

References

Blume, J., et al. (2013). Head-up display: Next generation with augmented reality. *ATZelektronik Worldwide, 8*, 4–7.

Bortz, J., & Schuster, C. (2010). Durchführung einer Clusteranalyse. In *Statistik für Human- und Sozialwissenschaftler* (pp. 462–466). Berlin/Heidelberg: Springer.

Cheng, Y. (1995). Mean shift, mode seeking, and clustering. *IEEE Transactions on Pattern Analysis and Machine Intelligence, 17*(8), 790–799.

Díaz, L. (2005). *Optical Aberrations in Head-Up Displays*. Doctoral thesis, Universidad Pontificia Comillas Madrid, escuela técnica superior de ingeniería.

Eichhorn, N., & Zink, O. (2012). Auto head-up displays: "View-Through" for drivers, www.gefasoft-regensburg.de

ITU-R. (2002). Methodology for the subjective assessment of the quality of television pictures, www.itu.int/dms_pubrec/itu-r/rec/bt/R-REC-BT.500-13-201201-I!!PDF-E.pdf. Recommendation ITU-R BT.500-11, 10–13.

Jordan, M. (2014). Mercedes-Benz Accessories mit Head-Up Display als Nachrüstlösung ab 2. Quartal 2014, http://blog.mercedes-benz-passion.com/2014/03/mercedes-benz-accessories-mit-head-up-display-als-nachruestloesung-ab-2-quartal-2014/

Kohn, W., & Öztürk, R. (2010). Kovarianz und Korrelationskoeffizient. In *Statistik für Ökonomen* (pp. 103–108). Berlin/Heidelberg: Springer.

Miličić, N. (2009). *Sichere und ergonomische Nutzung von Head-Up Displays im Fahrzeug*. Doctoral thesis, TU München, Fakultät für Elektrotechnik.

Neumann, A. (2012). *Simulationsbasierte Messtechnik zur Prüfung von Head-up Displays*. Doctoral thesis, TU München, Institut für Informatik.

Schürmann, J. (1996). Polynomial regression. In *Pattern classification: A unified view of statistical and neural approaches* (pp 102–186). New York: Wiley.

SMIA. (2004). TV distortion. In *Camera characterisation specification*. SMIA 1.0 Part 5, 61–63, http://wenku.baidu.com/view/33ff23bec77da26925c5b03f.html

Minimizing Redundancy Among Genes Selected Based on the Overlapping Analysis

Osama Mahmoud, Andrew Harrison, Asma Gul, Zardad Khan, Metodi V. Metodiev, and Berthold Lausen

Abstract For many functional genomic experiments, identifying the most characterizing genes is a main challenge. Both the prediction accuracy and interpretability of a classifier could be enhanced by performing the classification based only on a set of discriminative genes. Analyzing overlapping between gene expression of different classes is an effective criterion for identifying relevant genes. However, genes selected according to maximizing a relevance score could have rich redundancy. We propose a scheme for minimizing selection redundancy, in which the Proportional Overlapping Score (POS) technique is extended by using a recursive approach to assign a set of complementary discriminative genes. The proposed scheme exploits the gene masks defined by POS to identify more integrated genes in terms of their classification patterns. The approach is validated by comparing its classification performance with other feature selection methods, Wilcoxon Rank Sum, mRMR, MaskedPainter and POS, for several benchmark gene expression datasets using three different classifiers: Random Forest; k Nearest Neighbour; Support Vector Machine. The experimental results of classification error rates show that our proposal achieves a better performance.

O. Mahmoud (✉)
Department of Mathematical Sciences, University of Essex, Colchester, UK

Department of Applied Statistics, Helwan University, Cairo, Egypt
e-mail: ofamah@essex.ac.uk

A. Harrison • Z. Khan • B. Lausen
Department of Mathematical Sciences, University of Essex, Colchester, UK
e-mail: harry@essex.ac.uk; zkhan@essex.ac.uk; blausen@essex.ac.uk

A. Gul
Department of Mathematical Sciences, University of Essex, Colchester, UK

Department of Statistics, Shaheed Benazir Bhutto Women University Peshawar,
Khyber Pukhtoonkhwa, Pakistan
e-mail: agul@essex.ac.uk

M.V. Metodiev
School of Biological Sciences/Proteomics Unit, University of Essex, Colchester, UK
e-mail: mmetod@essex.ac.uk

© Springer International Publishing Switzerland 2016
A.F.X. Wilhelm, H.A. Kestler (eds.), *Analysis of Large and Complex Data*, Studies in Classification, Data Analysis, and Knowledge Organization,
DOI 10.1007/978-3-319-25226-1_24

275

1   Introduction

Microarray technology, as well as other high-throughput functional genomics experiments, has become a fundamental tool for gene expression analysis in recent years. A major challenge with microarray data is the problem of dimensionality; tens of thousands of genes' expressions are observed in a small number, tens to few hundreds, of observations. For a particular classification task, microarray data are inherently noisy since most genes are irrelevant and uninformative to the given classes (phenotypes).

Performing a supervised classification based on expressions of discriminative genes, identified by an effective gene selection technique, leads to improved prediction accuracy, as well as interpretation of the biological relationship between genes and the considered clinical outcomes. This procedure of pre-selection of informative genes also helps in avoiding over-fitting problem and building a faster model by providing only the features that contribute most to the considered classification task. Identification of discriminative genes for their use in classification has been investigated in many studies (e.g., Apiletti et al. 2012; Mahmoud et al. 2014a). Various approaches have been proposed including Best Individual Genes (Su et al. 2003), Max-Relevance and Min-Redundancy based approaches (Peng et al. 2005), Set Covering Machines (Kestler et al. 2006), MaskedPainter (Apiletti et al. 2012) and Proportional Overlapping Score (POS) approach (Mahmoud et al. 2014a). Different criteria have been used in order to detect the most informative genes including: p-values of statistical tests, e.g. t-test or Wilcoxon Rank Sum test (Lausen et al. 2004); ranking genes using statistical impurity measures, e.g. information gain, Gini index and max minority (Su et al. 2003); selecting genes based on overlapping analysis (Apiletti et al. 2012; Mahmoud et al. 2014a).

Analyzing the overlap between gene expression measures for different classes is an effective criterion for identifying discriminative genes for a considered classification task. Mahmoud et al. (2014a) developed a procedure specifically designed to select genes based on their overlapping degree across different classes. This procedure, named *Proportional Overlapping Score* (POS), calculates a relevance score for each gene. For binary class situations, this score estimates the overlapping degree between the expression intervals of both classes taking into account three factors that form the characteristics of classes' overlapping. It has been defined to provide higher scores for genes with lower discriminative power. Genes are then ranked in ascending order according to their scores. POS method characterizes each gene by means of a *gene mask* that represents the capability of a gene to unambiguously assign training observations to their correct classes. Characterization of genes using training observation masks with their overlapping scores allows the detection of a minimum subset of genes that provides the best classification coverage on a training set of observations. A final gene set is then provided by combining the minimum gene subset with the top ranked genes according to the estimated scores. Feature selection produced by POS is robust against outliers, since gene masks are defined based on the interquartile range of gene's expressions. However, the top ranked genes, given based on POS relevance score, may provide a classifier with redundant information.

In this article, we propose an extended version of POS method, called POSr, that can exploit detection of the minimum subset of genes in a recursive way in order to mitigate redundancy in the final gene selection.

The article is organized as follows. Section 2 shows the main idea of POS and explains the proposed method. The results of proposal are compared with some other gene selection techniques in Sect. 3. Section 4 concludes the article.

2 Methods

2.1 Overlapping Analysis for Binary Class Problems

Microarray data are usually presented in the form of a gene expression matrix, $X = [x_{ij}]$, such that $X \in \mathbb{R}^{P \times N}$ and x_{ij} is the observed expression value of gene i for observation (tissue sample) j where $i = 1, \ldots, P$ and $j = 1, \ldots, N$. Each observation is also characterized by a target class label, y_j, representing the phenotype of the observation being studied. Let $Y \in \mathbb{R}^N$ be the vector of class labels such that its jth element, y_j, has a single value c which is either 1 or 2.

Analyzing the overlap between expression intervals of a gene for different classes can provide a classifier with an important aspect of a gene's characteristic. The idea is that a certain gene i can assign observations to class c because their gene i expression interval in that class is not overlapping with gene i interval of the other class. In other words, gene i has the ability to correctly classify observations for which their gene i expressions fall within the expression interval of a single class.

POS method, proposed by Mahmoud et al. (2014a), initially exploits the interquartile range approach to robustly define gene masks that report the discriminative power of genes avoiding outlier effects. Construction of these masks can be described as follows.

2.1.1 Core Intervals and Gene Masks

For a certain gene i, two expression intervals, one for each class, can be defined for that gene. The cth class core interval for gene i can be defined in the form:

$$I_{i,c} = [a_{i,c}, b_{i,c}], \quad i = 1, \ldots, P, \quad c = 1, 2, \tag{1}$$

such that:

$$a_{i,c} = Q_1^{(i,c)} - 1.5\,\mathrm{IQR}^{(i,c)}, b_{i,c} = Q_3^{(i,c)} + 1.5\,\mathrm{IQR}^{(i,c)}, \tag{2}$$

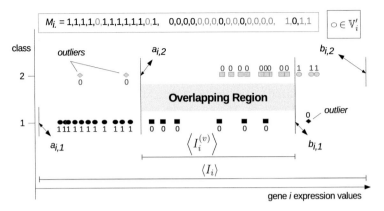

Fig. 1 Core intervals with gene mask. An example for core expression intervals of a gene with 18 and 14 observations belonging to class 1, in *black colour*, and class 2, in *grey colour*, respectively, with its associated mask elements. Elements of the non-overlapped observations set are represented by *circles*

where $Q_1^{(i,c)}$, $Q_3^{(i,c)}$ and $\mathrm{IQR}^{(i,c)}$ denote the first, third empirical quartiles, and the interquartile range of gene i expression values for class c, respectively. The multiplier value of 1.5 is the default value that commonly used with the interquartile range approach for detecting outliers (Tukey 1977).

For each gene, a mask is defined based on its observed expression values and constructed core intervals. Gene i mask is represented by a vector of length equal to the total number of observations. It reports the observations that gene i can unambiguously assign to their correct target classes. Thus, gene masks can represent the capability of genes to classify correctly each observation, i.e. it represents a gene's classification power. For a particular gene i, element j of its mask is set to 1 if the corresponding expression value x_{ij} belongs only to core expression interval I_{i,c_j} of the single class c_j, where c_j is the target class of observation j. Otherwise, it is set to zero.

Figure 1 shows the constructed core expression intervals $I_{i,1}$ and $I_{i,2}$ associated with a particular gene i along-with its gene mask. The non-overlapped observations are represented by circles. The gene mask is sorted corresponding to the observations ordered by increasing expression values.

A matrix of gene masks $M = [m_{ij}]$ can be produced such that the mask of gene i is presented by $M_{i.}$ (the ith row of M) and gene mask element m_{ij} is defined as:

$$m_{ij} = \begin{cases} 1 & if \quad j \in \mathbb{V}'_i \\ 0 & otherwise \end{cases}, \tag{3}$$

where $\mathbb{V}'_i$ is the set that includes non-outliers observations whose observed expressions fall into the non-overlapping region such that $i = 1, \ldots, P$ and $j = 1, \ldots, N$.

2.1.2 Proportional Overlapping Score

An overlapping measure, called Proportional Overlapping Score (*POS*), is developed to estimate the overlapping degree between different expression intervals taking into account three factors: (1) length of the overlapping region; (2) number of overlapped observations; (3) the proportion of classes' contribution to the overlapped observations (Mahmoud et al. 2014a). For each gene i, POS_i is estimated as follows:

$$
POS_i = 4 \frac{\left\langle I_i^{(v)} \right\rangle}{\langle I_i \rangle} \frac{v_i}{\ell_i} \left(\prod_{c=1}^{2} \theta_c \right),
\tag{4}
$$

where $\left\langle I_i^{(v)} \right\rangle$ is the length of the overlapping region, and $\langle I_i \rangle$ is the length of the total core interval which is given by the region between the global minimum and global maximum boundaries of core intervals for both classes, see Fig. 1. Whereas v_i and ℓ_i represent number of observations whose observed expressions of gene i fall within the overlapping region and number of non-outlier observations, respectively, while θ_c is the proportion of class c observations among overlapped observations. Hence, θ_c can be defined as:

$$
\theta_c = \frac{v_{i,c}}{v_i},
\tag{5}
$$

where $v_{i,c}$ represents number of the overlapped observations belonging to class c. The factor 4 is included in (4) in order to scale *POS* values within the interval $[0, 1]$ (Mahmoud et al. 2014a). According to (4) and (5), the value of *POS* measure for gene i shown in Fig. 1 is $4 \cdot \frac{15}{29} \cdot \left(\frac{6}{15} \cdot \frac{9}{15} \right) \cdot \frac{\left\langle I_i^{(v)} \right\rangle}{\langle I_i \rangle} = \frac{72}{145} \cdot \frac{\left\langle I_i^{(v)} \right\rangle}{\langle I_i \rangle}$.

2.2 *Recursive Minimum Sets for Minimizing Redundancy (POSr)*

POS gives its final selection by combining a minimum gene subset produced using gene masks, defined in (3), with the top ranked genes according to the estimated *POS* scores, defined in (4). It is an effective feature selection method for identifying discriminative genes for a considered classification task.

However, POS selections may provide a classifier with redundant information since the set of top ranked genes is likely to have redundancy among its members. Such a redundancy increases the model complexity since it increases the dimensionality without adding further information. Moreover, redundancy may affect classification prediction accuracy as well as interpretation of the underlying biological relationship between the features and considered clinical outcomes.

A gene mask reflects the capability of the gene to correctly classify each observation to its target class. Genes with higher number of 1 bits in their masks are more informative to the considered classification problem [see (3)]. When two genes classify in the same way the same observations, then their masks should be identical. Genes with complementary masks, on the other hand, can provide diverse information to the classifier model.

In this article, we propose an extended version of POS, called POSr, in which gene masks along-with *POS* measure are exploited to identify minimum subsets of genes in a recursive way in order to mitigate the potential redundancy in the final gene selection. The subset is designated to be the minimum one that correctly classifies the maximum number of observations in a given training set, avoiding the effects of expression outliers.

Let $\mathbb{G}_z$ be a set of remaining genes at the zth iteration given by excluding the selected subset of genes at the $(z - 1)$th iteration, such that $\mathbb{G}_1$ is the full set of all genes (i.e., $|\mathbb{G}_1| = P$). Also, let $\overline{M}(\mathbb{G}_z)$ be its aggregate mask which is defined as the logical disjunction *(logic OR)* among all masks corresponding to genes that belong to the set $\mathbb{G}_z$. It can be expressed as follows:

$$\overline{M}(\mathbb{G}_z) = \bigvee_{i \in \mathbb{G}_z} M_{i.} \tag{6}$$

At iteration z, our objective is to search the set, $\mathbb{G}_z$, for the minimum subset, denoted by $\mathbb{G}_z^*$, for which $\overline{M}(\mathbb{G}_z^*)$ equals to the aggregate mask of the corresponding set of genes, $\overline{M}(\mathbb{G}_z)$. In other words, our minimum subset of genes should satisfy the following statement:

$$\underset{\mathbb{G}_z^* \subseteq \mathbb{G}_z}{\arg \min} \left(|\mathbb{G}_z^*| \,\middle|\, \left(\overline{M}(\mathbb{G}_z^*) = \bigvee_{i \in \mathbb{G}_z^*} M_{i.} = \overline{M}(\mathbb{G}_z) \right) \right). \tag{7}$$

This procedure is performed in a recursive way and ends when the required number of genes, set by the user, is selected.

The pseudo code of our procedure, POSr, is reported in Algorithm 3. Its inputs are: the matrix of gene masks, M; *POS* scores for all genes; number of genes to be selected, r. It produces the sequence of selected genes, $\mathbb{T}^*$, as output.

At the initial step ($z = 0$), we let $\mathbb{T} = \emptyset$ (line 2); where $\mathbb{T}$ is a set created to contain the successively selected minimum subsets of genes. Then at each iteration, z, the following steps are performed:

1. We let $k = 0$, $\mathbb{G}_z^* = \emptyset$ and $\overline{M}(\mathbb{G}_z^*) = \mathbf{0}_N$ (lines 5–7) to initialize individual selection within the minimum subset $\mathbb{G}_z^*$, where $\overline{M}(\mathbb{G}_z^*)$ is the aggregate mask of the set $\mathbb{G}_z^*$, see (6). Then at each sub-iteration, k, the following sub-steps are performed:

   a. Among genes of the set $\mathbb{G}_z$, the one(s) with the highest number of mask bits assigned to 1 is (are) chosen to form the set $\mathbb{S}_{zk}$ (line 10).

Algorithm 3: POSr method: recursive minimum subsets

---

1 **Inputs:** M, *POS* scores and number of required genes (r).
2 **Output:** Sequence of the selected genes $\mathbb{T}^*$.

1: $z = 0$ {Initialization}
2: $\mathbb{T} = \emptyset$
3: **while** $|\mathbb{T}| < r$ **do**
4: $\quad z = z + 1$
5: $\quad k = 0$ {Initialization of individual selection}
6: $\quad \mathbb{G}_z^* = \emptyset$
7: $\quad \overline{M}\left(\mathbb{G}_z^*\right) = \mathbf{0}_N$
8: $\quad$ **while** $\overline{M}\left(\mathbb{G}_z^*\right) \neq \overline{M}\left(\mathbb{G}_z\right)$ **do**
9: $\quad\quad k = k + 1$
10: $\quad\quad S_{zk} = \underset{i \in \mathbb{G}_z}{argmax}\left(\sum_{j=1}^{N} I\left(m_{ij}^{(k)} = 1\right)\right)$

$\quad\quad$ {Assign gene set whose masks have max. bits of 1}
11: $\quad\quad g_{zk} = \underset{i \in S_{zk}}{argmin}\,(POS_i)$ {Select the candidate with the best score among the assigned set}

12: $\quad\quad \mathbb{G}_z^* = \mathbb{G}_z^* + g_{zk}$ {Update the target set by adding the selected candidate}
13: $\quad\quad$ **for all** $i \in \mathbb{G}_z$ **do**
14: $\quad\quad\quad M_{i.}^{(k+1)} = M_{i.}^{(k)} \wedge \overline{M}'\left(\mathbb{G}_z^*\right)$ {update gene masks such that the uncovered observations are only considered}
15: $\quad\quad$ **end for**
16: $\quad$ **end while**
17: $\quad \mathbb{T} = \mathbb{T} + \mathbb{G}_z^*$
18: $\quad \mathbb{G}_{z+1} = \mathbb{G}_z - \mathbb{G}_z^*$
19: **end while**
20: $\mathbb{T}^*$ is the sequence whose members are the first r genes in $\mathbb{T}$
21: **return** $\mathbb{T}^*$

---

   b. The gene with the lowest *POS* score among genes in S_{zk}, if there are more than one, is then selected (line 11). It is denoted by g_{zk}.

   c. The set $\mathbb{G}_z^*$ is updated by adding the selected gene, g_{zk} (line 12).

   d. All masks of genes in $\mathbb{G}_z$ are also updated by performing the logical conjunction (*logic AND*) with negated aggregate mask of set $\mathbb{G}_z^*$ (line 14). Note that $M_{i.}^{(k)}$ represents updated mask of gene i at the kth iteration such that $M_{i.}^{(1)}$ is its original gene mask whose elements are computed according to (3).

   e. This sub-procedure is successively iterated and ends when all masks of genes in $\mathbb{G}_z$ have no one bits anymore, i.e. the selected genes cover the maximum number of observations. This situation is accomplished iff $\overline{M}\left(\mathbb{G}_z^*\right) = \overline{M}\left(\mathbb{G}_z\right)$.

2. The set $\mathbb{T}$ is updated by adding the detected minimum subset of genes, $\mathbb{G}_z^*$ (line 17).

3. Genes within the selected minimum subset, $\mathbb{G}_z^*$, are then removed from the set of genes, $\mathbb{G}_z$ (line 18).

4. The procedure is successively iterated and ends when the size of the set $\mathbb{T}$ is greater than or equal the number of required genes, r. Then, the target sequence of selected genes, $\mathbb{T}^*$, is produced by selecting the first r genes in $\mathbb{T}$ (lines 20, 21).

Thus, this approach combines recursively the detected minimum subsets of genes that provide the best classification coverage for a given training set. Selection of the minimum subsets based on the updated gene masks allows to minimize redundancy among the final selection list.

Table 1 Description of used gene expression datasets

Dataset	Genes	Observations	Class-sizes	Est. error	Source
Leukaemia	7129	72	47/25	0.049	Golub et al. (1999)
Breast	4948	78	34/44	0.369	Michiels et al. (2005)
Srbct	2308	54	29/25	0.0008	Statnikov et al. (2005)
Lung	12533	181	150/31	0.003	Gordon et al. (2002)
GSE24514	22215	49	34/15	0.0406	Alhopuro et al. (2012)
GSE4045	22215	37	29/8	0.2045	Laiho et al. (2007)
GSE14333	54675	229	138/91	0.4141	Jorissen et al. (2009)

3 Results and Discussion

For evaluating a feature selection method, one can assess the accuracy of a classifier applied after the feature selection process. Such an assessment can verify the efficiency of gene selections. In this article, our experiment is conducted using seven publicly available gene expression datasets in which the POSr method is validated by comparison with three well-known gene selection techniques along-with POS method. The performance is evaluated by obtaining the classification error rates from three different classifiers: Random Forest (RF), k Nearest Neighbour (kNN), Support Vector Machine (SVM).

Table 1 summarizes the characteristics of the datasets. The estimated classification error rate is based on the Random Forest classifier with the full set of features, without pre-selection.

Fifty repetitions of tenfold cross validation analysis were performed for each combination of dataset, feature selection algorithm, and a given number of selected genes, up to 50, with the considered classifiers. For each experimental repetition, the split seed was changed while the same folds and training datasets were kept for all feature selection methods. To avoid bias, gene selection algorithms have been performed only on the training sets. For each fold, the best subset of genes has been selected according to the Wilcoxon Rank Sum (Wil-RS) technique, Minimum Redundancy Maximum Relevance (mRMR) method (Peng et al. 2005), MaskedPainter (MP) (Apiletti et al. 2012), POS (which is implemented in *propOverlap* R package (MAHMOUD et al. 2014b), along-with our proposed method. The expressions of the selected genes as well as the class labels of the training observations have then been used to construct the considered classifiers. The classification error rates on the test sets are separately reported for each classifier and the average error rate over all the 50 repetitions is then computed.

To highlight the entire performances of the compared methods against our proposed approach, a comparison between the minimum error rates achieved by each method was conducted. Table 2 summarizes these results. Each row shows the minimum error rate (along-with its corresponding set size, shown in brackets) for a specific dataset, reported in the first column. In addition, the error rates of the corresponding classifiers with the full set of features, without feature selection, are

Table 2 Comparison between the minimum error rates yielded by the feature selection methods using RF, *k*NN and SVM classifiers

Dataset	Classifier	Wil-RS	mRMR	MP	POS	POSr	Full set
Leukaemia	RF	0.030 (20)	0.118 (40)	0.015 (9)	0.0002 (40)	**0.000 (9)**	0.049
	*k*NN	0.074 (6)	0.135 (50)	0.019 (1)	**0.005 (1)**	**0.005 (1)**	0.109
	SVM	0.047 (8)	0.126 (50)	0.022 (1)	**0.005 (1)**	**0.005 (1)**	0.131
Lung	RF	0.040 (30)	0.016 (48)	0.008 (46)	0.007 (48)	**0.006 (48)**	0.003
	*k*NN	0.203 (12)	0.027 (49)	0.017 (17)	0.011 (12)	**0.002 (40)**	0.0005
	SVM	0.066 (50)	0.026 (50)	0.021 (19)	0.010 (47)	**0.008 (38)**	0.024
Breast	RF	0.371 (50)	0.407 (48)	0.354 (48)	**0.308 (45)**	0.317 (48)	0.369
	*k*NN	0.405 (11)	0.404 (50)	0.346 (19)	0.332 (11)	**0.328 (11)**	0.405
	SVM	0.401 (39)	0.407 (50)	0.359 (21)	0.313 (22)	**0.303 (37)**	0.438
Srbct	RF	0.069 (24)	0.074 (46)	0.009 (32)	0.003 (48)	**0.002 (44)**	0.0008
	*k*NN	0.157 (3)	0.098 (48)	0.005 (26)	**0.005 (22)**	0.008 (32)	0.034
	SVM	0.131 (50)	0.124 (49)	0.010 (21)	**0.003 (8)**	0.004 (47)	0.079
GSE4045	RF	0.134 (24)	0.187 (37)	0.137 (21)	0.114 (27)	**0.105 (33)**	0.205
	*k*NN	0.166 (43)	0.207 (38)	0.137 (50)	0.142 (3)	**0.112 (6)**	0.103
	SVM	0.134 (24)	0.187 (37)	0.095 (47)	0.114 (29)	**0.085 (47)**	0.214
GSE14333	RF	**0.421 (10)**	–	0.438 (31)	0.437 (34)	0.442 (44)	0.414
	*k*NN	**0.420 (8)**	–	0.455 (23)	0.450 (34)	0.448 (47)	0.438
	SVM	0.427 (9)	–	**0.412 (1)**	0.431 (1)	0.431 (1)	0.407
GSE24514	RF	0.054 (47)	0.063 (50)	0.036 (48)	**0.032 (24)**	0.034 (26)	0.041
	*k*NN	**0.032 (20)**	0.041 (50)	0.036 (50)	0.039 (50)	0.038 (49)	0.041
	SVM	0.041 (40)	0.059 (50)	0.037 (40)	**0.034 (30)**	0.036 (43)	0.070

Boldface numbers indicate the lowest classification error rates (highest accuracy among compared methods) achieved using the corresponding classifier. The numbers in brackets represent the size of the gene sets that corresponding to the minimum error rate

reported in the last column. Due to limitations of the R package "mRMRe" (De Jay et al. 2013), mRMR selections could not be conducted for datasets having more than "46340" features. Therefore, mRMR method is excluded from the analysis of the "GSE14333" dataset.

Table 2 demonstrates that the proposed approach, POSr, provides the minimum error rates (the highest accuracy) for all used classifier models with most of the used datasets. In particular, for the "Leukaemia", "Lung" and "GSE4045" datasets, it outperforms the other methods using all different classifiers. For the "Breast" and "Srbct" datasets, POSr provides the best performance using kNN and SVM, for the "Breast" dataset, and RF, for the "Srbct" dataset. While for the "GSE14333" and "GSE24514" datasets, Wil-RS and POS methods, respectively, outperformed the other compared methods.

Figure 2 shows that our proposed approach provides less classification error rates than other compared gene selection methods on the "Breast" and "Lung" datasets at different selected gene set sizes. The stability index proposed by Lausser et al. (2013) is used to measure the stability of the compared method at different set sizes

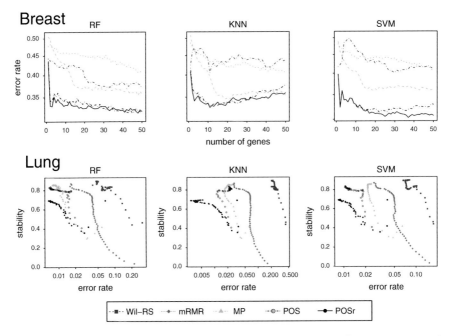

Fig. 2 Averages of classification error rates and stability-accuracy plots: (*first row*) averages of classification error rates for the "Breast" dataset using RF, kNN and SVM classifiers, (*second row*) stability-accuracy plots for the "Lung" dataset

of features. The relation between the accuracy and stability has been depicted for the "Lung" dataset. Different dots for the same gene selection method correspond to different set sizes of genes. For all classifiers, POSr achieves a good trade-off between accuracy and stability for "Lung" data, see the second row panels of Fig. 2.

4 Conclusion

A gene selection method, POSr, is proposed as an extension of POS technique. The proposed approach detects minimum subsets of genes in a successive way. The final selection is then produced by combining these subsets in order to reduce the redundancy among selected genes. It is designed for binary class situations. The classification error rates achieved by Random Forest, k Nearest Neighbour and SVM classifiers for POSr were compared with Wilcoxon Rank Sum, Maximum Relevance Minimum Redundancy, MaskedPainter and POS on seven benchmarked gene expression datasets. The relation between classification accuracy and selection stability is also outlined. The proposed method performed better than compared methods on most datasets for all classifiers. It is an effective approach in enhancing the prediction classification performance of the considered classifier models using less number of features compared to the other studied gene selection methods. Furthermore, POSr approach provides good stability scores at small as well as large sets of selected genes.

References

Alhopuro, P., Sammalkorpi, H., Niittymäki, I., Biström, M., Raitila, A., Saharinen, J., et al. (2012). Candidate driver genes in microsatellite-unstable colorectal cancer. *International Journal of Cancer, 130*(7), 1558–1566.

Apiletti, D., Baralis, E., Bruno, G., & Fiori, A. (2012). Maskedpainter: Feature selection for microarray data analysis. *Intelligent Data Analysis, 16*(4),717–737.

De Jay, N., Papillon-Cavanagh, S., Olsen, C., El-Hachem, N., Bontempi, G., & Haibe-Kains, B. (2013). mRMRe: An R package for parallelized mRMR ensemble feature selection. *Bioinformatics, 29*(18), 2365–2368.

Golub, T. R., Slonim, D. K., Tamayo, P., Huard, C., Gaasenbeek, M., Mesirov, J., et al. (1999). Molecular classification of cancer: Class discovery and class prediction by gene expression monitoring. *Science, 286*(5439), 531–537.

Gordon, G., Jensen, R., Hsiao, L., Gullans, S., Blumenstock, E., Ramaswamy, S., et al. (2002). Translation of microarray data into clinically relevant cancer diagnostic tests using gene expression ratios in lung cancer and mesothelioma. *Cancer Research, 62*(17), 4963–4967.

Jorissen, R. N., Gibbs, P., Christie, M., Prakash, S., Lipton, L., Desai, J., et al. (2009). Metastasis-associated gene expression changes predict poor outcomes in patients with Dukes stage B and C colorectal cancer. *Clinical Cancer Research, 15*(24), 7642–7651.

Kestler, H., Lindner, W., & Müller, A. (2006). Learning and feature selection using the set covering machine with data-dependent rays on gene expression profiles. In F. Schwenker & S. Marinai (Eds.), *Artificial neural networks in pattern recognition (ANNPR 06) volume LNAI 4087* (pp 286–297). Heidelberg: Springer.

Laiho, P., Kokko, A., Vanharanta, S., Salovaara, R., Sammalkorpi, H., Järvinen, H., et al. (2007). Serrated carcinomas form a subclass of colorectal cancer with distinct molecular basis. *Oncogene, 26*(2), 312–320.

Lausen, B., Hothorn, T., Bretz, F., & Schumacher, M. (2004). Assessment of optimal selected prognostic factors. *Biometrical Journal, 46*(3), 364–374.

Lausser, L., Müssel, C., Maucher, M., & Kestler, H. A. (2013). Measuring and visualizing the stability of biomarker selection techniques. *Computational Statistics, 28*(1), 51–65.

Mahmoud, O., Harrison, A., Perperoglou, A., Gul, A., Khan, Z., & Lausen, B. (2014b). propOverlap: Feature (gene) selection based on the proportional overlapping scores. R package version 1.0, http://CRAN.R-project.org/package=propOverlap

Mahmoud, O., Harrison, A., Perperoglou, A., Gul, A., Khan, Z., Metodiev, M., et al. (2014a). A feature selection method for classification within functional genomics experiments based on the proportional overlapping score. *BMC Bioinformatics, 15*, 274.

Michiels, S., Koscielny, S., & Hill, C. (2005). Prediction of cancer outcome with microarrays: A multiple random validation strategy. *The Lancet, 365*(9458), 488–492.

Peng, H., Long, F., & Ding, C. (2005). Feature selection based on mutual information criteria of max-dependency, max-relevance, and min-redundancy. *IEEE Transactions on Pattern Analysis and Machine Intelligence, 27*(8), 1226–1238.

Statnikov, A., Aliferis, C. F., Tsamardinos, I., Hardin, D., & Levy, S. (2005). A comprehensive evaluation of multicategory classification methods for microarray gene expression cancer diagnosis. *Bioinformatics, 21*(5), 631–643.

Su, Y., Murali, T., Pavlovic, V., Schaffer, M., & Kasif, S. (2003). Rankgene: Identification of diagnostic genes based on expression data. *Bioinformatics, 19*(12), 1578–1579.

Tukey, J. (1977). *Exploratory data analysis*. Reading, Mass. Addison-Wesley.

The Identification of Relations Between Smart Specialization and Sensitivity to Crisis in the European Union Regions

Beata Bal-Domańska

Abstract The purpose of the article is an attempt to measure and assess the sensitivity to crisis of the European Union regional economies having considered their sector structure. The research results presented in literature references indicate that the differences in sector structure of particular economies were the main reason of diverse crisis consequences. The study covered the NUTS-2 level regions in the period 2005–2011. Econometric models for panel data with adequate estimation techniques are used for the assessment of the EU regions' sensitivity to the effects of 2008 crisis. The application of panel data allows for including in the analysis also the specific, non-measurable, individual effects for particular regions and time, what seems a particularly useful tool for the description of regional economies growth in the crisis.

1 Introduction

The effects of 2008 financial crisis, initiated at the American housing market, were experienced by many European economies and manifested by the prosperity downturn, which resulted in the decrease of production and reduced economic growth rate.

The research results, presented in the subject literature Groot et al. (2011), Gorzelak (2009), indicate that the differences in sector structure of particular economies were the main reason for the occurred, diverse consequences of the crisis. Special resistance to crisis is associated with "modern" sectors and also the ones related to non-material services the development of which is based on knowledge and innovation (Markowska 2013). According to Eurostat based study about high and medium-high technology industries, authored by Jaegers et al. (2013), it was confirmed that a manufacturing sector segment was the powerhouse of growth. Since 2005 the industrial production index and other short-term statistical indicators

B. Bal-Domańska (✉)
Department of Regional Economics, Wroclaw University of Economics, Nowowiejska 3, 58-500 Jelenia Góra, Poland
e-mail: beata.bal-domanska@ue.wroc.pl

© Springer International Publishing Switzerland 2016
A.F.X. Wilhelm, H.A. Kestler (eds.), *Analysis of Large and Complex Data*, Studies in Classification, Data Analysis, and Knowledge Organization,
DOI 10.1007/978-3-319-25226-1_25

were developing much more favourably for the EU-27 high-tech manufacturing than for the entire industry. Despite the financial and economic crisis, high technology manufacturing production increased by 26 % between the first quarter of 2005 and the third quarter of 2012. Medium-high technology industries showed the growth by 7 %, whereas medium-low technology and low-tech production recorded even a decrease (−5 and −6 %). As opposed to the entire industry the level of production in 2012 was almost the same as in 2005.

The need for smart growth is particularly emphasized in the long-term strategy Europe 2020 European Commission (2010). It is to be achieved through the development of knowledge-based economy and innovation. It can be assumed that smart growth, representing the set of instruments (indicators) stimulating growth, can be defined in terms of three pillars: creativity, innovation and smart specialization. The presented analysis is based on one of the pillars Markowska and Strahl (2013), i.e. smart specialization, which emphasizes the actual volume and role of knowledge, as well as high and mid-tech sector in region's economy.

The purpose of the article is an attempt to measure and assess the sensitivity to crisis of the European Union regional economies having considered their sector structure in terms of smart specialization in manufacturing and service sector.

2 Research Procedure and Data

The 2008 economic crisis was visible in many spheres and had impact on the deteriorating socio-economic situation of countries and regions. Several variables could be used to characterize the crisis effects in economy, of which two were selected for the final analysis: *GDP*—gross domestic product per capita by purchasing power standard (PPS), *EMPL*—employment rate in % or$\Delta EMPL$ employment changes value in % (in the group aged 25–64). In order to analyse the regions Eurostat database was used. The data for employment come from the EU Labour Force Survey (LFS). The conducted analysis was based on the panel covering 268 EU regions (except French regions of Guadeloupe, Martinique, Guyane and Réunion) at NUTS-2 level in the period 2005–2011.

The first part of the analysis refers to the assessment of economic crisis effects' diversification in the space of EU regions and in the defined classes of regions regarding the level of smart specialization in manufacturing and service sector. This part applies the selected descriptive statistics and the underlying classifications.

Smart specialization emphasizes the actual volume and role of knowledge-based sectors (manufacturing and services) in the employment structure of particular countries. Two diagnostic indicators were defined for smart specialization: *KIS*—employment in knowledge-intensive services as the share of total employment (%)—the classification of knowledge-intensive services is based on the share of tertiary education graduates at NACE 2-digit level, HMMS —employment in high and medium-high technology manufacturing as the share of total employment (%)—high and medium-high technology manufacturing covers the subset of manufacturing industries in which expenditure on research and development is higher than 8 % (high) or 2 % (medium) of revenues.

The following groups of regions, distinguished on the basis of median value calculated as the percentage of employment in *KIS/HMMS* against the total employment in 2008 in the above-defined sectors, were defined: (1) ALL—reference group consisting of all analysed regions, (2) low/(3) high KIS—each of them covers 134 regions which recorded the value of *KIS* indicator below/above median (median = 37.2), (4) low / (5) high HMMS —covers 134 regions which recorded the value of *HMMS* indicator below/above median (median = 5.0).

The spatial distributions of regions featuring high KIS and high HMMS are not equal—the differences are quite visible. KIS regions, in which the employment in *KIS* sector is over 37.2 % of total employment, are located mainly in central (France, The Netherlands, Denmark, some regions of Germany) and northern (Scandinavian countries) Europe and also GB. The regions where HMMS are of big importance (*HMMS* presents minimum 5.0 % of total employment) belong to Germany, northern Italy, France, GB, western Poland, Czech Republic, Slovakia, Romania, Austria, Denmark, southern regions of Sweden and Finland. The regions characterized by low employment level in both knowledge sectors (services and manufacturing) are represented by Portuguese, Spanish, Greek, Baltic countries' regions (Latvia, Lithuania, Estonia), north-eastern Poland and Romania.

The second part of the article analyses the impact of changes in employment level and human capital quality on the level of regional development. The relationship assessment was performed based on the model structure referring to an extended neoclassical growth model according to Mankiw et al. (1992)—MRW. The analysed model can be described in the following way:

$$\ln \text{GDP}_{it} = \alpha_i + \beta_{jlk} \ln \text{TETR}_{it} + \beta_{jlk} \ln \Delta\text{EMPL}_{it} + \varepsilon_{it} \qquad (1)$$

where: $\ln \text{GDP}_{it}$—gross domestic product (*GDP*) per capita in PPS in i-th region ($i = 1, 2, \ldots, N$) and t-th year ($t = 1, 2, \ldots T$), β_{jlk}—parameter defining the impact of j-th ($j = \ln\text{TETR}, \ln \Delta\text{EMPL}$) variable in l-th group of regions ($l = \text{KIS}$, HMMS) in k-th period (the period 2005–2007 before the crisis; the period 2008–2011 after the crisis); α_i—specific for each region, fixed in time, individual effects; ε_{it}—random term, *TETR*—employment among workers with tertiary education as the percentage of total employment (aged 25–64). The analysis is focused on major factors particularly sensitive to economic crisis, such as changes in employment rate level. Moreover, a variable illustrating human capital resources was introduced in the model. This variable does not show strong relationship with the effects of crisis, its purpose was to introduce variability into the model resulting from different levels of knowledge and innovation advancement in particular regions. Such model specification allowed assessing to what extent the factors directly related to labour market are capable of explaining GDP variability.

In order to estimate structural parameter values, adequate estimation techniques, typical for panel data, were applied (Maddala 2006; Wooldridge 2002; Verbeek 2000; Greene 2003; Danska 2000). Two estimators: Least Squares with Dummy

Variable (fixed effects, FE) and GLS (random effect, RE) model were applied in the study. FE remove the effect of individual effects and hence the estimated coefficients of models cannot be biased due to the omitted time-invariant characteristics and therefore the net effect of the predictors on the depended variable can be assessed. The random effects approach (GLS) is identified with zero correlation between the unobservable effect α_i and the values of observable explanatory variables x_{it}. However, if the fixed effects occur, it actually means allowing the possibility of individual effects correlation α_i with the explanatory variables x_{it}. A critical approach to the assumption about the absence of correlation between α_i and x_{it} Mundlak's effect (Mundlak 1961) can be found in subject literature since, in practice, it is difficult to hold. The assumption about the independence of individual effects and explanatory variables can be verified by Hausman's test, which checks whether the results obtained based on FE and GLS estimators are significantly different. The null hypothesis is that the preferred model is random effects vs. the alternative the fixed effects. In accordance with the idea of this test, having assumed that $E(c_i x_{it}) = 0$, estimator GLS is more effective than FE estimator, the obtained assessments should not be significantly different. Robust estimator HAC Arellano (2003) was applied in order to avoid negative consequences of heteroscedasticity and autocorrelation for model estimates. In order to assess whether model specification is correct and the introduction of individual effects is founded F test for FE model and Breuch-Pagan Lagrange Multiplier test for RE model were used (Greene 2003). F test allows checking the joint significance of artificial variables regarding individual effects for each object (regions) of the study. Zero hypothesis referring to constant intercepts (individual effects) can be presented in the following way: $H_0 = \alpha_i = \alpha = $ const., where $i = 1, 2, \ldots, N$.

Breuch-Pagan tests whether there is no significant difference across units (i.e. no panel effect). The null hypothesis in the BP test states that variances across objects (regions) are zero. R^2 determination coefficient (only for FE model—two version FE with dummy variable or within) was used as the quality measure of model adjustment to empirical data. It informs about the extent to which the variability of an explained variable is presented by the model which ranges from 0 to 1 (the closer the value to unity the more y variability was explained in the model).

3 The Selected Economy Phenomena Sensitivity to Crisis in Groups of Region Distinguished by the Level of Smart Specialization

The presentation of research results starts from two basic indicators, i.e.: the level of *GDP* per capita (by PPS) and also employment rate in the period of 2005 and 2011. Figure 1 presents the average values and the distribution of these macroeconomic indicators in the two groups of regions, i.e. KIS and HMMS divided by the median,

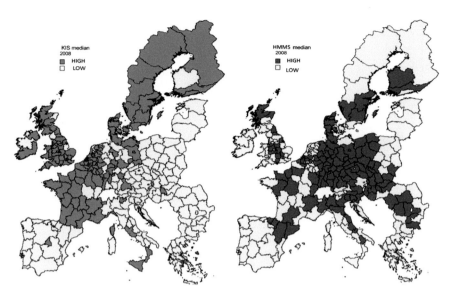

Fig. 1 The EU regions characterized by high and low employment level in KIS and HMMS knowledge sectors in 2008. *Source*: author's compilation

to regions featuring low, below 37.3 % employment rate in knowledge-intensive services (left) and 5.2 % employment rate in HMMS sector (right).

The differences between the average level of *GDP* in the regions presenting low and high level of KIS/HMMS are significant. High KIS region reached 40–50 % higher *GDP* (average) than low KIS region. In HMMS regions the difference is only 15–17 % in favour of high-tech regions. One more conclusion results from the analysis of the above presented figures, i.e. the visible outliers refer to the regions featuring an exceptionally high *GDP* level. These regions are characterized by the highly developed service sector and rather poorly developed HMMS sector.

Figure 2 presents the same set of information for employment rate. In this case the average level of the discussed phenomenon was also higher in "high" regions and lower in the "low" ones. The differences between employment rates are smaller in case of *GDP* level. In KIS regions they reached from 4 to 6 percentage points. In HMMS regions these differences were insignificant in the first analysed period and amounted to less than 1 pp in 2005. In the following years the distance between high and low HMMS regions kept growing to reach 5 pp. in 2011 (regarding mean value).

In the period of 2008 crisis and also in 2009 the significant drops in the level of regional macroeconomic indicators (*GDP* see Fig. 2 and employment rate see Fig. 3) were recorded. In accordance with *GDP* per capita, the regions (regarding mean value) returned to the level preceding 2008. In terms of employment rate only the highly developed regions (KIS and HMMS) returned to the level preceding 2008 (especially the regions featuring a well-developed industry sector). In case of

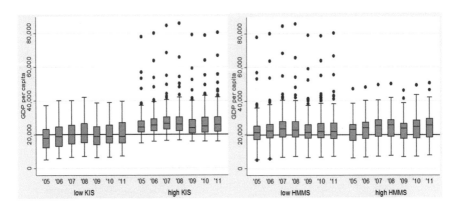

Fig. 2 The average values and the distribution of *GDP* in KIS (*left*) and HMMS regions (*right*) in the period of 2005–2011. *Source*: author's compilation

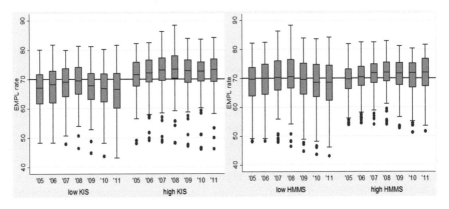

Fig. 3 The average values and the distribution of employment rate in KIS (*left*) and HMMS (*right*) regions in the period of 2005–2011. *Source*: author's compilation

high-tech manufacturing the Eurostat study concludes that the "recovery was driven by pharmaceuticals and air and spacecraft machinery. The decline in the production of high-tech businesses between the second quarter of 2008 and the first quarter of 2009 was mainly due to a fall in the production of computers, electronic and optical products" (where the competition from developing countries is minimal)— Jaegers et al. (2013). The regions characterized by the low level of knowledge development (in services and manufacturing) recorded an ongoing decrease in average employment rate until 2008 and in 2011 reached the lowest employment rate in the entire studied period.

Human capital, apart from knowledge, represents one of the more important growth factors. Figure 4 illustrates human capital changes measured by the share of workers with tertiary education in the total number of workforce in the defined groups of regions, i.e. low/high KIS/HMMS. In accordance with the presented data human capital resources were growing year after year in both groups.

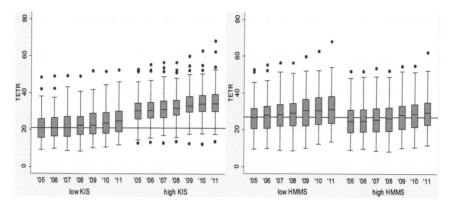

Fig. 4 The average values and the distribution of TETR in KIS (*left*) and HMMS (*right*) regions in the period of 2005–2011. *Source*: author's compilation

It is worth noticing that the regions featuring high level of KIS sector present higher human capital level against the ones classified in the low KIS group (which results from the methodology for identifying which services are included in KIS based on education level). In case of the group of regions characterized by a well-developed high and mid-tech sector the regions covered by high HMMS group present smaller resources of educated workers. Therefore, one can conclude that the number of workers with tertiary education is not of key significance for the development of high and medium-high technology sector.

4 Results of Econometric Analysis of Economic Growth Models in Groups of Region Distinguished by the Level of Smart Specialization in the Period Before and After 2008 Crisis

Table 1 and Figs. 5, 6, and 7 present the results of Solow model estimations. All calculations were done in STATA 11 and GRETL program. The model estimates the impact of two factors related to work: the growth of human resources $\Delta EMPL$ and human capital *TETR* comparing to *GDP* per capita in the periods prior to (2005–2007) and after the crisis (2008–2011). The analyses were conducted for five defined groups of regions: ALL, low and high KIS, low and high HMMS. Summarizing the estimation results (Table 1) the following observations can be presented: (1) in each case the individual effects characteristic for each region turned out to be statistically significant (as confirmed by F test for FE models and Breuch-Pagan test for RE models), (2) factors included in the model explained the *GDP* per capita variability to a small extent, the *GDP* variance is due to differences across panels (rho equals over 95 % in each model), (3) in accordance with Hausman's test results: in the

Table 1 The results of growth models estimation for the EU regions in the period before and after 2008 crisis

Specification	Before crisis (2005–2007)		After crisis (2008–2011)	
	FE	GLS (RE)	FE	GLS (RE)
ALL				
R^2 (FE/within)	0.987/0.097	–	0.990/0.095	–
F/B	$F = 110.8(0.000)$	$B = 740.0(0.000)$	$F = 218.9(0.000)$	$B = 1438.9(0.000)$
Hausman (H)	$H = 10.0(0.007)$		$H = 86.2(0.000)$	
Low KIS				
R^2 (FE/within)	0.984/0.134	–	0.99/0.166	–
F/B	$F = 114.6(0.000)$	$B = 369.5(0.000)$	$F = 257.8(0.000)$	$B = 733.9(0.000)$
Hausman (H)	$H = 2.6(0.273)$		$H = 17.1(0.000)$	
High KIS				
R^2 (FE/within)	0.979/0.074	–	0.983/0.054	–
F/B	$F = 76.6(0.000)$	$B = 361.2(0.000)$	$F = 146.5(0.000)$	$B = 727.8(0.000)$
Hausman (H)	$H = 47.7(0.021)$		$H = 27.2(0.000)$	
Low HMMS				
R^2 (FE/within)	0.987/0.121	–	0.991/0.09	–
F/B	$F = 119.6(0.000)$	$B = 402.8(0.000)$	$F = 248.6(0.000)$	$B = 781.4(0.000)$
Hausman (H)	$H = 5.22(0.073)$		$H = 54.3(0.000)$	
High HMMS				
R^2 (FE/within)	0.986/0.081	–	0.989/0.151	–
F/B	$F = 101.2(0.000)$	$B = 336.3(0.000)$	$F = 197.0(0.000)$	$B = 661.1(0.000)$
Hausman (H)	$H = 4.8(0.091)$		$H = 33.9(0.000)$	

Source: author's compilation

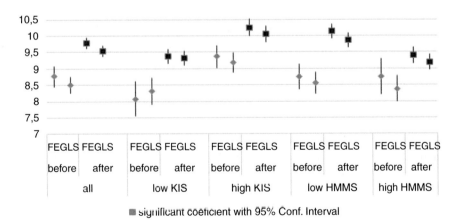

Fig. 5 The results of growth models estimation for the EU regions in the period before and after 2008 crisis for common for all regions constant term (the beginning level of ln*GDP*). *Source*: author's compilation

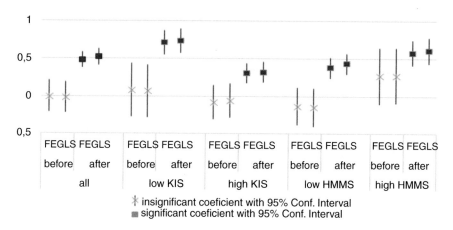

Fig. 6 The results of growth models estimation for the EU regions in the period before and after 2008 crisis for human resources (ln$\Delta EMPL$). *Source*: author's compilation

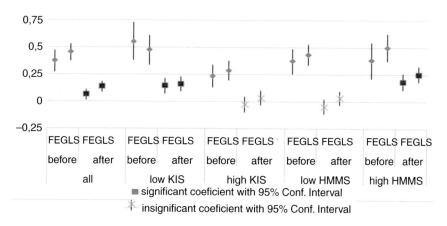

Fig. 7 The results of growth models estimation for the EU regions in the period before and after 2008 crisis for human capital (ln*TETR*). *Source*: author's compilation

period before crisis GLS at 0.02 significance level was the more effective estimator (except for ALL group); in the period after the crisis FE turned out to be the proper estimator at any significance level.

In the period before the crisis human capital (described by ln*TETR*) only was recognized as the statistically significant factor in ln*GDP* per capita creation, whereas employment growth did not show any statistically significant relation. In the groups of regions distinguished in terms of knowledge sectors development (low and high HMMS) human capital significance for *GDP* per capita changes was at similar level and amounted to 0.433 % in low HMMS regions and 0.499 in high HMMS regions. Human capital presented a much higher flexibility level in the low KIS group regions and amounted to 0.555 %, whereas in high KIS regions only

0.237 %. It can be interpreted that in the regions featuring low development level (Fig. 2.) the increase in human capital is significantly related to ln*GDP* growth, while in the regions characterized by high development level and, at the same time, high human capital saturation (Fig. 4.) further increase in its resources results in smaller effects (one should look for develop indicators in others factors).

The situation changed in the period after the crisis. The increase in human resources (ln$\Delta EMPL$) in each of the analysed groups was recognized as the significant factor enhancing regional growth. Human capital (ln*TETR*) represented a statistically significant factor only in low KIS and high HMMS groups, i.e. in the regions with the relatively low level of human capital. The values of structural parameters indicate impact of human capital on ln*GDP* per capita changes amounting to 0.187 % in high HMMS and 0.145 % in low KIS. In the groups of regions where the statistically significant impact of human capital on ln*GDP* per capita was recorded the much higher parameter values of workforce growth impact on *GDP* per capita were also observed (0.577 % in high HMMS and 0.714 % in low KIS against 0.376 % in low HMMS and 0.302 % in high KIS).

5   Conclusions

The most important conclusions resulting from the presented analysis are as follows: The level of KIS and HMMS sectors development influences the level of regional development manifested by higher *GDP* value and higher level of employment rate in the regions featuring better developed KIS and high and mid-tech industry sectors (HMMS), however, the differences in development level are significantly higher in the regions with the well-developed service sector. The regions were reacting differently to crisis in terms of production (*GDP*) and employment rate. After the crisis the employment rate in the regions presenting low KIS and low HMMS groups was decreasing. In case of *GDP* per capita after the crisis having mainly considered the final analysed period 2010–2011 the highest growth rate was recorded in high HMMS and low KIS regions (against the, respectively, low HMMS and high KIS). The increase in human resources and human capital explain *GDP* per capita level changes only to a small extent, especially in the post-crisis period and for low KIS regions other regional factors, unobservable in the model structure, were of significant importance for regional development. Human capital increase represented the most important and the only growth factor in each of the groups of regions before the crisis, however, in the regions characterized by the highest development level, measured by *GDP* per capita (high KIS), its significance for *GDP* per capita changes was lower. The increase in workforce number represented the most important growth factor after the crisis along with the human capital in low KIS and high HMMS regions, however, the changes recorded in human capital were of smaller significance than in the period before the crisis.

Acknowledgements Project has been financed by the Polish National Centre for Science, decision DEC-2013/09/B/HS4/00509.

References

Arellano, M. (2003). *Panel data econometrics.* Oxford: Oxford University Press.

Danska, B. (2000). *Przestrzenno-czasowe modelowanie zmian w działalności produkcyjnej w Polsce. Zatosowanie modeli panelowych.* Absolwent, Łódź.

European Commission. (2010). *A strategy for smart, sustainable and inclusive growth.* Communication from the Commission EUROPE 2020, Brussels, 3.3.2010.

Gorzelak, G. (2009). *Geografia polskiego kryzysu. Kryzys peryferii czy peryferia kryzysu [The geography of Polish crisis. The crisis of periphery or the periphery of crisis].* Warszawa: Regional Studies Association.

Greene, W. H. (2003). *Econometric analysis.* Upper Saddle River, NJ: Pearson Education International.

Groot, S. T. P., Mohlmann, J. L., Garretsen, J. H., & de Groot, H. L. F. (2011). The crisis sensitivity of European countries and regions: Stylized facts and spatial heterogeneity. *Cambridge Journal of Regions, Economy and Society, 4*(3), 437–456.

Jaegers, T., Lipp-Lingua, C., & Amil, D. (2013). *High-technology and medium-high technology industries main drivers of EU-27's industrial growth.* Eurostat, Statistics in Focus 1.

Maddala, G. S. (2006). *Ekonometria.* Warszawa: PWN.

Mankiw, N., Pomer, D., & Weil, D. (1992). A contribution to the empirics of economic growth. *The Quarterly Journal of Economics, 107*(2), 407–437.

Markowska, M. (2013). *Ocena zależności miedzy rozwojem inteligentnym a odpornoscia na kryzys ekonomiczny w wymiarze regionalnym – przeglad badań,* (The assessment of relations between smart growth and resistance to economic crisis in regional dimension—research review, The Conference entitled *Local economy in theory and practice*), Mysłakowice.

Markowska M., & Strahl D. (2013), Klasyfikacja europejskiej przestrzeni regionalnej ze względu na filar inteligentnego rozwoju. In Pawełek, B. (red). *Zastosowanie metod iloś- ciowych i jakościowych w modelowaniu i prognozowaniu zjawisk społeczno-gospodarczych.* (s. 201–219). Kraków: Wyd. Uniwersytetu Ekonomicznego w Krakowie.

Mundlak, Y. (1961). *Empirical production functions free of management bias. Journal of Farm Economics, 43*(1), 44–56. doi:10.2307/1235460.

Verbeek, M. (2000). *A guide to modern econometric.* New York: Wiley.

Wooldridge, J. M. (2002). *Econometric analysis of cross section and panel data.* Cambridge: Massachusetts Institute of Technology.

Part VII
Data Analysis in Marketing

Market Oriented Product Design and Pricing: Effects of Intra-Individual Varying Partworths

Stephanie Löffler and Daniel Baier

Abstract Conjoint analysis is a widespread method for modeling and measuring preferences for multi-attributed products in marketing: A sample of customers are asked to evaluate (fictive) offers (attribute-level-combinations). From these individual responses partworths are estimated and used to design and price offers that maximize, e.g., market share, sales, or profit. However, it can be theoretically and empirically argued that partworths not only vary across individuals but also within them. In this paper, we discuss an approach that respects these variations. Partworths are situation-specific modeled at the individual level. The empirical partworth distributions are estimated using Bayesian procedures. The approach is applied to waterpark design and pricing using simulated and real data. It is shown that taking these variations into account influences the maximization.

1 Introduction

Marketing managers use conjoint analysis since the 1970s to design and price competing multi-attributed goods or services. Maybe the best known early application was Marriot's introduction of the new hotel chain "Courtyard by Marriot" in 1982 (see Wind et al. 1989): To support the hotel chain design, 50 hotel chain attributes and 167 levels were selected during workshops. A sample of business and nonbusiness travelers were asked to evaluate (fictive) offers characterized by these attributes and levels with respect to their willingness to stay overnight. The responses were analyzed using regression-like procedures, the resulting partworths then formed the basis for market share predictions and the development of design and pricing recommendations. The approach was so successful that nowadays, conjoint analysis is widespread in many application fields.

Selka and Baier (2014) describe 1899 commercial applications over the last 10 years alone in Germany. In most cases design (59 %) and pricing problems (88 %)

S. Löffler (✉) • D. Baier
Institute of Business Administration and Economics, BTU Cottbus-Senftenberg, Senftenberg, Germany
e-mail: stephanie.loeffler@tu-cottbus.de; daniel.baier@tu-cottbus.de

© Springer International Publishing Switzerland 2016
A.F.X. Wilhelm, H.A. Kestler (eds.), *Analysis of Large and Complex Data*, Studies in Classification, Data Analysis, and Knowledge Organization,
DOI 10.1007/978-3-319-25226-1_26

were in the focus (multiple assignments allowed). Most of the applications used choice-based conjoint analysis (94 %) and web interviewing (74 %). However, Selka and Baier (2014) also found out that the average validity of these applications has deteriorated over time. Selka et al. (2014) ascribe this deterioration partly to the advancement of online data collection: Sampled customers perform their evaluation tasks in changing environments with many distraction possibilities and therefore are not able to give unambiguous answers. As one possible solution to this problem often (e.g., Baier 2014; Baier and Polasek 2003) the usage of Bayesian procedures has been proposed where the respondents' partworths are modeled in a more flexible (stochastic and multi-modal) manner. In this paper, we discuss this proposition in more detail and apply it to simulated and real data in tourism.

2 Conjoint Analysis Based Market Simulation

2.1 Traditional Data Collection and Partworth Estimation

Conjoint analysis is a well-known and widespread method for measuring customers' preferences with respect to multi-attributed goods and services with many methodological variants (see, e.g., Green et al. 2001; Selka et al. 2014). All of them have in common, that sampled customers (in the following: respondents) are confronted with (fictive) offers described conjointly by attribute-levels. From these evaluations partworths for the attribute-levels are estimated and used to predict choice decisions. The conjoint analysis variants differ with respect to the data collection formats and estimation procedures. However, since many years, the choice-based conjoint analysis variant (CBC, see Desarbo et al. 1995; Louviere and Woodworth 1983) is the most popular one (see, e.g., Selka and Baier 2014).

CBC—another label is discrete choice analysis (see, e.g., Ben-Akiva and Lerman 1985)—received its popularity due to the widespread CBC software system (Sawtooth Software 2014a). CBC supports the data collection format that I respondents ($i = 1, \ldots, I$) are subsequently confronted with J_i choice sets ($j = 1, \ldots, J_i$) of K_{ij} offers ($k = 1, \ldots, K_{ij}$). The respondents are asked to select a most preferred offer in each set. In the following, q_{ijk} denotes the results of these selections: $q_{ijk} = 1$ if k was the preferred offer in set j for respondent i, $q_{ijk} = 0$ if not. The sets are constructed in a balanced manner with respect to pre-defined attributes and levels. Typical settings for CBC data collection are three to five competing offers within one set (sometimes with an additional "no choice" attribute and a corresponding offer in each set). The respondents are confronted with, e.g., 10 to 15 choice sets. The offers itself are characterized using, e.g., four up to ten attributes that can take two up to ten attribute-levels (see Sawtooth Software 2014a for examples). Typically, the attributes are nominally scaled (e.g., "low" and "high" as levels for "good quality"), but—for partworth estimation—are converted to intervally scaled attributes using a dummy coding with respect to—say—M dummy-coded variables. Here, $\mathbf{x}_{jk}$ denotes the dummy coded description of offer k in choice set j.

For partworth estimation, now, the observed selections in each choice set must be predicted using the M-dimensional vectors of model parameters $\boldsymbol{\beta}_i$ ($i = 1, \ldots, I$) for the (unknown) partworths of customer i. Following the multinomial logit approach with an assumed independently, identically type I extreme distributed additional error in the (overall) utilities, the observed selections are modeled using

$$p_{ijk} = \frac{\exp(\mathbf{x}_{jk}{}'\boldsymbol{\beta}_i)}{\sum_{k'=1}^{K_{ij}} \exp(\mathbf{x}_{jk'}{}'\boldsymbol{\beta}_i)} \quad \forall i = 1, \ldots, I, j = 1, \ldots, J_i, k = 1, \ldots, K_{ij} \qquad (1)$$

as the probability that customer i selects offer k in choice set j. The model parameters are estimated by maximizing the data likelihood, but—due to a typical mismatch between the number of observations per respondent (J_i) and the number of model parameters (M)—CBC usually assumes identical partworths and error distributions across all respondents. The data of all respondents are used to derive an M-dimensional mean partworth vector $\boldsymbol{\beta}$.

2.2 Estimation of Inter- and Intra-Individual Varying Partworths

Bayesian procedures for conjoint analysis now differ from the above described modeling insofar that the model parameters $\boldsymbol{\beta}_i$ ($i = 1, \ldots, I$) are not assumed to be deterministic (with unknown values) but to be distributed. The usual assumption (see, e.g., Sawtooth Software 2014b; Baier et al. 2015) is that they follow a multivariate normal distribution with expected values $\boldsymbol{\mu}$ and covariance matrix $\mathbf{H}$, i.e.

$$\boldsymbol{\beta}_i \sim \text{Normal}(\boldsymbol{\mu}, \mathbf{H}) \quad \forall i = 1, \ldots, I. \qquad (2)$$

In order to estimate the model parameters, a hierarchical modeling is used that assumes two modeling layers:

- At the higher layer, respondents' partworths $\boldsymbol{\beta}_i$ are described by a normal distribution according to formula (2).
- At the lower layer the respondents' probabilities of selecting an offer are governed by the multinomial logit model according to formula (1).

The parameters are estimated by an iterative process where in each of these steps one set of parameters is reestimated conditionally, given current values of the other sets. As a result we receive from each iteration a draw of all parameters. The draws across all iterations form joint empirical distributions of the model parameters (see, e.g., Sawtooth Software 2014b; Baier et al. 2015 for details). So, e.g., when L iterations were used for estimation, we receive with $\boldsymbol{\beta}_{il}(l = 1, \ldots, L)$ the empirical distribution of respondent i's partworths.

Here, in this paper, a more complex modeling is assumed: The respondents' partworths $\boldsymbol{\beta}_i$ $(i = 1, \ldots, I)$ are described by a mixture of T multivariate normal distributions with expected values $\boldsymbol{\mu}_t$ and covariance matrix $\mathbf{H}_t$ for component t $(t = 1, \ldots, T)$ and mixing parameters η_{it} $(i = 1, \ldots, I, t = 1, \ldots, T)$, i.e.

$$\boldsymbol{\beta}_i \sim \sum_{t=1}^{T} \eta_{it} \text{Normal}(\boldsymbol{\mu}_t, \mathbf{H}_t) \quad \forall i = 1, \ldots, I. \tag{3}$$

The underlying idea for this extension ($T = 1$ is the already discussed special case as implemented, e.g., in Sawtooth Software 2014b) is that now the partworths of the respondents are allowed to vary situation-specific over choices. So, e.g., when asked to evaluate a choice set of waterpark offers, a respondent can have different situations in mind, e.g., a short visit to go swimming alone or a day trip with the family. Depending on the situation in mind, the partworths vary. So, e.g., in the first situation the "low distance" has a higher and the "good quality" of the saunas and fun pools has a lower importance. The components reflect these different situations, the normal distribution allows to model ambiguous evaluations. The approach is similar to a market segmentation assumption that extends the formulation by Baier and Polasek (2003) and was also discussed by Otter et al. (2004), but here, the focus is on the assumption that individuals are not allocated to market segments but have situation-specific intra-individual varying partworths.

The details for deriving empirical distributions of the model parameters via Bayesian estimation procedures (e.g., $\boldsymbol{\beta}_{il}$ with $l = 1, \ldots, L$ as an index for the draws) are not discussed here, but it should be mentioned that Baier (2014) has shown in a similar setting for Bayesian procedures in metric conjoint analysis that the generalized version (with $T > 1$) can—in many cases—be approximated by the more simple approach (with $T = 1$, implemented, e.g., in Sawtooth Software 2014b) without a major loss of validity. This is due to the flexibility of the Bayesian estimation procedures.

2.3 Usage of the Empirical Partworth Distributions for Predictions

The empirical distribution of the intra-individual varying partworths can now be used to predict choice probabilities for the respondents in an assumed market scenario with K^* competing offers with descriptions $\mathbf{x}_k^*$ $(k = 1, \ldots, K^*)$. For this,

$$p_{ik}^* = \frac{1}{L} \sum_{l=1}^{L} \frac{\exp(\mathbf{x}_k^{*'} \boldsymbol{\beta}_{il})}{\sum_{k'=1}^{K^*} \exp(\mathbf{x}_{k'}^{*'} \boldsymbol{\beta}_{il})} \quad \forall = 1, \ldots, I, k = 1, \ldots, K^* \tag{4}$$

(a similar formulation to 1) is used as the probability that respondent i selects offer k in this scenario. Using some weighted average across the respondents (for taking varying demand into account) these probabilities can be aggregated to predict market shares and—with additional buying intensity and cost information—sales and profit for the competing offers (statistical projection).

3 Application to Simulated Data

3.1 Generation of Empirical Partworth Distributions

The first application is used to demonstrate the usefulness of the new approach. The simulated application domain is waterpark design and pricing. We assume that the competing offers can be described by three dummy-coded attributes ("low distance," "good quality," "low price"). So, e.g., the level "1" of the attribute "low price" indicates that a waterpark has a pre-defined low entry fee (say, e.g., 1.5 money units) whereas "0" has a pre-defined high entry fee (say, e.g., 5 money units). The level "1" of the attribute "good quality" is associated with more service that causes additional costs per visit (say, e.g., 1.5 money units) whereas the level "0" reflects standard service without additional costs per visit (say, e.g., 0 money units).

Further, empirical partworth distributions for $I = 20$ respondents are generated that assume that they come from four customer segments: Segment 1 represents "sport enthusiasts" (who usually go swimming but sometimes like also to visit a waterpark for relaxation), 2 "families with children," 3 "working singles" (who like to swim in the morning but visit a waterpark on weekends), and 4 "retired persons" (with a small income who like to swim).

For each segment two usage situations are assumed, one with a focus on "swim" and one with a focus on a longer "visit." As discussed in the last section, we assume that the partworths across segments and usage situations may differ, so, the empirical distributions are generated by drawing values from mixtures of multivariate normal distributions with $M = 3$ dimensions and $T = 4 * 2 = 8$ components. Further assumptions reflect the different sizes of the segments (segment 1: 6 respondents, 2: 6 respondents, 3: 4 respondents, 4: 4 respondents), the proportions of the components, and the component-specific mean partworths and covariances. Table 1 describes these settings.

One can easily see that, e.g., all segments in the usage situation "swim" have a stronger focus on the attribute-level "low distance" and that, e.g., the segment 2 ("family with children") in the usage situation "visit" has a stronger focus on "low price." The weights reflect the proportions of the eight components across the sample, e.g. the segments "sports enthusiasts" and "family with children" have in both usage situations higher demands than "working singles." As additional distributional assumptions for the components a standard deviation of 1 was

Table 1 Mean partworths and weights across segments and usage situations

Segment	1		2		3		4	
Usage situation	Swim	Visit	Swim	Visit	Swim	Visit	Swim	Visit
Component	1	2	3	4	5	6	7	8
Low distance	0.714	0.091	0.714	0.111	0.143	0.143	0.455	0.455
Good quality	0.143	0.455	0.143	0.333	0.143	0.714	0.091	0.091
Low price	0.143	0.455	0.143	0.566	0.143	0.143	0.455	0.455
Weight	0.18	0.12	0.12	0.18	0.10	0.10	0.12	0.08

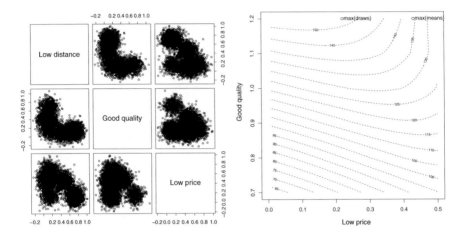

Fig. 1 *Left*: Empirical partworth distributions in the application with simulated data (all draws); *right*: derived indifference map (based on all draws) for predicted relative sales (100 reflects the status quo) depending on the attributes "low price" and "good quality"; max(draws) gives the maximum in this map; max(means) gives the maximum when using the averaged draws per respondent for prediction

assumed. According to these settings, for each respondent an empirical distribution with $L = 250$ draws was generated. The results are shown on the left side of Fig. 1.

3.2 Usage of Empirical Partworth Distributions to Design and Price

The generated empirical partworth distributions are now used for market simulation and to design and price a waterpark. We assume a scenario as described in Table 2. The table contains the actual values $\mathbf{x}_k^*$ ($k = 1, \ldots, K^*$) of four competing waterparks, for one waterpark (offer 4) a new pricing is looked for. Basing on the market prediction formula (3) sales predictions can be made for varying levels of the attributes "low price" and "good quality" with respect to the sample of respondents now.

Table 2 Waterpark market scenario with four competing offers

Waterpark	1	2	3	4
Low distance	0.9	0.6	1.0	0.1
Good quality	0.1	0.6	0.5	0.8
Low price	0.9	0.1	0.3	0.3

The indifference map on the right side of Fig. 1 reflects these predictions in percent of the status quo sales (with x_k^* values of the four competing waterparks according to Table 2 where the values for our waterpark were 0.8 for "good quality" and 0.3 for "low price"). The 100 % isoquantel line shows that the actual sales could also be achieved, e.g., with values 0.9 for "good quality" and 0.0 for "low price," this means that a price uplift from actual 4 money units to 5 money units could be compensated by a respective increase of services (Note that additional costs for the increase of services are already modeled.). The indifference map also gives some hints with respect to improve the sales: Values 1.2 for "good quality" and 0.11 for "low price" [see the point "max(draws)"] allow to increase the sales to 160 % of the actual sales. The management has to decide whether this change of strategy could be an alternative. The map also gives the results when using the mean partworths instead of the empirical distribution of the partworths with inter-individual variation for sales prediction: Here, the sales could be maximized for values 1.2 for "good quality" and 0.35 for "low price" [see the point "max(means)"]. However, as also can be seen in the map, these values would lead to a suboptimal solution when taking the empirical distribution into account with only an increase of the sales to 137 % of the actual sales.

4 Application to Real Data

4.1 Data Collection and Estimation of Partworth Distributions

For a regional waterpark in Eastern Germany the authors performed—in cooperation with the waterpark management—a conjoint study to understand the customers' preferences. The waterpark market in Germany is very competitive and so, the management was very interested in revising the waterpark's design and pricing, what was successfully done on the basis of the results of the study. In the following—for confidentiality reasons—the attribute and levels as well as the results of the study are alienated, but nevertheless the main idea of the study and results with respect to the modeling of intra-individual varying partworths are discussed.

The attributes and levels for a CBC web interviewing approach were selected by applying focus group interviews with customers and waterpark managers. Additionally, an internet-based analysis of all near-by waterparks and their offers as well as a literature overview on relevant studies were performed. The respondents were sampled using regional social networks and banners on websites where sports activities were promoted. A quota controlling was performed according to the waterpark management's demands. So, e.g., the respondents were mainly sampled

Table 3 Means and standard deviations (SD) of the partworth distributions (across all "draws" and solely across the respondent-specific "means")

Attribute	Level	Mean partworth	SD (draws)	SD (means)
Low distance	0: 31 vs. 1: 4 min	0.766	1.654	1.271
Low price	0: 26 vs. 1: 15 Euro	2.084	2.056	1.680
Fun pool	0: no vs. 1: yes	−0.730	1.358	1.062
Outside bath area	0: no vs. 1: yes	0.754	1.718	1.183
Brine bath	0: no vs. 1: yes	0.836	1.890	1.426
Many saunas	0: 5 vs. 1: 14 saunas	1.065	2.030	1.708
Recommended	0: no vs. 1: yes	1.015	1.506	1.161
Calm	0: no vs. 1: yes	1.963	3.325	3.071
No choice	0: no vs. 1: yes	−5.719	3.520	2.719

from residents of a near-by major town that forms the main customer reservoir for the waterpark. To fulfill the quota with respect to elderly people, also personal interviews were conducted. Each respondent had to deliver 16 choice tasks, each with three competing and a "no choice" alternative. Also, the respondents' usage intensity of waterparks was collected to distinguish heavy from light users by allocating weights. So, e.g., an adult who goes swimming one time a year got a weight of 1 whereas a family with two adults and more than one child that go swimming several times a weak received a weight of 300. The data collection took the different usage situations of the respondents into account. All in all, a total of 201 interviews formed the basis for the analysis.

For partworth estimation, Sawtooth Software's CBC/HB system was used that allows to derive for each respondent an empirical partworth distribution (Note that here—as already mentioned—the version with $T = 1$ is used to simulate the estimation of the situation-specific modeling approach.). The data validity was tested by using the averaged root likelihood (RLH) value. The collected data showed an averaged RLH value of 0.613, which is clearly superior to 1/4—the value of a naive model. Table 3 gives the resulting means and standard deviations of the partworths (across all draws for taking the inter- and intra-individual variation into account and across the respondents for taking only the inter-individual variation into account). One can easily see the high importance of the attribute "low price" but also—as in the simulated data application from above—the reduced information when only taking the inter-individual variation into account.

4.2 Usage of Empirical Partworth Distributions to Design and Price

Now, again as in the simulated application, we can predict market shares and sales in an assumed market scenario of competing waterparks as defined in Table 4. Again,

Table 4 Waterpark market scenario with three competing and a "no choice" alternative

Attribute/Waterpark	1	2	3	No choice
Low distance	0.0	1.0	0.161	0.0
Low price	0.237	1.0	0.0	0.0
Fun pool	0.7	0.6	0.0	0.0
Outside bath area	0.5	0.5	0.5	0.0
Brine bath	0.7	0.0	1.0	0.0
Many saunas	1.0	0.357	0.5	0.0
Recommended	0.1	0.0	1.0	0.0
Calm	1.0	0.0	1.0	0.0
No choice	0.0	0.0	0.0	1.0

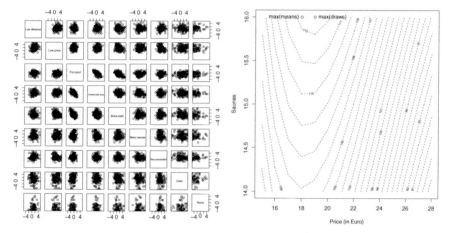

Fig. 2 *Left*: Empirical partworth distributions in the application with real data (averaged draws per respondent); *right*: indifference map (based on all draws) of predicted relative sales (100 reflects the status quo) depending on the (recoded) attributes "price (in Euro)" and "saunas"; max(draws) gives the maximum in this map; max(means) gives the maximum when using the averaged draws per respondent for prediction

the table contains the actual x_k^* ($k = 1, \ldots, K^*$) values of four competing waterparks in the region, again, for our waterpark (offer 1 in Table 4) a new pricing is looked for. As can be seen from Table 4, our waterpark is far away from the target population ('low distance' has level 0), not very cheap (32 Euro) but provides some interesting attribute-levels (e.g., w.r.t. "brine bath," "many saunas," and "calm"). Waterpark 2 is nearby the target population, relatively cheap (15 Euro) but has less interesting attribute-levels. Waterpark 3 has advantages w.r.t. "brine bath" or "recommended" but is even more expensive. Now, again, basing on the market prediction formula (3) sales predictions can be made for varying levels of the attributes "low price" and "many saunas" (as a proxy for quality improvement). Figure 2 gives the distribution of the partworths (on the left) and the respective improvements in contrast to the actual sales (right). Again, there are some improvement possibilities (up to 113 %) and the results show that taking the means would lead to a suboptimal result compared to the full draws.

5 Conclusion and Outlook

The paper has presented a new approach to market oriented product design and pricing taking inter- and intra-individual varying partworths into account. The application to waterpark design using simulated and real data shows promising results. But—of course—more analyses are necessary, applying, e.g., a Monte Carlo setting to analyze when the modes of the empirical distributions (the mean partworths per respondents) are especially inferior in prediction.

References

Baier, D. (2014). Bayesian methods for conjoint analysis-based predictions: Do we still need Latent Classes? In *German-Japanese interchange of data analysis results. Studies in classification, data analysis, and knowledge organization* (Vol. 47, pp. 103–113). Berlin: Springer.

Baier, D., Pelka, M., Rybicka, A., & Schreiber, S. (2015). TCA/HB compared to CBC/HB for predicting choices among multi-attributed products. *Archives of Data Science, 1*(1), 1–11.

Baier, D., & Polasek, W. (2003). Market simulation using Bayesian procedures in conjoint analysis. In *Exploratory data analysis in empirical research. Studies in classification, data analysis, and knowledge organization* (Vol. 23, pp. 413–421). Berlin: Springer.

Ben-Akiva, M., & Lerman, S. R. (1985). *Discrete choice analysis: Theory and application to travel demand.* Cambridge: MIT Press.

Desarbo, W. S., Ramaswamy, V., & Cohen, S. H. (1995). Market segmentation with choice-based conjoint analysis. *Marketing Letters, 6*(2), 137–147.

Green, P. E., Krieger, A. M., & Wind, Y. (2001). Thirty years of conjoint analysis: Reflections and prospects. *Interfaces, 31*(3), 56–73.

Louviere, J. J., & Woodworth, G. (1983). Design and analysis of simulated consumer choice or allocation experiments: An approach based on aggregate data. *Journal of Marketing Research, 20*(4), 350–367.

Otter, T., Tüchler, R., & Frühwirth-Schnatter, S. (2004). Capturing consumer heterogeneity in metric conjoint analysis using Bayesian mixture models. *International Journal of Research in Marketing, 21*(3), 285–297.

Sawtooth Software. (2014a). *The CBC system for choice-based conjoint analysis version 8.* Orem, UT: Sawtooth Software, Inc.

Sawtooth Software. (2014b). *The CBC/HB system for hierarchical Bayes estimation version 5.* Sequim, WA: Sawtooth Software, Inc.

Selka, S., & Baier, D. (2014). Kommerzielle Anwendung auswahlbasierter Verfahren der Conjointanalyse: Eine empirische Untersuchung zur Validitätsentwicklung. *Marketing ZFP – Journal of Research and Management, 36*(1), 54–64.

Selka, S., Baier, D., & Kurz, P. (2014). The validity of conjoint analysis: An investigation of commercial studies over time. In *Data analysis, machine learning and knowledge discovery. Studies in classification, data analysis, and knowledge organization* (Vol. 48, pp. 227–234). Berlin: Springer

Wind, J., Green, P. E., Shifflet, D., & Scarbrough, M. (1989). Courtyard by Marriott: Designing a hotel facility with consumer-based marketing models. *Interfaces, 19*(1), 25–47.

The Use of Hybrid Predictive C&RT-Logit Models in Analytical CRM

Mariusz Łapczyński

Abstract Predictive models in analytical CRM (customer relationship management) are closely related to the customer's life cycle. Prediction of binary dependent variable refers to the most common areas such as customer acquisition, customer development (cross-selling and up-selling), and customer retention (churn analysis). While building static predictive models one usually applies decision trees, logistic regression, support vector machines or ensemble methods such as different algorithms of boosted decision trees or random forest. Recently one can observe increasing use of hybrid models in the analytical CRM, i.e. those that combine several different analytical tools, e.g. cluster analysis with decision trees, genetic algorithms with neural networks, or decision trees with logistic regression. The purpose of this paper is to compare the results obtained by using hybrid predictive CART-logit models with single decision tree models and logistic regression models. All analyses have been conducted on the basis of data sets relating to analytical CRM.

1 Introduction

Analytical customer relationship management (CRM) refers to all activities related to the gathering and analysis of data and the construction of descriptive and predictive models. The research areas of its domain include: customer acquisition, customer development/retention and measurement of customer lifetime value. Three research areas listed above made it possible to develop a model called ACURA (Christopher et al. 2008), which is an acronym for acquisition, cross-selling, up-selling, retain, and advocacy. This model is consistent with the popular concept of "customer life cycle," which consists in passing from one stage of the cycle to another with changing relationships between parties.

For each of these phases predictive models are built that improve the managerial decision-making process. As far as customer acquisition models are concerned, one

M. Łapczyński (✉)
Cracow University of Economics, Kraków, Poland
e-mail: lapczynm@uek.krakow.pl

© Springer International Publishing Switzerland 2016
A.F.X. Wilhelm, H.A. Kestler (eds.), *Analysis of Large and Complex Data*, Studies in Classification, Data Analysis, and Knowledge Organization,
DOI 10.1007/978-3-319-25226-1_27

311

can use various analytical tools such as: discriminant analysis, logistic regression, decision trees, probit models, or diffusion models. With regard to cross-sell and up-sell models analysts can additionally utilize association rules, collaborative filtering, sequence analysis, Markov models, or neural networks. In the case of static churn models, random forest, boosted trees, support vector machines were also used, and in the case of dynamic ones—survival analysis.

A characteristic feature of all static models in analytical CRM is the frequent occurrence of a binary dependent variable, whose categories can refer to response/no response, buy/no buy, churn/no churn. In recent times some attempts of predicting such variables by using hybrid approaches that combine various algorithms and analytical tools have been appearing in the literature. Although this combination results in an increased duration of computing, it is sometimes compensated by a better performance of models, and may overcome the class imbalance problem or enrich the interpretation of the model. The hybrid predictive CART-logit model presented in this paper is suitable for such variables and sometimes delivers better results than if these methods were utilized separately.

2   Hybrid Predictive C&RT-Logit Model

During the construction of predictive models for relationship marketing purposes one focuses not only on prediction of phenomena but also on understanding of the nature of relationships among analyzed variables. If a model serves both predictive and descriptive purposes at the same time, it allows not only to choose the right target group but also to create effective marketing campaigns. The combination of decision trees with logistic regression meets both of these research objectives.

The hybrid C&RT-logit model used in this study combines C&RT (classification and regression trees) algorithm with logistic regression. The STATISTICA software was used for the data analysis and therefore the abbreviation CART (a registered trademark of Salford Systems company) was replaced with the acronym C&RT. CART (Breiman et al. 1984) is considered as one of the most advanced decision tree algorithms. The features of that method that distinguish it from the logistic regression model include: automatic selection of the best predictors, no need for the transformation of variables, automatic detection of interaction effects, resistance to outliers, utilizing surrogate variables while classifying cases with missing data, and minimal supervision of the researcher while building the model.

The construction of logistic regression models, in turn, requires the supervision of an experienced analyst and frequently takes much longer than the construction of the decision tree. Logit models are sensitive to outliers and require imputation of missing data (cases with missing data are removed from the analysis). The big advantage of this approach is the ability to calculate the unique probability of belonging to a class (category dependent variable) for each case. On the other hand, decision trees provide as many probabilities as many leaves they have for each terminal node and cases belonging to it.

The combination of decision trees (CHAID algorithm) with logistic regression carried out by Lindahl and Winship (1994) was probably the first attempt to build this kind of a hybrid model. Hybridization was based on the sequential use of these analytical tools. After the initial exploration of data set by using CHAID algorithm cases were divided into terminal nodes. In the second step of the procedure a separate logistic regression model was built for each leaf.

Another concept of hybridization was proposed a few years later (in Steinberg and Cardell 1998). It combined decision trees (CART algorithm) with logit models. This time it was also a two-step procedure, however, the set of independent variables in the logit model was supplemented with an additional variable whose categories informed about the terminal node to which the case was assigned. The new variable was transformed into a set of dummy variables. CART model from the first stage of the procedure was based on the same set of independent variables, and each leaf took into account the interaction between the predictors. The authors pointed out that such hybridization is more effective, because the partition of the data set into subsets according to the first concept is connected with a reduction of sample size (instances are divided into terminal nodes) and the loss of information (it can happen that the sets of independent variables for each logit model will be slightly different). Moreover, patterns discovered by logit models are local (limited to the leaves) and the variability of the dependent variable and variance of predictors is lower in terminal nodes than in the entire data set. The advantages of the CART-logit approach include a higher predictive accuracy of the hybrid model, a faster detection of interactions by the CART algorithm and, in general, no need to replace missing data.

So far the hybrid CART-logit models have been utilized in the analysis of medical data sets (in Su 2007) and in the analysis of the causes of the Asian currency crisis (Ait-Saadi and Jusoh 2011). According to the author's knowledge, no one else has used the CART-logit approach in the analysis of data sets related to analytical CRM. After 1998 several researchers attempted to combine decision trees with logistic regression. They developed new hybrid approaches known as LOTUS (Logistic Tree with Unbiased Selection) (Chan and Loh 2004), LMT (Logistic Model Tree) (Landwehr et al. 2005), or PLUTO (Penalized, Logistic Regression, Unbiased Splitting, Tree Operator) (in Zhang and Loh 2014).

3 Experiments

During the experiment three data sets were used. All were obtained from popular repositories and related to the areas of analytical CRM. The first model and the second model pertain to the churn analysis, while the third model refers to a direct marketing campaign carried out by a Portuguese bank. Due to the problem of imbalanced classes random under-sampling was utilized in all data sets used in this experiment in order to modify the learning sample.

3.1   The First Approach to Hybridization: Churn Model

The data set used in this experiment refers to churn analysis. It was obtained from http://www.dataminingconsultant.com/DMMM.htm. The sample size and distribution of the target variable are shown in Table 1.

Prior to building the hybrid model, the size of decision tree was reduced to four terminal nodes (Fig. 1). In leaf ID 3 there are customers for whom the daily number of minutes of calls is higher than 246.6. The leaf ID 5 includes customers for whom the "total day minutes" is fewer than or equal to 246.6 and the number of calls to the call center exceeds 3. The leaf ID 6 consists of buyers having international plan calls for whom the daily number of minutes is fewer than or equal to 246.6 and the number of calls to the call center is fewer than or equal to 3. In the terminal node ID 7 there are customers who do not have international plan, for whom the daily number of minutes is fewer than or equal to 246.6 and the number of calls to the customer service center is fewer than or equal to 3.

The hybrid model was supplemented with an additional variable "terminal node," which was transformed to dummies with the reference category "leaf ID 3". Table 2 presents the results of the hybrid approach.

Table 1 Sample size and distribution of dependent variable (the first model)

Data set	Sample size	Number and percentage of churners in dependent variable
Entire data set	5000	707 (14.40 %)
Learning sample (random under-sampling)	1687	506 (29.99 %)
Test sample	1506	201 (13.35 %)

Fig. 1 Reduced decision tree model

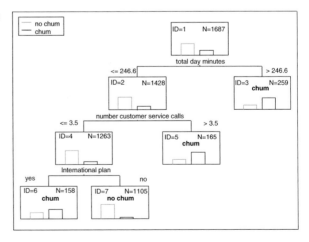

Table 2 Results of hybrid C&RT-logit model (limited to independent variables that have positive effects with respect to the response)

Variable	Estimate	Standard error	p value	Odds ratio
Intercept	−3.786	0.732	0.000	0.02
International plan	1.205	0.418	0.004	3.34
Total day minutes	0.005	0.002	0.004	1.01
Total eve minutes	0.007	0.001	0.000	1.01
Total night minutes	0.006	0.001	0.000	1.01
Total intl charge	0.417	0.099	0.000	1.52
State NJ	0.830	0.402	0.039	2.29
State MT	1.293	0.482	0.007	3.64
State CA	1.664	0.693	0.016	5.28
Leaf ID 5	0.939	0.327	0.004	2.56

In the hybrid approach there are several predictors that significantly contributed to the model and have high positive effects with respect to the response (increase the probability of churn):

- "international plan"—the probability of churning among customers with the international calling plan is almost 3.5 times higher than the probability among customers without that plan, one can notice a decrease in the value of the odds ratio in comparison with the basic logistic model;
- "total intl charge"—for every unit increase in "total intl charge" the odds of churning increase by approximately 52 %;
- "state NJ"—clients living in the state of New Jersey have approximately 2 times greater odds of churning than clients from the states not included in the model (increase in the odds ratio from 2.17 in the basic model to 2.29 in the hybrid model);
- "state MT"—customers from Montana have 3.64 times higher probability of churning than customers from the states not included in the model (the odds ratio value in the basic logistic model is equal to 3.65);
- "state CA"—customers living in California have about 5 times greater odds of churning than customers from the states not included in the model (decrease in the odds ratio from 6.09 to 5.28);
- leaf ID 5—the probability of churning among the clients from the terminal node ID 5 (those for whom the daily number of minutes is lower than or equal to 246.6 and the number of calls to call center exceeds 3) is about 156 % higher than the probability among customers who talk longer (reference category—leaf ID 3).

Terminal nodes that significantly contributed to the model adjusted other predictors from the basic logistic model. It turned out that the independent variable "number customer service calls" did not contribute to the hybrid model. On the other hand, it enriched the interpretation of the model by detecting interactions of predictors. In the basic logistic model (the main effects model) the relationship

Table 3 Performance of models (the first model)

Measure	C&RT	Logistic model	Hybrid C&RT-logit
Accuracy	0.837*	0.523	0.689
Recall	0.806	0.915*	0.866
Precision	0.439*	0.208	0.282
TNR	0.841*	0.462	0.661
G-mean	0.823*	0.650	0.757
F measure	0.568*	0.339	0.426
Lift (decile 1/decile 2)	5.14*/4.04*	3.70/2.97	4.49/3.75

Table 4 Sample size and distribution of dependent variable (the second model)

Data set	Sample size	Number and percentage of churners in dependent variable
Entire data set	50,000	3672 (7.34 %)
Learning sample (random under-sampling)	14,666	2932 (20.00 %)
Test sample	10,087	740 (7.34 %)

between "total day minutes" and churning was almost unobservable (the odds ratio was equal to 1.01).

To evaluate the models' performance several popular measures such as accuracy, recall, precision, true negative rate (TNR), G-mean, and F-measure were used (see details in Table 3). The best results are marked with an asterisk (*). As one can easily see, the C&RT model outperforms other methods (except for recall). In general, the hybrid C&RT-logit model turned out to be better than the basic logistic model.

The lift measures in the first and in the second decile allow to draw additional conclusions. One can see the higher predictive power of the decision tree model in comparison with the hybrid approach and the basic logistic model. This means in this case that hybridization allowed only to enrich the interpretation of the model and to detect quickly the interaction between the variables. The researcher can interpret the odds ratios which are not directly available in the C&RT model, however losing the predictive properties of the model. It should be noted that in terms of the lift measure, the difference between the models becomes smaller and smaller starting from the third decile.

3.2 The Second Approach to Hybridization: Churn Model

The second set of observations also refers to churn modeling and was used during the KDD Cup in 2009. The information about the sample size and distribution of dependent variable is shown in Table 4.

Table 5 Performance of models (the second model)

Measure	C&RT	Logistic model	Hybrid C&RT-logit
Accuracy	0.536*	0.106	0.143
Recall	0.584	0.989*	0.969
Precision	0.090*	0.075	0.077
TNR	0.532*	0.037	0.078
G-mean	0.557*	0.190	0.275
F measure	0.156*	0.140	0.142
Lift (decile 1/decile 2)	1.55/1.38	2.35*/1.82*	2.26/1.72

The procedure for building hybrid model is identical to that carried out previously. In the first step, decision tree was reduced to four terminal nodes. In the second step, new dummy variables (leaves of the decision tree) were introduced to the hybrid model. All terminal nodes significantly contributed to the model and have positive effects with regard to the response. Table 5 presents the models' performance evaluated by using measures that were used in the first hybridization. The best results are marked with asterisk (*). Again, it was found that the C&RT model outperforms other methods (except for recall). In general, the hybrid approach turned out to be better than the main effects model. As far as lift measure is concerned, the main effects model outperforms the other approaches, however, it is only slightly better than the hybrid model.

3.3 The Third Approach to Hybridization: Acquisition Model

The third data set was related to the direct marketing campaign of a Portuguese bank (Moro et al. 2011). The distribution of dependent variables and sample sizes is presented in Table 6.

Prior to building the hybrid model, the size of decision tree was reduced to three terminal nodes. In the next step of the procedure an additional variable "terminal node" was introduced to the hybrid C&RT-logit model. It was transformed to dummies with the reference category "leaf ID2". Apart from independent variables that were present in the basic logistic model, two terminal nodes from decision tree significantly contributed to the hybrid model. Both have positive effects with respect to the response.

It is worth noting that the terminal nodes of decision tree are categorized continuous predictor "call duration," which—in its original form—had a minimal influence on the response. The odds ratio was equal to 1.003, which means that for every increase of one second in the "call duration" the odds of acquiring a new customer increase by approximately 0.003 (0.3 %). The odds ratios for the terminal nodes are equal to 2.49 and 5.22. As far as other predictors that have positive effects

Table 6 Sample size and distribution of dependent variable (the third model)

Data set	Sample size	Number and percentage of churners in dependent variable
Entire data set	45,211	5289 (11.70 %)
Learning sample (random under-sampling)	18,595	3719 (20.00 %)
Test sample	13,659	1570 (11.49 %))

Table 7 Performance of models (the third model)

Measure	C&RT	Logistic model	Hybrid C&RT-logit
Accuracy	0.721*	0.696	0.710
Recall	0.807	0.905*	0.885
Precision	0.265	0.262	0.269*
TNR	0.710*	0.669	0.687
G-mean	0.757	0.778	0.780*
F measure	0.399	0.406	0.412*
Lift (decile 1/decile 2)	3.51/2.32	4.81*/3.61*	4.77/3.59

with respect to the response are concerned, their odds ratios are comparable with the odds ratio from the basic (the main effects) model.

The six measures that are presented in Table 7 are based on misclassification matrices. The best results are marked with asterisk (*). It turned out that decision tree model delivered the highest accuracy and true negative rate. The basic logistic model once again outperforms other methods with regard to recall, and hybrid approach delivers the highest values of precision, G-mean and F measure. The lift measures in the first and in the second decile allow to draw additional conclusions. The basic logistic model performs much better than decision tree, however, only slightly better than in the case of the hybrid approach.

3.4 The Comparison of Logistic Regression Model and C&RT-Logit Model

Comparing the logistic regression model with the hybrid model it can be noticed that (Table 8) the values of Cox and Snell's R2 were higher for C&RT-logit models than for the main effects models. These results were confirmed by other measures, such as Nagelkerke's R2, the Akaike information criterion (AIC), and the Bayesian information criterion (BIC). Apart from recall and sometimes lift, the hybrid approach outperforms the basic logistic model. In each C&RT-logit model the terminal nodes from the decision tree appeared, which usually resulted in removing from the model the variables that were included in the rules describing these nodes. The odds ratios for these variables (tree leaves) were noticeably higher than the odds ratios for other independent variables.

Table 8 The comparison of logistic regression model and the hybrid approach

Data set	Logistic model	Hybrid C&RT-logit
Churn 1	Cox and Snell's R2 = 0.27	Cox and Snell's R2 = 0.38
	AIC = 1567, BIC = 1643	AIC = 1295, BIC = 1398
	Higher recall	Higher accuracy, precision, TNR, G-mean
		F-measure and lift
		1 leaf as predictor with positive effect
		(combines 3 predictors)
Churn 2	Cox and Snell's R2 = 0.04	Cox and Snell's R2 = 0.06
	AIC = 14,068, BIC = 14,242	AIC = 13,869, BIC = 14,013
	Higher recall and lift	Higher accuracy, precision, TNR, G-mean
		and F-measure
		3 leaves as predictors with positive effects
		(combine 3 predictors)
Acquisition	Cox and Snell's R2 = 0.29	Cox and Snell's R2 = 0.30
	AIC = 12,246, BIC = 12,411	AIC = 12,017, BIC = 12,197
	Higher recall and lift	Higher accuracy, precision, TNR, G-mean
		and F-measure
		2 leaves as predictors with positive effects
		(categorized continuous predictor "duration")

4   Conclusions

Prior to the building of the hybrid C&RT-logit model one counted on benefits from combining advantages of both analytical tools. To summarize the three models presented in this paper it can be noted that the performance measures for hybrid model have never been the worst. The C&RT-logit approach outperformed at least one basic model (logistic or decision trees) and in some cases even both. The drawback of the proposed approach is certainly a more time-consuming procedure. On the other hand, it is compensated by the automatic detection of interaction effects and enriched interpretation of the studied domain. While comparing the hybrid approach to the decision tree one can observe an unquestionable advantage, which is assigning unique probabilities for cases from the test sample. The experiment certainly should be extended to other data sets with binary dependent variable relating to analytical CRM. It would also be interesting to make a comparison of the CART-logit approach with other hybrid models mentioned earlier, i.e. LOTUS, LMT and PLUTO. Twenty years of attempts to hybridize decision trees with logistic regression models indicate an attractiveness of this approach and still unexplored areas for researchers.

Acknowledgements The project was financed by a grant from National Science Centre (DEC—2011/01/B/HS4/04758).

References

Ait-Saadi, I., & Jusoh, M. (2011). What we know, what we still need to know: The Asian currency crisis revisited. *Asian-Pacific Economic Literature, 25*(2), 21–37.

Breiman, L., Friedman, J., Stone, C. J., & Olshen, R. A. (1984). *Classification and regression trees* (3rd ed.). New York: Chapman and Hall.

Chan, K.-Y., & Loh, W.-Y. (2004). LOTUS: An algorithm for building accurate and comprehensible logistic regression trees. *Journal of Computational and Graphical Statistics, 13*, 826–852.

Christopher, M., Payne, A., & Ballantyne, D. (2008). *Relationship marketing. creating stakeholder value* (5th ed.). Oxford: Butterworth Heinemann.

Landwehr, N., Hall, M., & Frank, E. (2005). Logistic model trees. *Machine Learning, 59*(1–2), 161–205.

Lindahl, W. E., & Winship, C. (1994). A logit model with interactions for predicting major gift donors. *Research in Higher Education, 35*(6), 729–743.

Moro, S., Laureano, R., & Cortez, P. (2011). Using data mining for bank direct marketing: An application of the CRISP-DM methodology. In P. Novais, et al. (Eds.), *Proceedings of the European Simulation and Modelling Conference - ESM'2011, Guimarães, Portugal* (pp. 117–121).

Steinberg, D., & Cardell, N. S. (1998). The hybrid CART-logit model in classification and data mining (online at http://www.salford-systems.com)

Su, X. (2007). Tree-based model checking for logistic regression. *Statistics in Medicine, 26*, 2154–2169.

Zhang, W., & Loh, W.-Y. (2014). PLUTO: Penalized unbiased logistic regression trees (online at arXiv:1411.6948v1).

Part VIII
Data Analysis in Finance

Excess Takeover Premiums and Bidder Contests in Merger & Acquisitions: New Methods for Determining Abnormal Offer Prices

Wolfgang Bessler and Colin Schneck

Abstract We investigate for mergers and acquisitions in Europe and the USA whether the size of the takeover premium offered by the first bidder prevents a second bidder from making a competing offer. Previous studies find only mixed evidence for the relationship between the size of a takeover premium and the occurrence of a takeover contest. Because the size of the premium varies over time and between merger waves and usually differs between countries and industries, it is essential to use the excess premium instead of the standard premium. We introduce and compare different methods for calculating the excess premium and test for the 1990–2012 period whether or not bidders can prevent a takeover contest when the initial offer includes an excess takeover premium. We calculate the excess premium as the percentage (a) above the pre-offer market value of the target, (b) over the industry mean, or (c) over the country mean. We then analyze whether these different methods provide results more consistent with the expected effect of excess premiums on the occurrence of takeover contests. The results suggest that the method used to calculate the excess premium significantly affects the size of the excess premium in takeover contests. We provide empirical evidence that when using the industry excess premiums, offering an above average premium reduces the probability of a takeover contest, especially in cash deals, whereas the standard method does not correctly discriminate between average and excess premiums. Consequently, only excess premiums are adequate for properly testing the effects of the premium size on the occurrence of takeover contests.

1 Introduction

The objective of our study is to provide empirical evidence for the factors that affect the occurrence or prevention of takeover contests. In most mergers and acquisitions (M&As), only one single bidder submits an offer for a target firm and if successful,

W. Bessler (✉) • C. Schneck
Center for Finance and Banking, Justus-Liebig-University Giessen, Licher Strasse 74, 35394 Giessen, Germany
e-mail: wolfgang.bessler@wirtschaft.uni-giessen.de; colin.schneck@wirtschaft.uni-giessen.de

© Springer International Publishing Switzerland 2016   323
A.F.X. Wilhelm, H.A. Kestler (eds.), *Analysis of Large and Complex Data*, Studies in Classification, Data Analysis, and Knowledge Organization,
DOI 10.1007/978-3-319-25226-1_28

both parties sign the deal within a certain period. However, subsequent to the initial offer a rival bidder competes in some cases by making another bid for the same target, resulting in a takeover contest. One important factor explaining why takeover contests occur is the size of the initially offered takeover premium. However, the literature so far presents only mixed results regarding the relationship between the size of the premium and the occurrence of takeover contests (Officer 2003; Jeon and Lion 2011). Thus, it is important to investigate whether the method employed for calculating takeover premiums has any effect on the empirical evidence regarding the occurrence of takeover contests.

Because takeover premiums differ between industries, countries and vary over time (Madura et al. 2012) it is inadequate to compare premiums across industries, countries as well as over time without controlling for these effects. Figure 1 shows the mean final premium for the manufacturing industry and the healthcare industry adjusted by the mean final premium over all industries. It reveals that premiums differ between industries and are time-varying, especially in recent years. We find a similar pattern for countries and the method of payment. The literature suggests that preemptive bidding, i.e., offering a high premium often containing a substantial cash part, could deter the occurrence of contests (Fishman 1988; Officer 2003). Thus, the initial bidder has a first-mover advantage in choosing the conditions of the initial

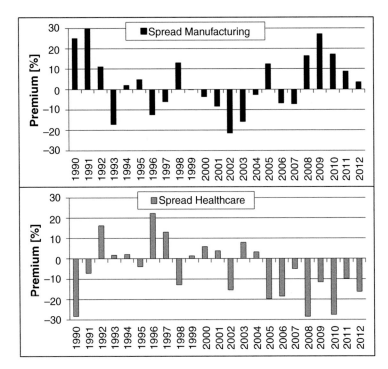

Fig. 1 Mean industry premiums per year for the manufacturing and healthcare industry adjusted with the mean premium over all industries per year

bid to avoid attracting other bidders (Jennings and Mazzeo 1993). Consequently, with respect to the initial premium, we hypothesize that only a premium above the expected premium deters rival bidders and consequently prevent takeover contests from occurring.

Two main research questions result from this discussion. First, how to correctly calculate excess premiums, and second do excess premiums reduce the probability that a takeover contest occurs. Therefore, the objective of this study is to analyze and compare different methods for calculating excess takeover premiums and then to test whether there is any interaction between these measures and the occurrence of takeover contests.

2 Literature Review

We provide a brief review of the factors that determine the size of the premium and explain how this premium relates to other variables before introducing our excess premium measures in the next section. Several factors determine the size of the takeover premium, which differ with respect to bid characteristics as well as industry and macroeconomic factors. First, the premium size depends on the bid characteristics. The takeover premium reflects expected benefits from the deal, which suggests that expected synergies and growth opportunities should positively correlate with the premium and depend on the bidders valuation of the target firm. Furthermore, the method of payment is another important factor as stock deals are typically associated with higher premiums than cash deals. Initial premiums are higher in stock deals compared to cash deals because often bidder firms are believed to pay with stock to exploit the overvaluation of its own shares (Eckbo 2009). Another important factor is the aggressiveness of the takeover bid, i.e., premiums for hostile takeovers are higher than in friendly mergers (Schwert 2000). Moreover, cross-border deals are also associated with higher premiums compared to domestic takeovers (Eckbo 2009). Second, industry effects, i.e., premiums paid for recent deals within an industry, also affect the premium for the most recent deal (Madura et al. 2012). We use the premiums paid in the last deals in an industry as a benchmark for evaluating the recent offer or for comparing them with the expected synergies. In addition, premiums are usually higher if the target provides important assets or access to resources that offers the bidder a competitive edge in its industry (Akdoğu 2009). Furthermore, industry shocks may intensify the competition for targets resulting in higher premiums as well. Third, macroeconomic shocks, i.e., high capital liquidity and GDP increases, lead to higher premiums (Madura et al. 2012).

3   Methodology

This section presents the methodology to calculate excess takeover premiums. Before developing our own excess premium measures, we discuss the important interactions of takeover premiums with several factors. Premiums are clustered over industries, time, and merger waves and depend on the premiums recently paid in the USA, especially for global takeovers (Madura et al. 2011; Madura and Ngo 2008). Furthermore, the size of the premium of a new offer correlates with the size of the premiums recently paid for deals within the same industry, method of payment, and takeover form (merger or tender offer). These links with previous premiums exist over a short time period of 3 months rather than over longer periods of 6 or 12 months. Differences in premiums over time relate more to previous premiums in the same industry than to target or bidder characteristics (Madura and Ngo 2008). By not including these differences, but using only the size of the premium, a comparison may be ambiguous. Therefore, we introduce excess takeover premium measures where we adjust the expected premium on different dimensions.

The previous literature suggests calculating takeover premiums *(P)* as the target offer price (OP) relative to the market value (Value) at the time prior to the announcement of a firm *i* (Madura and Ngo 2008). Often the shortest period of one day prior to the announcement is used, but employing a longer period of 4 weeks prior to the announcement seems appropriate as well, for example, to include the usual run-up and information leakage (Schwert 2000). To capture all value changes from the beginning to the end of the takeover process, an alternative approach calculates the premium as the target value at the completion date relative to the target value 4 weeks prior to the announcement. Nevertheless, none of these premium measures adjusts for the expected premium derived from similar takeovers. Hence, we extend the literature by introducing a new method for analyzing the effect of excess takeover premiums.

$$P_{(i)} = \frac{OP_{(i)} - \text{Value}_{(i)}}{\text{Value}_{(i)}} \qquad (1)$$

Often managers and shareholders of bidder and target firms use premiums paid in prior comparable deals as a benchmark when determining the initial premium for a new deal. We follow this approach but define the excess takeover premium (EP) as the initial premium (IP) offered in a takeover adjusted for the mean premium paid in comparable previous deals. We call this value our adjustment benchmark (AB), and we use different methods to determine this adjustment benchmark. Starting with the recent deal *i*, we include all takeovers *j* that took place within a fixed period of 3, 6, or 12 months prior to the announcement of the most recent deal and calculate as expected premium the mean final premium (FP) in these deals. Alternatively, we use a fixed number of takeovers such as the last 10, 20, or 30 deals that occurred prior to the announcement of the recent deal for determining the benchmark.

$$EP_{(i)} = IP_{(i)} - AB_{(j)} \quad \text{with} \quad AB_{(j)} = \frac{1}{N} \sum_{j=1}^{n} FP_{(j)} \tag{2}$$

Because M&A activity varies over time, both the fixed period and the fixed size excess takeover premium approaches have some inherent problems. Since the fixed period excess premiums are calculated over a fixed period before the recent deal is announced, the number of deals within the benchmark can differ due to a changing M&A frequency. In contrast, fixed size period excess premiums determine the adjustment benchmark using the last fixed number of deals before the recent deal announcement. Thus, the estimation period differs. All these problems result in changing variances in the adjustment benchmark portfolios. To address these issues we first combine fixed period and fixed size excess premiums, and second we use standardized excess premiums as a robustness check. Combining fixed period and fixed size approaches results in calculating the adjustment benchmark over a period that guarantees the minimal benchmark size. The combination solves the problem of increasing variance in the adjustment benchmark due to insufficient comparable deals in the fixed period. The advantage of using standardized premiums is that they are easily comparable but their interpretation is challenging. In our analysis, we calculate standardized excess takeover premium (SEP) as follows:

$$SEP_{(i)} = \frac{IP_{(i)} - \left[\frac{1}{N} \sum_{j=1}^{n} FP_{(j)} \right]}{Sd \left(\frac{1}{N} \sum_{j=1}^{n} FP_{(j)} \right)} \tag{3}$$

For all approaches, we calculate the benchmark over deals within the same industry (Fama-French 12 industry portfolios), the same country or region, and for the same method of payment. In line with the literature, we define a bid as cash deal (stock deal) if the bidder offers at least 80 % cash (stock). Finally, we perform regression on premium size with the following independent variables: method of payment, industries, countries, years, hostility, cross border, toehold, and target size. Thus, we control for several factors simultaneously and calculate the expected premium with the coefficients of these variables. We use all deals in our sample and rerun the regression using only deals that took place before the recent deal (rolling).

In the second step, we test the relevance of our excess premium measures for explaining the occurrence of takeover contests by estimating several probit models. In the probit model (4), the dummy variable (Y) takes the value of one if a takeover contest occurs and zero otherwise. The function $\Phi(.)$ denotes the standard normal distribution function. The estimated probability y is interpreted as the likelihood that a takeover contest occurs.

$$Pob(Y = 1|x) = \int_{-\infty}^{x'\beta} \phi(t)dt = \Phi(x'\beta) \tag{4}$$

4 Data

For our empirical analysis and the calculation of excess premium measures, we use all M&As available in the Thomson One Global Mergers and Acquisitions Database that meet the following criteria: all acquisition announcements occur between 1990 and 2012. The target firm is located in Europe or the USA. The bidder firm owns less than 50 % of the target shares at the announcement and seeks to own at least 50 % of the target shares after the transaction. All deals are completed. This results in a final sample of 6604 takeovers. For each takeover, we determine whether any other bidder made a public takeover offer for the same target within a period of 6 months. Deals with more than one bidder for the target we denote as takeover contest. All other deals are denoted as single bidder M&As without contest. Thus, our focus is on empirically investigating the occurrence of takeover contests with respect to excess takeover premium measures.

5 Empirical Results

In this section we present our empirical results. We first focus on the excess premium measures (Sect. 5.1) and then analyze the effects on takeover contests (Sect. 5.2).

5.1 Excess Premium Measures

Figure 2 displays the cumulative density functions of the standard premium and the excess premium measures. The various methods for calculating excess premiums

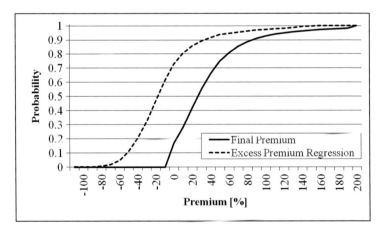

Fig. 2 Cumulative density function of standard premium and excess takeover premium measures

Table 1 Selected excess takeover premium measures (EP) in takeover contests and takeovers without competition for countries, industries, and method of payment (MOP)

	No contest (1)			Contest (2)			Diff. (1)–(2)	
Premium	N	Mean	Median	N	Mean	Median	Mean	Median
Initial premium	5915	44.50	35.14	350	46.46	39.10	−1.97	−3.97
EP country 6 months	5913	−1.45	−9.30	350	−5.75	−8.42	4.30*	−0.87*
EP industry 6 months	5871	−1.38	−8.42	347	−6.35	−10.17	4.97**	1.75
EP MOP 6 months	5844	−1.33	−9.32	349	−6.89	−10.73	5.56**	1.41
EP country 20 deals	5882	−1.37	−8.70	345	−5.98	−8.58	4.61**	−0.12
EP industry 20 deals	5726	−1.53	−8.62	333	−7.04	−9.46	5.51**	0.84
EP MOP 20 deals	5796	−1.30	−8.72	342	−6.42	−11.06	5.12**	2.35
EP country 6M, 20D	5882	−1.43	−9.31	346	−5.73	−8.42	4.30*	−0.89
EP industry 6M, 20D	5704	−1.58	−8.81	333	−6.64	−9.42	5.06**	0.61
EP MOP 6M, 20D	5796	−1.53	−9.38	342	−6.70	−11.29	5.34**	1.91
EP regr. all	5915	−1.71	−8.88	350	−6.53	−10.10	4.82**	1.22
EP regr. rolling	5908	−1.62	−7.78	348	−7.55	−10.01	5.93***	2.23*

***, **, and * denote significance at the 1, 5, and 10 % level, respectively.

result in different distributions compared to the standard calculation measure. For example, a premium of 10 % in the standard calculation measure indicates a low premium, but this relatively low premium can result in a positive excess premium if we adjust for the expected premium in comparable deals. Table 1 illustrates the results of excess premium calculations differentiated by bidders in contests and bidders without contests. The standard initial premium in the first row suggests that there is no significant difference between the size of the initial premiums in contests and takeovers without contests. Because we are interested in explaining the occurrence of contests, we expect that a low initial premium could increase the risk of a takeover contest. The next rows in Table 1 show the descriptive results of several excess premium measures. We use fixed period and fixed size approaches to determine the excess premiums and a combination of both approaches to address the problem of changing variances in the adjustment benchmark in times of lower M&A activity. In addition, we use a regression approach that controls for several factors simultaneously. As expected, we find that fixed period, fixed size excess premiums, and the combination of both are significantly lower in contests compared to takeovers without competition. Our results for regression-based premiums are strongest. Overall, excess takeover premiums are negative for offers in contests, which indicate that the mean initial offer is lower than the expected premium. In takeovers without contests, the excess premium differs only slightly from the expected premium whereas in contests the initial offer differs significantly from the expected premium. Therefore, we expect that the excess premium measures are superior in explaining the occurrence of takeover contests compared to the standard method.

5.2 Occurrence of Takeover Contests

The descriptive results of the excess premium measures indicate that these premiums are significant lower for bidders in contests compared to bidders in takeovers without competition. Therefore, we now test whether the excess premium measures are significantly in explaining the occurrence of takeover contests in contrast to the standard method. To test the impact of the excess premium measures we estimate probit models to determine the probability that a takeover contest occurs. We use a dummy variable as the dependent variable that takes a value of one if a rival bidder makes an offer for the same target after the initial offer, and zero if there is no competing offer. Furthermore, we use a set of variables to control for deal and target characteristics. We also include year dummies to control for year fixed effects and cluster standard errors by the 12 Fama-French industries.

Table 2 presents the results of our probit models. Model 1 includes the standard premium measure (initial premium) where we find an insignificantly negative effect on the occurrence of takeover contests. The further Models include the fixed period excess premium (Models 7–9), the fixed size excess premium (Models 10–12) estimated over industries, countries, and the method of payment. Models 2–4 include a combination of fixed period and fixed size excess premium. Models 5 and 6 contain the regression based excess premium. The results suggest that our measures are significant in explaining the occurrence of takeover contests in contrast to the standard method. However, our findings are limited to cash deals. In stock deals, the size of the premium varies due to the bidders stock price movements and therefore affects the offered premium during the takeover process. In more detail, fixed period excess premiums reduce the probability of a takeover contest if calculated over comparable deals within a country or the same method of payment. Offering a high excess premium, i.e., a premium above the mean for the country or the method of payment reduces the probability of a takeover contest. Moreover, we find that the fixed period premiums are widely robust to standardization (not tabulated). Country and method of payment fixed period excess premiums significantly reduces the probability of contests. Fixed size excess premiums are with some exceptions significant if they are determined within the same industry or country. The results further indicate that the number of deals in the adjustment benchmark is crucial and should contain at least 20 deals. The combination of fixed period and fixed size approach improves the results for the industry excess premium measure. We find the strongest effect on the occurrence of contests in the rolling regression based excess premiums. This indicates that several factors affect the premium size and should be included in calculating the expected and the excess premium. Overall, these results suggest that our excess premium measures are valuable in explaining the occurrence of takeover contests in contrast to the standard method.

In sum, our empirical analysis of takeover contests offers the following results. The multivariate analysis suggests that, at least for cash deals, fixed period and fixed size excess premiums significantly reduce the probability of a takeover contest.

Table 2 Probit occurrence of takeover contests

DV: contest	Model 1	Model 2	Model 3	Model 4	Model 5	Model 6
Initial premium	−0.0013					
EP industry 6M,20D		−0.0018**				
EP country 6M,20D			−0.0017*			
EP MOP 6M,20D				−0.0016*		
EP regr. all					−0.0015*	
EP regr. rolling						−0.0022**
Const.	−0.8639	−0.9857	−0.9021	−0.9102	−0.8716	−0.9199
R²	0.1453	0.1465	0.1498	0.1497	0.1464	0.1510
F-test	993.2756	546.2869	899.7584	870.6365	1072.1448	867.4884
N	3573	3457	3551	3554	3573	3566

DV: Contest	Model 7	Model 8	Model 9	Model 10	Model 11	Model 12
EP industry 6M	−0.0011					
EP country 6M		−0.0017*				
EP MOP 6M			−0.0017**			
EP industry 20D				−0.0018**		
EP country 20D					−0.0019**	
EP MOP 20D						−0.0013
Const.	−0.8862	−0.8643	−0.8688	−0.9617	−0.8987	−0.9054
R²	0.1502	0.1470	0.1470	0.1462	0.1482	0.1494
F-test	1088.1929	1048.3911	977.1911	542.4609	873.7536	1123.3046
N	3543	3571	3572	3470	3550	3554

***, **, and * denote significance at the 1, 5, and 10 % level, respectively.

Because the size of the premium in stock deals varies, the stock price of the bidder influences the premium during the takeover process. Industry adjusted premiums are important if they are calculated over a longer time period or over a larger number of deals, indicating that the number of deals in the adjustment benchmark should be sufficiently large. Furthermore, we find that fixed period measures are widely robust to standardization. Overall, the suggested measures are significant in explaining the occurrence of contests in contrast to the standard method. Offering a positive excess premium significantly reduces the probability that a contest occurs.

6 Conclusion

This study analyzes various approaches to calculate excess takeover premiums and tests empirically whether these measures are significant in explaining the occurrence or prevention of takeover contests. The literature so far provides only mixed evidence for the influence of premiums on takeover contests. Probably, the initial bidder has a first-mover advantage in choosing the conditions of his initial bid (Jennings and Mazzeo 1993) and therefore has the means to reduce the probability that a contest occurs. Therefore, we argue that only a premium above the expected premium is useful in deterring rival bidders and in preventing takeover contests in contrast to the standard calculation of premiums. Hence, we introduce different methods to calculate excess takeover premiums, given that premiums differ across industries, countries, methods of payment, and over time. We determine excess premiums by using different adjustment benchmarks (industry, country, and method of payment). Our results reveal significant differences in excess premiums between contest bidders and bidders without contests. Furthermore, our findings suggest that bidders in takeovers without a subsequent contest offer an initial premium that is only slightly below the expected premium. In contrast, initial takeover offers resulting in bidder contests use initial premiums that are significantly below the expected premium. Thus, we observe a negative correlation between the excess premium measures and the occurrence of takeover contests. These empirical findings suggest that our measures are valuable in explaining the occurrence of takeover contests.

References

Akdoğu, E. (2009). Gaining a competitive edge through acquisitions: Evidence from the telecommunications industry. *Journal of Corporate Finance, 15*, 99–112.
Eckbo, B. E. (2009). Bidding strategies and takeover premiums: A review. *Journal of Corporate Finance, 15*, 149–178.
Fishman, M. J. (1988). A theory of preemptive takeover bidding. *The RAND Journal of Economics, 19*, 88–101.

Jennings, R. H., & Mazzeo, M. A. (1993). Competing bids, target management resistance, and the structure of takeover bids. *Review of Financial Studies, 6*, 883–909.

Jeon, J. Q., & Lion, J. A. (2011). How much is reasonable? The size of termination fees in mergers and acquisitions. *Journal of Corporate Finance, 17*, 959–981.

Madura, J., & Ngo, T. (2008). Clustered synergies in the takeover market. *Journal of Financial Research, 31*, 333–356.

Madura, J., Ngo, T., & Viale, A. M. (2011). Convergent synergies in the global market for corporate control. *Journal of Banking & Finance, 35*, 2468–2478.

Madura, J., Ngo, T., & Viale, A. M. (2012). Why do merger premiums vary across industries and over time? *Quarterly Review of Economics and Finance, 52*, 49–62.

Officer, M. S. (2003). Competing bids, termination fees in mergers and acquisitions. *Journal of Financial Economics, 69*, 431–467.

Schwert, G. W. (2000). Hostility in takeovers: In the eyes of the beholder? *Journal of Finance, 55*, 2599–2640.

Firm-Specific Determinants on Dividend Changes: Insights from Data Mining

Karsten Luebke and Joachim Rojahn

Abstract This paper aims at investigating the performance of state-of-the-art Data Mining techniques in identifying important firm-specific determinants of dividend changes. Since announcements of dividend changes are said to be informative and likely to affect stock prices, an accurate prediction of dividend changes is of vital interest. Therefore, we compare Data Mining techniques like Classification Trees, Random Forests or Support Vector Machines with classical methods like Multinomial Logit or Linear Discriminant Analysis. This comparison is done on data of the dividend payout of German Prime Standard Issuers during the years 2007–2010, as in this phase of financial turmoil many dividend changes can be observed. To our best knowledge this is the first application of Data Mining techniques in this research field concerning the German Stock Market.

1 Introduction

There is a growing body of literature dealing with the application of machine learning techniques for financial market predictions. This paper deals with the application of machine learning techniques in the problem of predicting dividend changes. As announcements of dividend changes are said to have significant effects on stock prices (see, e.g., Docking and Koch 2005), a deeper analysis of the firm-specific determinants of dividend changes is of some importance, e.g., for investors but also for researchers in the field of management decision processes. Though, most papers dealing with this research question focus on single classification techniques, see, e.g., Payne (2011) analysing the firm-specific determinants of

K. Luebke (✉)
FOM Hochschule für Oekonomie und Management, c/o B1st software factory, Rheinlanddamm 201, 44139 Dortmund, Germany
e-mail: karsten.luebke@fom.de

J. Rojahn
FOM Hochschule für Oekonomie und Management, Leimkugelstraße 6, 45141 Essen, Germany
e-mail: joachim.rojahn@fom.de

© Springer International Publishing Switzerland 2016   335
A.F.X. Wilhelm, H.A. Kestler (eds.), *Analysis of Large and Complex Data*, Studies in Classification, Data Analysis, and Knowledge Organization,
DOI 10.1007/978-3-319-25226-1_29

dividend initiations by using a multiple discriminant analysis or Li and Lie (2006) running a multinominal logistic regression.

However, diverse classification methods may lead to different predictions of a class and varying performance in terms of misclassification error rates (e.g. Dietterich 1998). Additionally, depending on the method different variables may be important for the classification (see, e.g., Bolón-Canedo et al. 2013 for a recent review on variable selection methods).

Whereas classical multivariate methods used in econometrics like the Multinomial Logit or Linear Discriminant Analysis (LDA) use all variables in estimation and prediction simultaneously, the hierarchical Classification Tree approach not only includes an intrinsic variable selection but also may allow deeper insights in the management decision process. Random Forests (an ensemble of trees) are quite successful in terms of prediction and it is also possible to assess the importance of a variable in the prediction (see, e.g., Varian 2014). The classification performance of Support Vector Machines (SVM) is also usually quite good and may outperform the classical methods, see, e.g., Bennett and Campbell (2000). So these methods are compared in the fields of interpretation, prediction performance and variable importance in the application of announcements of dividend changes.

We analyse the payout policy of German issuers, as it is declared to be more flexible than in the USA or UK (Goergen et al. 2005). We focus on the period 2007–2010, as in this phase of financial turmoil many dividend changes can be observed. To reach for a high level of transparency in the decision making of dividend changes, as many explanatory variables as available are investigated and the effect and importance of each is assessed by means of the different classification methods mentioned above.

The paper is organized as follows: in the next section the classification methods are shortly discussed. In Sect. 3 the different variable importance measures are given. The data set and the analysed explanatory variables are introduced in Sect. 4 followed by the results of the different classification methods and variable importance measures in Sect. 5. Finally our findings are summarized and a short outlook is given.

2   Classification Methods

In this section we give a short introduction on the methods used to model and predict the changing behaviour of dividend payout. This dependent variable Y is on an ordinal scale: decrease, unchanged, increase of the dividend payout. So with K being the number of groups or classes in the application $K = 3$. We focus on the assumption on the independent variables random vector X with realizations $x = (x_1, x_2, \ldots, x_d)$, where d is the number of variables used, as well as the consequences in interpreting the model. More mathematical details are given in textbooks on Data-Mining, Machine-Learning or Multivariate-Analysis like Hastie et al. (2009).

2.1 Multinomial Logit

The Multinomial Logit (MN) is an extension for a multinomial (instead of a binomial) dependent variable of the well-known Logistic Regression. It is derived by modelling the log odds of each class against a reference class (unchanged in our application):

$$\log \frac{P(Y = i | X = x)}{P(Y = k | X = x)} = \beta_{i0} + \beta_i^T x$$

There are no distributional assumptions on x, especially using nominal explanatory variables is via dummy coding no problem. The sign of β_{ij} indicates whether an increase of x_j will raise the probability of class i—compared to the reference class k. Statistical testing procedures for the hypotheses $H_0 : \beta_{ij} = 0$, i.e. the variable X_j has no effect on the probability of class i, are available. Allocation is done according to an arg max rule, i.e. assign an observation to the class with the highest probability. The decision boundaries are linear in the explanatory variables.

2.2 Linear Discriminant Analysis

The LDA is similar to the Multinomial Logit Model, but uses the assumption of a multivariate normal distribution of X with a common covariance matrix Σ within the classes. With $\mu_j = E(X | K = j)$ and prior probabilities π_j allocation via

$$\arg\max_i \left(x^T \Sigma^{-1} \mu_i - \frac{1}{2} \mu_i^T \Sigma^{-1} \mu_i + \log \pi_i \right).$$

So the differences in the classes are analysed by the differences in the mean vector of each class—weighted by the inverted covariance matrix. If the assumptions are met, the LDA is optimal in terms of the misclassification rate. Like the Multinomial Logit Model the decision boundaries are linear.

2.3 Classification Tree

In a Classification Tree (Tree) analysis the space of the explanatory variables is partitioned into a set of rectangles and the same class y is assigned to each rectangle. To build up the rectangles feasible variables are split at suitable points. It is a hierarchic process from the root of the tree to the final nodes, where the building and pruning process of the tree depends on the split criteria which is based on a appropriate impurity measure (e.g. Gini index) of the nodes—and some further parameters like minimum number of observations in the terminal nodes.

The main advantage of a Classification Tree—as long as it is not too complex— is the easy interpretation of the classification rules which may reassemble in some ways the management decision process. It has no assumption on the distribution of the explanatory variables, but unfortunately Classification Trees have a high variance and are therefore unstable. Also the performance in terms of misclassification rate may be high compared to other methods.

2.4 Random Forests

In order to overcome the instability of a single Classification Tree many of such trees can be built and the generated forest of trees can be used for a bagged classifier: a Random Forest (RF) is generated by a bootstrapped sample for which a full tree is build where a random sub-sample of the d explanatory variables are the candidates for each split. Allocation of an observation is then done by a majority vote of all, e.g. 500 Classification Trees.

The algorithm is quite easy to understand and applied, and the performance in terms of misclassification rate is generally quite good. However, as classification is based on a large number of trees and rules, results are hard to interpret.

2.5 Support Vector Machines

SVM aim at construction large margin separating hyperplanes in a high dimensional feature-space. By using a so-called kernel trick the original variables can be transformed into some high dimensional space where the classes may be separated linearly by hyperplanes—even though the classes cannot be separated (linearly) in the space of the original variables.

SVM have some very nice mathematical properties and (depending on the choice of the parameters, especially the kernel) can perform quite well in terms of the misclassification rate, but interpretation is not straightforward. Nevertheless they return the Support Vectors, i.e. those observations that are either close to the decision boundary and therefore needed to construct the hyperplane or are misclassified.

3 Variable Importance Measures

Which firm-specific financial characteristics do affect the decision to increase or decrease the dividend payout? And which variables are most important for classification? To indicate the importance of a variable for the classification rule a wide range of methods is available. In this study mostly method-intrinsic measures are used to identify these key factors, i.e. variables:

For the Multinomial Logit Model the mean absolute value of the t-test statistic of $H_0 : \beta_{ij} = 0$ for $i = 1, \ldots, K - 1$ is used. As the LDA can be linked to Multivariate Analysis of Variance (see, e.g., Mardia et al. 1979), we employ the forward selection based on the P-value of Wilk's lambda to asses the variable importance for classification there.

For the importance of a variable in a Classification Tree simply the number of observations split by the variable can be counted. Therefore variables that split on top of the tree get more weight than variables that are only needed for terminal nodes with much less observations. In Random Forest the data x_j is permuted and the decrease in the prediction performance after permutation averaged over all trees is measured for each explanatory variable (Strobl et al. 2007).

The variable importance within the Support Vector Classifier is measured by the mean of the area under the ROC. The area under curve is averaged over each binary classification problem and averaged for all variables x_j, $j = 1, \ldots, d$, see Hand and Till (2001).

4   Data Set

The initial sample consists of all companies listed at the German Prime Standard at least once during the sample period from 2007 to 2010.

Since our investigation approach requires dividend data in at least two subsequent years, we have to eliminate observations because of missing data in the previous period due to IPOs or upgrades from other German stock market segments or missing data in the subsequent period because of de-listings. Additionally, we drop REITs and financials according to the GICS classification as dividend policy is subject to regulatory constraints. Finally, we exclude issuers following a zero-dividend policy, so that the final sample consists of $n = 609$ observations.

The following independent variables are used to model and predict the dependent variable dividend payout (increase, unchanged, decrease):

- Standardized net income: net income scaled by book equity
- Loss-Dummy: net income negative $= 1$, 0 otherwise
- Standardized FCF: free cash flow scaled by turnover
- LN(Size): natural logarithm of total assets
- Debt Ratio: total debt to total assets
- Cash to total assets
- Free float: fraction of shares which is frequently traded
- Price to book ratio: proxy for perceived undervaluation and/or growth opportunities
- Turnover growth rate

- Index: dummy for index membership in DAX, MDAX, TecDAX and SDAX, = 1 index member, 0 otherwise
- Year: 2007–2010. The period of financial turmoil with many dividend changes. Additionally, the German tax system experienced a major change as a flat tax system became effective in 2009, so that capital gains get less attractive for investors

For scaled variables, i.e. net income, free cash flow, debt ratio, cash to total asset, free float and price to book ratio, we also include the differences from period $t-1$ to period t (denoted by "Delta"). Also included are the relative changes in lN(size), denoted by "Percentage Change". The data is collected from Bloomberg.

5 Results

The results of the Multinomial Logit are shown in Table 1. Those factors which are significant at 0.05 level are marked by a (*). It can be seen that firms are more likely to increase their dividend payments, if they are profitable, have high turnover growth rates, are index members and if their debt ratio is low. The probability that firms cut their dividends is rising when losses occur, net income is shrinking and the

Table 1 Estimated t-values for dividend change in multinomial logit model

Variable	Reduction	Increase
(Intercept)	1.00	0.36
Standardized net income	2.36 (*)	3.20 (*)
Loss dummy	3.07 (*)	−0.51
Standardized FCF	−1.12	0.12
LN(Size)	−2.30 (*)	0.22
Debt ratio	−0.42	−2.12 (*)
Cash to total assets	−1.11	−0.86
Free float	−0.02	−0.61
Price to book ratio	−2.65 (*)	−1.13
Turnover growth rate	0.15	3.75 (*)
Index	1.88	3.12 (*)
Year 2008	−0.53	−0.38
Year 2009	1.36	−3.22 (*)
Year 2010	0.96	−2.39 (*)
Delta scaled net income	−3.62 (*)	−0.70
Delta scaled FCF	1.29	−0.46
Percentage change in total assets	−0.88	1.74
Delta scaled debt ratio	1.24	−1.79
Delta scaled cash reserves	−1.56	1.51
Delta free float	−0.63	−0.35
Delta price to book ratio	1.47	1.24

Table 2 Estimated class means in the linear discriminant analysis for dividend change

Variable	Reduction	Unchanged	Increase
Standardized net income	0.00	0.12	0.21
Loss dummy	0.30	0.04	0.01
Standardized FCF	0.02	0.05	0.06
LN(Size)	6.44	6.47	7.00
Debt ratio	0.22	0.20	0.18
Cash to total assets	0.11	0.14	0.13
Free float	0.62	0.61	0.65
Price to book ratio	1.57	1.96	2.65
Turnover growth rate	−0.01	0.03	0.13
Index	0.56	0.49	0.70
Year 2008	0.12	0.23	0.34
Year 2009	0.41	0.29	0.18
Year 2010	0.39	0.29	0.16
Delta scaled net income	−0.17	−0.01	0.02
Delta scaled FCF	−0.00	0.00	0.00
Percentage change in total assets	0.02	0.06	0.15
Delta scaled debt ratio	0.03	0.01	−0.00
Delta scaled cash reserves	−0.01	−0.00	0.00
Delta free float	−0.03	−0.02	−0.02
Delta price to book ratio	−0.18	−0.22	−0.28

price to book ratio is low. These results are in line with the findings of Li and Lie (2006), which referred the negative impact of price to book ratio on the probability to cut dividends to poor performance. Furthermore the results reveal the impact of the financial market crisis on payout behaviour in the periods 2009 and 2010.

These findings are coherent with the results of the LDA, displayed in Table 2. The ordering of the class wise means is also in line to the theory as well as to the order of the dependent variable.

For the result of the Classification Tree (see Fig. 1) it is interesting to note that the first node is the yearly change of the scaled net income. If this is below −0.043, the observations are classified as Reduction, especially if the net income is low or the debt ratio is high. If the yearly change of the scaled net income is above −0.043, the dividend payout is classified mainly as Increase, only if also the price to book ratio is <1.48 and the turnover growth rate is <0.83 the dividend payout is classified as unchanged.

As the result, i.e. classification of a Random Forest is more like a black-box only the classification performance is reported (see below). The same is true for a Support Vector Machine (here with a Radial Basis Kernel) but note that out of the 609 observations in the data set 455 are support vectors.

Without any fine tuning of parameters—all calculations are done with help of the statistical computing software R and the applicable packages out of the box—

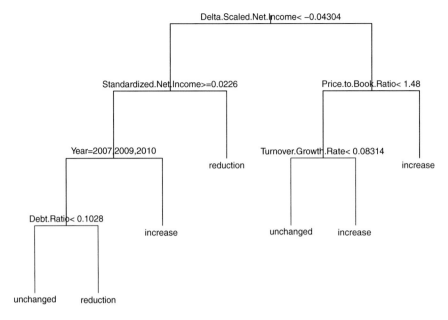

Fig. 1 Classification Tree for dividend change

Table 3 Apparent misclassification rate for dividend change

Method	MN	LDA	Tree	RF	SVM
Error rate	0.35	0.36	0.32	0.34	0.25

Table 4 Ex-Post-Ante Error Rate for dividend change

Year	MN	LDA	Tree	RF	SVM
2008	0.35	0.30	0.42	0.37	0.36
2009	0.53	0.53	0.52	0.51	0.57
2010	0.46	0.53	0.60	0.45	0.54
Mean	0.45	0.45	0.51	0.46	0.51

the apparent error rate, i.e. estimation/learning and evaluation are performed on the same data, are given in Table 3. One can see that about 30 % of the observations are misclassified, with the Support Vector Machine slightly outperforming the other methods.

Investors are mainly interested in the prediction performance of the methods, i.e. if the method is estimated on previous data, how well will it perform on new data. Therefore we also calculated the Ex-Post-Ante Error: e.g. use data 2007–2008 for estimation/learning, use data 2009 for prediction and testing (Weihs and Luebke 2009). Concerning this prediction oriented error rate on average (see Table 4) the Multinomial Logit Model, LDA and Random Forest perform similar, with the Classification Tree and SVM slightly worse. Please note the increase in the misclassification rate with the manifestation of the financial turmoil in 2009.

Table 5 Variable importance ranking for dividend change

	MN	LDA	Tree	RF	SVM
Top 1	Standardized net income	Loss dummy	Delta scaled net income	Delta scaled net income	Delta scaled net income
Top 2	Index	Turnover growth rate	Price to book ratio	Price to book ratio	Standardized net income
Top 3	Year 2009	Year 2009	Standardized net income	Standardized net income	Turnover growth rate
Top 4	Delta scaled net income	Year 2010	Turnover growth rate	Turnover growth rate	Price to book ratio
Top 5	Turnover growth rate	Index	Year	Year	Year

The five most important variables measured as described in Sect. 3 are given in Table 5. The turnover growth rate is ranked highly by all methods. Instead of the net income—which is ranked highly by all other methods—LDA selects the binary loss dummy. Delta scaled net income is the most important variable for the Data Mining methods (Tree, RF, SVM) and top 4 for the Multinomial Logit Model but not by LDA. The Data Mining methods also select the price to book ratio, which is also important, i.e. significant in the Multinomial Logit (see Table 1 for a decrease of the dividend payout but not for an increase.

6 Conclusion and Outlook

Despite the differences over the classification methods our analysis reveals four predominant financial characteristics affecting the decision to change dividends of German Prime Standard issuers in the years 2007–2010: the yearly change of the net income, the net income itself, the turnover growth rate and the price to book ratio. Interestingly, firms are more likely to increase dividends, if turnover growth rates are high. No trade-off between growth and dividend increases can be detected so there seem to be no dividend puzzle.

Nevertheless there are high error rates despite a quite large number of explanatory variables which may be due to missing variables or because of some kind of randomness within the process. Also it is possible that the reasons and actions may vary over time as this is suggested by the variation of the Ex-Post-Ante Error Rate as well as by the importance of the variable Year.

Data Mining techniques (especially without any fine-tuning of parameters) are not per-se outperforming established econometric methods at least for this data, but the conditional structure of the Classification Tree can enhance business understanding.

In a future work other potential variables should be investigated as well as a longer time period. Under the aspect of meta learning it will be interesting to understand why the classification performance varies and which data characteristics influence this.

References

Bennett, K. P., & Campbell, C. (2000). Support vector machines: Hype or hallelujah? *ACM SIGKDD Explorations Newsletter, 2*(2), 1–13.

Bolón-Canedo, V., Sánchez-maroño, N., & Alonso-Betanzos, A. (2013). A review of feature selection methods on synthetic data. *Knowledge and Information Systems, 34*(3), 483–519.

Dietterich, T. G. (1998). Approximate statistical tests for comparing supervised classification learning algorithms. *Neural Computation, 10*(7), 1895–1923.

Docking, D. S., & Koch, P. D. (2005). Sensitivity of investor reaction to market direction and volatility: Dividend change announcements. *Journal of Financial Research, 28*(1), 21–40.

Goergen, M., Renneboog, L., & Da Silva, C. (2005). When do German firms change their dividend? *Journal of Corporate Finance, 11*(1), 375–399.

Hand, D. J., & Till, R. J. (2001). A simple generalisation of the area under the ROC curve for multiple class classification problems. *Machine Learning, 45*(2), 171–186.

Hastie, T., Tibshirani, R., & Friedman, J. (2009). *The elements of statistical learning*. New York: Springer.

Mardia, K. V., Kent, J. T., & Bibby, J. M. (1979). *Multivariate analysis*. New York/London: Academic Press.

Li, W., & Lie, E. (2006). Dividend changes and catering incentives. *Journal of Financial Economics, 80*(2), 293–308.

Payne, B. C. (2011). On the financial characteristics of firms that initiated new dividends during a period of economic recession and financial market turmoil. *Journal of Economics and Finance, 35*(2), 149–163.

Strobl, C., Boulesteix, A. L., Zeileis, A., & Hothorn, T. (2007). Bias in random forest variable importance measures: Illustrations, sources and a solution. *BMC Bioinformatics, 8*(1), 25.

Varian, H. R. (2014). Big data: New tricks for econometrics. *The Journal of Economic Perspectives, 28*(2), 3–27.

Weihs, C., & Luebke, K. (2009). Prediction optimal classification of business phases. In: A. Wagner (Ed.), *Empirische Wirtschaftsforschung heute* (pp. 149–156). Stuttgart: Schäffer-Poeschel.

Selection of Balanced Structure Samples in Corporate Bankruptcy Prediction

Mateusz Baryła, Barbara Pawełek, and Józef Pociecha

Abstract Selecting samples is one of the methodological issues in the case of corporate failure prediction. In practice, when constructing a sample of balanced structure (50 % of failed and 50 % of non-failed firms), the most popular approach is based on subjective pairings of available bankrupt companies with non-bankrupt ones. This method is sometimes called pair-matched sampling. However, samples obtained in this way are not independent random ones. For this reason, other techniques should be taken into consideration. The simplest solution seems to be the employment of random sampling with replacement. The paper presents a comparative study of prognostic capabilities of four types of bankruptcy prediction models obtained as a result of applying two sampling techniques (pair-matched sampling and random sampling with replacement). The conducted analysis is based on financial data of Polish manufacturing companies.

1 Introduction

In corporate bankruptcy modelling, one looks for models which will be able to predict, most effectively, whether a company collapses or not. Although various types of models have been proposed so far, it cannot be definitely determined which of them are characterized by the best prognostic capabilities. Some researchers prefer data mining techniques (e.g. neural networks, classification trees), whereas others claim that statistical methods, such as linear discriminant analysis or logistic regression, could give better results. Since it is difficult to choose the most appropriate type of model, the question about possible sources of errors committed during the process of bankruptcy prediction arises. Pawełek and Pociecha (2012) indicate that one such error may be a method for selecting samples. The application of a sampling technique results in obtaining either a balanced or non-balanced sample. In the paper, special attention is given to pair-matched sampling which leads to the creation of a sample of balanced structure.

M. Baryła (✉) • B. Pawełek • J. Pociecha
Cracow University of Economics, 27 Rakowicka Street, 31-510 Cracow, Poland
e-mail: mateusz.baryla@uek.krakow.pl; barbara.pawelek@uek.krakow.pl;
jozef.pociecha@uek.krakow.pl

© Springer International Publishing Switzerland 2016   345
A.F.X. Wilhelm, H.A. Kestler (eds.), *Analysis of Large and Complex Data*, Studies
in Classification, Data Analysis, and Knowledge Organization,
DOI 10.1007/978-3-319-25226-1_30

In the classical approach of, e.g., discriminant analysis, frequently used in this kind of research, samples are randomized. Additionally, taking into consideration the fact that the populations of bankrupt and non-bankrupt companies are not usually very large (especially the population of firms that collapsed), the selection would require random sampling with replacement. However, in practice, samples are not drawn randomly. Information on insolvent companies is based on all the failures filed by court registers in a given period of time, so the entire population of bankrupt companies is considered, not a sample. Additionally, bankrupt companies are matched with well-performing ones with similar parameters, such as size or industry. Although pair-matched sampling is commonly used (Altman 1968; Boritz et al. 2007; Charitou et al. 2004; Wu et al. 2007; Zhang et al. 1999), it causes two main difficulties. Not only is it a non-random technique, but also it does not provide samples in their primary meaning.

The main aim of the paper is to compare two methods of sampling, that is pair-matched sampling and random sampling with replacement. In addition to this, an attempt to find the answer to the following question is made: does the method of selecting samples from the population of bankrupt and non-bankrupt companies have an impact on the prognostic capabilities of models? In order to answer the question, the case of balanced samples was taken into account. Operating on samples of the same type for both considered sampling techniques involved the necessity of making a reliable comparative analysis of prognostic abilities of models.

2 Characterization of Database and Variants of Study

To carry out the intended analysis, the authors used a database containing 7329 records that included financial information about 1852 manufacturing companies operating in Poland (133 of them were bankrupts). The data came from the period between 2005 and 2009 from the Legal Gazette of the Polish Government Part B and EMIS base (Emerging Markets Information Service). Each company was described by 33 financial indicators grouped into: liquidity ratios (4 variables), liability ratios (10 variables), profitability ratios (8 variables), productivity ratios (11 variables), and zero-one variable that equals "1" if a company was bankrupt between 2007 and 2010, and "0" if a company was non-bankrupt from 2005 to 2010. Financial ratios included in the database are listed in Table 1.

Assuming that financial data regarding the same company in various years is treated as information on different firms, the records of the database can be identified with various companies. Therefore, the data set contained the financial data of 7329 companies (182 of them went bankrupt, whereas the others did not fail).

Considering the most efficient use of the database, three variants of the analysis were regarded. The first variant (variant V_1) focused on bankruptcy prediction taking into account financial information from 2006 to 2009 and 1 year horizon of forecasting. In this case, the population of bankrupt companies consisted of 59

Table 1 Financial ratios appearing in the database and their definitions

Liquidity ratios	Profitability ratios
$R_{01} = \dfrac{\text{Current assets}}{\text{Short term liabilities}}$	$R_{15} = \text{EBITDA}$
$R_{02} = \dfrac{\text{Current assets} - \text{Inventories}}{\text{Short term liabilities}}$	$R_{16} = \dfrac{\text{EBITDA}}{\text{Total assets}}$
$R_{03} = \dfrac{\text{Current assets} - \text{Inventories} - \text{Short term receivables}}{\text{Short term liabilities}}$	$R_{17} = \dfrac{100 * \text{Gross profit (loss)}}{\text{Sales revenues}}$
$R_{04} = \dfrac{\text{Current assets} - \text{Short term liabilities}}{\text{Total assets}}$	$R_{18} = \dfrac{100 * \text{Net profit (loss)}}{\text{Sales revenues}}$
	$R_{19} = \dfrac{100 * \text{Net profit (loss)}}{\text{Shareholders' equity}}$
	$R_{20} = \dfrac{100 * \text{Net profit (loss)}}{\text{Total assets}}$
	$R_{21} = \dfrac{\text{Operating profit (loss)}}{\text{Total assets}}$
	$R_{22} = \dfrac{\text{Operating profit (loss)}}{\text{Sales revenues}}$

(continued)

Table 1 (continued)

Liability ratios	Productivity ratios
$R_{05} = \dfrac{\text{Long term liabilities} + \text{Short term liabilities}}{\text{Total assets}}$	$R_{23} = \dfrac{\text{Sales revenues}}{(\text{Short term receivables}(t) + \text{Short term receivables}(t-1))/2}$
$R_{06} = \dfrac{\text{Long term liabilities} + \text{Short term liabilities}}{\text{Shareholders' equity}}$	$R_{24} = \dfrac{\text{Sales revenues}}{(\text{Fixed assets}(t) + \text{Fixed assets}(t-1))/2}$
$R_{07} = \dfrac{\text{Long term liabilities}}{\text{Shareholders' equity}}$	$R_{25} = \dfrac{\text{Sales revenues}}{(\text{Total assets}(t) + \text{Total assets}(t-1))/2}$
$R_{08} = \dfrac{\text{Shareholders' equity}}{\text{Total assets}}$	$R_{26} = \dfrac{\text{Sales revenues}}{\text{Total assets}}$
$R_{09} = \dfrac{\text{Short term liabilities}}{\text{Total assets}}$	$R_{27} = \dfrac{\text{Short term liabilities}}{\text{Operating costs}}$
$R_{10} = \dfrac{\text{Fixed assets}}{\text{Total assets}}$	$R_{28} = \dfrac{\text{Inventories}}{\text{Sales revenues}}$
$R_{11} = \dfrac{\text{Net profit (loss)} + \text{Depreciation}}{\text{Long term liabilities} + \text{Short term liabilities}}$	$R_{29} = \dfrac{\text{Inventories}}{\text{Operating costs}}$
$R_{12} = \dfrac{\text{Shareholders' equity}}{\text{Long term liabilities} + \text{Short term liabilities}}$	$R_{30} = \dfrac{\text{Short term receivables}}{\text{Sales revenues}}$
$R_{13} = \dfrac{\text{Gross profit (loss)}}{\text{Short term liabilities}}$	$R_{31} = \dfrac{\text{Operating costs}}{\text{Short term liabilities}}$
$R_{14} = \dfrac{\text{Shareholders' equity} + \text{Long term liabilities}}{\text{Fixed assets}}$	$R_{32} = \dfrac{\text{Sales revenues}}{\text{Short term receivables}}$
	$R_{33} = \dfrac{100 * \text{Operating costs}}{\text{Sales revenues}}$

firms, while the population of non-bankrupt companies amounted to 5943 firms. The second variant (variant V_2) was based on data from 2005 to 2008 and concerned a 2-year-ahead failure prediction. The populations of bankrupts and non-bankrupts were composed of 123 and 5922 firms, respectively. The last variant of study (variant V_3) referred to data from 2007 and involved a 2-year forecasting horizon. In the case of this variant, 63 firms constituted the population of bankrupts, and 1644 firms formed the population of well-performing companies.

3   Research Procedure Description

To construct a sample of balanced structure for each variant of the study, two techniques of sampling were employed, that is pair-matched sampling and random sampling with replacement. Using the first mentioned method, all companies that failed in a given period of time constituted a sample of bankrupts. This period of time was directly connected with the applied variant of the analysis. Next, the bankrupt companies were paired up with non-bankrupt ones, taking into account the same industry code and similar value of assets. For variants V_1 and V_2, when matching companies in pairs, financial data came from the same year within every pair of firms. It meant that if a company went bankrupt, e.g. in 2008, it was paired with a firm which in the year did not fail. Thus, three balanced samples were created. They comprised: 118 companies (variant V_1), 246 companies (variant V_2) and 126 companies (variant V_3).

Three subsequent balanced samples were constructed with the use of another sampling technique, i.e. random sampling with replacement. So as to create them, the specific numbers of firms were drawn independently from the populations of bankrupts and non-bankrupts. The sizes of three samples which were obtained by means of random sampling with replacement remained in accordance with the sizes of three samples chosen by applying pair-matched sampling. The same number of companies constituting samples was necessary to guarantee a reliable comparative analysis between the two described methods of sampling. When companies were drawn independently from the populations, criteria for matching firms in pairs (e.g. similar size of companies) were not taken into consideration.

Before constructing bankruptcy prediction models for Polish manufacturing companies and comparison of their prognostic capabilities, some additional assumptions were made. They regarded: the type of applied models, the division of a sample into training data and testing data, as well as the technique used in the selection of variables. Many various types of bankruptcy prediction models have been proposed so far. Their detailed classification is presented, e.g. by McKee (2000). In the research, four commonly used methods of building such models were used, i.e.: linear discriminant analysis, logistic regression, neural networks (only multi-layer perceptrons) and classification trees (based on the CART algorithm). It is worth stressing that the first two mentioned methods belong to the group of statistical techniques, whereas the last two are data mining methods.

Table 2 Sample division
into training set and test set

Split type	Variant	Training set	Test set	Total
7:3	V_1	82	36	118
	V_2	172	74	246
	V_3	88	38	126
6:4	V_1	70	48	118
	V_2	148	98	246
	V_3	76	50	126

In bankruptcy modelling, a sample is usually divided into two subsamples. The first one, called a training set, is used to estimate a model, while the second subsample, known as a test set, is used to check the prognostic abilities of a built model. In the study, two splits were taken into account. According to one of them, 70 % of companies formed training data and 30 %—testing data. The other split was connected with dividing data into a training group and testing group in a ratio of 6:4. In each case, in a training set and test set, the number of bankrupt companies equalled the number of non-bankrupt ones. The exact amount of firms in a training group and testing group on account of the applied data split and variant of the analysis is presented in Table 2.

Selecting variables is another crucial issue while constructing models. For statistical models, two techniques were used; that is forward stepwise analysis and backward stepwise analysis within linear discriminant analysis and logistic regression. In order to create classification trees, the CART algorithm was used, which simultaneously triggers variables reduction. Variables obtained as a result of the previously mentioned techniques were also used to construct neural networks. In Polish and foreign literature on bankruptcy prediction, authors often obtain models with no more than six predictors. For this reason, models with a number of variables greater than six were not considered.

In the study, the following procedure was applied. For a particular variant of the analysis, particular sampling technique and particular data split, a sample had been divided randomly into training group and testing group until ten models of a given type were obtained. These models had to meet some conditions. Firstly, they were models with the maximum number of predictors equalling six. Secondly, values of type I error (percentage of bankrupt companies that were incorrectly classified as non-bankrupt ones) and type II error (percentage of non-bankrupt companies that were incorrectly classified as bankrupt ones) for both a training set and test set were less than 50 %. Thirdly, all parameters of statistical models were significant at the 0.05 level. Next, from every group of ten such models, the best model with the highest predictive power was chosen according to the following criterion. In the first place, the lowest value of type I error for testing data was considered. If it occurred that some models had the same value of type I error, the one with the minimum value of type II error for testing data was selected as the best model.

It should be emphasized that the choice of type I error as the first criterion of models selection was legitimate. In the literature on bankruptcy modelling, it is generally agreed that type I error is more important than type II error (Bellovary et al. 2007). From the perspective of financial institutions, such as banks, greater significance has an error being a consequence of granting a loan for a company that will go bankrupt (type I error) than an error resulting from not giving a loan to a firm that will be able to repay it (type II error). In the first case, a bank suffers a financial loss, whereas the second situation is connected with an alternative cost which is equal to the profit that a bank would gain if it granted a loan.

All the calculations were made by the use of STATISTICA 10. In the case of neural networks, three-layer perceptrons (an input and an output layer with one hidden layer) were considered. The BFGS algorithm was used for training neural networks. To transform the activation level of a neuron into an output signal, the following functions were taken into account: identity function, logistic function, hyperbolic tangent function, exponential function and softmax function. Classification trees were constructed by means of the CART algorithm. The Gini Index was employed as the impurity measure so as to assess the quality of the obtained splits of objects (companies) in the nodes. The tree pruning was based on the cost-complexity criterion which is characteristic of the CART method. More details regarding the implemented algorithms in STATISTICA (also for linear discriminant analysis and logistic regression which were used in the study) can be found in Statsoft (2013). In the case of the logit model, the classification of companies was based on the predicted probabilities, assuming the cut-off point at the level of 0.5. If the probability was greater than or equal to 0.5, a company was classified as a bankrupt one; otherwise a company was classified as a non-bankrupt one.

4 Empirical Results

The implementation of the previously described procedure led to the selection of 48 models with the highest prognostic capabilities. Half of them were formulated by the use of the non-random sampling method, and the rest—by the random sampling technique.

Table 3 shows a ranking of the best bankruptcy prediction models when pair-matched sampling was applied. The last three columns of the table contain information about the values of type I error, type II error, as well as total error (percentage of bankrupts and non-bankrupts that were incorrectly classified) calculated for companies which formed testing groups. Places in the presented rankings were allotted to models according to their predictive power (in the first place, a value of type I error was considered and after that, if it was necessary, a value of type II error).

Table 3 Rankings of the best models in the case of applying pair-matched sampling

Variant	Split type	Place in ranking	Model type	Type I error	Type II error	Total error
V_1	6:4	1	NN_1	4.17	16.67	10.42
		2	D_1	4.17	25.00	14.58
		3	CT_1	4.17	29.17	16.67
		4	L_1	8.33	25.00	16.67
	7:3	1	NN_2	5.56	27.78	16.67
		2	L_2	11.11	16.67	13.89
		3	D_2	11.11	22.22	16.67
		4	CT_2	11.11	33.33	22.22
V_2	6:4	1	NN_3	16.33	30.61	23.47
		2	CT_3	16.33	36.73	26.53
		3	L_3	22.45	30.61	26.53
		4	D_3	32.65	28.57	30.61
	7:3	1	NN_4	18.92	18.92	18.92
		2	CT_4	18.92	24.32	21.62
		3	L_4	27.03	27.03	27.03
		4	D_4	29.73	45.95	37.84
V_3	6:4	1	NN_5	20.00	36.00	28.00
		2	D_5	20.00	40.00	30.00
		3	L_5	24.00	40.00	32.00
		4	CT_5	24.00	48.00	36.00
	7:3	1	NN_6	5.26	36.84	21.05
		2	L_6	15.79	42.11	28.95
		3	D_6	15.79	47.37	31.58
		4	CT_6	36.84	31.58	34.21

Note: D denotes discriminant model, L logit model, NN neural network, CT classification tree, numbers from 1 to 6 are used in the subscript to distinguish between the same type of models

Similar rankings of the best models (on account of their predictive power) for three variants of the study and two considered types of data split are presented in Table 4. In this case, the models were built and their prognostic capabilities tested on the basis of samples created with the use of random sampling with replacement.

To check whether the applied method of sampling has an influence on the classification of companies from a testing group or not, a comparison of pairs of models was made. When comparing models, the same kind of model was considered as well as the same type of sample division and the same variant of study, but different technique of samples selection. A better model from every group of two compared models turned out to be the one with the lowest value of type I error for a test set. If both models had the same value of the error, then values of type II error for testing data were taken into consideration. The obtained results are shown in Table 5.

Table 4 Rankings of the best models in the case of applying random sampling with replacement

Variant	Split type	Place in ranking	Model type	Type I error	Type II error	Total error
V_1	6:4	1	NN_7	0.00	8.33	4.17
		2	CT_7	0.00	20.83	10.42
		3	D_7	8.33	0.00	4.17
		4	L_7	12.50	33.33	22.92
	7:3	1	NN_8	0.00	16.67	8.33
		2	D_8	0.00	27.78	13.89
		3	CT_8	5.56	11.11	8.33
		4	L_8	5.56	16.67	11.11
V_2	6:4	1	CT_9	10.20	38.78	24.49
		2	NN_9	12.24	32.65	22.45
		3	L_9	18.37	38.78	28.57
		4	D_9	26.53	24.49	25.51
	7:3	1	NN_{10}	10.81	35.14	22.97
		2	CT_{10}	16.22	32.43	24.32
		3	D_{10}	18.92	37.84	28.38
		4	L_{10}	24.32	27.03	25.68
V_3	6:4	1	NN_{11}	12.00	24.00	18.00
		2	D_{11}	16.00	32.00	24.00
		3	CT_{11}	24.00	28.00	26.00
		4	L_{11}	32.00	24.00	28.00
	7:3	1	NN_{12}	0.00	21.05	10.53
		2	L_{12}	15.79	26.32	21.05
		3	CT_{12}	21.05	10.53	15.79
		3	D_{12}	21.05	10.53	15.79

Note: in the fourth column, numbers from 7 to 12 are used in the subscript to distinguish between the same type of models

Comparing 24 pairs of bankruptcy prediction models, it can be stated that random sampling with replacement provides better forecasts in as many as twenty cases. With reference to variant V_2, independently of the applied sample split, the random technique of sampling always led to the creation of models with higher predictive power. For both types of data split within variant V_3, in three cases out of four, random sampling with replacement ensured better prognostic capabilities of models. It occurred that the random method of sample selection also provided better results for all cases of variant V_1, when a sample was split into training data and testing data in a ratio of 7:3. Only in one situation (variant: V_1, type of data split: 6:4), pair-matched sampling equaled random sampling with replacement on account of the obtained forecasts in the group of four considered kinds of bankruptcy prediction models.

Table 5 Outcomes of models comparisons on account of applied sampling method

Variant	Split type	Number of winning comparisons	Result of comparison	Winning models
V_1	6:4	2	In favour of *rswr*	NN_7, CT_7
		2	In favour of *p-ms*	D_1, L_1
	7:3	4	In favour of *rswr*	NN_8, D_8, CT_8, L_8
		0	In favour of *p-ms*	–
V_2	6:4	4	In favour of *rswr*	CT_9, NN_9, L_9, D_9
		0	In favour of *p-ms*	–
	7:3	4	In favour of *rswr*	$NN_{10}, CT_{10}, D_{10}, L_{10}$
		0	In favour of *p-ms*	–
V_3	6:4	3	In favour of *rswr*	NN_{11}, D_{11}, CT_{11}
		1	In favour of *p-ms*	L_5
	7:3	3	In favour of *rswr*	NN_{12}, L_{12}, CT_{12}
		1	In favour of *p-ms*	D_6

Note: *rswr* random sampling with replacement, *p-ms* pair-matched sampling

It is worth noting that whenever pair-matched sampling contributed to the creation of models with higher predictive power, it referred only to some statistical models, namely: D_1, L_1, L_5, D_6. In the case of models in the form of neural networks and classification trees, the outcomes always argued for random sampling with replacement.

5   Conclusions

Empirical results of the conducted analysis led to the conclusion that the type of applied technique of sample selection has an influence on model prognostic abilities in the group of four considered types of models. As far as statistical models are concerned (linear discriminant models and logit models), the use of random sampling with replacement provided better forecasts in most cases. In eight out of twelve compared pairs of statistical models, a more accurate classification of firms from the testing groups was observed for models which were tested on random samples drawn independently from the populations of bankrupts and non-bankrupts. It also turned out that in the case of neural networks and classification trees, the implementation of random sampling with replacement always guaranteed models with higher predictive power. The outcomes showed that random sampling with replacement can be perceived as an alternative to pair-matched sampling.

Undoubtedly, assumptions which were made in the research had an impact on the results. It is worth recalling some of them. Firstly, the procedure of generating new models of the same type was repeated until the building of ten models which complied with the defined rules. Of course, creating new models could last to the moment of getting more than ten such models. Secondly, the choice of a model with the highest predictive power from groups of ten estimated models was made in accordance with a certain criterion. The minimum value of type I error was considered in the first place, and then the minimum value of type II error. The same rule found its implementation when rankings were created and comparison of pairs of models was made. Despite the fact that other criteria could be taken into account, it does not change the fact that for credit institutions, more important is type I error than type II error. Thirdly, the conducted analysis could be extended by including other types of bankruptcy prediction models, different kinds of sample division into training and testing sample, or alternative techniques of variables selection.

It should also be emphasized that the prognostic abilities of the created models were checked on the basis of test sets which were subsamples of selected samples. This raises the question: how accurate will the classification of companies be if they are from other samples than the tested one; for example, chosen from the population of firms that belong to the same sector of industry and the same period of time? Additionally, the problem of checking the predictions for their stability remains: how would the outcomes change when a new sample was drawn? Due to many assumptions which were made when doing the research, the issue of influence of employed sampling technique leading to the creation of balanced structure samples on predictive power of models still stays open.

Acknowledgements The authors would like to express their appreciation for the support provided by the National Science Centre (NCN, grant No. N N111 540 140).

References

Altman, E. I. (1968). Financial ratios, discriminant analysis and the prediction of corporate bankruptcy. *The Journal of Finance, 23*(4), 589–609.

Bellovary, J., Giacomino, D., & Akers, M. (2007). A review of bankruptcy prediction studies: 1930 to present. *Journal of Financial Education, 33*, 1–42.

Boritz, J. E., Kennedy, D. B., & Sun, J. Y. (2007). Predicting business failures in Canada. *Accounting Perspectives, 6*(2), 141–165.

Charitou, A., Neophytou, E., & Charalambous, C. (2004). Predicting corporate failure: Empirical evidence for the UK. *European Accounting Review, 13*(3), 465–497.

Mckee, T. E. (2000). Developing a bankruptcy prediction model via rough sets theory. *International Journal of Intelligent Systems in Accounting, Finance and Management, 9*(3), 159–173.

Pawełek, B., & Pociecha, J. (2012). General SEM model in researching corporate bankruptcy and business cycles. In: J. Pociecha & R. Decker (Eds.), *Data analysis methods and its applications* (pp. 215–231). Warsaw: C.H. Beck.

Statsoft, Inc. (2013). *Electronic statistics textbook.* Tulsa: StatSoft, http://www.statsoft.com/textbook/

Wu, C.-H., Tzeng, G.-H., Goo, Y.-J., & Fang, W.-C. (2007). A real-valued genetic algorithm to optimize the parameters of support vector machine for predicting bankruptcy. *Expert Systems with Applications, 32*, 397–408.

Zhang, G., Hu, M. Y., Patuwo, B. E., & Indro, D. C. (1999). Artificial neural networks in bankruptcy prediction: General framework and cross-validation analysis. *European Journal of Operational Research, 116*, 16–32.

Facilitating Household Financial Plan Optimization by Adjusting Time Range of Analysis to Life-Length Risk Aversion

Radoslaw Pietrzyk and Pawel Rokita

Abstract The article presents a concept of two-person household model with an original approach to expressing life-length risk aversion, allowing, at the same time, to simplify financial plan optimization. The technique uses (with improvements and corrections) concepts introduced in some earlier works by the authors, but it has not been presented so far as the main subject nor discussed in details. Moreover, financial plans for two persons treated as a household are compared here with a sum of two single individuals. This enriches the presentation of the model by an analysis of advantages of the joint (household) approach.

1 Introduction

The aim of this research is to propose a household financial planning framework that simplifies optimization of two-person household financial plan, but may be easily augmented, still preserving the same concept. Secondary aims are: demonstrating how the model works for stylized data and indicating the significance of household effect by comparing the results obtained for a household and for a sum of two separate individuals.

This piece of research expands and specifies more precisely some elements of the concept presented by Feldman et al. (2014a).

Classical approaches, on the pattern of Yaari (1965) model, allow to optimize consumption for a single person rather than a household. This was not earlier than the last two decades of the twentieth century when research by Kotlikoff and Spivak (1981) pointed out the significance of the fact that a vast part of life-long financial decision makers are married couples. Hurd (1999) gave an analytical solution of optimal consumption problem for a couple. Brown and Poterba (2000) continued the current of research initiated by Kotlikoff and Spivak. They investigated life annuities designed for married couples, providing examples showing advantages of joint annuities. Albeit this article does not discuss joint annuities, the aforementioned

R. Pietrzyk (✉) • P. Rokita
Wroclaw University of Economics, Wroclaw, Poland
e-mail: radoslaw.pietrzyk@ue.wroc.pl; pawel.rokita@ue.wroc.pl

© Springer International Publishing Switzerland 2016
A.F.X. Wilhelm, H.A. Kestler (eds.), *Analysis of Large and Complex Data*, Studies in Classification, Data Analysis, and Knowledge Organization,
DOI 10.1007/978-3-319-25226-1_31

357

advantages may serve as an argument for analyzing joint properties of household finance.

The model discussed in this article is designed to facilitate building easily applicable decision supporting tools. It is shown in its basic version, which in the phase of concept presentation is rather a merit than shortfall.

This is a discrete time, two-person household model. More persons are allowed, but treated just as part of financial situation of the main two. The approach is general in this sense that it does not impose any particular consumption model, utility function, nor underlying survival model. Despite the fact that some strong simplifying assumptions (comp. Sects. 2.1 and 4) are used, relaxing them would not entail changes of the very concept. Depending on the intended application and research objectives, the model may be modified and augmented, like, for instance, its variants used by Feldman et al. (2014b) or Pietrzyk and Rokita (2014). The model allows also to analyze advantages of household (joint) approach. In order to perform this analysis, a so-called *disjoint variant* is introduced for comparison.

There are two main contributions of this article. The first is combining reduction of the number of survival scenarios with an original and, at the same time, intuitive way of expressing risk aversion (*range of concern*). The scenario reduction, even though it is a simplification, makes the model rather more than less realistic, cutting off scenarios that are hardly plausible from psychological point of view (see Sect. 2.3). The second contribution is facilitating decision making thanks to a straightforward graphical interpretation of the financial planning results (expected cumulated surplus trajectories—comp. Sect. 2.1). Each financial plan has a corresponding shape of expected cumulated surplus. The model also allows to take advantage of internal risk sharing effect within the household (comp. Sects. 3 and 4), and, as it has already been mentioned, is highly elastic and easily extendible.

The article is composed as follows: Sect. 2 presents assumptions and general concept of the model, Sect. 3 introduces the so-called disjoint variant which is designed to be comparable with household approach in order to identify advantages of a joint treatment of household finance, and in Sect. 4 a numerical example is presented. By the example also some differences between a household and disjoint variant are shown. Section 5 concludes.

2   The Model

This is a cash-flow based, discrete time, life-long financial plan model for two-person household with pre-declared consumption needs.

2.1 General Concept and Assumptions

The model assumes, so far, two financial goals: retirement and bequest. The second is not set as a typical financial goal (defined in terms of time and magnitude) but it is taken into account by utility of residual wealth.

Household financial characteristics at the start of the plan include:

1. Income of Person 1 and Person 2.
2. Consumption of the household.
3. Retirement dates of Person 1 and Person 2.
4. Expected retirement gap of Person 1 and Person 2.
5. Growth rates of income and consumption, expressed in real terms.
6. Expected rates of return on investments (risky and risk free) in real terms.
7. Some other parameters, like expected inflation.
8. Mortality model for the two persons, or data from life expectancy tables.

All the components listed above, but survival, are represented by their expected values. Consistently, only life-length risk is explicitly present. To help adjust the plan to changes of other variables, regular revisions are performed.

Household consumption is composed of three elements: (a) common consumption—fixed and not attributed to any particular person, (b) consumption of Person 1, and (c) consumption of Person 2.

Household members pay contributions to private pension programs on regular basis. This is the only form of investment (but investing ongoing surplus at the risk free rate). The programs are separated, but if a person dies before retirement age, the amassed capital is transferred to that other one.

Investments of the household are divided into:

1. Systematic investment program assigned to Person 1.
2. Systematic investment program assigned to Person 2.
3. "Uninvested" and unconsumed surplus (invested at some low rate).

It is assumed that the surplus is invested at a rate not higher than the risk free one, though in fact it might be. It is just the matter of cautiousness (all the more so because cumulated surplus plays also the role of an immediately accessible liquid reserve). Simplifying the discussion somewhat, the cumulated surplus may be treated as *risk free investment*, and what is here referred to as systematic investment program assigned to Person 1 or 2—as *risky investment*.

Retirement and bequest preferences are strictly contradictive in the model $\alpha = 1 - \beta$, where α, β—consumption and bequest preference, respectively).

Consumption preference influences also the amount of capital required to fulfill the retirement goal. It is postulated here that consumption increases in real terms, both in pre-retirement and retirement phase. As a special case, consumption may be constant in real terms (inflation indexed).

The value function is constructed so that two subtypes of life-length risk are taken into account: premature death risk and longevity risk.

To the key parameters of the household belong the so-called *critical dates*. There are four such dates distinguished—comp. Feldman et al. (2014a): $R1$—retirement of Person 1, $R2$—retirement of Person 2, $E(D1)$—expected date of death of Person 1, $E(D2)$—expected date of death of Person 2.

The only two random factors of the model are dates of death: $D1$—date of death of Person 1, $D2$—date of death of Person 2.

A pair $Sc = (D1, D2)$ is a survival scenario. Some survival scenarios give very similar final financial outputs and some differ a lot. The differences consist in location of $D1$ and $D2$ amongst critical dates.

2.2 The Task

The choice of an optimal financial plan consists in maximization of the value function with respect to the following two decision variables:

1. Consumption-investment proportion,
2. Division of investments into the part assigned to Person 1 and Person 2.

At the moment 0 there is no surplus generated. All household income is spent either on consumption or investment. Thus, initial consumption rate (c_0) determines consumption-investment proportion at start. Then, income and consumption change deterministically (growth rates). In all subsequent periods the difference between income and consumption growth rates may be a source of surplus. The initial level of consumption determines not only consumption-investment proportion, but also investment-surplus proportion in the next periods. Investment-surplus proportion may be here also interpreted as proportion between risky and risk-free investments (comp. Sect. 2.1). Of course, the proportions are fixed only at the start of the plan.

Division of household investment into parts assigned to Person 1 and Person 2 determines what part of household investment contributions goes to programs maturing on the retirement date of Person 1 (υ) and Person 2 $(1 - \upsilon)$. It is assumed that capital accumulated in the investment may be used to buy an individual annuity (joint annuities are not taken into account).

The optimization task consists in maximization of a goal function described in Sect. 2.4 (value function of the household), with the aforementioned decision variables. The constraints are: the budget constraint (no unrecoverable shortfall allowed) and minimum acceptable level of consumption. Each pair of values of decision variables (c_0, υ) stands for a particular financial plan.

2.3   Range of Concern

In order to simplify application, it is proposed that the risk aversion parameters should be expressed in terms of periods before and after $E(Di)$, in years, delimiting what will be further referred to as the *range of concern*. Household members define the scope of survival scenarios that they find worth their concern. As regards premature death, they define some number of years before unconditional expected date of death (parameter γ^*). With respect to longevity risk, they define a number of years after expected date of death (parameter δ^*). The range of concern is thus defined as [Eq. (1)]:

$$(D_1^*, D_2^*) \in \left[E(D1) - \gamma^*; E(D1) + \delta^* \right] \times \left[E(D2) - \gamma^*; E(D2) + \delta^* \right]. \qquad (1)$$

The range of concern should not be confused with domain of optimization. It is not a range of decision variable values, but a set of scenarios to be used.

This procedure differs significantly from the most commonly used ones. In classical approaches consumption is optimized across the whole life cycle or the whole retirement period (if the model is designed to optimize decumulation)—comp. Yaari (1965), Kotlikoff and Spivak (1981), Hurd (1999), Brown and Poterba (2000), Huang et al. (2012), Blake et al. (2013), etc.

Taking into account all possible scenarios might result in excess saving and amassing too much retirement capital. The household would have to decrease its consumption in early years, in order to fulfill optimization constraints for each combination of individual survival scenarios. Moreover, some of these scenarios, even though mathematically probable (small but nonzero probability), may be hardly plausible in real life. For example, let us consider the following survival scenario: a men dies at the age of 25 and his wife lives on to 95. Treating her further life just as continuation of the original plan does not seem reasonable. Thirdly, attempting to find optimal solution for all survival scenarios is computationally tough, because their number increases proportionally to square of the number of years taken into account.

2.4   Value Function

The goal function [Eq. (2)] is an expansion of that proposed by Feldman et al. (2014a). And it is a corrected version of that presented by Feldman et al. (2014b). Its components are expected discounted utilities of (a) consumption in pre-retirement phase, (b) consumption in retirement, and (c) bequest. Consumption and bequest are calculated for bivariate survival scenarios. Probabilities of survival scenarios may be derived from virtually any survival model. It may be a combination of two independent survival models like univariate (Gompertz 1825), or a joint bivariate model (direct implementation of stochastic force of mortality, like proposed by Huang et al. 2012 may be, however, problematic).

Utility of consumption is calculated for whole paths of the survival process, beginning at the start of the financial plan ($t = 0$), up to the moment when the last household member dies under the scenario ($t = \max\{D_1^*, D_2^*\}$). Utility of bequest is taken for each scenario only once, namely at the moment when the survival trajectory ends under this scenario ($t = \max\{D_1^*, D_2^*\}$).

In a household model with more than one person, financial categories assigned to the persons are interconnected even if the underlying survival processes are assumed to be independent. There are also more survival states than just "alive" and "not alive," and in each state of the type "Person i is alive and Person j is not alive" it is important when exactly Person j died. Critical dates are of special significance, particularly dates of retirement. Moreover, some of important financial categories, like cumulated investment and cumulated surplus follow path-dependent processes, due to their cumulative nature. All these properties make hardly applicable for households the traditional models, in which the underlying survival process is expressed by means of conditional probability of survival. That inconvenience is overcome by taking whole scenarios (process trajectories) with their unconditional probabilities.

The value function of a household is given by the formula (2):

$$
V(c_0, \upsilon)
$$

$$
= \sum_{D_2^*=E(D2)-\gamma^*}^{E(D2)+\delta^*} \sum_{D_1^*=E(D1)-\gamma^*}^{E(D1)+\delta^*} P_{(D_1^*,D_2^*)} \cdot \left(\alpha \cdot \left(\sum_{t=0}^{\max\{D_1^*,D_2^*\}} \frac{1}{(1+r_C)^t} u\left(C(t; D_1^*, D_2^*)\right)(\gamma(t)+\delta(t)) \right) + \beta \cdot \frac{1}{(1+r_B)^{\max\{D_1^*,D_2^*\}}} u\left(B(\max\{D_1^*,D_2^*\}; D_1^*, D_2^*)\right) \right),
$$

$$(2)$$

where: c_0—consumption rate at the moment 0; υ—proportion of Person 1 investment in joint one-period contribution of the household ($\upsilon \equiv \upsilon_1$, $\upsilon_1 = 1 - \upsilon_2$); δ^*—longevity risk aversion parameter; γ^*—premature-death risk aversion parameter; r_C, r_B—discount rates for consumption and bequest, respectively; $P_{(D_1^*,D_2^*)}$—(unconditional) probability of such scenario that $\left(D1 = D_1^*, D2 = D_2^*\right)$; α—consumption preference; β—bequest preference; $\max\{D_1^*, D_2^*\}$—time of household end, under a scenario of $\left(D1 = D_1^*, D2 = D_2^*\right)$; $C(t; D_1^*, D_2^*)$—consumption at time t for $\left(D1 = D_1^*, D2 = D_2^*\right)$ scenario; $B(t; D_1^*, D_2^*)$—cumulated surplus at time t for $\left(D1 = D_1^*, D2 = D_2^*\right)$ scenario (available bequest if $t = \max\{D_1^*, D_2^*\}$).

In the double summation loop of the formula (2) there are taken only such survival scenarios (pairs of dates of death) that belong to the range of concern. When calculating discounted utility of consumption, the whole trajectory of consumption is taken for each such scenario (till the end, that is—$\max\{D_1^*, D_2^*\}$). The argument of the consumption function, $C(t; D_1^*, D_2^*)$, is time (t). Scenario information (D_1^*, D_2^*) is treated as parameters. The arguments and parameters taken by the bequest function, $B(t; D_1^*, D_2^*)$, are the same, with the difference that the time argument is for each scenario just the end of this scenario ($t = \max\{D_1^*, D_2^*\}$). Because whole paths of consumption (and cumulated surplus) are used, it is always easy to determine when exactly they end. Scenarios are then weighted with their unconditional probabilities.

Functions $\delta(t)$ and $\gamma(t)$ play the role of multipliers, distinguishing between utility of consumption before and after expected date of household end ($E(D) = \max\{E(D1), E(D2)\}$). They additionally modify utility, introducing a time-dependent risk aversion factor, which is not household-specific. Whereas γ^* and δ^* are risk aversion parameters, $\gamma(t)$ and $\delta(t)$ may be thought of as risk aversion measures. They express severity of a potential instance of death. The more distant the period from expected time of death, the higher severity (in financial sense). For $t = E(D)$ the multipliers are of the value 1 (neutral as utility modifiers). Before $E(D)$ the $\gamma(t)$ function is decreasing and convex, it falls to unity at $E(D)$, and is zero after $E(D)$. The $\delta(t)$ function is of value zero before $E(D)$, starts with the value one at $E(D)$ and then is convex and increasing. The slope of $\delta(t)$ is higher than of $\gamma(t)$—the intuition is that longevity has more severe consequences than premature death, due to less possibilities of financial recovery in an old age. Formula (3) is just a proposition of how the $\delta(t)$ and $\gamma(t)$ functions might be defined:

$$\gamma(t) = \begin{cases} \left(\frac{1}{1+\gamma^*}\right)^{\left(\frac{t-E(D)}{E(D)}\right)} & t \leq E(D) \\ 0 & t > E(D) \end{cases}$$

$$\delta(t) = \begin{cases} (1+\delta^*)^{\left(\frac{t-E(D)}{\delta^*}\right)} & t > E(D) \\ 0 & t \leq E(D) \end{cases} \qquad (3)$$

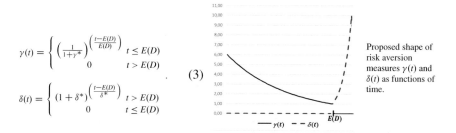

Proposed shape of risk aversion measures $\gamma(t)$ and $\delta(t)$ as functions of time.

3 Joint vs. Disjoint Treatment

To show the difference in financial plans prepared for a two-person household and two separate individuals, a disjoint model was created. For comparativeness, all cash flows from financial plans of the two separate individuals are summed up and presented also jointly. Moreover, household common costs, that are not assigned individually (common consumption), are just divided into two equal parts. The main difference between joint and disjoint variant of the plan is that in the disjoint variant the persons may invest only in their own retirement capital and then buy life annuity for themselves individually. Also no part of income of one person may be spent on consumption of that other one. Comparison of the joint and disjoint variant is presented in Table 1.

Table 1 Household (joint) vs. disjoint variant

	Household	Disjoint variant
Common costs (not-assigned consumption)	Joint	Divided into two equal parts
Individual costs	Individually assigned, covered jointly	Individually assigned and covered
Income	Individually assigned but spent jointly	Individually assigned
Investment	The difference between joint income and joint consumption of the household	The difference between individual income and individual consumption
Private pension plan	Individually assigned but the accumulated capital may be used for purchasing life annuity for any person	Individually assigned and the accumulated capital is used to purchase life annuity for this particular person
Private retirement	Individually assigned but spent jointly	Individually assigned

4 Numerical Example

It is assumed that the household has common and fixed utility function of $\sqrt{X}$ (where X stands for consumption or bequest).

The stylized example of this section is constructed to illustrates the idea of the model and the risk-sharing mechanism by internal transfer of capital. The household model might be also more complicated (children, elderly parents of the main household members, other models of consumption, and labor income evolution, etc.), but this would not make the example more comprehensive.

Let the household be composed of two persons: a 34-year-old man, earning annually 31,000 monetary units, with life expectancy (at t_0) of 74 years, and a woman, 32 year old, earning 19,000, with life expectancy (at t_0) of 82 years. Other parameters (e.g., growth rates) are set so that they resembles the reality.

The effect of financial plan optimization is presented using plots of cumulated net cash flow for three financial plans. The plans are further referred to as **Plan 0**, **Plan 1**, and **Plan 2**. They differ only by the range of concern. **Plan 0** assumes that $\gamma^* = 0$, $\delta^* = 0$, which means no risk aversion, and optimization for the expected scenario ($D1 = E(D1), D2 = E(D2)$) only. **Plan 1** is built for parameters: $\gamma^* = 5$, $\delta^* = 5$. **Plan 2** assumes even wider range of concern, with: $\gamma^* = 7$, $\delta^* = 9$. All results are presented for already optimized plans (each optimized given its range of concern).

Cumulated surplus of these plans is analyzed under three scenarios (the expected one and two other than expected). Let **Scenario (0,0)** denote the expected scenario, (0,0) meaning that there is no deviation from expectations. **Scenario (−5,5)** denotes a scenario under which Person 1 dies 5 years before and Person 2 dies 5 years after her or his expected date of death. And in the **Scenario (−7,9)** it is 7 years before and 9 years after, respectively.

In Figs. 1, 2, and 3 there are shown cumulated surplus time structures of the plans under different scenarios. The bar plots present the result for the household (joint) variant and the linear plots—for the disjoint variant.

In Fig. 1 there is shown a result of **Plan 0** optimization. This plan is optimized only for one, expected scenario. In the subplot 1(a) cumulated surplus for the expected **Scenario (0,0)** is presented. This plan is very sensitive to any deviation from expectations. Subplot 1(b) presents cumulated surplus trajectory for **Scenario (−5,5)**. The plan suffers under this scenario a serious shortfall. There is also observed no advantage of household approach over disjoint variant for this plan (compare the bar plot and the solid line).

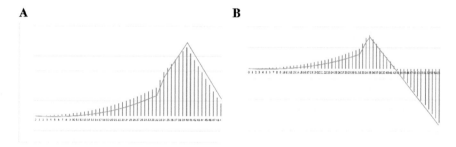

Fig. 1   Plan 0. Realized scenarios: **(a) Scenario (0,0)**, **(b) Scenario (−5,5)**

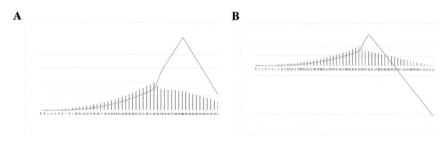

Fig. 2   Plan 1. Realized scenarios: **(a) Scenario (0,0)**, **(b) Scenario (−5,5)**

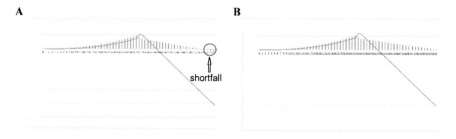

Fig. 3   Subplot **(a) Plan 1, Scenario (−7,9)**. Subplot **(b) Plan 2, Scenario (−7,9)**

In Fig. 2 behavior of a more conservative **Plan 1** is shown. The subplot 2(a) presents cumulated surplus trajectory for the expected **Scenario (0,0)**. Under this scenario both the household and disjoint variant give stable financial liquidity. The advantage of the household approach becomes apparent under the more adverse **Scenario (−5,5)**, shown in the subplot 2(b). For the household (joint) variant there is a stable financial situation also for this scenario, whereas in the disjoint variant a deep unrecoverable shortfall occurs.

The effect of even more adverse **Scenario (−7,9)** is shown in Fig. 3. Subplot 3(a) presents performance of the **Plan 1**. Since the scenario falls beyond the range of concern of this plan, the shortfall at the end is not a surprise. If the household wanted to be protected against such scenarios, it should optimize its financial plan for a broader range of concern.

Let us assume that the household has declared: $\gamma^* = 7$, $\delta^* = 9$. This gives **Plan 2**. The subplot 3(b) presents **Plan 2.** outcome under the **Scenario (−7,9)**. The plan guarantees stable financial liquidity also for this adverse scenario. The last refers only to the household (joint) variant.

5 Conclusions

The article presents a simplified, yet general, household financial planning framework that allows to overcome some modeling problems. Narrowing the bunch of survival scenarios (the range of concern) is here the basic concept. This is strictly connected with an easily applicable interpretation of risk aversion (the idea already used by Feldman et al. 2014b and Pietrzyk and Rokita 2014, but then yet without the comparison how the concept works for a household and disjoint variant).

Further research in this field might concentrate on a more detailed insight in the difference between the joint (household) and disjoint variant, the conditions under which the joint variant outperforms disjoint one, and the potential of reducing costs of private pension investment in the joint variant.

Another direction is taking into account other types of risk, other financial goals, introducing also ways of financing specific to particular types of goals.

Acknowledgements The research project was financed by The National Science Centre (NCN) grant, on the basis of the decision no. DEC-2012/05/B/HS4/04081.

References

Blake, D., Cairns, A. J. G., & Dowd, K. (2013). Pensionmetrics 2: Stochastic pension plan design during the distribution phase. *Insurance: Mathematics and Economics, 33*, 29–47. Discussion Paper PI-0103, The Pensions Institute, Cass Business School, City University, London.

Brown, J. R., & Poterba, J. M. (2000). Joint life annuities and annuity demand by married couples. *The Journal of Risk and Insurance, 67*(4), 527–554.

Feldman, L., Pietrzyk, R., & Rokita, P. (2014a). A practical method of determining longevity and premature-death risk aversion in households and some proposals of its application. In M. Spiliopoulou, L. Schmidt-Thieme, & R. Janning (Eds.), *Data analysis, machine learning and knowledge discovery* (pp. 255–264). Berlin/Heidelberg: Springer.

Feldman, L., Pietrzyk, R., & Rokita, P. (2014b, January). Multiobjective optimization of financing household goals with multiple investment programs. *Statistics in Transition. New Series, 15*(2), 243–268.

Gompertz, B. (1825). On the nature of the function expressive of the law of human mortality, and on a new mode of determining the value of life contingencies. *Philosophical Transactions of the Royal Society of London, 115*, 513–585.

Huang, H., Milevsky, M. A., & Salisbury, T. S. (2012). Optimal retirement consumption with a stochastic force of mortality. *Insurance: Mathematics and Economics, 51*(2), 282–291.

Hurd, M. (1999). *Mortality Risk and Consumption by Couples*. NBER working paper series, 7048. http://www.nber.org/papers/w7048.pdf

Kotlikoff, L. J., & Spivak, A. (1981). The family as an incomplete annuities market. *Journal of Political Economy, 89*(2), 372–391.

Pietrzyk, R., & Rokita, P. (2014). Optimization of financing multiple goals with multiple investment programs in financial planning for households. In *Conference Proceedings of the 32nd International Conference Mathematical Methods in Economics* (pp. 795–800). Olomouc: Palacky University.

Yaari, M. E. (1965). Uncertain lifetime, life insurance and theory of the consumer. *The Review of Economic Studies, 32*(2), 137–150.

Dynamic Aspects of Bankruptcy Prediction Logit Model for Manufacturing Firms in Poland

Barbara Pawełek, Józef Pociecha, and Mateusz Baryła

Abstract Many types of bankruptcy prediction models have been formulated by business theory and practice. Among them, a wide group is composed of classification models, which can divide firms' population into two groups: bankrupts and non-bankrupts. The current bankruptcy prediction models for firms in Poland are usually based on the company's internal financial factors which mainly have a static character. The aim of the paper is to present the possibility of introducing into the bankruptcy prediction logit model a time factor which represents dynamic changes in external economic environment. The proposal of time factor inclusion in this type of model was tested on data concerning manufacturing companies in Poland from 2005 to 2008.

1 Introduction

In corporate bankruptcy prediction, one uses models (including logit model) which are built on the basis of data drawn from financial statements of bankrupt and non-bankrupt companies. Financial information is often taken from several years. Connecting data from various periods is the consequence of the impossibility of collecting a large enough data set which would contain information from only 1 year (e.g. García et al. 2014).

In literature, the discussion of some issues arising while building models for a binary dependent variable (also for logit model), taking into account data from diverse periods can be found (e.g. Beck et al. 1998). The proposal of replacing static models (e.g. single-period logit model) with models including changes of observed values over time (e.g. multi-period logit model) appears in scientific papers (e.g. Shumway 2001).

B. Pawełek (✉) • J. Pociecha • M. Baryła
Cracow University of Economics, 27 Rakowicka Street, 31-510 Cracow, Poland
e-mail: barbara.pawelek@uek.krakow.pl; jozef.pociecha@uek.krakow.pl; mateusz.baryla@uek.krakow.pl

© Springer International Publishing Switzerland 2016   369
A.F.X. Wilhelm, H.A. Kestler (eds.), *Analysis of Large and Complex Data*, Studies in Classification, Data Analysis, and Knowledge Organization,
DOI 10.1007/978-3-319-25226-1_32

The goal of corporate bankruptcy prediction is, among other things, to build a model which is characterized by good predictive power. Using data which reflects the financial condition of companies in various years, and a different economic situation in a certain country (e.g. Trabelsi et al. 2014), the following question arises: can a model with parameters which were estimated without including changes in companies' economic environment be a reliable tool for failure prediction? The answer to this question is not simple. One of the potential sources of errors being committed during bankruptcy forecasting can be the unstable nature of the investigated population (e.g. Pawełek and Pociecha 2012).

Models presented in literature are constructed on the basis of data usually taken from several years, and their usefulness for collapse prediction is assessed according to a certain criterion assumed by authors. Some models find their practical implementation for the several subsequent years after the time-frame analysed in the case of research sample. So, does not a bankruptcy risk depend on the nationwide economic situation? Can a model which was estimated for financial data from the period of economic boom retain its predictive ability at the time of the economic crisis?

When analysing models appearing in literature, which were created taking into consideration data from various periods, one can notice their diversity on account of employed financial ratios as well as estimated values of parameters for individual variables (financial ratios) in different models. The observed diversity may be a consequence of a different economic situation at the time from which the data was taken. It can be expected that a model created on the basis of financial information from the period of boom (slump) in the economy keeps its ability to predict bankruptcy in the period of boom (slump) in the economy. However, it is hard to assume that such a model will be able to predict correctly if a company fails at the time of economic situation reverse to the one which was observed during sample selection.

The aim of this paper is to discuss the possibility of introducing a time factor into failure prediction models, which can represent changes in firms' economic environment. In addition to this, the article also presents obtained empirical outcomes resulting from the implementation of proposed modifications of the logit model.

During global financial crisis, the following research question gains in importance: how can a time factor, which shows changes in the economy of a certain country, be included in a bankruptcy prediction model? Introducing this factor, which represents alterations in the companies' economic environment, into the bankruptcy prediction logit model is an attempt to make the model dynamic.

2 Data

Two balanced samples were used in order to verify the usefulness of the logit model which was expanded by dummy variables for bankruptcy prediction of manufacturing companies in Poland. Each sample consisted of 246 firms.

The first one (sample S_1) was created as a consequence of matching non-bankrupt companies with bankrupt ones taking into account the same industry code and similar value of assets. The second sample (sample S_2) was drawn independently from the populations of failed and non-failed firms so that the number of bankrupts can equal the number of non-bankrupts. Data was taken from the period between 2005 and 2008. Bankrupt companies were described by the values of financial ratios from the 2 years preceding the firms' failure. In the study, 33 variables (financial ratios), which are presented in Table 1, were considered.

Both samples of companies were divided into a training group and testing group [variant I: 60 % of firms and 40 % of firms (S_{11} and S_{21}) or variant II: 70 % of firms and 30 % of firms (S_{12} and S_{22})]. In the case of symbol S_{sd} ($s = 1, 2; \ d = 1, 2$), subscript s indicates the method of sample selection (1—pair-matched sampling, 2—random sampling with replacement), and subscript d denotes the variant of sample splitting into training data and testing data (1—6:4 division, 2—7:3 division).

3 Proposals of Logit Model Extension

The logit model was used to make an attempt to include dynamic changes in the economy so as to predict the failure of manufacturing companies in Poland. The logit model can be written as follows:

$$P(y_i = \text{bankrupt}|\ \mathbf{x_i}) = \frac{\exp(\mathbf{x_i}\boldsymbol{\beta})}{1 + \exp(\mathbf{x_i}\boldsymbol{\beta})}, \tag{1}$$

where: $\mathbf{x_i}$—vector of values of independent variables for the ith object, $\boldsymbol{\beta}$—vector of parameters.

It was assumed that the economic situation in a given country has an influence on the financial standing of companies. Financial ratios are used to assess this standing and, therefore, they are the base for building bankruptcy prediction models. Changes of financial ratios values are partially caused by changes in companies' economic environment. It was also assumed that the significance of financial ratios depends on the overall economic situation. Thus, information about the year which the financial statement comes from is a link between financial ratios and the state of economy.

The main goal of the presented methodological proposal, which consists in using interactions between qualitative variables (binary variables which identify years which financial data are from) and quantitative variables (financial ratios) for failure forecasting of companies in Poland, is to overcome some difficulties which arise

Table 1 Financial ratios appearing in the database and their definitions

Liquidity ratios	Profitability ratios
$R_{01} = \dfrac{\text{Current assets}}{\text{Short term liabilities}}$	$R_{15} = \text{EBITDA}$
$R_{02} = \dfrac{\text{Current assets} - \text{Inventories}}{\text{Short term liabilities}}$	$R_{16} = \dfrac{\text{EBITDA}}{\text{Total assets}}$
$R_{03} = \dfrac{\text{Current assets} - \text{Inventories} - \text{Short term receivables}}{\text{Short term liabilities}}$	$R_{17} = \dfrac{100 * \text{Gross profit (loss)}}{\text{Sales revenues}}$
$R_{04} = \dfrac{\text{Current assets} - \text{Short term liabilities}}{\text{Total assets}}$	$R_{18} = \dfrac{100 * \text{Net profit (loss)}}{\text{Sales revenues}}$
	$R_{19} = \dfrac{100 * \text{Net profit (loss)}}{\text{Shareholders' equity}}$
	$R_{20} = \dfrac{100 * \text{Net profit (loss)}}{\text{Total assets}}$
	$R_{21} = \dfrac{\text{Operating profit (loss)}}{\text{Total assets}}$
	$R_{22} = \dfrac{\text{Operating profit (loss)}}{\text{Sales revenues}}$

Liability ratios

$$R_{05} = \frac{\text{Long term liabilities} + \text{Short term liabilities}}{\text{Total assets}}$$

$$R_{06} = \frac{\text{Long term liabilities} + \text{Short term liabilities}}{\text{Shareholders' equity}}$$

$$R_{07} = \frac{\text{Long term liabilities}}{\text{Shareholders' equity}}$$

$$R_{08} = \frac{\text{Shareholders' equity}}{\text{Total assets}}$$

$$R_{09} = \frac{\text{Short term liabilities}}{\text{Total assets}}$$

$$R_{10} = \frac{\text{Fixed assets}}{\text{Total assets}}$$

$$R_{11} = \frac{\text{Net profit (loss)} + \text{Depreciation}}{\text{Long term liabilities} + \text{Short term liabilities}}$$

$$R_{12} = \frac{\text{Shareholders' equity}}{\text{Long term liabilities} + \text{Short term liabilities}}$$

$$R_{13} = \frac{\text{Gross profit (loss)}}{\text{Short term liabilities}}$$

$$R_{14} = \frac{\text{Shareholders' equity} + \text{Long term liabilities}}{\text{Fixed assets}}$$

Productivity ratios

$$R_{23} = \frac{\text{Sales revenues}}{(\text{Short term receivables}(t) + \text{Short term receivables}(t-1))/2}$$

$$R_{24} = \frac{\text{Sales revenues}}{(\text{Fixed assets}(t) + \text{Fixed assets}(t-1))/2}$$

$$R_{25} = \frac{\text{Sales revenues}}{(\text{Total assets}(t) + \text{Total assets}(t-1))/2}$$

$$R_{26} = \frac{\text{Sales revenues}}{\text{Total assets}}$$

$$R_{27} = \frac{\text{Short term liabilities}}{\text{Operating costs}}$$

$$R_{28} = \frac{\text{Inventories}}{\text{Sales revenues}}$$

$$R_{29} = \frac{\text{Inventories}}{\text{Operating costs}}$$

$$R_{30} = \frac{\text{Short term receivables}}{\text{Sales revenues}}$$

$$R_{31} = \frac{\text{Operating costs}}{\text{Short term liabilities}}$$

$$R_{32} = \frac{\text{Sales revenues}}{\text{Short term receivables}}$$

$$R_{33} = \frac{100 * \text{Operating costs}}{\text{Sales revenues}}$$

Source: authors' own study

when traditional methods are used for corporate bankruptcy modelling. Similar attempts to make bankruptcy prediction models dynamic can be found in literature. The models created for the Italian economy using data from the years between 1995 and 1998 (De Leonardis and Rocci 2008), as well as 1999 and 2005 (De Leonardis and Rocci 2014) can be quoted as an example.

Searching for the possibilities for the time factor implementation into the logit model, it has been assumed that the most important criterion for the evaluation of usefulness of a given methodological proposal will be, above all, sensitivity (percentage of bankrupt companies which were correctly classified as bankrupt ones) calculated for the model on the basis of a test set. The classification of companies was based on probabilities given by Eq. (1), assuming the cut-off point at the level of 0.5. Objects for which the probability equalled at least 0.5 were classified as bankrupts.

Three following solutions were considered:

- variant I: the creation of models for each year separately,
- variant II: the introduction of binary variables in the form of:

$$Y^t = \begin{cases} 1 & \text{if} \quad \text{year} = t \\ 0 & \text{if} \quad \text{year} \neq t \end{cases} \quad (t = 2006,\ 2007,\ 2008)\ , \tag{2}$$

identifying the year which a financial statement is taken from,
- variant III: the introduction of both binary variables identifying the year of a particular financial statement as well as qualitative-quantitative ones in the form of:

$$R_i^t = \begin{cases} R_i & \text{if} \quad \text{year} = t \\ 0 & \text{if} \quad \text{year} \neq t \end{cases} \quad (i = 01,\ \ldots,\ 33;\ t = 2006,\ 2007,\ 2008)\ , \tag{3}$$

which reflect fluctuating significance of financial ratios for corporate bankruptcy prediction over time.

Binary variables Y^t as well as qualitative-quantitative variables R_i^t are referred to as dummy variables (Maddala 2001).

The aforementioned variants are associated with certain practical problems. In the case of variant I, the problem with a small number of bankrupts to be considered in separate years was encountered. Whereas the application of variant II implied that it was possible to observe the influence of time factor only on the intercept of a model. Ultimately, it was decided to utilize the third solution which allows to influence not only the intercept, but also the coefficients of the logistic regression model. Further in the article, the results of empirical studies based on variant III are shown.

4 Empirical Verification of Usefulness of Logit Model with Dummy Variables

The best model among traditional logit models (that is, models based only on financial ratios—L_j, where: j denotes the model number), $L_1(R_{11}, R_{12})$ constructed on sample S_{11} [where: $L = \text{logit}(p) = \ln\left(\frac{p}{1-p}\right)$ and p is the probability of success], was characterized by the following results for testing data: sensitivity = 77.55 % (that is 38 out of 49 bankrupts were correctly classified as bankrupt companies), specificity = 69.39 % (that is 34 out of 49 non-bankrupts were correctly classified as non-bankrupt companies), and correctly classified = 73.47 % (that is 72 out of all 98 companies were correctly classified). The aim of the conducted research was to verify whether the inclusion of dummy variables in the logit model may lead to the prognostic quality improvement of a model, measured mainly by sensitivity for a test set. The logit models were built with the use of backward stepwise analysis within logistic regression.

The search for a prognostic model initially based on ratios from all four groups as well as binary and qualitative-quantitative variables led to the creation of L_1^M model which contained variables R_{12} and R_{20}^{2008} belonging to the group of liability and profitability ratios (p-values are given below estimates of the parameters):

$$L_{i,1}^M = \text{logit}(p_i) = \ln\left(\frac{p_i}{1-p_i}\right) = \underset{(0.0020)}{0.7408} - \underset{(0.0003)}{0.6703}\,R_{i,12} - \underset{(0.0437)}{0.1132}\,R_{i,20}^{2008}, \qquad (4)$$

where: $p_i = \hat{P}(y_i = \text{bankrupt}|\,\mathbf{x_i})$.

The estimated model L_1^M informs that the increase in the value of liability ratio R_{12} in a given year reduced the probability of bankruptcy in 2 years' time, *ceteris paribus*. Additionally, the increase in the value of profitability ratio R_{20} in 2008 caused the decrease in the probability of failure in 2010, *ceteris paribus*.

The prognostic capability of this model measured by sensitivity (85.71 %) was the highest from among other considered models. However, its specificity (59.18 %) and being correctly classified (72.45 %) were lower than in the case of other analysed models.

The ROC (receiver operating characteristic) curve was used to illustrate the classification accuracy of modified logit models (that is, models which also contain dummy variables—L_j^M, where j denotes the model number) (e.g. Birdsall 1973; Stein 2005; Verikas et al. 2010). The curve is shown in a square of the surface area that equals one. Values calculated as 1 minus specificity are presented on the horizontal axis. The vertical axis shows in turn sensitivity of a model. Therefore, points on the ROC curve have coordinates 1-specificity and sensitivity for different cut-off values.

In the analysis of the ROC curve, a very important role is played by the area under the curve (AUC), which is used as a measure of classification accuracy of a model. The value of AUC at the level of 0.5 signifies that the classification accuracy

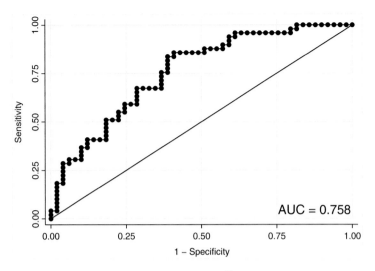

Fig. 1 ROC curve for testing data in the case of model L_1^M

of a considered model corresponds to the random allocation of objects to groups. In turn, AUC $= 1$ indicates excellent classification accuracy of an analysed model.

The classification accuracy of logit model L_1^M for the test set was at a satisfactory level (AUC $= 0.758$, Fig. 1—created using Stata/IC 12.0), slightly higher than for the training set (AUC $= 0.755$). The value of AUC for model L_1, calculated on the basis of the test set, equalled 0.806. Therefore, the classification accuracy of the logit model with dummy variables (measured by AUC) was lower than the accuracy of the traditional model.

In the variant based on sample S_{21} which consists of companies selected by means of random sampling with replacement, the traditional logit model $L_2(R_{02}, R_{11}, R_{13}, R_{16})$ was obtained. Relying on financial ratios included in model L_2 and dummy variables, a modified logit model $L_2^M(R_{11}, R_{13}, R_{02}^{2006})$ was created with the use of backward stepwise analysis:

$$L_{i,2}^M = \underset{(0.0003)}{0.9747} - \underset{(0.0001)}{6.7781}\, R_{i,11} + \underset{(0.0333)}{2.0917}\, R_{i,13} - \underset{(0.0013)}{1.3769}\, R_{i,02}^{2006}. \qquad (5)$$

The model L_2^M informs that the increase in the value of liability ratio R_{11} in a given year influenced the reduction in the probability of bankruptcy in 2 years' time, *ceteris paribus*. However, the growth of the value of liability ratio R_{13} in a given year caused the increase in the probability of failure in 2 years' time, *ceteris paribus*. Additionally, the increase in the value of liquidity ratio R_{02} in 2006 affected the decrease in the probability of bankruptcy in 2008, *ceteris paribus*.

The prognostic abilities of the logit model L_2^M were higher than the abilities of traditional logit model L_2 and they can be presented as follows: sensitivity $= 89.80\,\%$, specificity $= 69.39\,\%$, correctly classified $= 79.59\,\%$.

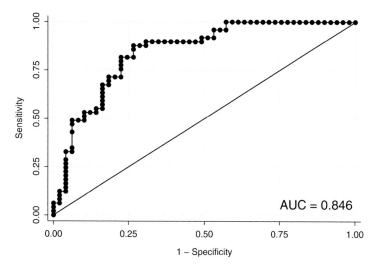

Fig. 2 ROC curve for testing data in the case of model L_2^M

The classification accuracy of logit model L_2^M for the test set was at a good level (AUC = 0.846, Fig. 2—created using Stata/IC 12.0), higher than for the training set (AUC = 0.806). The value of AUC for model $L_2(R_{02}, R_{11}, R_{13}, R_{16})$, calculated on the basis of the test set, equalled 0.776. Modified logit model L_2^M had, therefore, higher classification accuracy (measured by AUC) than its corresponding traditional model L_2.

A similar analysis was conducted for samples S_{12} and S_{22}. The following modified logit models were obtained:

$$L_{i,3}^M = \underset{(0.0001)}{1.2721} - \underset{(0.0002)}{1.8485}\,R_{i,02} - \underset{(0.0001)}{2.5527}\,R_{i,11} + \underset{(0.0184)}{1.0971}\,R_{i,02}^{2007}, \tag{6}$$

$$L_{i,4}^M = \underset{(0.0004)}{0.9051} - \underset{(0.00001)}{12.9975}\,R_{i,11} + \underset{(0.0038)}{4.3793}\,R_{i,13} + \underset{(0.0024)}{9.5844}\,R_{i,11}^{2007} - \underset{(0.0377)}{4.6146}\,R_{i,13}^{2007}. \tag{7}$$

Taking into account the parameters estimates for model L_3^M, it can be stated that the increase in the value of liability ratio R_{11} in a given year had an impact on the decrease in the probability of bankruptcy in 2 years' time, *ceteris paribus*. Moreover, the growth in the value of liquidity ratio R_{02} in a given year affected the reduction in the probability of failure in 2 years' time, *ceteris paribus*. However, the influence of the increase in the value of ratio R_{02} in 2007 on the decline in the probability of bankruptcy in 2009 was lower than in the other analysed years, *ceteris paribus*.

The estimated model L_4^M shows that the increase in the value of liability ratio R_{11} in a given year caused the decrease in the probability of failure in 2 years' time, *ceteris paribus*. However, in 2007 this impact on the decline in the probability of bankruptcy in 2009 was lower than in the other considered years, *ceteris paribus*.

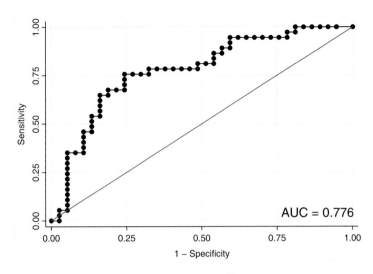

Fig. 3 ROC curve for testing data in the case of model L_3^M

Additionally, the growth in the value of liability ratio R_{13} in 2005, 2006 and 2008 had an effect on the rise in the probability of failure in 2 years' time, *ceteris paribus*. The year 2007 was an exception. In this year, the influence of this ratio on the bankruptcy probability in 2009 was reverse, *ceteris paribus*.

The prognostic capability of the model L_3^M measured by sensitivity (75.68 %) and being correctly classified (74.32 %) was higher than in the case of model $L_3(R_{02}, R_{11})$. Specificity (72.97 %) in turn was at the same level as for model L_3. The classification accuracy of logit model L_3^M for the testing data was at a satisfactory level (AUC = 0.776, Fig. 3—created using Stata/IC 12.0), but lower than for model L_3 (AUC = 0.804).

The model L_4^M was characterized by a higher sensitivity (81.08 %) than the traditional model L_4, however, it had a lower value of specificity (67.57 %). The percentage of correctly classified companies for the model L_4^M was the same as in the case of the model L_4 (74.32 %). The classification accuracy of the logit model L_4^M for the test set was at a good level (AUC = 0.811, Fig. 4—created using Stata/IC 12.0), higher than for model L_4 (AUC = 0.784).

Recapitulating the results of the conducted empirical research, it can be stated that the inclusion of dummy variables in the logit model, which is used to predict the failure of manufacturing companies in Poland, may lead to the improvement of prognostic capability of a model, measured by sensitivity for a test set. Models estimated using financial data from the period between 2005 and 2008, which includes the beginning of the worldwide financial crisis, point at the changing significance of financial ratios groups employed in bankruptcy prediction of manufacturing companies in Poland. While forecasting failures 2 years in advance for 2008, special attention was paid to liquidity ratios from 2006, that is from the period of an overall good economic situation. In the bankruptcy prediction for

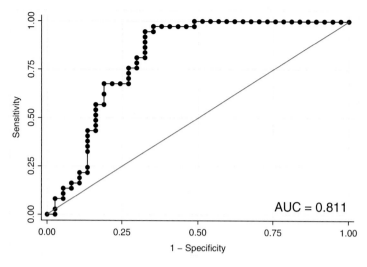

Fig. 4 ROC curve for testing data in the case of model L_4^M

2009, an important role was played by liquidity and liability ratios from 2007, that is from the beginning of financial crisis in Poland. In the case of making predictions for 2010, profitability ratios from 2008 gained in importance. The year 2008 was characterized by a worsening economic situation in Poland. The economic environment of manufacturing companies in Poland was then unstable. The degree of bankruptcy threat depended on various factors connected with different fields of business activity. It appears that conclusions drawn on the basis of estimated logit models with dummy variables are in accordance with the knowledge of how companies function during the economic boom and a slump in the economy.

5 Simulation Verification of Usefulness of Logit Model with Dummy Variables

Because of the small size of the test sets, the bootstrap technique (Efron 1979) was used in order to verify empirical values of sensitivity, specificity and AUC of the discussed models. For each pair of models L_j and L_j^M ($j = 1, 2, 3, 4$), 10,000 bootstrap test samples were generated on the basis of the test samples (considered in Sect. 4) resulting from two types of data division (6:4 or 7:3). The generated bootstrap test samples consisted of 98 or 74 observations.

The outcomes shown in Table 2 confirm the conclusions presented in the previous section. The values of sensitivity for the modified logit models are higher than in the case of the traditional models. The values of specificity and AUC for the modified

Table 2 Bootstrapping estimates of sensitivity, specificity and AUC

Sample		Sensitivity		Specificity		AUC	
S_{11}		L_1	L_1^M	L_1	L_1^M	L_1	L_1^M
	Mean	0.7764	0.8581	0.6940	0.5921	0.8062	0.7580
	St.dev.	0.0594	0.0499	0.0664	0.0711	0.0439	0.0485
S_{21}		L_2	L_2^M	L_2	L_2^M	L_2	L_2^M
	Mean	0.8165	0.8980	0.6120	0.6937	0.7758	0.8452
	St.dev.	0.0552	0.0432	0.0683	0.0656	0.0454	0.0395
S_{12}		L_3	L_3^M	L_3	L_3^M	L_3	L_3^M
	Mean	0.7287	0.7562	0.7300	0.7295	0.8032	0.7757
	St.dev.	0.0735	0.0708	0.0725	0.0732	0.0515	0.0550
S_{22}		L_4	L_4^M	L_4	L_4^M	L_4	L_4^M
	Mean	0.7567	0.8112	0.7288	0.6752	0.7832	0.8103
	St.dev.	0.0700	0.0647	0.0732	0.0776	0.0561	0.0540

Source: own calculations by means of R packages "base", "stats", "verification" (R Core Team 2013; NCAR 2014)

models can be perceived as higher or similar in two out of four instances (one for each of two splittings—6:4 and 7:3).

The bootstrapping estimates of standard deviation are worth considering. The obtained results indicate the smaller diversity of the sensitivity, specificity and AUC values which are characterized by higher mean value.

6 Summary

The problem with the way of utilizing logit models with dummy variables, connected with the period between 2006 and 2008, in the case of possessing financial data from years after 2008, stays open.

Dummy variables included in the logit model can be viewed as variables reflecting the meaning of some financial ratios for bankruptcy prediction of manufacturing companies in Poland at the time of the economic boom or recession. The year 2005 may be regarded as a boom in the Polish economy and models without dummy variables correspond to this period. The year 2007 is considered to be the beginning of worldwide financial crisis. 2008 is the year of a slowdown in the Polish economy and in the case of such an economic situation, model terms which allow taking into consideration the economic environment of companies should be used.

One of the solutions to the problem consists in the application of several logit models both traditional and modified. This approach assumes our lack of knowledge whether the year of financial data is characterized by an economic boom or slump. Also, the obtained results need to be juxtaposed, and e.g., the voting method might be employed.

The second solution recommends preceding the corporate bankruptcy prediction by the analysis of similarities between the Polish economic situation in the years between 2005 and 2008, and the economic situation in the period which the financial data comes from. The analysis of similarities between the economic situation in the chosen years can be conducted on the basis of GDP changes. Financial data used as a basis for estimation of the presented models came from financial statements for years of both a boom and slowdown in the Polish economy. Thus, there is the possibility of selecting appropriate logit models for bankruptcy prediction of manufacturing companies in Poland 2 years in advance depending on the assessment of the Polish economic situation.

In the literature on corporate bankruptcy prediction, one can also find proposals that consist in the implementation of chosen market or macroeconomic ratios in the traditional models (e.g. Chava and Jarrow 2004; Shumway 2001). In the discussion, especially ratios whose values reflect the changes in the economic situation of a given country are taken into account. On the basis of the research results in which an attempt was made to apply macroeconomic ratios to bankruptcy prediction models, it can be presumed that the usefulness of this approach depends, among other things, on the length of the time-frame which the data was taken from (e.g. De Leonardis and Rocci 2008, 2014).

The determined direction of further research, regardless of all the problems arising while collecting data (among other things changes in the principles of making and publishing financial statements in Poland), which is aimed at being the basis for modelling and bankruptcy prediction of companies in Poland, is interesting and deserves special attention. In the future, the authors plan to extend their research to include also non-balanced samples.

Acknowledgements The authors would like to express their appreciation for the support provided by the National Science Centre (NCN, grant No. N N111 540 140).

References

Beck, N., Katz, J. N., & Tucker, R. (1998). Taking time seriously: Time-series–cross-section analysis with a binary dependent variable. *American Journal of Political Science, 42*(4), 1260–1288.

Birdsall, T. G. (1973). *The theory of signal detectability: ROC curves and their character.* Technical Report, No. 177, Cooley Electronics Laboratory, Department of Electrical and Computer Engineering, The University of Michigan, Ann Arbor, MI.

Chava, S., & Jarrow, R. A. (2004). Bankruptcy prediction with industry effects. http://dx.doi.org/10.2139/ssrn.287474.

De Leonardis, D., & Rocci, R. (2008). Assessing the default risk by means of a discrete-time survival analysis approach. *Applied Stochastic Models in Business and Industry, 24*, 291–306. interscience.wiley.com. doi:10.1002/asmb.705.

De Leonardis, D., & Rocci, R. (2014). Default risk analysis via a discrete-time cure rate model. *Applied Stochastic Models in Business and Industry, 30*(5), 529–543.

Efron, B. (1979). Bootstrap methods: Another look at the jackknife. *The Annals of Statistics, 7*(1), 1–26.

García, V., Marqués, A. I., & Sánchez, J. S. (2014, September). An insight into the experimental design for credit risk and corporate bankruptcy prediction systems. *Journal of Intelligent Information Systems*. doi:10.1007/s10844-014-0333-4.

Maddala, G. S. (2001). *Introduction to econometrics*, 3rd edn. West Sussex: Wiley.

NCAR - Research Applications Laboratory. (2014). Verification: Weather forecast verification utilities. R package version 1.37. http://CRAN.R-project.org/package=verification. Accessed 20 Dec 2014.

Pawełek, B., & Pociecha, J. (2012). General SEM model in researching corporate bankruptcy and business cycles. In J. Pociecha & R. Decker (Eds.), *Data analysis methods and its applications* (pp. 215–231). Warsaw: C.H. Beck.

R Core Team. (2013). *R: A language and environment for statistical computing*. Vienna: R Foundation for Statistical Computing. http://www.R-project.org/. ISBN:3-900051-07-0. Accessed 20 Dec 2014.

Shumway, T. (2001). Forecasting bankruptcy more accurately: A simple hazard model. *The Journal of Business, 74*(1), 101–124.

Stein, R. M. (2005). The relationship between default prediction and lending profits: Integrating ROC analysis and loan pricing. *Journal of Banking and Finance, 29*, 1213–1236.

Trabelsi, S., He, R., He, L., & Kusy, M. (2014, January). A comparison of Bayesian, hazard, and mixed logit model of bankruptcy prediction. *Computational Management Science*. doi:10.1007/s10287-013-0200-8.

Verikas, A., Kalsyte, Z., Bacauskiene, M., & Gelzinis, A. (2010). Hybrid and ensemble-based soft computing techniques in bankruptcy prediction: A survey. *Soft Computing, 14*(9), 995–1010.

Part IX
Data Analysis in Medicine and Life Sciences

Estimating Age- and Height-Specific Percentile Curves for Children Using GAMLSS in the IDEFICS Study

Timm Intemann, Hermann Pohlabeln, Diana Herrmann, Wolfgang Ahrens, and Iris Pigeot, on behalf of the IDEFICS consortium

Abstract In medical diagnostics age-specific reference values are needed for assessing the health status of children. However, for many clinical parameters such as blood cholesterol or insulin reference curves are still missing for children. To fill this gap, the IDEFICS study provides an excellent data base with 18,745 children aged 2.0–10.9 years. The generalised additive model for location, scale and shape (GAMLSS) was used to derive such reference curves while controlling for the influence of various covariates on the parameters of interest. GAMLSS, an extension of the LMS method, is able to model the influence of more than one covariate. It is also able to model the kurtosis using different distributions. The Bayesian information criterion (BIC), Q-Q plots and wormplots were applied to assess the goodness of fit of alternative models. GAMLSS has proven to be a useful tool to model the influence of more than one covariate when deriving age- and sex-specific percentile curves for clinical parameters in children. This will be demonstrated exemplarily for the bone stiffness index (SI) where age- and height-specific percentile curves were calculated for boys and girls based on the model which showed the best goodness of fit.

1 Introduction

Most medical tests used in diagnostics are based on clinical parameters with established thresholds. Preferably, such thresholds should be related to the risk of subsequent diseases. For adults, for instance, cut-off values are defined to diagnose hypertension: A systolic blood pressure of at least 140 mm Hg and/or a diastolic blood pressure of at least 90 mm Hg are considered as indicators of hypertension which in turn may cause heart attack or stroke. Another well-known example is the

T. Intemann • H. Pohlabeln • D. Herrmann • W. Ahrens • I. Pigeot (✉), on behalf of the IDEFICS consortium
Leibniz Institute for Prevention Research and Epidemiology - BIPS, Achterstraße 30, 28359 Bremen, Germany
e-mail: intemann@bips.uni-bremen.de; pohlabeln@bips.uni-bremen.de; herrmann@bips.uni-bremen.de; ahrens@bips.uni-bremen.de; pigeot@bips.uni-bremen.de

© Springer International Publishing Switzerland 2016
A.F.X. Wilhelm, H.A. Kestler (eds.), *Analysis of Large and Complex Data*, Studies in Classification, Data Analysis, and Knowledge Organization,
DOI 10.1007/978-3-319-25226-1_33

385

classification of obesity in adults where according to the World Health Organization definition a body mass index (BMI) greater than or equal 25 (30) classifies a person as overweight (obese).

Such single cut-off values are, however, not useful in children since the distributions of clinical parameters in children are heavily age-dependent due to their physical development during growth. Thus, paediatricians need reference curves in their daily practice. Such curves exist for the assessment of childhood obesity (see e.g. Cole and Lobstein 2013 for BMI) but are still missing for most clinical parameters. Because measurements obtained in clinical studies typically do not represent a healthy population a large population-based sample is required to derive reference curves. The IDEFICS study (Identification and prevention of Dietary- and lifestyle-induced health EFfects In Children and infantS; Ahrens et al. 2011) provides such a sample of healthy European children. Its aim is to enhance the knowledge of health effects of a changing diet and altered social environment and lifestyle of children in Europe.

Based on the rich data base of the IDEFICS study age- and sex-specific reference values for several clinical parameters have been derived and published in a supplement volume of the International Journal of Obesity (Ahrens et al. 2014). One of these parameters is the bone stiffness index (SI) as an indicator for bone health. It is measured by quantitative ultrasound (QUS) of the calcaneus which has been shown to predict the risk of bone fractures in adults (Krieg et al. 2006; Maggi et al. 2006). However, QUS has been poorly investigated in children and adolescents, especially with regard to the prediction of health endpoints such as bone fracture or osteopenia (Jaworski et al. 1995; Herrmann et al. 2014). SI reference curves may help to assess skeletal development in early life and to detect deviations from a normal growth already at an early stage. They may also guide therapeutic decisions in children treated with medications affecting bone density, e.g. during cancer treatment.

The generalised additive model for location, scale and shape (GAMLSS) is used to derive such reference curves. In the literature, various approaches to calculate percentile curves can be found. The probably most often used method has been the so-called LMS method introduced by Cole and Green (1992). The basic idea of LMS is to estimate the distribution of the response variable by estimating the mean μ, the coefficient of variance σ and the Box–Cox transformation parameter v which accounts for skewness by smooth curves depending on one covariate. The LMS method is a special case of the more recent GAMLSS proposed by Rigby and Stasinopoulos (2005). Nowadays, GAMLSS has gained importance in percentile curve estimation because of its ability to not only account for the mean, variation and skewness of a response variable but also for the kurtosis if necessary. Additionally, GAMLSS is not restricted to one covariate and was used, for example, to derive age- and height-specific reference ranges for spirometry (Cole et al. 2009). Finally, GAMLSS fulfills the primary criteria defined in Borghi et al. (2006) to derive growth curves: GAMLSS has the ability to estimate outer percentiles precisely, to estimate percentiles simultaneously, avoiding percentiles to cross, to estimate z-scores and percentiles using direct formulae, to apply continuous age smoothing and to account for skewness and kurtosis. This method will be described in detail here, using SI in children as an example.

The GAMLSS method, the study population and details of model selection are described in Sect. 2, followed by a graphical presentation of the results in Sect. 3 and a conclusion in Sect. 4.

2 Methods

2.1 Generalised Additive Model for Location, Scale and Shape

A semiparametric GAMLSS is a regression model assuming $Y_n, n = 1, \ldots, N$, to be independent random variables following a distribution D_{θ^n} with K distribution parameters $(\theta_{n1}, \ldots, \theta_{nK}) = \boldsymbol{\theta}^n$ and a probability density function $f(y_n | \boldsymbol{\theta}^n)$. For $k = 1, \ldots, K$ the link function g_k links the parameter $\boldsymbol{\theta}_k = (\theta_{1k}, \ldots, \theta_{Nk})^{\mathsf{T}}$ to covariates by

$$g_k(\boldsymbol{\theta}_k) = \mathbf{X}_k \boldsymbol{\beta}_k + \sum_{j_k=1}^{J_k} h_{j_k k}(\mathbf{x}_{j_k k}), \tag{1}$$

where $\mathbf{X}_k$ is a known design matrix, containing the covariate information, $\boldsymbol{\beta}_k$ is a parameter vector and the function $h_{j_k k} : \mathbb{R}^N \to \mathbb{R}^N$, $j_k = 1, \ldots, J_k$, is an unknown function of covariates which can be estimated, e.g., by splines. GAMLSS allows to consider a wide range of different distributions D_{θ^n}. In contrast to the generalised additive model the assumption of an underlying exponential family is relaxed to a more general class of distributions including highly skewed and kurtotic distributions.

The only requirements regarding the distribution to be fitted originates from the implementation in the statistical software **R**: the probability density function and its first, second and cross derivatives with respect to each of the parameters have to be computable.

Two important examples of distributions that may be fitted within the framework of GAMLSS are the Box–Cox Cole and Green distribution (BCCG) and the Box–Cox power exponential (BCPE) distribution. Both distributions result from the Box–Cox transformation

$$Z = \begin{cases} \frac{1}{\sigma \nu} \left[\left(\frac{Y}{\mu} \right)^{\nu} - 1 \right] & , \nu \neq 0 \\ \frac{1}{\sigma} \log \left(\frac{Y}{\mu} \right) & , \nu = 0 \end{cases}.$$

The random variable Y follows a BCCG(μ, σ, ν) distribution if the random variable Z follows a normal distribution. If Z follows a standard power exponential (PE) distribution then Y follows a BCPE(μ, σ, ν, τ) distribution. The parameter τ originates from the PE distribution and accounts for the kurtosis which was the main reason for introducing the BCPE distribution by Rigby and Stasinopoulos (2004). Within the GAMLSS framework models including natural cubic splines are estimated by maximising the penalised likelihood function

$$l_p = l - \frac{1}{2} \sum_{k=1}^{K} \sum_{j_k=1}^{J_k} \lambda_{j_k k} \mathbf{h}_{j_k k}^{\mathsf{T}} \mathbf{K}_{j_k k} \mathbf{h}_{j_k k},$$

where for $k = 1, \ldots, K$ and $j_k = 1, \ldots, J_k$, l denotes the log likelihood, $\lambda_{j_k k}$ is the smoothing parameter, $\mathbf{h}_{j_k k}$ is a vector which evaluates $h_{j_k k}$ at $\mathbf{x}_{j_k k}$ and $\mathbf{K}_{j_k k}$ is a known penalty matrix depending on $\mathbf{x}_{j_k k}$. Two algorithms are available for fitting a GAMLSS: the CG algorithm based on an algorithm from Cole and Green (1992) and the RS algorithm from Rigby and Stasinopoulos (2005).

2.2 An Application of GAMLSS

2.2.1 Study Population

The IDEFICS survey includes 18,745 2.0–10.9 year old children (recruited from 2007 to 2010) in eight European countries. The study population is described in detail in Ahrens et al. (2014). The SI was measured in a subsample. The analysis group that was used to derive reference curves for the SI included 5,412 boys and 5,379 girls. A more specific description of the analysis group can be found in Herrmann et al. (2014).

2.2.2 Modelling

The following analysis will be exemplarily conducted for boys. The analysis for girls proceeded analogously. We used the GAMLSS package (version 4.2–6) of the statistical software **R** (version 3.0.1; R Core Team 2013). For the application of the semiparametric GAMLSS all components of Model (1) for the response variable must be determined, these include:

- the distribution of the response variable, here the SI,
- link functions g_k of the distribution parameters,
- terms depending on covariates for modelling the distribution parameters and
- appropriate covariates.

Descriptive analysis of the distribution of the SI showed an effect of both age and height on the SI (Fig. 1). Therefore, age and height were considered as covariates.

Fig. 1 Histograms of the SI for boys aged 5.0–5.9 years: (**a**) below median height and (**b**) above median height; aged 10.0–10.9 years: (**c**) below median height and (**d**) above median height. The *black bar* displays the respective mean

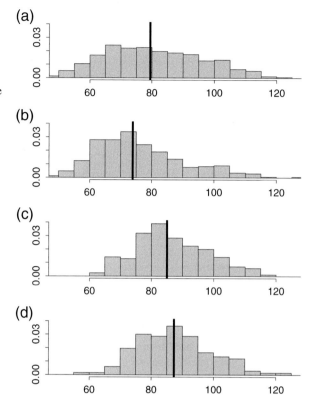

We used cubic splines (cs), linear functions (both depending on age and height) and constant terms to model the distribution parameters.

For the link functions we only used the default link functions depending on the distribution and distribution parameters since, on the one hand, the link functions had no influence on the Bayesian information criterion (BIC) and, on the other hand, the default approach ensured that the requirements on the distribution parameters are fulfilled, e.g. the log link for σ of the BCPE distribution ensured that $\sigma > 0$.

We fitted several distributions, ranging from symmetric (normal and logistic distribution) and skewed (gamma and inverse Gaussian distribution) two parameter distributions to distributions with three parameters which account for kurtosis (power exponential and t-family distribution) or skewness (BCCG) to four parameter distributions which account for skewness and kurtosis (BCPE and Box–Cox t distribution).

We used a stepwise model selection for each distribution mentioned above. Models were fitted for both sexes separately. The particular distribution parameters were modelled in terms of age and height. The BIC was used to assess different models. The model selection included a forward approach (step 1–4) and a backward elimination (step 5–7) to minimise the BIC. In step 1 a model only for μ was built. Given the model selected in this step a model was built for σ in step 2. Given the model selected in step 2 a model was built for ν in step 3 and given the model selected in step 3 a model for τ was built in step 4. Given the model selected in step 4 it was checked if the terms for ν were needed (step 5). Given the model selected in step 5 it was checked if the terms for σ were needed in step 6 and finally given the model selected in step 6 it was checked if the terms for μ were needed (step 7).

The results of this procedure are summarised in Table 1: For example, for the BCPE distribution μ was modelled as a linear function of age and a cubic spline of height, $\log(\sigma)$ and ν as a cubic spline of height and $\log(\tau)$ was modelled as a constant. Models based on the BCPE and BCCG distribution resulted in the smallest BIC values. The residuals of these models provided further details regarding the goodness of fit. Table 2 summarises the quantile residuals which indicated that both models fitted the data well since the residual distributions were very close to a standard normal distribution. This was also confirmed by the Q-Q plots and the density estimations of the residuals which are depicted in Fig. 2.

Additionally we inspected wormplots (van Buuren and Fredriks 2001) to decide between the two models (Fig. 3). Wormplots are useful to check if the first four moments are well captured by the model. The wormplots indicated that in both models there was no need to account for kurtosis because the wormplots exhibited no S-shape. Thus we considered the BCCG model as sufficient and we discarded

Table 1 Results of the stepwise model selection in boys for the BCCG, BCPE and normal distribution

Distribution	Parameters μ	$\log(\sigma)$	ν	$\log(\tau)$	BIC
BCCG	Age+cs(height)	Cs(height)	Cs(height)	–	42,652
BCPE	Age+cs(height)	Cs(height)	Cs(height)	1	42,661
Normal	Age+cs(height)	Cs(height)	–	–	43,154

Table 2 Summary of quantile residuals

Model	BCCG	BCPE
Mean	-2.56×10^{-5}	4.64×10^{-5}
Variance	1.00019	1.00018
Coefficient of skewness	-0.00319	-0.00193
Coefficient of kurtosis	3.01162	2.96746

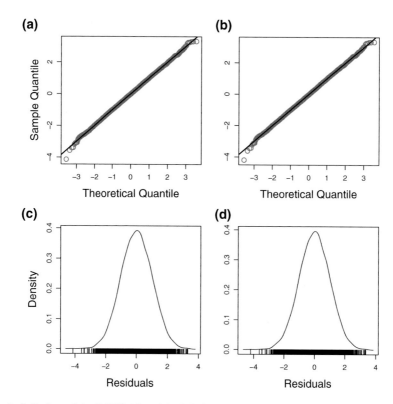

Fig. 2 Q-Q plots of the BCCG (**a**) and the BCPE model (**b**); density estimate of BCCG (**c**) and the BCPE model (**d**). Figure (a) from Herrmann et al. (2014)

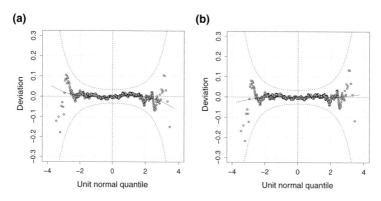

Fig. 3 Wormplots of the BCCG (**a**) and BCPE model (**b**)

the more complex four-parametric BCPE distribution. Percentile curves based on the BCCG model were derived for boys. Our model selection procedure for girls also led to a BCCG model.

3 Results

As already mentioned, the reference curves for the SI depend on age and height. Thus, three-dimensional curves stratified by sex are required to adequately represent the functional relationship. One straightforward approach to present this relationship is a wireframe plot using the covariates age and height as *y*- and *x*-coordinates and one percentile of the SI as *z*-coordinate in a three-dimensional Euclidean space (Fig. 4a). Using all dimensions a contour plot could be an alternative (Fig. 4b). For the sake of clarity, improbable combinations of age and height should be omitted. For this we used the age-specific 3rd and 97th height percentiles as cut-offs. In Herrmann et al. (2014), we decided to provide easy-to-read tables and figures which can be useful in daily practice, see, e.g., Fig. 5. Here, two-dimensional plots are presented where percentile curves for only three height percentiles of major clinical relevance are plotted. In younger children, the 1st and 50th percentile of the SI showed a negative association with age and height. In older children, a positive association was observed for all percentiles and their height had a smaller effect.

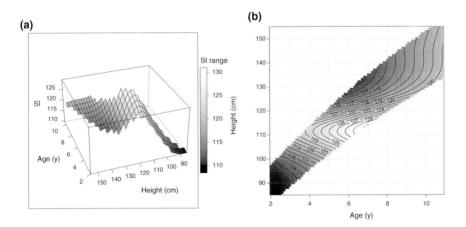

Fig. 4 The 99th percentile of the SI in a wireframe plot (**a**) and in a contour plot (**b**)

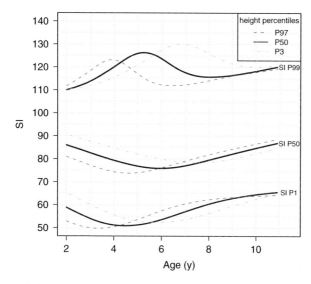

Fig. 5 The 99th, 50th, 1st percentile curve of the SI in boys depending on age and three different height percentiles (P97, P50 and P3) from Herrmann et al. (2014)

4 Conclusion

We demonstrated the application of GAMLSS to derive age- and height-specific percentile curves for children. We also showed how to find appropriate models performing a model selection based on BIC and how to compare different models by means of model diagnostic tools. Especially wormplots seem to be an adequate tool in this context because they permit to assess the need to account for mean, variance, skewness and particularly kurtosis. Finally, we presented some examples how to display the resulting percentile curves.

Although a model assuming a BCCG distribution was chosen, which corresponds to the LMS method, two advantages of GAMLSS became apparent. First, it is possible to derive percentile curves that account for more than one covariate which is useful for clinical parameters not only depending on age but also on height. Second, it is possible to consider different models and to compare them to decide whether to account for skewness and kurtosis. Thus, the chosen model and the derived percentile curves result from a more comprehensive evaluation of various models which justifies the additional effort for model selection related to GAMLSS compared to the LMS method.

Taking SI as an example, GAMLSS has proven to be useful to generate reference curves for children. For many clinical parameters, including SI, reference curves were derived to guide clinical practice (Ahrens et al. 2014). These and further applications of GAMLSS to clinical parameters, where reference values are missing, will help to fill the gap in medical diagnostics that currently exists for children.

Acknowledgements This work was done as part of the IDEFICS study (www.idefics.eu). We gratefully acknowledge the financial support of the European community within the Sixth RTD Framework Programme Contract No. 016181 (FOOD).

Disclaimer The information in this document reflects the author's view and is provided as is.

References

Ahrens, W., Bammann, K., Siani, A., Buchecker, K., De Henauw, S., Iacoviello, L., et al. (2011). The IDEFICS cohort: Design, characteristics and participation in the baseline survey. *International Journal of Obesity, 35*(Suppl 1), S3–S15.

Ahrens, W., Moreno, L., & Pigeot, I. (2014). Filling the gap: International reference values for health care in children. *International Journal of Obesity, 38*, S2–S3.

Borghi, E., De Onis, M., Garza, C., Van Den Broeck, J., Frongillo, E.A., Grummer-Strawn, L., et al. (2006). Construction of the World Health Organization child growth standards: Selection of methods for attained growth curves. *Statistics in Medicine, 25*, 247–265.

Cole, T. J., & Green, P. J. (1992). Smoothing reference centile curves: The LMS method and penalized likelihood. *Statistics in Medicine, 11*, 1305–1319.

Cole, T. J., & Lobstein, T. (2012). Extended international (IOTF) body mass index cut-offs for thinness, overweight and obesity. *Pediatric Obesity, 7*, 284–294.

Cole, T. J., Stanojevic, S., Stocks, J., Coates, A.L., Hankinson, J. L., & Wade, A. M. (2009). Age- and size-related reference ranges: A case study of spirometry through childhood and adulthood. *Statistics in Medicine, 28*, 880–898.

Herrmann, D., Intemann, T., Lauria, F., Mårild, S., Molnár, D., Moreno, L. A., et al. (2014). Reference values of bone stiffness index and C-terminal telopeptide in healthy European children. *International Journal of Obesity, 38*, S76–S85.

Jaworski, M., Lebiedowski, M., Lorenc, R. S., & Trempe, J. (1995). Ultrasound bone measurement in pediatric subjects. *Calcified Tissue International, 56*, 368–371.

Krieg, M. A., Cornuz, J., Ruffieux C., Van Melle, G., Büche, D., Dambacher, M. A., et al. (2006). Prediction of hip fracture risk by quantitative ultrasound in more than 7000 Swiss women ≥ 70 years of age: Comparison of three technologically different bone ultrasound devices in the SEMOF study. *Journal of Bone and Mineral Research, 21*, 1457–1463.

Maggi, S., Noale, M., Giannini, S., Adami, S., Defeo, D., Isaia, G., et al. (2006). Quantitative heel ultrasound in a population-based study in Italy and its relationship with fracture history: The ESOPO study. *Osteoporosis International, 17*, 237–244.

R Core Team. (2013). *R: A language and environment for statistical computing*. Vienna: R Foundation for Statistical Computing. http://www.R-project.org/

Rigby, R. A., & Stasinopoulos, D. M. (2004). Smooth centile curves for skew and kurtotic data modelled using the Box-Cox power exponential distribution. *Statistics in Medicine, 23*, 3053–3076.

Rigby, R. A., & Stasinopoulos, D. M. (2005). Generalized additive models for location scale and shape. *Journal of the Royal Statistical Society: Series C (Applied Statistics), 54*, 507–554.

van Buuren, S., & Fredriks, M. (2001). Worm plot: A simple diagnostic device for modelling growth reference curves. *Statistics in Medicine, 20*, 1259–1277.

An Ensemble of Optimal Trees for Class Membership Probability Estimation

Zardad Khan, Asma Gul, Osama Mahmoud, Miftahuddin Miftahuddin, Aris Perperoglou, Werner Adler, and Berthold Lausen

Abstract Machine learning methods can be used for estimating the class membership probability of an observation. We propose an ensemble of optimal trees in terms of their predictive performance. This ensemble is formed by selecting the best trees from a large initial set of trees grown by random forest. A proportion of trees is selected on the basis of their individual predictive performance on out-of-bag observations. The selected trees are further assessed for their collective performance on an independent training data set. This is done by adding the trees one by one starting from the highest predictive tree. A tree is selected for the final ensemble if it increases the predictive performance of the previously combined trees. The proposed method is compared with probability estimation tree, random forest and node harvest on a number of bench mark problems using Brier score as a performance measure. In addition to reducing the number of trees in the ensemble,

Z. Khan (✉)
Department of Mathematical Sciences, University of Essex, Colchester, UK

Department of Statistics, Abdul Wali Khan University, Mardan, Pakistan
e-mail: zkhan@essex.ac.uk; zardadkhan@awkum.edu.pk

A. Gul
Department of Mathematical Sciences, University of Essex, Colchester, UK

Department of Statistics, Shaheed Benazir Bhutto Women University Peshawar,
Khyber Pukhtoonkhwa, Pakistan
e-mail: agul@essex.ac.uk

M. Miftahuddin • A. Perperoglou • B. Lausen
Department of Mathematical Sciences, University of Essex, Colchester, UK
e-mail: mmifta@essex.ac.uk; aperpe@essex.ac.uk; blausen@essex.ac.uk

O. Mahmoud
Department of Mathematical Sciences, University of Essex, Colchester, UK

Department of Applied Statistics, Helwan University, Cairo, Egypt
e-mail: ofamah@essex.ac.uk

W. Adler
Department of Biometry and Epidemiology, University of Erlangen-Nuremberg, Erlangen,
Germany
e-mail: werner.adler@fau.de

© Springer International Publishing Switzerland 2016   395
A.F.X. Wilhelm, H.A. Kestler (eds.), *Analysis of Large and Complex Data*, Studies
in Classification, Data Analysis, and Knowledge Organization,
DOI 10.1007/978-3-319-25226-1_34

our method gives better results in most of the cases. The results are supported by a simulation study.

1 Introduction

The usual task of pattern recognition or discrimination is to make a simple statement about the group membership of an individual. For example, this simple statement about a tumour patient could be that he/she is having a malignant or a benign tumour. This might also be of interest to know the class membership probability of the individual which is an important biomedical application. It is usually required by surgeons, oncologists, pathologists, professionals involved in internal medicine and human genetics and paediatricians (Malley et al. 2012). For instance, carrier probabilities are calculated in genetic counseling and treatment response probability is estimated in personalized medicine of every patient (Kruppa et al. 2012, 2014b).

The logistic regression model is the standard and classical approach for estimating individual probabilities (Kruppa et al. 2012, 2014a). A major problem with the logistic regression is the requirement of correct and full specification of the model. Misspecified model will give biased and inconsistent results.

Machine learning methods, on the other hand, can be used as non-parametric alternatives to the classical logistic regression models to avoid the assumptions involved and to overcome the problem of misspecification. These methods have been utilized in various biomedical applications (Kruppa et al. 2012, 2014a; Malley et al. 2012). Most of these methods are based on the idea of combining multiple models to build a strong model (Ali and Pazzani 1996; Hothorn and Lausen 2003). Studies have shown that the generalization error can be reduced by combining the outputs of multiple models (Maclin and Opitz 2011). In this paper, the possibility of creating an ensemble of optimal trees for class membership probability estimation is considered that is motivated by Breiman's (2001) upper bound for the overall prediction error of a random forest ensemble which is given by

$$PE^* \leq \bar{\rho} PE_j, \tag{1}$$

where $j = 1, 2, 3, \ldots, T$. T is the total number of trees in the forest, PE^* is the overall prediction error of a random forest, $\bar{\rho}$ is the weighted correlation between residuals from two independent trees and PE_j is the prediction error of tree j in the forest. This relation indicates that individually accurate and diverse trees could make an efficient forest. Based on this intuition, trees are selected from a total of T trees grown on bootstrap samples drawn from a given learning data set. A similar approach is proposed in Gul et al. (2015) where the idea of random feature set selection and bagging is used with k-nearest neighbours classifiers for the issue of non-informative features in the data. We compare the method with k-nearest neighbours, tree, random forest (RF), node harvest (NH) (Meinshausen 2010) and support vector machines (SVMs) for probability estimation. The rest of the paper is arranged as follows: Sect. 2 discusses the methods mentioned before; Sect. 3

describes the Brier score; Sect. 4 introduces our method; Sect. 5 gives experiments and results and conclusion is given in Sect. 6.

2 Probability Machines

Machine learning techniques that are used to give estimates of probability for the group membership in binary class problems are named probability machines by Malley et al. (2012). Here we briefly explain how kNN, tree, RF, NH and SVM could be used for estimating class membership probabilities before introducing our method, the optimal trees ensemble (*OTE*).

2.1 Probability Estimation Trees

To find the conditional probability, $P(Y|X)$, of an individual belonging to a particular class, the steps are

1. On a bootstrap sample from the training data $\mathcal{L} = (\mathbf{X}, \mathbf{Y})$, grow a classification or regression tree.
2. Filter a test observation through the tree until it reaches to a leaf node Q'.
3. The proportion $p_i, i = 1, 2$ of an observations of a particular class in Q' is determined which is the required probability, where

$$p_i = \frac{\text{\# of } i\text{th class observations in } Q'}{\text{\# of observations in } Q'}.$$

2.2 Random Forest as Probability Machine

The Breiman (2001) random forest can effectively be used for estimating the conditional probability function $P(\mathbf{Y}|\mathbf{X})$ (Liaw and Wiener 2002). To find the group membership probability $P(\mathbf{Y}|\mathbf{X})$, take the following steps:

1. Draw T bootstrap samples from the given training data $\mathcal{L} = (\mathbf{X}, \mathbf{Y})$ and grow T probability estimation trees (PETs).
2. A test observation is filtered through each tree until it reaches a leaf node.
3. The estimate of class probability is the average proportion of a class observations in the leaf nodes of all the trees where the test observation resides.

2.3 Node Harvest as Probability Machine

Node harvest, proposed by Meinshausen (2010), is a tree-based algorithm that takes a large set of nodes as an initial ensemble and selects the most useful nodes for the final decision. Class membership probability of an observation is estimated as follows:

1. Take a sufficiently large number of nodes from an initial tree ensemble.
2. Allow non-negative weights that take on values in the continuous interval [0,1] and select those nodes that are assigned the highest weights.
3. Remove nodes that are identical.
4. The estimate of class probability is the average proportion of a class observations in the selected nodes where the test observation resides.

2.4 k-Nearest Neighbours as Probability Machine

To estimate class membership probability of a test observation via *k*NN, the steps are as follows:

1. Compute the distance of a test observation from all the training instances.
2. Find *k* nearest instances to the test point according to the distance.
3. The estimate of the probability is the proportion of instances of a class in the *k* nearest neighbours.

2.5 SVMs for Probability Estimation

Given a training data set $\mathscr{L} = (\mathbf{X}, \mathbf{Y})$, SVMs can be used to produce estimates of class membership probability instead of class labels. This is done by the implementation of Platt's posteriori probabilities (Platt 2000) in several R packages, where the following sigmoid function is used.

$$p(y|\mathbf{X}) = \frac{1}{1 + \exp(Af(\mathbf{X}) + B)}, \text{ where } f(\mathbf{X}) \text{ is a decision function.} \tag{2}$$

A and *B* are the parameters to be estimated. For further information on this, see Platt (2000).

Before introducing the proposed ensemble, we explain the performance measure used in the algorithm and its comparison to other methods.

3 Assessment of the Probability Machines

We use the Brier score as performance measure which is generally used when the true probabilities are not available (Malley et al. 2012). Gneiting and Raftery (2007) argued that the Brier score is a proper score and its minimum value can only be obtained if the estimated probabilities are taken exactly equal to the true unknown probabilities. It means that any probability machine having the smallest Brier score is estimating class probabilities in the best possible way. The Brier score is represented by the following equation:

$$BS = \mathbb{E} \left(Y - P(Y|X) \right)^2, \qquad (3)$$

where Y is the state of the response variable in the 0,1 form for the two classes and $P(Y|X)$ is the true unknown probability for the binary response given the features. An estimator for the above score is

$$\hat{BS} = \frac{\sum_{i=1}^{\text{\# of test cases}} \left(y_i - \hat{P}(y_i|X) \right)^2}{\text{total \# of test cases}}, \qquad (4)$$

where y_i is the state of the response for observation i in the 0,1 form and $\hat{P}(y_i|X)$ is the estimate of probability for the binary response given the features.

4   The Ensemble of Optimal Trees

For obtaining the ensemble of best (accurate and diverse) trees, divide the given training data $\mathscr{L} = (\mathbf{X}, \mathbf{Y})$ randomly into two non overlapping parts, $\mathscr{L}_{\mathscr{B}} = (\mathbf{X_B}, \mathbf{Y_B})$ and $\mathscr{L}_{\mathscr{V}} = (\mathbf{X_V}, \mathbf{Y_V})$. Grow T trees on T bootstrap samples from $\mathscr{L}_{\mathscr{B}} = (\mathbf{X_B}, \mathbf{Y_B})$. Accurate and diverse trees are selected as follows:

1. Estimate the error of each tree (growing by random forest without pruning) by using the out-of-bag (OOB) observations (observations left out from a bootstrap sample) as the validation data.
2. Arrange the trees in ascending order with respect to the prediction errors and take the first M trees.
3. To find diverse trees, the second best tree out of the M trees is combined with the best tree to get an ensemble of size two and see how they perform on $\mathscr{L}_{\mathscr{V}} = (\mathbf{X_V}, \mathbf{Y_V})$. Then the third best tree is added and the performance is measured and so on until the final Mth tree is added.
4. Tree $\hat{L}_k, k = 1, 2, 3, \ldots, M$ is selected if its addition to the ensemble without the kth tree fulfils the following criterion.

   - Let $BS^{(k-1)}$ be the Brier score of the ensemble without the kth tree and $BS^{(k)}$ be the Brier score of the ensemble including the kth tree, then tree $\hat{L}_k$ is selected if

$$BS^{(k)} < BS^{(k-1)}. \qquad (5)$$

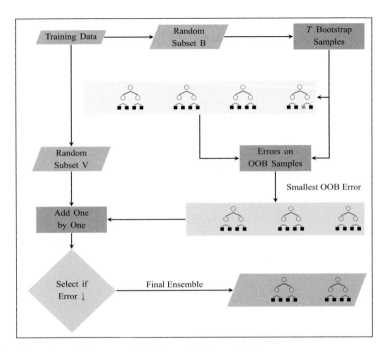

Fig. 1 A flow chart of the steps of *OTE* for probability estimation

To estimate class probability of an observation, apply steps 2 and 3 of random forest on the M selected trees.

A simple illustrative flow chart of the steps is given in Fig. 1.

5 Experiments and Results

5.1 Simulation

We simulate data consisting of various structures to make the recognition problem slightly difficult for simple classifiers, kNN and PET for example. We aimed our method to perform better than the simple classifier and compete with the complex and powerful classifiers, SVM, random forest and node harvest in our study, in finding the structures. To this end we generate four models with a different number of tree components where all the components are partitioning the data set on a subset of the feature space. For each model we consider four different cases/complexity levels by altering the weights η_{ijk} of the tree nodes to move from highly non-uniform distributions (low entropy) to distributions with high entropy. Thus we get four different values of the Bayes error where the lowest Bayes error means a data set with meaningful patterns and the highest Bayes error indicates a data set with no patterns. Table 1 lists the various values of η_{ijk} used in models 1, 2, 3 and 4.

Table 1 Node weights, η_{ijk}, used in simulation models where i is tree number, j is node number in each tree and k is denoting a variant of the weights

		Model 1						Model 2						Model 3						Model 4			
		k						k						k						k			
i	j	1	2	3	4	i	j	1	2	3	4	i	j	1	2	3	4	i	j	1	2	3	4
1	1	0.9	0.8	0.7	0.6	1	1	0.9	0.8	0.7	0.6	1	1	0.9	0.9	0.9	0.8	1	1	0.9	0.9	0.9	0.8
	2	0.1	0.2	0.3	0.4		2	0.1	0.2	0.3	0.4		2	0.1	0.1	0.1	0.2		2	0.1	0.1	0.1	0.2
	3	0.1	0.2	0.3	0.4		3	0.1	0.2	0.3	0.4		3	0.1	0.1	0.1	0.2		3	0.1	0.1	0.1	0.2
	4	0.9	0.8	0.7	0.6		4	0.9	0.8	0.7	0.6		4	0.9	0.9	0.9	0.8		4	0.9	0.9	0.9	0.8
2	1	0.9	0.8	0.7	0.6	2	1	0.9	0.8	0.7	0.6	2	1	0.9	0.9	0.9	0.8	2	1	0.9	0.9	0.9	0.8
	2	0.1	0.2	0.3	0.4		2	0.1	0.2	0.3	0.4		2	0.1	0.1	0.1	0.2		2	0.1	0.1	0.1	0.2
	3	0.1	0.2	0.3	0.4		3	0.1	0.2	0.3	0.4		3	0.1	0.1	0.1	0.2		3	0.1	0.1	0.1	0.2
	4	0.9	0.8	0.7	0.6		4	0.9	0.8	0.7	0.6		4	0.9	0.9	0.9	0.8		4	0.9	0.9	0.9	0.8
3	1	0.9	0.8	0.7	0.6	3	1	0.9	0.8	0.7	0.6	3	1	0.9	0.8	0.7	0.7	3	1	0.9	0.9	0.9	0.8
	2	0.1	0.2	0.3	0.4		2	0.1	0.2	0.3	0.4		2	0.1	0.2	0.3	0.3		2	0.1	0.1	0.1	0.2
	3	0.1	0.2	0.3	0.4		3	0.1	0.2	0.3	0.4		3	0.1	0.2	0.3	0.3		3	0.1	0.1	0.1	0.2
	4	0.9	0.8	0.7	0.6		4	0.9	0.8	0.7	0.6		4	0.9	0.8	0.7	0.7		4	0.9	0.9	0.9	0.8

(continued)

Table 1 (continued)

Model 1

Model 2

4	1	0.9	0.8	0.7	0.6
	2	0.1	0.2	0.3	0.4
	3	0.1	0.2	0.3	0.4
	4	0.9	0.8	0.7	0.6

Model 3

4	1	0.9	0.8	0.7	0.7
	2	0.1	0.2	0.3	0.3
	3	0.1	0.2	0.3	0.3
	4	0.9	0.8	0.7	0.7
5	1	0.9	0.8	0.7	0.7
	2	0.1	0.2	0.3	0.3
	3	0.1	0.2	0.3	0.3
	4	0.9	0.8	0.7	0.7

Model 4

4	1	0.9	0.8	0.7	0.7
	2	0.1	0.2	0.3	0.3
	3	0.1	0.2	0.3	0.3
	4	0.9	0.8	0.7	0.7
5	1	0.9	0.8	0.7	0.6
	2	0.1	0.2	0.3	0.4
	3	0.1	0.2	0.3	0.4
	4	0.9	0.8	0.7	0.6
6	1	0.9	0.8	0.7	0.6
	2	0.1	0.2	0.3	0.4
	3	0.1	0.2	0.3	0.4
	4	0.9	0.8	0.7	0.6

Node weights for getting the four complexity levels are given in four columns of the table for $k = 1, 2, 3, 4$, for each model. All the four models are derived from the following equation for producing class probabilities of the Bernoulli response $\mathbf{Y} = \text{Bernoulli}(p)$ given the $n \times 3T$ dimensional vector $\mathbf{X}$ of n iid observations from Uniform$(0, 1)$, T being the total number of trees.

$$p(y|\mathbf{X}) = \frac{\exp\left(b \times \left(\frac{\zeta_m}{T} - a\right)\right)}{1 + \exp\left(b \times \left(\frac{\zeta_m}{T} - a\right)\right)}, \quad \text{where } \zeta_m = \sum_{t=1}^{T} \Upsilon_t. \tag{6}$$

$a, b \in \mathbb{R}$ are any arbitrary constants, $m = 1, 2, 3, 4$ is the model number and ζ_m's are $n \times 1$ vector of probabilities. T is the total number of trees in a model and Υ_t's are probabilities for a particular class in the response $\mathbf{Y}$ generated by different tree structures as follows:

$$\Upsilon_1 = \eta_{11k} \times \mathbb{1}(x_1 \leq 0.5 \ \& \ x_3 \leq 0.5) + \eta_{12k} \times \mathbb{1}(x_1 \leq 0.5 \ \& \ x_3 > 0.5)$$
$$+ \eta_{13k} \times \mathbb{1}(x_1 > 0.5 \ \& \ x_2 \leq 0.5) + \eta_{14k} \times \mathbb{1}(x_1 > 0.5 \ \& \ x_2 > 0.5),$$

$$\Upsilon_2 = \eta_{21k} \times \mathbb{1}(x_4 \leq 0.5 \ \& \ x_6 \leq 0.5) + \eta_{22k} \times \mathbb{1}(x_4 \leq 0.5 \ \& \ x_6 > 0.5)$$
$$+ \eta_{23k} \times \mathbb{1}(x_4 > 0.5 \ \& \ x_5 \leq 0.5) + \eta_{24k} \times \mathbb{1}(x_4 > 0.5 \ \& \ x_5 > 0.5),$$

$$\Upsilon_3 = \eta_{31k} \times \mathbb{1}(x_7 \leq 0.5 \ \& \ x_8 \leq 0.5) + \eta_{32k} \times \mathbb{1}(x_7 \leq 0.5 \ \& \ x_8 > 0.5)$$
$$+ \eta_{33k} \times \mathbb{1}(x_7 > 0.5 \ \& \ x_9 \leq 0.5) + \eta_{34k} \times \mathbb{1}(x_7 > 0.5 \ \& \ x_9 > 0.5),$$

$$\Upsilon_4 = \eta_{41k} \times \mathbb{1}(x_{10} \leq 0.5 \ \& \ x_{11} \leq 0.5) + \eta_{42k} \times \mathbb{1}(x_{10} \leq 0.5 \ \& \ x_{11} > 0.5)$$
$$+ \eta_{43k} \times \mathbb{1}(x_{10} > 0.5 \ \& \ x_{12} \leq 0.5) + \eta_{44k} \times \mathbb{1}(x_{10} > 0.5 \ \& \ x_{12} > 0.5),$$

$$\Upsilon_5 = \eta_{51k} \times \mathbb{1}(x_{13} \leq 0.5 \ \& \ x_{14} \leq 0.5) + \eta_{52k} \times \mathbb{1}(x_{13} \leq 0.5 \ \& \ x_{14} > 0.5)$$
$$+ \eta_{53k} \times \mathbb{1}(x_{13} > 0.5 \ \& \ x_{15} \leq 0.5) + \eta_{54k} \times \mathbb{1}(x13 > 0.5 \ \& \ x_{15} > 0.5),$$

$$\Upsilon_6 = \eta_{61k} \times \mathbb{1}(x_{16} \leq 0.5 \ \& \ x_{17} \leq 0.5) + \eta_{62k} \times \mathbb{1}(x_{16} \leq 0.5 \ \& \ x_{17} > 0.5)$$
$$+ \eta_{63k} \times \mathbb{1}(x_{16} > 0.5 \ \& \ x_{18} \leq 0.5) + \eta_{64k} \times \mathbb{1}(x_{16} > 0.5 \ \& \ x_{18} > 0.5),$$

where $0 < \eta_{ijk} < 1$ are weights given to the nodes of the trees, $k = 1, 2, 3, 4$. The four models use the following specifications for using (6)

5.1.1   Model 1

This model consists of 3 tree components each based on 3 variables. Therefore, $T = 3$, $\zeta_1 = \sum_{t=1}^{3} \Upsilon_t$ and $\mathbf{X}$ becomes a $n \times 9$ dimensional vector. A tree used in this model is shown in Fig. 2.

Fig. 2 A tree used in
simulation model 1

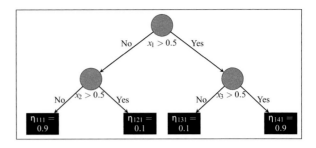

5.1.2 Model 2

For this model we take $T = 4$ trees where $\zeta_2 = \sum_{t=1}^{4} \Upsilon_t$ and $\mathbf{X}$ becomes a $n \times 12$ dimensional vector.

5.1.3 Model 3

This model is based on $T = 5$ trees such that $\zeta_3 = \sum_{t=1}^{5} \Upsilon_t$ and $\mathbf{X}$ becomes a $n \times 15$ dimensional vector.

5.1.4 Model 4

This model consists of 6 tree components with $T = 6$, $\zeta_4 = \sum_{t=1}^{6} \Upsilon_t$ and $\mathbf{X}$ becomes a $n \times 18$ dimensional vector.

We see in Table 2 that tree, kNN, NH and SVM gave consistently poor performance as compared to RF and *OTE*. *OTE* gave comparable results with RF in most of the cases. Comparable/better results can be seen in the first of the four cases of all the remaining models. From these results, it follows that the proposed method can produce comparable results to random forest with a significant reduction in the ensemble size (given in the last column of Table 2) if there are some meaningful patterns in the data.

5.2 Bench Mark Problems

We considered 20 bench mark problems taken from various open sources. Dystrophy and Glaucoma data sets are taken from "ipred" R package, Musk from "kernlab" R package and Body data set is from "gclus" R package. Appendicitis and SAHeart are from http://sci2s.ugr.es/keel/dataset.php?cod=183. Oil-Spill data is from http://openml.org/d?from=180. All the rest of the data sets are from UCI machine learning repository http://archive.ics.uci.edu/ml/. A brief description of these data is given in the first four columns of Table 3 where n is sample size and d is the number of features.

Table 2 Brier scores of kNN, tree, RF, NH, SVM and *OTE* on simulated data

Model	d	n	Bayes error	kNN	Tree	RF	NH	SVM (Radial)	SVM (Linear)	SVM (Bessel)	SVM (Laplacian)	OptTreesEns	Reduction in ensemble size (%)
Model 1	9	1000	0.09	0.16	0.09	0.10	0.12	0.13	0.13	0.14	0.13	0.08	90.7
			0.14	0.18	0.12	0.12	0.14	0.16	0.16	0.16	0.16	0.13	89.5
			0.17	0.22	0.13	0.12	0.14	0.19	0.18	0.19	0.18	0.12	89.5
			0.33	0.27	0.23	0.22	0.22	0.23	0.23	0.23	0.22	0.23	90.8
Model 2	12	1000	0.21	0.19	0.16	0.13	0.16	0.16	0.16	0.19	0.16	0.13	89.9
			0.24	0.21	0.18	0.15	0.17	0.17	0.17	0.20	0.17	0.15	89.7
			0.28	0.24	0.21	0.18	0.2	0.20	0.20	0.22	0.20	0.19	89.7
			0.3	0.25	0.22	0.21	0.21	0.21	0.21	0.23	0.21	0.21	89.2
Model 3	15	1000	0.15	0.21	0.17	0.14	0.18	0.16	0.16	0.25	0.16	0.14	90.7
			0.18	0.21	0.18	0.15	0.18	0.17	0.17	0.25	0.17	0.16	89.1
			0.21	0.22	0.18	0.16	0.18	0.18	0.18	0.25	0.18	0.16	91.1
			0.24	0.24	0.2	0.19	0.2	0.19	0.19	0.25	0.19	0.18	89.9
Model 4	18	1000	0.21	0.22	0.2	0.16	0.19	0.17	0.17	0.19	0.18	0.16	89.8
			0.22	0.23	0.2	0.16	0.19	0.18	0.18	0.20	0.19	0.17	89.3
			0.25	0.25	0.22	0.18	0.2	0.20	0.20	0.21	0.22	0.18	90.5
			0.26	0.26	0.22	0.19	0.2	0.21	0.21	0.22	0.24	0.19	90.2

The last column is the percentage reduction in ensemble size of *OTE* compared to RF

Table 3 Data sets summary (FT means feature type with R: real, I: integer and N: nominal number of features) and Brier scores of kNN, tree, random forest, node harvest, SVM and OTE

Data set	n	d	FT (R/I/N)	kNN	Tree	RF	NH	SVM (Radial)	SVM (Linear)	SVM (Bessel)	SVM (Laplacian)	OptTreesEns
Mammographic	830	5	(0/5/0)	0.1412	0.1229	0.1288	**0.1207**	0.1340	0.1252	0.1313	0.1354	0.1366
Dystrophy	209	5	(2/3/0)	0.1051	0.1344	0.0947	0.1161	0.0831	0.0872	0.0802	**0.0792**	0.0864
Monk3	122	6	(0/6/0)	0.0886	0.0687	0.0657	0.1817	0.0695	0.1570	0.0663	0.0938	**0.0610**
Appendicitis	106	7	(6/0/0)	0.1263	0.1354	**0.1199**	0.1165	0.1360	0.1257	0.1156	0.1178	0.1242
SAHeart	462	9	(5/3/1)	0.2092	0.2074	0.1895	0.1880	0.1850	**0.1794**	0.1966	0.1816	0.2006
Tic-tac-toe	958	9	(0/0/9)	0.2279	0.1467	**0.0408**	0.1997	0.1483	0.2188	0.1200	0.1972	0.0437
Heart	303	13	(1/12/0)	0.2226	0.1683	**0.1231**	0.1441	0.1442	0.1278	0.1235	0.1247	0.1286
House vote	232	16	(0/0/16)	0.0655	0.0323	0.0293	0.0656	0.0299	0.0345	0.1580	0.0386	**0.0290**
Bands	365	19	(13/6/0)	0.2231	0.2549	0.1878	0.2240	0.1991	0.2028	0.2230	0.2107	**0.1814**
Hepatitis	80	20	(2/18/0)	0.3105	0.1378	0.0970	0.0950	0.0964	0.1042	0.1158	0.0894	**0.0883**
Parkinson	195	22	(22/0/0)	0.1151	0.1138	0.0676	0.0930	0.0763	0.1195	0.1544	0.0931	**0.0636**
Body	507	23	(22/1/0)	0.0190	0.0734	0.0311	0.0553	0.0124	**0.0120**	0.2377	0.0219	0.0295
Thyroid	9172	27	(3/2/22)	0.0305	0.0104	0.0084	0.0161	0.0388	0.0321	0.0572	0.0382	**0.0079**
WDBC	569	29	(29/0/0)	0.0541	0.0643	0.0311	0.0425	0.0266	**0.0212**	0.2034	0.0283	0.0308
WPBC	198	32	(30/2/0)	0.1825	0.2131	0.1679	0.1686	0.1603	**0.1542**	0.1806	0.1626	0.1653
Oil-spill	937	49	(40/9/0)	0.0395	0.0334	0.0282	0.0293	0.0326	0.0373	0.0331	0.0364	**0.0274**
Spam base	4601	57	(55/2/0)	0.1744	0.0948	0.0383	0.0906	0.0730	0.0618	0.2407	0.0814	**0.0374**
Glaucoma	196	62	(62/0/0)	0.1365	0.1095	**0.0890**	0.0916	0.0941	0.1239	0.2193	0.1193	0.0904
Nki 70	144	76	(71/5/0)	0.1458	0.1410	0.1465	0.1473	0.1675	0.2024	0.2349	0.1832	**0.1329**
Musk	476	166	(0/166/0)	0.1420	0.1884	0.0963	0.1746	0.0956	0.1107	0.2470	0.1886	**0.0871**

The best result is shown in bold

5.3   Experimental Setup and Results for Bench Mark Problems

The data sets are divided into two parts. The training part consisted of 90 % of observations (of which 90 % is used for bootstrapping and 10 % for diversity check) and the remaining part is taken as the testing part. A total of 1000 runs are performed to calculate the average Brier score on all the data sets. The results are given in Table 3 where the average Brier scores of kNN, tree, random forest, node harvest, SVM and *OTE* are given against each data set. Four kernels; Radial, Linear, Bessel and Laplacian, are considered for SVM with the rest of parameters on their default values in the "kernlab" R package. Tenfold cross validation is used for tuning the parameters of kNN, tree and RF. kNN is tuned for $k = 1, \ldots, 10$. For finding the optimal number of splits and the minimal optimal depth of the trees, values $(5, 10, 15, 20, 25, 30)$ are tried. For tuning the node size of RF, we tried values $(1, 5, 10, 15, 20, 25, 30)$, for *ntree*, $(500, 1000, 1500, 2000)$ and for tuning *mtry*, we tried $(\text{sqrt}(d), d/5, d/4, d/3, d/2)$ where d is the total number of features. Number of nodes in the initial set for NH is fixed at 1500. The result of the best performing method is given in bold. R package, version 3.1.0 is used in all the experiments (R Core Team 2014). It is clear from Table 3 that *OTE* outperforms all the other methods on most of the data sets. The new method is giving the smallest Brier scores on 10 out of 20 data sets. On 4 data sets random forest gave the smallest Brier scores. On 1 data set, node harvest gave the best result while SVM gave the best performance on 5 data sets. A large number of trees in the initial set can be recommended under the available computational resources. For $T > 1000$ the results of the proposed method are invariant and the method converges afterwards for the data sets considered. This can be seen in Fig. 3a. As shown in Fig. 3b, class membership probability estimations by using *OTE* is unaffected by varying

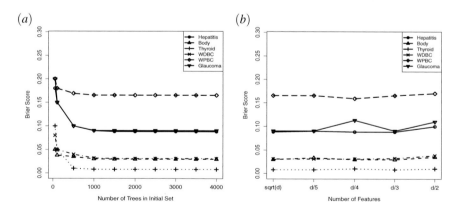

Fig. 3 (**a**) The effect of the number of trees in the initial set on *OTE*. (**b**) The effect of the number of features selected at random for splitting the nodes of the trees on *OTE*

the number of features selected at random for splitting the nodes of the trees. This means that growing trees for the initial set through random forest or simple bootstrap technique might lead to very similar final ensembles.

6   Conclusion

We have proposed an ensemble of optimal trees, *OTE*, as a non-parametric method for estimating class membership probabilities in binary class problems. We compared PETs, random forest, node harvest and the proposed *OTE* on a number of bench mark and simulated data sets. The proposed method outperformed *k*NN, tree, random forest, node harvest and SVM on most of the data sets. We also used tree style simulation models to generate data sets with several structures. The proposed method is observed to use fewer accurate and diverse trees and hence could be very helpful in reducing the number of trees in tree ensembles which might increase interpretability. The method is observed to be unaffected by varying the number of features selected at random for splitting the nodes of the trees and they could simply be grown using the simple bagging technique. The method is implemented in an R package *OTE*. The proposed method could better be used, in conjunction with some feature selection method, (Mahmoud et al. 2014a,b, for example) in high dimensional settings.

References

Ali, K. M., & Pazzani, M. J. (1996). Error reduction through learning multiple descriptions. *Machine Learning, 24*, 173–202.

Breiman, L. (2001). Random forests. *Machine Learning, 45*, 5–32.

Gneiting, T., & Raftery, A. E. (2007). Strictly proper scoring rules, prediction, and estimation. *Journal of the American Statistical Association, 102*, 359–378.

Gul, A., Khan, Z., Mahmoud, O., Perperoglou, A., Miftahuddin, M., Adler, W., et al. (2015). Ensemble of k-nearest neighbour classifiers for class membership probability estimation. In *The Proceedings of European Conference on Data Analysis, 2014.*

Hothorn, T., & Lausen, B. (2003). Double-bagging: Combining classifiers by bootstrap aggregation. *Pattern Recognition, 36*, 1303–1309.

Kruppa, J., Liu, Y., Biau, G., Kohler, M., Konig, I. R., Malley, J. D., et al. (2014a). Probability estimation with machine learning methods for dichotomous and multicategory outcome: Theory. *Biometrical Journal, 56*, 534–563.

Kruppa, J., Liu, Y., Diener, H. C., Weimar, C., Konig, I. R., & Ziegler, A. (2014b). Probability estimation with machine learning methods for dichotomous and multicategory outcome: Applications. *Biometrical Journal, 56*, 564–583.

Kruppa, J., Ziegler, A., & Konig, I. R. (2012). Risk estimation and risk prediction using machine-learning methods. *Human Genetics, 131*, 1639–1654.

Liaw, A., & Wiener, M. (2002). Classification and regression by random forest. *R News, 2*, 18–22.

Maclin, R., & Opitz, D. (2011). Popular ensemble methods: An empirical study. *Journal of Artificial Research, 11*, 169–189.

Mahmoud, O., Harrison, A., Perperoglou, A., Gul, A., Khan, Z., & Lausen, B. (2014b). propOverlap: Feature (Gene) selection based on the proportional overlapping scores. R package version 1.0. http://CRAN.R-project.org/package=propOverlap

Mahmoud, O., Harrison, A., Perperoglou, A., Gul, A., Khan, Z., Metodiev, M. V., et al. (2014a). A feature selection method for classification within functional genomics experiments based on the proportional overlapping score. *BMC Bioinformatics, 15*, 274.

Malley, J., Kruppa, J., Dasgupta, A., Malley, K., & Ziegler, A. (2012). Probability machines: Consistent probability estimation using nonparametric learning machines. *Methods of Information in Medicine, 51*, 74–81.

Meinshausen, N. (2010). Node harvest. *The Annals of Applied Statistics, 4*, 2049–2072.

Platt, J. C. (2000). Probabilistic outputs for support vector machines and comparison to regularized likelihood methods. In A. J. Smola, P. Bartlett, B. Schölkopf, & D. Schuurmans (Eds.), *Advances in large margin classifiers* (pp. 61–74). Cambridge, MA: MIT Press.

R Core Team. (2014). R: A language and environment for statistical computing. http://www.R-project.org/

Ensemble of Subset of k-Nearest Neighbours Models for Class Membership Probability Estimation

Asma Gul, Zardad Khan, Aris Perperoglou, Osama Mahmoud, Miftahuddin Miftahuddin, Werner Adler, and Berthold Lausen

Abstract Combining multiple classifiers can give substantial improvement in prediction performance of learning algorithms especially in the presence of non-informative features in the data sets. This technique can also be used for estimating class membership probabilities. We propose an ensemble of k-Nearest Neighbours (kNN) classifiers for class membership probability estimation in the presence of non-informative features in the data. This is done in two steps. Firstly, we select classifiers based upon their individual performance from a set of base kNN models, each generated on a bootstrap sample using a random feature set from the feature space of training data. Secondly, a step wise selection is used on the selected learners, and those models are added to the ensemble that maximize its predictive performance. We use bench mark data sets with some added non-informative features for the evaluation of our method. Experimental comparison of the proposed method with usual kNN, bagged kNN, random kNN and random forest shows that it leads to high predictive performance in terms of minimum Brier score on most of the data sets. The results are also verified by simulation studies.

A. Gul (✉)
Department of Mathematical Sciences, University of Essex, Colchester, UK

Department of Statistics, Shaheed Benazir Bhutto Women University Peshawar, Khyber Pukhtoonkhwa, Pakistan
e-mail: agul@essex.ac.uk

Z. Khan • A. Perperoglou • M. Miftahuddin • B. Lausen
Department of Mathematical Sciences, University of Essex, Colchester, UK
e-mail: zkhan@essex.ac.uk; aperpe@essex.ac.uk; mmifta@essex.ac.uk; blausen@essex.ac.uk

O. Mahmoud
Department of Mathematical Sciences, University of Essex, Colchester, UK

Department of Applied Statistics, Helwan University, Cairo, Egypt
e-mail: ofamah@essex.ac.uk

W. Adler
Department of Biometry and Epidemiology, University of Erlangen-Nuremberg, Erlangen, Germany
e-mail: werner.adler@fau.de

© Springer International Publishing Switzerland 2016   411
A.F.X. Wilhelm, H.A. Kestler (eds.), *Analysis of Large and Complex Data*, Studies in Classification, Data Analysis, and Knowledge Organization,
DOI 10.1007/978-3-319-25226-1_35

1 Introduction

In numerous real-life applications, class membership probabilities of individuals are required in addition to their class labels. For example, in safety-critical domains such as surgery, oncology, internal medicine, pathology, paediatrics and human genetics, these probabilities are needed. In all the aforementioned areas, probability estimates are more useful than simple classification as they provide a measure of reliability of the decision taken about an individual (Lee et al. 2010; Malley et al. 2012; Kruppa et al. 2012, 2014a,b). Machine learning techniques used mainly for classification can be used as non-parametric methods for class membership probability estimation in order to avoid the assumptions imposed in parametric models used for the estimation of these probabilities (Kruppa et al. 2012; Malley et al. 2012).

In many real-life problems, one often encounters imprecise data such as data with non-informative features. These features dramatically decrease the prediction performance of the algorithms (Nettleton et al. 2010). Feature selection methods that investigate the most discriminative features from the original features are usually recommended to mitigate the effect of such non-informative features (Mahmoud et al. 2014a,b). However, different feature selection methods result in different feature subsets for the same data set thus varying feature relevancy. This encourages combining the results of several best feature subsets.

It has been investigated in the last two decades that combining the outputs of multiple models, known as ensemble techniques, results in improved prediction performance (Breiman 1996; Hothorn and Lausen 2003; Kuncheva 2004) and are more resilient to non-informative features in the data than using an individual model (Melville et al. 2004). Recently, an ensemble of optimal trees has been suggested for class membership probability estimation by Khan et al. (2015).

k-Nearest Neighbour (kNN) learning algorithm is one of the simplest and oldest methods. It classifies an unknown observation to the class of majority among its k-Nearest Neighbour points in the training data as measured by a distance metric (Cover and Hart 1967). Despite its simplicity, kNN gives competitive results and in some cases even outperforms other complex learning algorithms. However, kNN is vulnerable to non-informative features in the data. Attempts have been made by researchers to improve the performance of Nearest Neighbour algorithm by ensemble techniques (Bay 1998; Li et al. 2011; Samworth 2012). In this manuscript, we propose an ensemble of subset of kNN classifiers (ESkNN) for the task of estimating class membership probability, particularly in the presence of non-informative features in the data set and compare the results with those of simple kNN, bagged kNN (BkNN), random kNN (RkNN) and random forest (RF).

2 Proposed Ensemble of Subset of kNN Algorithm

To construct the ensemble of subset of kNN models (ESkNN), a two stage strategy is implemented. Consider a training data set of $n \times (d + 1)$ dimensions, consisting of data points $\mathscr{L} = (\mathbf{x}, \mathbf{y})$, an instance is characterized by d features along with the corresponding class label. The training data set $\mathscr{L}$ is randomly divided into two sets, a learning set and validation set and the ensemble is developed in the following steps.

1. Draw m random feature sets of size l from d input features, $l < d$ and draw m bootstrap samples on these feature sets from the learning set.
2. Build m kNN models and select h of the m models that give the highest accuracy on the out-of-bag observations (observations that are left out from the bootstrap sample).
3. In the next stage add the selected h models one by one starting from the best model and assess its collective performance on the validation set. The process is repeated until all the h models are evaluated in the ensemble.
4. A model is selected if it gives minimum Brier score (BS) on the validation set. Let $BS^{(r-1)}$ be the Brier score of the ensemble without the rth model and $BS^{(r)}$ is the Brier score after adding the model, the rth model is selected if

$$BS^{(r)} < BS^{(r-1)}. \tag{1}$$

5. The group membership probability estimate of the test instance is the averaged probability estimate over all t selected models.

A flow chart of the procedure of ESkNN is shown in Fig. 1.

3 Performance Measure of the Methods

As a performance measure, we use Brier score introduced by Brier (1950). It provides a measure of accuracy of the predicted probabilities. It is the most common and appropriate criterion for binary class outcome and can be used for the evaluation of predicted probabilities by a machine learning algorithm, in situations where the true probabilities are unknown (Malley et al. 2012). Gneiting and Raftery (2007) stated that the Brier score is a proper measure and its minimum value can only occur if the calculated probabilities are taken as the true probabilities which are unknown. It follows that a machine learning technique that has the smallest value of the Brier score will be performing best in estimating group membership probabilities. The Brier score is given as:

$$BS = E(y_i - p(y_i|\mathbf{x}))^2, \tag{2}$$

where $y_i \in \{0, 1\}$ and $p(y_i|\mathbf{x})$ is the true but unknown probability of the state of the outcome for y_i given the features. An estimator for the above score for t test

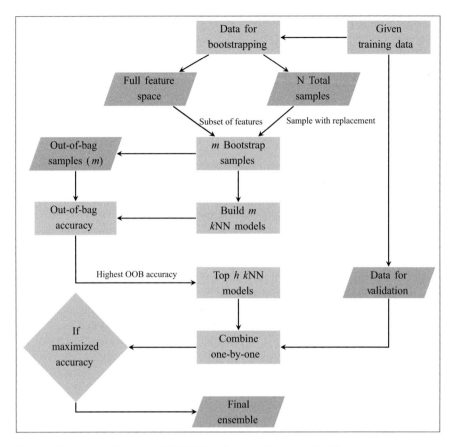

Fig. 1 A flow chart of the steps of *ESkNN* for class membership probability estimation

observations is:

$$\widehat{BS} = \frac{\sum_{i=1}^{t} (y_i - \hat{p}(y_i|\mathbf{x}))^2}{t}.$$ (3)

4 Results and Discussion

4.1 Experiments and Discussion on Bench Mark Problems

The performance of the ESkNN in terms of the Brier score, is evaluated on a total
of 25 data sets taken from UCI, KEEL databases and from within R-Libraries,
mlbench, mboost, ipred, gclus and mmst. Summary of the data sets is given in
Table 1.

Table 1 Summary of the data sets

Data Sets	Sample size	Features Feature type	(Continuous/Discrete/ Catagorical)
Adenocarcinoma	76	9869	(9869/0/0)
Prostate	102	6033	(6033/0/0)
Breast2	77	4869	(4869/0/0)
Leukaemia	38	3052	(3052/0/0)
Colon	61	2000	(2000/0/0)
nki70 breast cancer	144	77	(72/1/4)
Glaucoma M	198	62	(62/0/0)
Wpbc	194(198)	33	(31/2/0)
Body	507	24	(24/0/0)
Biopsy	683(699)	9	(0/9/0)
SAheart	462	9	(5/3/1)
Diabetes	768	8	(8/0/0)
Appendicitis	106	7	(7/0/0)
Bupa	345	6	(1/5/0)
Dystrophy	194	5	(2/3/0)
Mammographic	830(961)	5	(0/5/0)
Transfusion	748	4	(2/2/0)
Hepatitis	80	19	(2/17/0)
Indian liver patients	583	10	(5/4/1)
Haberman	306	3	(0/3/0)
Phoneme	1000	5	(5/0/0)
Two norms	1000	20	(20/0/0)
German credit	1000	20	(0/7/13)
House voting	435	16	(0/0/16)
Bands	365	19	(13/6/0)
Sonar	208	60	(60/0/0)

The first five data sets are from microarray studies

The ES*k*NN is assessed in two scenarios. Firstly, all the methods are applied on the data sets with their original features and secondly, the feature space of all the data sets are extended by adding 500 randomly generated non-informative features. The results for both the cases are given in Tables 2 and 3. Each of the data sets is divided into test and training parts where the training part consists of 90 % of observations and the remaining 10 % is reserved for testing. The methods are applied on each data set in a total of 1000 runs and are evaluated on the testing data set in each run. The final Brier score is the average of Brier scores of the 1000 runs. The experiments are carried out using the R-Program. The values of the hyper parameters for the methods are selected by using the "tune" function within the R-Package "e1071".

The results from Table 2 reveal that ES*k*NN is giving the smallest Brier scores on 23 data sets out of the total 25 data sets among all *k*NN-based methods, whereas on Bands data set it gives better probability estimate than *k*NN and B*k*NN and

Table 2 Brier scores on the data sets on five methods

Data Sets	kNN	BkNN	RkNN	ESkNN	RF
Haberman	0.199	0.197	0.181	**0.171**	0.199
Dystrophy	0.105	0.102	0.098	0.097	0.096
Mammographic	0.141	0.140	0.127	**0.115**	0.129
Transfusion	0.168	0.167	0.164	**0.160**	0.172
Bupa	0.221	0.215	0.217	0.215	**0.190**
Appendicitis	0.126	0.119	0.109	**0.105**	0.119
Diabetes	0.177	0.172	0.173	0.168	**0.156**
Biopsy	0.024	0.024	0.025	**0.021**	0.025
SAheart	0.209	0.207	0.203	0.200	**0.189**
Bands	0.235	0.231	0.207	0.208	**0.183**
German credit	0.216	0.214	0.201	0.178	**0.159**
Body	0.019	0.019	0.036	**0.012**	0.031
Wpbc	0.182	0.182	0.176	0.172	**0.168**
Sonar	0.179	0.179	0.109	**0.092**	0.127
Glaucoma M	0.147	0.144	0.142	0.130	**0.089**
Indian liver	0.191	0.189	0.179	**0.163**	0.174
Phoneme	0.130	0.128	0.130	0.121	**0.105**
Two norms	0.067	0.068	0.084	**0.029**	0.062
Hepatitis	0.310	0.259	0.209	0.221	**0.195**
House voting	0.065	0.065	0.065	0.044	**0.030**
Colon	0.145	0.144	0.139	0.138	**0.129**
Leukaemia	0.030	0.030	0.062	**0.027**	0.054
Breast2	0.243	0.241	0.233	0.230	**0.210**
Prostate	0.138	0.137	0.145	0.100	**0.084**
Adenocarcinoma	0.126	0.119	0.119	**0.114**	0.125

Smallest Brier scores are indicated by bold numbers

comparable to RkNN. When comparing to random forest it gives low Brier scores on 10 data sets.

In case of non-informative features in the data sets from Table 3, the ESkNN outperforms kNN-based methods on most of the data sets. Comparing to random forest it gives low Brier scores on 10 data sets. These results indicate that the ESkNN is better than the kNN and kNN-based methods and comparable to random forest.

ESkNN is evaluated for various values of k, the number of nearest neighbours and m, the number of models in the initial ensemble. Figure 2 reveals varied behaviour of ESkNN on different data sets for the choice of k and m. It is recommended to fine tune the value of k by cross validation, for example. Figure 2b shows that a very small number of models are not reasonable and a very large number of models might be computationally expensive hence a moderate number of models is recommended.

Table 3 Brier scores of the methods for the data sets with added non-informative features

Data sets	*k*NN	B*k*NN	R*k*NN	ES*k*NN	RF
Haberman	0.204	0.202	0.196	**0.191**	0.196
Dystrophy	0.158	0.172	0.220	0.149	**0.118**
Mammographic	0.153	0.160	0.231	0.139	**0.123**
Transfusion	0.187	0.186	0.180	**0.160**	0.166
Bupa	0.229	0.228	0.243	**0.222**	0.230
Appendicitis	0.143	0.142	0.145	0.139	**0.132**
Diabetes	0.240	0.236	0.225	0.216	**0.173**
Biopsy	0.053	0.052	0.067	0.048	**0.029**
SAheart	0.252	0.247	0.228	0.225	**0.218**
Bands	0.237	0.235	0.222	**0.213**	0.221
German credit	0.218	0.216	0.210	0.208	**0.182**
Body	0.082	0.082	0.107	0.078	**0.065**
Wpbc	0.196	0.190	0.180	**0.179**	0.181
Sonar	0.164	0.139	0.201	**0.104**	0.193
Glaucoma M	0.157	0.156	0.212	**0.121**	0.135
Indian liver	0.198	0.199	0.201	**0.183**	0.189
Phoneme	0.174	0.170	0.236	**0.154**	0.163
Two norms	0.126	0.084	0.203	**0.082**	0.124
Hepatitis	0.239	0.230	0.239	**0.223**	0.234
House voting	0.135	0.134	0.212	0.127	**0.103**

Smallest Brier scores are indicated by bold numbers

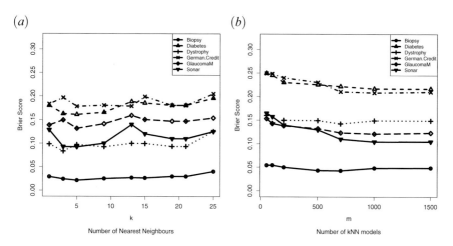

(*a*) (*b*)

Fig. 2 Performance of ES*k*NN in presence of non-informative features in the data for; (**a**): different values of *k*, (**b**): different values of *m*

4.2 Simulation Study

We evaluate the predictive performance of ES*k*NN by simulation study in addition to the benchmark data sets. We used two examples in our simulation study. The models proposed in our simulation study involve several variations to gain an understanding of the behaviour of the methods under different situations.

4.2.1 Simulation Model 1

In the first model, Model 1, binary class data is generated on 20 features. The features for class 1 are generated from $\mathcal{N}(\mathbf{2}, w\Psi)$, while those of class 2 are generated from $\mathcal{N}(\mathbf{1}, \mathbf{1})$. The values considered for w in class 1 are 3, 5, 10, 15 and 20. The predictive performance of the algorithms are investigated by adding 50, 100, 200 and 500 non-informative features, generated from normal distribution, to the data.

$$\Psi = \begin{pmatrix} \sigma_{1,1} & \varrho_{1,2} & , \ldots, & \varrho_{1,d} \\ \varrho_{2,1} & \sigma_{2,2} & , \ldots, & \varrho_{2,d} \\ \vdots & \vdots & \vdots & \vdots \\ \varrho_{d,1} & \varrho_{d,2} & , \ldots, & \sigma_{d,d} \end{pmatrix} \tag{4}$$

where ϱ_{ij} are the covariances between the features defined as:

$$\varrho_{ij} = (1/2)^{|i-j|}, i,j = 1, \ldots, d, \tag{5}$$

and $\sigma_{ij} = 1$ for $i = j$. The variables within class 1 are correlated among each other and are exhibiting negligible/no correlation with the features of class 2.

4.2.2 Simulation Model 2

The second simulation model developed here is a four-dimensional model, derived from the model proposed by Mease et al. (2007). The feature vector $\mathbf{x}$ is a random vector uniformly distributed over $[0, 100]$. The class is determined by the distance r, the distance of the feature vector $\mathbf{x}$ from the central point. The class probabilities given the features are:

$$p(y = 1 \mid \mathbf{x}) = \begin{cases} 1 & r < 110, \\ \frac{150-r}{140} & 110 \leq r \leq 140, \\ 0 & \text{otherwise.} \end{cases}$$

The binary response variable y is generated from the above distribution using a binomial random number generator. We extended the dimensions of the data by adding 50,100, 200 and 500 randomly generated non-informative feature.

4.3   Simulation Results and Discussion

The results from Table 4 reveal that ES*k*NN consistently outperform the other methods. In case of different values of w to the data in Model 1, as shown in Table 4, random forest outperforms all the other methods. However, in *k*NN-based methods the ES*k*NN consistently gives higher accuracy than *k*NN, B*k*NN and R*k*NN.

The Brier scores from Model 2 given in Table 5, show that ES*k*NN consistently outperforms *k*NN, B*k*NN, R*k*NN and RF for the data with original four features and added 50, 100, 200 and 500 features (Fig. 3).

Table 4 Brier score of the five methods with added non-informative features to the data set from Model 1 and different values of w on 70 features (20 + 50 non-informative) shown in first column

Features	*k*NN	B*k*NN	R*k*NN	ES*k*NN	RF
20	0.042	0.041	0.087	**0.039**	0.071
20+50	0.066	0.079	0.086	**0.060**	0.081
20+100	0.081	0.076	0.095	**0.061**	0.086
20+200	0.103	0.094	0.095	**0.062**	0.092
20+500	0.137	0.130	0.088	**0.061**	0.113
w	*k*NN	*Bk*NN	*Rk*NN	*ESk*NN	*RF*
3	0.198	0.151	0.155	0.102	**0.081**
5	0.221	0.191	0.136	0.101	**0.062**
10	0.222	0.186	0.099	0.081	**0.038**
15	0.251	0.172	0.089	0.057	**0.028**
20	0.256	0.159	0.062	0.043	**0.022**

The best result is highlighted in bold

Table 5 Brier score of the methods on the data from Model 2 with the added non-informative features

Features	*k*NN	B*k*NN	R*k*NN	ES*k*NN	RF
4	0.101	0.101	0.145	**0.090**	0.112
4+50	0.158	0.157	0.185	**0.146**	0.176
4+100	0.165	0.164	0.190	**0.152**	0.186
4+200	0.179	0.178	0.196	**0.162**	0.177
4+500	0.188	0.182	0.209	**0.151**	0.180

Results of the best performing method is highlighted in bold

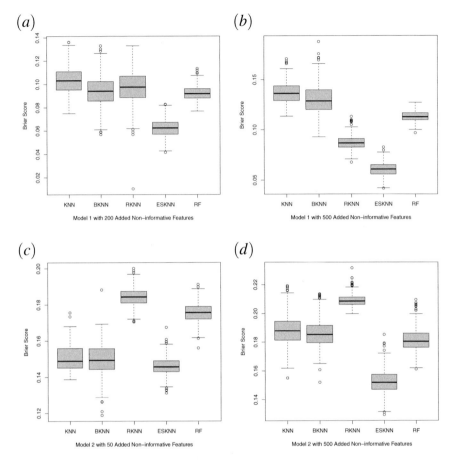

Fig. 3 Brier score, of simulated data from Model 1 (**a**, **b**) and Model 2 (**c**, **d**) for the five classifiers *k*NN, B*k*NN, R*k*NN, ES*k*NN and RF with added non-informative features to the data

5   Conclusion

We proposed an ensemble of subset of *k*NN models, ES*k*NN, for class membership probability estimation. The ES*k*NN improves the predictive performance of *k*NN-based methods. The ES*k*NN reveals better predictive performance than the *k*NN, bagged *k*NN and random *k*NN in most of the cases (both in bench marking and simulation) and gives comparable results to random forest. The performance of ES*k*NN is also evaluated in order to deal with the issue of non-informative features in the data. The results demonstrate that the ES*k*NN provides better estimates of class membership probability than the other methods considered in the presence of non-informative features in the data. Besides performance improvement, the

ES*k*NN as using *k*NN classifier is simple in implementation and interpretation. The ES*k*NN is implemented in an R-package, **ES*k*NN**.

References

Bay, S. (1998). Combining nearest neighbor classifiers through multiple feature subsets. In *Proceedings of the Fifteenth International Conference on Machine Learning* (Vol.3, pp. 37–45).

Breiman, L. (1996): Bagging predictors. *Machine Learning, 24*(2), 123–140.

Brier, G. W. (1950). Verification of forecasts expressed in terms of probability. *Monthly Weather Review, 78*, 1–3.

Cover, T., & Hart, P. (1967). Nearest nieghbor pattern classification. *IEEE Transaction on Information Theory, 13*, 21–27.

Gneiting, T., & Raftery, A. E. (2007). Strictly proper scoring rules, prediction, and estimation. *Journal of the American Statistical Association, 102* 359–378.

Hothorn, T., & Lausen, B. (2003). Double-bagging: Combining classifiers by bootstrap aggregation. *Pattern Recognition, 36*(9), 1303–1309.

Khan, Z., Perperoglou, A., Gul, A., Mahmoud, O., Adler, W., Miftahuddin, M., & Lausen, B. (2015). An ensemble of optimal trees for class membership probability estimation. In *Proceedings of European Conference on Data Analysis.*

Kruppa, J., Liu, Y., Biau, G., Kohler, M., Konig, I. R., Malley, J. D., et al. (2014a). Probability estimation with machine learning methods for dichotomous and multicategory outcome: Theory. *Biometrical Journal, 56*, 534–563.

Kruppa, J., Liu, Y., Diener, H. C., Weimar, C., Konig, I. R., & Ziegler, A. (2014b). Probability Estimation with machine learning methods for dichotomous and multicategory outcome: applications. *Biometrical Journal, 56*, 564–583.

Kruppa, J., Ziegler, A., & Konig, I. R. (2012). Risk estimation and risk prediction using machine-learning methods. *Human Genetics, 131*, 1639–1654.

Kuncheva, L. I.(2004). *Combining pattern classifiers. Methods and algorithms.* New York: Wiley.

Lee, B. K., Lessler, J., & Stuart, E. A. (2010). Improving propensity score weighting using machine learning. *Statistics in Medicine, 29*, 337–346.

Li, S., Harner, E. J., & Adjeroh, D. (2011). Random knn feature selection a fast and stable alternative to random forests. *BMC Bioinformatics, 12*(1), 450.

Mahmoud, O., Harrison, A., Perperoglou, A., Gul, A., Khan, Z., & Lausen, B. (2014b). Propoverlap: Feature (gene) selection based on the Proportional Overlapping scores. R package version 1.0, http://CRAN.R-project.org/package=propOverlap

Mahmoud, O., Harrison, A., Perperoglou, A., Gul, A., Khan, Z., Metodiev, M. V., et al. (2014a). A feature selection method for classification within functional genomics experiments based on the proportional overlapping score. *BMC Bioinformatics, 15*, 274.

Malley, J., Kruppa, J., Dasgupta, A., Malley, K., & Ziegler, A. (2012). Probability machines: Consistent probability estimation using nonparametric learning machines. *Methods of Information in Medicine, 51*, 74–81.

Mease, D., Wyner, A. J., & Buja, A. (2007). Boosted classification trees and class probability/quantile estimation. *The Journal of Machine Learning Research, 8*, 409–439.

Melville, P., Shah, N., Mihalkova, L., & Mooney, R. (2004). Experiments on ensembles with missing and noisy data. *Multiple Classifier Systems, 53*, 293–302.

Nettleton, D. F., Orriols-puig, A., & Fornells, A. (2010). A Study of the effect of different types of noise on the precision of supervised learning techniques. *Artificial Intelligence Review, 33*(4), 275–306.

Samworth, R. J. (2012). Optimal weighted nearest neighbour classifiers. *The Annals of Statistics, 40*(5), 2733–2763.

Part X
Data Analysis in Musicology

The Surprising Character of Music: A Search for Sparsity in Music Evoked Body Movements

Denis Amelynck, Pieter-Jan Maes, Marc Leman, and Jean-Pierre Martens

Abstract The high dimensionality of music evoked movement data makes it difficult to uncover the fundamental aspects of human music-movement associations. However, modeling these data via Dirichlet process mixture (DPM) Models facilitates this task considerably. In this paper we present DPM models to investigate positional and directional aspects of music evoked bodily movement. In an experimental study subjects were moving spontaneously on a musical piece that was characterized by passages of extreme contrasts in physical acoustic energy. The contrasts in acoustic energy caused surprise and triggered new gestural behavior. We used sparsity as a key indicator for surprise and made it visible in two ways. Firstly as the result of a positional analysis using a Dirichlet process gaussian mixture model (DPGMM) and secondly as the result of a directional analysis using a Dirichlet process multinomial mixture model (DPMMM). The results show that gestural response follows the surprising or unpredictable character of the music.

1 Introduction

Several authors suggested that humans perceive something as aesthetically interesting when there is a balanced mixture between recognition and surprise (Birbaumer et al. 1996). In 1933, Birkhoff was one of the first to present a mathematical theory for aesthetic measures, which he defined as the ratio of order(O) to complexity(C) (Birkhoff 1933). The idea that surprise is related to aesthetic feeling fully resonates with known theories of music processing and emotional arousal (Meyer 1956; Berlyne 1971; Huron 2006).

Surprise is often intended and in music it has a strong power to arouse listeners. Mayer-Kress et al. (1994) drew an analogy between musical structures and recurrence structures in chaotic systems. He stated that: "Perceived order and disorder, recurrence and complexity are common features observed in both

D. Amelynck (✉) • P.-J. Maes • M. Leman • J.-P. Martens
Ghent University, Gent, Belgium
e-mail: denis.amelynck@gmail.com; denis.amelynck@ugent.be; maes.pieterjan@gmail.com; marc.leman@ugent.be; jeanpierre.martens@ugent.be

© Springer International Publishing Switzerland 2016   425
A.F.X. Wilhelm, H.A. Kestler (eds.), *Analysis of Large and Complex Data*, Studies in Classification, Data Analysis, and Knowledge Organization,
DOI 10.1007/978-3-319-25226-1_36

chaos and music. These features can be perceived in music because the music has been intentionally designed to reveal them." An extreme example is the famous Symphony No. 94 in G major (Hoboken 1/94) written by J. Haydn, also known as the Surprise Symphony. Haydn was reputed for this type of surprises, and the Surprise Symphony is exemplary in that it contains a sudden fortissimo chord at the end of a piano opening theme in the variation-form second movement. The music then returns to normal and subsequent movements do not repeat the surprise. And this brings us to a key indicator of surprise and that is sparsity. Sparsity is a major attribute of many descriptions of surprise (e.g., Huron 2006; Margulis 2007; Itti and Baldi 2005; Keogh et al. 2002).

Based on the key insights that cognition is situated and embodied (Clark 1997; Leman 2008) we assume that the surprising character of the music gets embodied in the movement idiosyncrasies of subjects. A cognitive system, such as the human mind, is always interacting with its environment via its sensors that perceive, and effectors that produce actions. For listeners and dancers, surprises, or failures to anticipate, afford new opportunities (named gestural affordances) to move along with the music (Heylighen 2012; Godøy 2009).

The paper is organized as follows. In Sect. 2 we describe the experiment that is at the basis of our research. Section 3 describes the methods of analysis. The results are presented in Sect. 4. For a conclusion we refer to Sect. 5.

2 Experimental set-up

- *Subjects and Task.*

    Thirty-six subjects were participated in a music evoked body movement experiment [20 males and 16 females with a mean age of 24.2 year (SD=4.2)]. The experiment was set-up on a per individual basis. Before the actual execution of the experiment, the participant received the task of moving spontaneously to the music. This was formulated as: "Translate your experience of the music into free full-body movement. Try to become absorbed by the music that is presented and express your feelings into body movement. There is no good or wrong way of doing it. Just perform what comes up in you." The actual *motor-attuning* experiment took place in a motion capture space: an octagonal space with a diameter of 4 m enclosed by black curtains to separate the participant from the experimenters.

- *Stimuli.*

    The music was part of Johannes Brahms' *First Piano Concerto*, Opus 15 in D minor. This piece is characterized by passages articulating extreme contrasts in physical acoustic energy, symbol for the surprising character of the music. Based on this, we define two contrasting musical style categories which structure the main outline of the composition, namely a Heroic and Lyric style category. In the stimulus three Heroic passages were presented in alternation with three Lyric passages.

- *Data recording.*

  Registration of movement data for the complete upper body was realized at a sample rate of 100 Hz with an OPTITRACK infrared optical system. Participants were asked to wear a special jacket and cap with 22 infrared reflecting markers attached: four markers for hip, three markers each for head, chest, upper arms, and hands.

3 Analysis Method

3.1 Pre-Processing of the Data

Although we collected data from multiple markers, the analysis focused on the movement data from the hand as it is the body part with the highest degree of freedom (DOF). To eliminate the influences from other body parts (like translations and rotations of torso and/or shoulders) a new three dimensional axis system was defined as in Fig. 1:

The data for the directional analysis are based upon the velocity signals, calculated as derivatives from the positional data. To calculate these derivatives a local (linear) *derivation* filter is applied to the positional data. The size of the filter window is set at 0.175 s corresponding with a linear frequency response of the derivation filter in the useful frequency band of 0–4 Hz. The 0–4 Hz range is in line with the information from the spectrogram (Fig. 2).

3.2 Feature Space

The feature space for positional analysis consists of the positional coordinates (Cartesian coordinates x–y–z). The feature space for directional analysis is calculated from the velocity vectors. Velocity vectors are converted to spherical coordinates (radius, elevation, and azimuth), used to categorize directional information. Categorization is done with the help of two indicators. A first indicator comes from the elevation ($[\frac{-\pi}{2}, \frac{\pi}{2}]$) and divides the elevation range in four quadrants of $\frac{\pi}{4}$. A second indicator is derived from the azimuth ($[-\pi, \pi]$) dividing its range in eight octants of again $\frac{\pi}{4}$. The combination of these two indicators results in total in 32 categories. In addition, we create one category labeled "lack of movement." The criterion for lack of movement is based on low speed as indicated by the radius of the velocity vector. The decision border for low speed values was set per subject in such a way that 5 % of the values would be categorized as lack of movement.

Because of the degree of randomness or should we say chaos (Sprott 2003) in music evoked body movement, we do not look at directional categories at distinct time stamps but at category mixtures over a limited time interval. The time

Fig. 1 Relative axis system
used for hand representation

intervals are set at 3 s conforming to Pöppels' theory of the 3 s window of temporal integration (Pöppel 1989). To avoid artifacts we work with 50 % overlapping windows.

3.3 Dirichlet Process Models

The analysis uses Dirichlet process mixture (DPM) models to cluster the data. DPMs have an advantage that they learn the number of clusters from the data. This is in contrast with algorithms like K-means clustering or gaussian mixture models (GMMs) where the number of clusters has to be specified upfront or has to be determined by additional validation steps. For the positional analysis we fit a Dirichlet process gaussian mixture model (DPGMM). For the directional analysis a Dirichlet process multinomial mixture model (DPMMM) is applied. For readers not familiar with Dirichlet Process Models we refer to the existing literature (see,

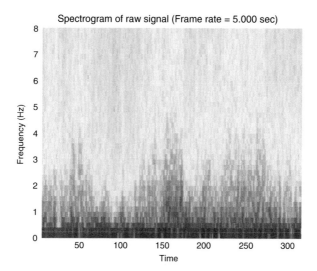

Fig. 2 Spectrogram of hand positional data

e.g., Teh 2010 and El-Arini 2008). The practical implementation was done in Matlab with the help of the demo programs from Yee Whye Teh (http://www.stats.ox.ac.uk/~teh/).

4   Results

The data from one subject were discarded due to technical problems during the recording.

4.1   DPGMM for Positional Analysis

The musical excerpt of Brahms was split-up in six fragments, namely, three heroic style fragments alternating with three lyric fragments. The DPGMM clustering analysis was performed for every combination of subject and fragment. For example Fig. 3 shows the data points collected for subject 2 fragment 2 (a lyric fragment). In this particular case the movement in terms of position can be described by a three cluster system.

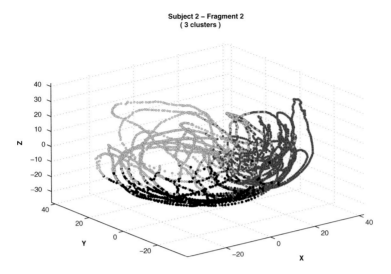

Fig. 3 Hand movement: Clustered data

4.1.1 Analysis of Small Data Clusters

Small clusters stand for sparse movement and might link to surprising, salient events
in the music. We defined a small cluster as clusters of maximum 300 data points.
This corresponds to a three second time interval if all the data points are adjacent in
time. Our assumption is that because these points are close in space (belonging to the
same cluster) they are close in time, as human movement is continuous and smooth.
This means that most of these small clusters represent small abnormal moves.

An interesting question is whether subjects made these moves at the same time.
That would point to some effect in the music that triggers these sudden (surprising)
events. We noticed four moments in time where this happened for at least five
subjects. Timestamps were at 5.2 s–57.1 s–102.4 s–300.6 s. For 5.2 s (warm-up?)
and 57.1 s we find no obvious explanation in the music but intriguing is that we
notice a similar event at 102.4 s and 300.6 s. There we localize a change in the
harmonic structure of the music with a major cord (happy) changing into a minor
chord (sad).

4.1.2 Analysis of Large Data Clusters

To understand sparsity we must also understand what is common. Therefore it is
instructive to study the large clusters as well. We defined clusters as large clusters if
they contained at least 10 % of the data points of a particular fragment.

Visual inspection (Fig. 4) learns that the centers of these clusters are located
on the surface of an ellipsoid. Unfortunately these locations are concentrated on

Fig. 4 The centers of the large clusters are near the surface of an ellipsoid

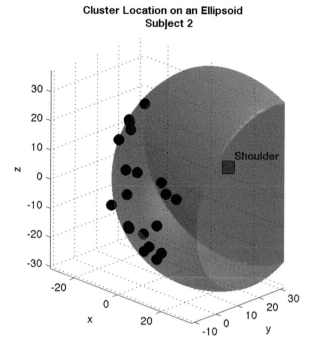

a limited area of the ellipsoid what makes it difficult for a fitting algorithm. If movement happens on an ellipsoid then only two coordinates are required for specification. According to the mathematical definition of dimension, this movement is then two dimensional and not three dimensional. It is our hypothesis that surprises in music can cause movement to suddenly enter a higher dimension but more research is required to confirm this.

Investigation of the covariance matrices learns that one eigenvalue is considerably smaller than the two others. Averaged per subject we see that this eigenvalue explains only about 8 % of the variance ($M = 8.10$, SD=1.20). In other words the movement of the hand is locally (centered at the cluster) rather two dimensional than three dimensional. The orientation of this two dimensional plane can be visualized by looking at the orientation of the eigenvector with the smallest eigenvalue as the two dimensional plane is perpendicular to this eigenvector. Figure 5 shows that the direction of these eigenvectors points to a central point near the shoulder. Variance in that direction corresponds with punching movements (from the body away and back). As this is the direction with the lowest variance (smallest eigenvalue) we can say that this type of movement was almost absent in our experiment.

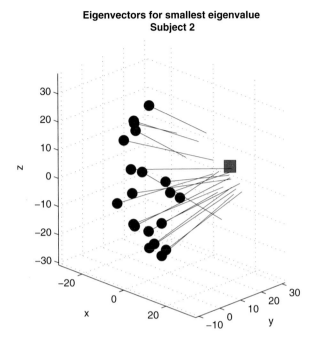

Fig. 5 The direction of the eigenvectors corresponding with the smallest eigenvalues are pointing to a central point close the shoulder (*square marker*)

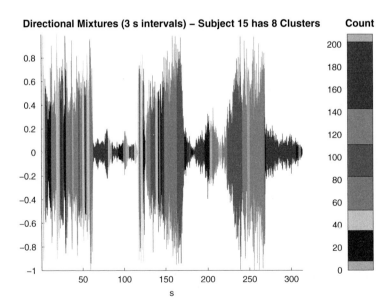

Fig. 6 Cluster assignment for subject 15, visualized on *top* of the musical amplitude. Each cluster stands for movement with the same directional mix

4.2   DPMMM for Directional Analysis

The task of the DPMMM analysis is to cluster the directional categorical mixtures as explained in Sect. 3.2. Figure 6 shows the clustering result for subject 15. As our interest lays in the relationship with music, the cluster assignment is displayed with the help of the musical amplitude.

For this subject, DPMMM assigned the time intervals to in total 8 different clusters. This means that the subject's movement style can be reduced to eight different ways of moving (direction-wise). In particular, the first and second heroic fragment show what we call sparse behavior. The clusters alternate there in a fast sequence. The third heroic interval however does not show this behavior and gives the impression that the subject is not anymore surprised by the music. These findings are based upon the results from a single subject. Question is, if we can generalize these results?

A way of consolidating is to bundle the results of all subjects in a single diagram, in a what we call a *directogram*. A *directogram* is a visual representation of the gestural affordances (direction-wise) in a musical excerpt. It represents a square matrix calculated as follows: If for example interval 17 and interval 24 belong for one subject to the same cluster we increase the value of element (17,24) of the square matrix by one. We loop then over all subjects and display the resulting matrix in a kind of a correlation plot (Fig. 7).

Lighter colors indicate that more subjects were moving their hand similarly (intra-subject) at the timestamps given by the horizontal and vertical index. Lighter colored areas appear in rectangles related to the musical structure of lyric and heroic style intervals. Rectangles across the diagonal depict a phenomenon that we define as persistency. It corresponds with continuous time intervals where for every subject a particular cluster (a direction-mix) dominates. Persistency is mostly present in the lyric style intervals. Next, we introduce the concept of consistency. Consistency is visible as off-diagonal high density areas. These areas appear here only at time intervals matching the lyric style intervals. This tells that all lyric time intervals are not only dominated by one single cluster but also that this cluster is also identical for all lyric intervals.

Consequently, darker colors stand for absence of movement structure and hence are indicators for the amount of surprise.

5   Conclusion

In this experiment, we analyzed a group of subjects moving spontaneously and individually to music. The idea was to search for sparsity, sparsity being a secondary indicator of surprise in music. The method applied Dirichlet Process Mixture (DPM) models and identified sparsity in positional and directional attributes of movement

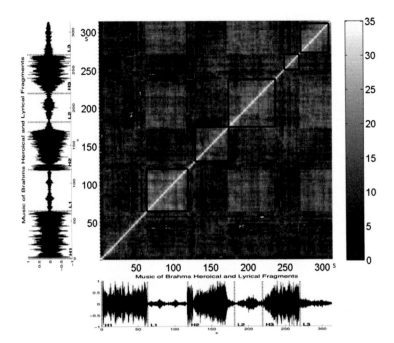

Fig. 7 Directogram: reveals "directional movement" characteristics of a musical excerpt. Persistency, along the diagonal and shown by *light colored areas* with *black borders*, answers questions like how long do we move similarly (in terms of direction). Consistency, off-diagonal *light colored areas*, compares remote intervals

data. The time stamps of sparsity could be linked to moments of surprise in the music.

The present experiment was executed with subjects moving on music of Brahms. Future work could include other, even modern musical styles.

Our methods are not limited to music evoked body movement but can be extended to other fields were sparsity (in movement) has to be measured. We think for example of applications in sports analysis and rehabilitation.

References

Birbaumer, N., Lutzenberger, W., Rau, H., Braun, C., & Mayer-Kress, G. (1996). Perception of music and dimensional complexity of brain activity. *International Journal of Bifurcation and Chaos, 6*(2), 267–278.

Birkhoff, G. D. (1933). *Aesthetic measure*. Cambridge, MA: Harvard University Press.

Berlyne, D. E. (1971). *Aesthetics and psychobiology*. New York: Appleton-Century-Crofts.

Clark, A. (1997). *Being there: Putting brain, body, and world together again.* Cambridge, MA: The MIT Press.

El-Arini, K. (2008): Dirichlet Process. A gentle tutorial. *Select Lab Meeting, 10*.

Godøy, R. I. (2009). Gestural affordances of musical sound. In R. I. Godøy & M. Leman (Eds.), *Musical gestures: Sound, movement, and meaning* (Vol. 5, pp. 103–125). New York: Routledge

Heylighen, F. (2012). Brain in a vat cannot break out. *Journal of Consciousness Studies, 19*(1–2), 1–2.

Huron, D. (2006). *Sweet anticipation: Music and the psychology of expectation*. Cambridge, MA: MIT Press.

Itti, L., & Baldi, P. F. (2005). Bayesian surprise attracts human attention. In *Advances in Neural Information Processing Systems, 2005* (pp. 547–554).

Keogh, E., Lonardi, S., & Chiu, B. Y. C. (2002). Finding surprising patterns in a time series database in linear time and space. In *Proceedings of the Eighth ACM SIGKDD International Conference on Knowledge Discovery and Data Mining* (pp.550–556). New York: ACM.

Leman, M. (2008). *Embodied music cognition and mediation technology*. Cambridge, MA: The MIT Press.

Margulis, E.H.(2007). Surprise and listening ahead: Analytic engagements with musical tendencies. *Music Theory Spectrum, 29*(2), 197–217.

Mayer-kress, G., Bargar, R., & Choi, I. (1994). Musical structures in data from chaotic attractors. In *Santa Fe Institute Studies in the Sciences of Complexity - Proceedings* (Vol. 18, pp. 341–341).

Meyer, L. B. (1956). *Emotion and Meaning in Music*. Chicago: The University of Chicago Press.

Pöppel, E. (1989): The measurement of music and the cerebral clock: A new theory. *Leonardo, 22*(1), 83–89.

Sprott, J. C. (2003): *Chaos and time-series analysis*. Oxford: Oxford University Press.

Teh, Y. W. (2010). Dirichlet process. In *Encyclopedia of machine learning* (pp. 280–287). New York: Springer

Comparing Audio Features and Playlist Statistics for Music Classification

Igor Vatolkin, Geoffray Bonnin, and Dietmar Jannach

Abstract In recent years, a number of approaches have been developed for the automatic recognition of music genres, but also more specific categories (styles, moods, personal preferences, etc.). Among the different sources for building classification models, features extracted from the audio signal play an important role in the literature. Although such features can be extracted from any digitised music piece independently of the availability of other information sources, their extraction can require considerable computational costs and the audio alone does not always contain enough information for the identification of the distinctive properties of a musical category. In this work we consider playlists that are created and shared by music listeners as another interesting source for feature extraction and music categorisation. The main idea is that the tracks of a playlist are often from the same artist or belong to the same category, e.g. they have the same genre or style, which allows us to exploit their co-occurrences for classification tasks. In the paper, we evaluate strategies for better genre and style classification based on the analysis of larger collections of user-provided playlists and compare them to a recent classification technique from the literature. Our first results indicate that an already comparably simple playlist-based classifiers can in some cases outperform an advanced audio-based classification technique.

1 Introduction

Many studies in the research field of music information retrieval (MIR) are aimed at the automated classification or categorisation of digital musical tracks. Having the available tracks automatically categorised allows us to build better applications which, e.g. recommend music that matches the user's favorite style, help users

I. Vatolkin (✉) • D. Jannach
Department of Computer Science, TU Dortmund, Dortmund, Germany
e-mail: igor.vatolkin@tu-dortmund.de; dietmar.jannach@tu-dortmund.de

G. Bonnin
LORIA, Nancy, France
e-mail: geoffray.bonnin@loria.fr

© Springer International Publishing Switzerland 2016
A.F.X. Wilhelm, H.A. Kestler (eds.), *Analysis of Large and Complex Data*, Studies
in Classification, Data Analysis, and Knowledge Organization,
DOI 10.1007/978-3-319-25226-1_37

organise their music collection based on genres, or are even capable to automatically extract semantic properties of individual musical pieces.

One of the most prominent classification scenarios is the recognition of genres and many efforts were spent on the improvement of such systems: Sturm (2014), for example, lists several hundred references. Other categorisation goals mentioned in the literature include the identification of emotions (Yang and Chen 2011), the recommendation of new music (Celma 2010), or the prediction of listener tags (Bertin-Mehieux et al. 2008); a number of further applications are described in Weihs et al. (2007).

1.1   The Music Classification Workflow

A typical algorithm chain for music categorisation comprises the following steps: (1) feature extraction, (2) feature processing, and (3) building classification models based on training examples.

Feature Extraction: As a first step, a set of typically numerical characteristics, or features, has to be chosen to represent the music data. The typical sources for the extraction of features for music data analysis are audio content, music score, music context, and user context (Serra et al. 2013).

Feature Processing: In the second step, the extracted features are further processed. These processing steps can serve different technically required purposes like data normalisation or the imputation of missing values. In addition, feature processing steps like feature selection or transforms to lower-dimensional spaces can aim at the improvement of the classification quality or at the reduction of computation costs.

Model Building: Finally, the resulting features can be used to build classification models on some training data (labels indicating the classes of some observations). Alternatively, unsupervised learning techniques can be applied to cluster the data based on the estimated distances between data instances in the feature space.

1.2   Using Playlists for Categorisation

Building classification models from audio features is probably the most common approach in the MIR literature. When using audio signals, the extractable characteristics often describe properties of time, spectrum, cepstrum, autocorrelation, phase, etc. Music classification with audio features was applied for example in Tzanetakis and Cook (2002) or Mierswa and Morik (2005); for an overview of commonly used features see, e.g., Theimer et al. (2008), or the regularly updated manual of the MIR Toolbox (Lartillot and Toivainen 2007).

Such approaches have the advantage that the features needed for the categorisation can be extracted from a digital music piece independently of the availability of any additional (meta-)information about it. However, relying only on audio features can have some disadvantages. First, the extraction of features from the musical signal can be computationally costly (Blume et al. 2008). Even if these computations have to be only done once and the task can be parallelised, the sheer size of today's music collections leaves this task still challenging. Furthermore, it is often still hard to robustly extract meaningful and "interpretable" properties of the musical tracks as sometimes music with similar audio characteristics is perceived as being different by the listeners, e.g. because of their cultural background. Alternative data sources for feature extraction mentioned in the literature include for example the musical score. Such data may however be hard to obtain for all considered tracks, in particular in the area of popular music.

The recent developments in the area of online music services and music- related platforms, however, opened new opportunities for researchers, as vast amounts, e.g. of user generated content annotations or listener preference information became available to be used in classification or music recommendation tasks (Hariri et al. 2012). The work presented in this paper continues these lines of research of using user-provided (Social Web) content. Specifically, we propose to use playlists that were created and shared by users on music platforms as a data source for the classification task and present a method that relies on artist co-occurrences in the playlists to derive labelled training data. These data vectors can then be used by various machine learning techniques to build models for music classification. To the best of our knowledge, the usage of user-created playlists as input for music classification has not been explored in the literature so far. To assess the classification quality, we compare our results with those that were obtained with a recent and optimised approach that relies on the audio signal for categorisation (Vatolkin 2013).

The paper is organised as follows. In Sect. 2, we describe the rationale and the technical details of our novel approach to use user-provided playlists as a source for music classification. Section 3 presents the design of our comparative evaluation and discusses the results that were observed for different musical genres and styles. In the final section, we provide an outlook on opportunities for future research in particular with respect to the combination of different data sources as was done, for example, in Lidy et al. (2007) or Mckay (2010).

2 Using Playlist Statistics for Feature Extraction

Our approach is based on the assumption that homogeneity is a major quality criterion for people creating playlists as discussed in Fields (2011) and that the tracks of a playlist are correspondingly somehow similar to each other. With respect to the classification problem, we therefore assume that the presence of a given music piece in a given playlist implies a higher probability that the other songs in this list belong to the same or a similar category.

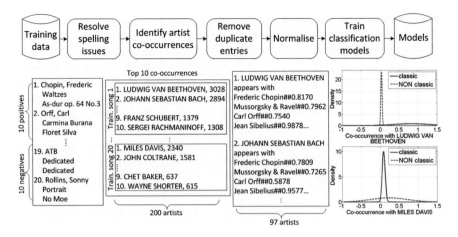

Fig. 1 Overview of algorithm steps for the extraction of playlist statistics

However, instead of relying on individual and possibly rare track co-occurrences, we propose to rather look at artist (composer, interpret) co-occurrences in the playlists. Given the artist of an unknown track, our goal is thus to use a machine learning model that is trained based on the information about frequently co-occurring artists for the categorisation of the track.

In the following, we describe a proposal of how to process a collection of user-provided playlists in a way that arbitrary classification algorithms like Support Vector Machines or Decision Trees can be applied. To achieve this goal, we have to derive *feature* vectors from the playlist data, which together with labelled training data points can be fed into supervised machine learning algorithms.

Figure 1 provides an overview of the steps required in our approach (top of the figure) and gives an example for the category "classic" (bottom of the figure). Our method has five steps: (1) Resolving spelling problems, (2) Identifying relevant artist co-occurrences, (3) Removing duplicates, (4) Normalisation and (5) Training of classification models.

2.1 Resolving Spelling Issues

A prerequisite to the computation of the co-occurrences of the tracks in the playlists is to correctly identify the tracks. As user-provided playlists often contain spelling mistakes, we applied a simple adaptation of the Smith–Waterman algorithm (Smith and Waterman 1981) on the artist and track spellings. This algorithm was originally designed for DNA sequence alignment and computes a distance between two sequences. Applying this algorithm, we could for instance match the track name "Fragile" of Sting to the following spellings: "How Fragile", "Sting Fragile", "How Fragile We Are", etc.

2.2 Identifying Relevant Artist Co-Occurrences

The next step is to count the artist co-occurrences in the playlists in order to determine a set of "informative" artists which co-occur with other artists frequently. To do so, we iterate over each artist a of a given training set which contains tracks belonging to a music category (*positive* examples) and not belonging to it (*negative* ones)[1] and count how often (tracks of) other artists co-occur with a in the playlists. For each training track, these numbers are then sorted in decreasing order. As shown in the example, pieces created by Ludwig van Beethoven appear most often together with pieces by Frederic Chopin (3028 times using Last.fm statistics), followed by Johann Sebastian Bach (2894 times), and so on. Given a negative training example track for the category "classic", pieces of the artist ATB (a DJ) appear most frequently together with tracks of Miles Davis (2340 times). Since not all co-occurrences are relevant and might introduce noise in our models, we store only the ten most frequent co-occurrences for each artist in the training dataset.[2]

2.3 Removing Duplicate Entries

After the previous step and as shown in Fig. 1 we end up with a set of informative artists, which co-occurred with the artists of the 20 tracks in the training dataset that was used in Vatolkin (2013). As the same artists may appear in the top co-occurring artists lists for several training tracks (in particular for positive examples which are expected to be more similar to each other), duplicate entries in the list are removed. For the concrete example of the recognition of the genre "classic", the number of artists and their co-occurring artists—which we will later on use as *features* in the classification models—is reduced from 200 to 97 as shown in Fig. 1. We would for example see that music pieces composed by Beethoven appear frequently not only together with Chopin, but also with decreasing frequency together with pieces by Mussorgsky, Ravel, Orff, Sibelius, etc.

2.4 Normalisation

We measure the relevance of each co-occurring artists using two standard approaches based on association rules (Han and Kamber 2006). The first approach

---

[1]More details of the training data will be given in Sect. 3.1.

[2]In a preliminary study, increasing this number to 20 did not lead to measurable improvements. Obviously, the optimal number depends on the category; this investigation is however beyond the scope of this first study.

is to use the *support* for normalisation:

$$\text{support}(\{a, b\}) = \frac{\text{count}(\{a, b\})}{N} \qquad (1)$$

where $count(\{a, b\})$ is the number of playlists that contain both artists a and b and N is the overall number of playlists. Since the support values are highly dependent on the general popularity of the musical pieces, we also use the *confidence* values as an alternative:

$$\text{confidence}(a \to b) = \frac{\text{support}(\{a, b\})}{\text{support}(\{a\})} \qquad (2)$$

2.5 Training of Classification Models

Based on the normalised co-occurrences with the artists from the training set (the co-occurrences values serve as features) and the given category assignments, classification models can be finally built using different machine learning approaches. For instance, Naive Bayes predicts classes based on feature distributions for positive and negative instances. An example of the density of the feature distribution is provided in the right hand side of Fig. 1. Tracks that do not belong to the "classic" genre appear very seldom together with Beethoven, which is indicated by the high peak of the density function for values close to zero. On the other hand, there are only a few classic pieces which appear together with tracks of Miles Davis.

At the end, after the models have been trained, they can be applied for the classification of unlabelled tracks for which the artist is known using the chosen machine learning technique.

3 Experiments

3.1 Experimental Setup

To be able to compare our playlist-based approach with a typical audio signal based one, we used the experimental setup from Vatolkin (2013), where the goal was to categorise music tracks into six genres (Classic, Jazz, Pop, etc.) and eight styles (e.g. ClubDance, HeavyMetal, Urban) using binary classifiers.

3.1.1 Dataset

For each of the 14 categories, the dataset comprises ten positive examples and ten negative ones. In addition, Vatolkin (2013) used an optimisation set of 120 tracks to apply an evolutionary feature selection technique in order to determine the most relevant audio features for learning. The models were then evaluated on a test set which also comprised 120 tracks and which had the same genre distribution as the optimisation set.

3.1.2 Audio Features

We use four sets of audio features after Vatolkin (2013). The first group describes 636 low-level audio signal characteristics. The second group consists of 566 high-level "semantic" descriptors, which are better interpretable, e.g. the recognised instruments, moods, harmonic properties, etc. The third group contains 13 Mel Frequency Cepstral Coefficients (MFCCs) which were developed for speech recognition but are commonly used in music classification (Meng et al. 2007). The fourth group contains the optimised feature sets after the application of an evolutionary feature selection strategy.

3.1.3 Playlist Features

For the four groups of playlist statistics, we used two datasets retrieved from public sources[3] and the two normalisation methods described in Eqs. (1) and (2).

3.1.4 Classification and Evaluation

As classification techniques, we used Decision Tree C4.5, Random Forest, Naive Bayes, and Support Vector Machines. In the following section, we report the results of the method that worked best for the specific classification task. Because the distribution across genres and especially across styles is not balanced, classification models are evaluated with the balanced relative classification error:

$$e_{\mathrm{BRE}} = \frac{1}{2}\left(\frac{\mathrm{FN}}{\mathrm{TP} + \mathrm{FN}} + \frac{\mathrm{FP}}{\mathrm{TN} + \mathrm{FP}} \right), \tag{3}$$

where *TP* denotes true positives, *TN* true negatives, *FP* false positives, and *FN* false negatives.

---

[3]The samples included about one million playlists from Last.fm and about 600,000 playlists from 8tracks, see also Bonnin and Jannach (2014).

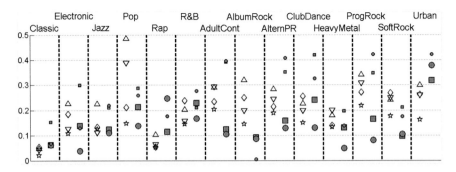

Fig. 2 Balanced relative classification errors for 14 music categories (labels above the figure) and eight feature sets. Audio features, signs with white background: *Downward-pointing triangles*: low-level features; *upward-pointing triangles*: high-level features; *diamonds*: MFCCs, *asterisks*: optimised feature sets. Playlist features, signs with *shaded background*: *rectangles*: 8tracks; *circles*: Last.fm; *larger signs*: normalisation with confidence; *smaller signs*: normalisation with support

3.2 Results

3.2.1 General Results

The classification errors obtained in the experiments for the eight feature sets and the 14 categorisation tasks using the classification method leading to the best results[4] are shown in Fig. 2. When looking on the audio-based approaches (symbols with white background), the feature optimisation method of Vatolkin (2013) not surprisingly worked best except for the category "Jazz" (for this category, the validation set contained more European Jazz and the optimisation set more American Jazz).

To some surprise, however, the comparably simple classification method based on playlist statistics and artist co-occurrences performs equally well and in many cases even better than the method based on optimised audio feature sets. The best variant of the playlist-based methods outperforms the best audio-based approach for 10 of 14 categories. This indicates that the computationally highly efficient and rather simple aggregation of playlist statistics can be indeed a good alternative for music classification. For some categories, however, audio features performed better. The MFCC-based feature set was for example particularly successful for the classification of Rap music. These results therefore suggest the use of hybrid strategies that combine the different approaches.

---

[4]The best performing method depends on the category. Moreover, the removal of a weaker classifier from ensemble of above mentioned methods led to a statistically significant reduction of performance in a previous study (Vatolkin et al. 2014).

3.2.2 Further Observations

The normalisation based on confidence generally performs better than when using the support statistic for the 8tracks data in 13 of 14 cases, and for Last.fm in 12 of 14 cases. Furthermore, the mean performance on the Last.fm dataset is generally higher than for 8tracks (in 10 of 14 cases). This can be simply explained by the larger amount of data that is available in the used playlist collection of Last.fm.

Another outcome of the study is that the obtained classification quality varies with the different classification methods. Playlist-based approaches seem to often perform slightly better if the models are trained with a Naive Bayes approach or Support Vector Machines. A systematic tuning of the hyperparameters of the classification methods has not yet been done but may help to further increase a classification performance. Another improvement could potentially be achieved if a feature optimisation strategy would also be applied to the playlist-based approach.

4 Conclusions and Outlook

In this work, we investigated how well two methods for the aggregation of playlist statistics are suited to build feature sets for genre and style classification. We compared the classification quality of using playlist statistics with the quality that can be achieved when using classification models based on optimised audio feature sets. Our results showed that playlist-based models were favourable over audio-based features sets for classification for more than half of the genres.

The choice of which features to use in real-world classification-based applications in our view strongly depends on the main guiding constraints in the goal of the particular application setting. Consider the following example scenarios.

1. If the application's goal is to derive interpretable harmonic and melodic properties, e.g. of user-defined personal categories, a music scientist would probably prefer automatic classification based on high-level audio features as playlist-based models do not operate on the basis of such features.
2. In case that the processing efficiency for the classification task is the main requirement, e.g. because huge music collections have to be analysed, one might prefer playlist-based models as they help to avoid the computationally costly extraction of features from the audio signal.
3. If the quality of the classification is the most important application requirement, a combination of audio features and features derived from playlist statistics might be the best choice.
4. Finally, for researchers, using playlist information in our view represents a comparably cheap way of developing classification approaches with competitive performance, because the musical tracks themselves do not have to be purchased or licensed for the analysis.

As part of our future work, we plan to examine the performance of combined feature sets where we also aim to apply feature selection techniques that simultaneously consider the feature sets of both sources. In addition, the validation of such an approach is planned using other public datasets.

Another promising direction for further research in our view is the development of further variants of our playlist-based classification methods and the evaluation of various parametrisations of the techniques. Specifically, this could involve the integration of statistics from other web sources, the systematic variation of individual parameters like the number of the stored top co-occurrences, the consideration of track and album co-occurrences, or the fine-tuning of the underlying classification methods.

References

Bertin-Mehieux, T., Eck, D., Maillet, F., & Lamere, P. (2008). Autotagger: A model for predicting social tags from acoustic features on large music databases. *Journal of New Music Research, 37*(2), 115–135.

Blume, H., Haller, M., Botteck, M., & Theimer, W. (2008). Perceptual feature based music classification - A DSP perspective for a new type of application. In *Proc. IC-SAMOS* (pp. 92–99).

Bonnin, G., & Jannach, D. (2014). Automated generation of music playlists: Survey and experiments. *ACM Computing Surveys, 47*(2), 26:1–26:35.

Celma, Ò. (2010). *Music recommendation and discovery: The long tail, long fail, and long play in the digital music space.* Berlin: Springer.

Fields, B. (2011). *Contextualize your Listening: The Playlist as Recommendation Engine.* PhD thesis, University of London.

Han, J., & Kamber, M. (2006). *Data mining: Concepts and techniques.* The Morgan Kaufmann Series in Data Management Systems. San Francisco, CA: Morgan Kaufmann.

Hariri, N., Mobasher, B., & Burke, R. (2012). Context-aware music recommendation based on latent topic sequential patterns. In *Proc. ACM RecSys 2013* (pp. 131–138).

Lartillot, O., & Toivainen, P. (2007). MIR in Matlab (II): A toolbox for musical feature extraction from audio. In *Proc. Int'l Conf. on Music Information Retrieval (ISMIR)* (pp. 127–130).

Lidy, T., Rauber, A., Pertusa, A., & Iñesta, J. M. (2007). Improving genre classification by combination of audio and symbolic descriptors using a transcription system. In *Proc. Int'l Conf. on Music Information Retrieval (ISMIR)* (pp. 61–66).

Mckay, C. (2010). *Automatic Music Classification with jMIR.* PhD thesis, McGill University.

Meng, A., Ahrendt, P., Larsen, J., & Hansen, L. K. (2007). Temporal feature integration for music genre classification. *IEEE Transactions on Audio, Speech, and Language Processing, 15*(5), 1654–1664.

Mierswa, I., & Morik, K. (2005). Automatic feature extraction for classifying audio data. *Machine Learning Journal, 58*(2–3), 127–149.

Serra, X., Magas, M., Benetos, E., Chudy, M., Dixon, S., Flexer, A., et al. (2013). *Roadmap for music information research.* Technical Report, The MIReS Consortium.

Smith, T., & Waterman, M. (1981). Identification of common molecular subsequences. *Journal of Molecular Biology, 147*, 195–197.

Sturm, B. (2014). A survey of evaluation in music genre recognition. In A. Nünberger, S. Stober, B. Larsen, & M. Detyniecki (Eds.), *Adaptive multimedia retrieval: Semantics, context, and adaptation. Lecture notes in computer science* (Vol. 8382, pp. 29–66). Cham:Springer.

Theimer, W., Vatolkin, I., & Eronen, A. (2008). *Definitions of audio features for music content description*. Technical Report TR08-2-001, TU Dortmund.

Tzanetakis, G., & Cook, P. (2002). Musical genre classification of audio signals. *IEEE Transactions on Speech and Audio Processing, 10*(5), 293–302.

Vatolkin, I. (2013). *Improving Supervised Music Classification by Means of Multi-Objective Evolutionary Feature Selection*. PhD thesis, Department of Computer Science, TU Dortmund.

Vatolkin, I., Bischl, B., Rudolph, G., & Weihs, C. (2014). Statistical comparison of classifiers for multi-objective feature selection in instrument recognition. In *Data analysis, machine learning and knowledge discovery* (pp. 171–178). Cham: Springer.

Weihs, C., Ligges, U., Mörchen, F., & Müllensiefen, D. (2007). Classification in music research. *Advances in Data Analysis and Classification, 1*(3), 255–291.

Yang, Y.-H., & Chen, H. H. (2011). *Music emotion recognition*. Boca Raton: CRC Press.

Duplicate Detection in Facsimile Scans of Early Printed Music

Christophe Rhodes, Tim Crawford, and Mark d'Inverno

Abstract There is a growing number of collections of readily available scanned musical documents, whether generated and managed by libraries, research projects, or volunteer efforts. They are typically digital images; for computational musicology we also need the musical data in machine-readable form. Optical Music Recognition (OMR) can be used on printed music, but is prone to error, depending on document condition and the quality of intermediate stages in the digitization process such as archival photographs. This work addresses the detection of one such error—duplication of images—and the discovery of other relationships between images in the process.

1 Introduction

1.1 Digitization and Early Music Online

Librarians have kept irreplaceable artifacts in trust for centuries. Now, with modern digital storage and networking technology, the opportunity has arisen to greatly widen access to heritage, and libraries and archives are taking this opportunity as and when resources permit. Normal digitization efforts involve taking pictures of sources; this is adequate for the most part, although in some cases (e.g. Henry Billingsley's 1570 translation of Euclid's Geometry, the first geometrical "pop-up" book printed in sixteenth-century England; see Swetz and Katz 2011) essential information is lost.

In *Early Music Online* (Rose 2011), a "Rapid Digitization" project funded by the Joint Information Systems Committee (JISC), over 320 printed volumes (35,000 pages) of music from sixteenth-century sources held in the British Library were digitized from microfilm, and made available to the community at large in the form of images, licensed for non-commercial use.

C. Rhodes (✉) • T. Crawford • M. d'Inverno
Goldsmiths, University of London, New Cross, London SE14 6NW, UK
e-mail: c.rhodes@gold.ac.uk; t.crawford@gold.ac.uk; dinverno@gold.ac.uk

© Springer International Publishing Switzerland 2016   449
A.F.X. Wilhelm, H.A. Kestler (eds.), *Analysis of Large and Complex Data*, Studies
in Classification, Data Analysis, and Knowledge Organization,
DOI 10.1007/978-3-319-25226-1_38

A photographic digitization process, as was carried out for *Early Music Online*, does not cause an immediate loss of information. The fact that digitization of the sources in *Early Music Online* was not from the originals but from microfilm has consequences for the published set of images—but the digitization also offers an extra opportunity: just as images of text could be further processed to make the text on those pages available, so we might want to make available not just the images of the musical source but also a representation of the musical content contained within it, in order to facilitate further analysis (by the human scholar, by automated processes, or most likely by a hybrid of the two).

However, we need to deal with the problem of images which, for one reason or another, are rescans of the same pages, as they must not be treated as distinct entities. These images are not precise digital duplicates of each other, and so must be detected through some approximate means. As well as duplicate scans, there are other forms of similarity present in the collection, such as musical relatedness and movable type reuse.

We present our work on developing and combining image-based near-duplicate detection, based on Scale-Invariant Feature Transform (SIFT) descriptors (Lowe 1999), with OMR-based musical content near-duplicate detection. We evaluate an order-statistic-based method for finding duplicate scans of pages, and additionally identify a number of distinct kinds of approximate similarity emergent from our distance measures: substantial reuse of graphical material; musical quotation; and title page detection.

1.2 Optical Music Recognition

Although Optical Music Recognition (OMR, by analogy with Optical Character Recognition for text) has been a subject of research since the 1960s (Pruslin 1966 and Prerau 1970; see Kassler 1972), it remains in general a difficult, unsolved problem (Rebelo et al. 2012). Partly this is because, unlike text, common musical notation is made up of a number of intersecting graphical elements; partly because, again unlike most text, the two-dimensional layout of the page is highly significant to the interpretation of the glyphs.

In our particular context, there is the additional difficulty that we are dealing with historical artifacts, from before the standardization of musical layouts—indeed, the *Early Music Online* collection is at the very start of printed music, when each printer would have had their own collection of movable type. Nevertheless, accuracy rates of around 90 % are achievable (Pugin and Crawford 2013) on the majority of the collection, with some sources allowing OMR to be performed with far greater precision and recall than others.

In the long term, we aim to overcome these difficulties, to allow full-music search and other algorithmic processing, just as OCR has allowed scholars to perform full-text search over the contents of documents, not just their metadata. This paper deals with one piece of the puzzle: namely, identifying portions of the source on which the results of OMR should *not* be included in any such automatic transcription, but rather flagged for a human expert to investigate. In the next section, we describe measures of similarity between images of musical notation; we then use these measures to characterize particular relationships between pages from three of the sources (475 pages) from *Early Music Online*.

2   Similarity Measurements

2.1   *Image Similarity*

As a basic measure of image similarity, we follow Lowe (1999) in computing SIFT descriptors for each image, which are invariant to (uniform) scaling and rotation, and robust against affine distortion and lighting changes. In order to compare the image similarity between a source image and a target, we compute for each descriptor in the source the *two* nearest (as measured by the Euclidean distance) descriptors in the target, and count a "hit" if the distance to the nearest is less than two-thirds of the distance to the second nearest. The overall similarity score for the pair of images is the sum of the "hits" from image to source, without reference to relative position or orientation. Note that this similarity is not necessarily symmetric, as the source and target images are treated differently.

2.2   *Musical Similarity*

We use the **Aruspix** software (Pugin 2006; Pugin and Crawford 2013) in untrained mode to extract musical data from images. Note that **Aruspix** will attempt to extract musical information no matter what the source image: for images containing no musical notation at all, this of course means that the output will be musically nonsensical, resulting from chance agglomerations of glyphs and graphical material which look "enough" like music to **Aruspix**'s recognizer. We convert the output of **Aruspix**, a representation of the musical data identified to strings representing either the diatonic melody or the diatonic intervals present on each staff, for example:

kind	string
melodic	SSQRSRPRQPNPONOPQRSTSSRTSRP
interval	-bAAabBaabBaaAAAAAAa-aBaab

The melodic string encodes the diatonic pitch (similar to chromatic pitch, but with seven notes per octave rather than 12, thus disregarding accidentals) as the ASCII character with code point 48 + the diatonic pitch. The interval string encodes the diatonic interval between successive diatonic pitches, with - indicating no change, capital letters representing movement upwards (A representing up one step, B up two) and lower-case letters movements downwards (a representing down one step, b down two, and so on)

We thus obtain one of these strings for each of the cases (melodic and interval) per line of music. We compute the similarity score of a source image to a target image by: first, taking the string for each line in the source image; second, finding and scoring the closest match in the target image using the Wu-Manber algorithm (Wu and Manber 1994, as implemented in `agrep`); and finally, summing those scores over all lines in the source page.

2.3 Outlier Analysis

We identify various possible scenarios for a scan X or a pair of scans (X,Y), which we encode as predicates:

music(X)   the scan X is primarily of musical notation
duplicate(X,Y)   the scans X and Y are near-duplicates of each other
musicsim(X,Y)   the scans X and Y contain substantially similar musical material
graphicsim(X,Y)   the scans X and Y contain substantially similar graphical material

Some of these predicates imply other relations:

- duplicate(X,Y) → graphicsim(X,Y)
- duplicate(X,Y) → (music(X) → musicsim(X,Y))
- duplicate(X,Y) → duplicate(Y,X);

the asymmetry arising from the fact that all scans contain graphical material, but not all scans contain musical material.

An ordered pair of scans (X,Y) will have two similarity scores associated with it: a similarity score based on image similarity, and a second score based on the musical similarity imputed from comparing the output of the OMR process. These similarity scores tell us nothing a priori; in order to extract meaning from them, we must compare them against thresholds. However, there is also no way of a priori deriving thresholds of similarity for "interestingness", so we use the distribution of similarity scores between X and all other scans as a way of establishing a threshold.

Specifically, we fit a lognormal distribution to the central 80 % of similarity scores, for each of the measures (image and music) separately; we then treat as a threshold the 0.5 % level of improbability, accepting the default thresholds from the implementation in the **extremevalues** R package (van der Loo 2010). This then gives us three possible diagnostics for each similarity measure:

- (X,Y) are unusually *similar* to each other;
- (X,Y) are unusually *dissimilar* to each other;
- (X,Y) have a similarity score which is not particularly distinctive.

These diagnostics, when the two similarity scores are combined, give a total of nine possible outcomes for each pair of scans.

2.4 Hypothesis

Our hypothesis is that we can use the combination of our music and image similarity measures to identify near-exact duplicates resulting from multiple images of the same pages on the microfilm.

Specifically, we invert the relationships in Sect. 2.3, and attempt to infer higher-level information from the low-level outlier information. If (X,Y) are unusually similar according to the music similarity measurement, we assert the music-sim(X,Y) relation; similarly with image similarity and graphicsim(X,Y); and we further infer duplicate(X,Y) from graphicsim(X,Y) ∧ musicsim(X,Y).

Other outlier cases (pairs where one similarity score is high but not the other, and pairs where at least one similarity score is low) are also potentially of interest, and we can attempt to characterize the relationships between pages that give rise to those scores more qualitatively in the results below.

3 Results

Our test collection is 475 images resulting from scans of three sets of partbooks of parody masses (mass settings based on a pre-existing piece of music) published in 1545–1546 by Tielman Susato in Antwerp. This is an interesting test set from the point of view of our similarity measures. Firstly, the nature of parody masses is that there will be significant reuse of musical content, within a single work (in the same voice and different mass section, and in the multiple voices) and between distinct works (for example, if there are multiple masses on the same original material, though this does not in fact occur in this set of images). Secondly, since the books were printed by the same printer there is the likelihood that graphical material might be reused without any musical similarity between the material on the pages.

Given this test collection, there are 225,625 pairwise comparisons between images, given that our definition of these comparisons is not symmetric, and including the comparison of a scan with itself. We would expect the identity comparison to show up as an outlier in both measures—indeed, this is useful as a consistency check—and at least 180,160 (80 % of the rest) to be considered as having uninteresting distances (since we are fitting the distribution to the central 80 %).

From Table 1, we can observe firstly that the lognormal fit is presumably working reasonably well: the number of non-outlier pairs is comfortably above the 180,160 which would be the minimum. This view of the aggregate data does not of course

Table 1 Counts of similarity judgments between all pairs of pages in our dataset, for both similarity measures

Similarity	Low (graphic)	Medium (graphic)	High (graphic)
Low (music)	1083	3215	7
Medium (music)	6091	213,122	1592
High (music)	0	18	497

Outliers according to the lognormal fit are labelled "low" and "high", while "medium" indicates a non-outlier

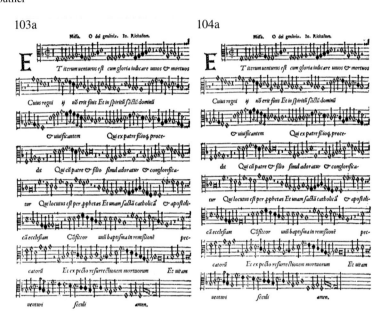

Fig. 1 Two pages with high image and musical similarity, from Susato (1545): these are most likely successive photographic shots of the same physical page

preclude there being individual cases for which the lognormal fit was inappropriate; however, on the dataset as a whole it appears to be justifiable.

Secondly, the number of high-melodic/high-image similarity pairs is 497, 22 above the 475 identity matches. From just this table it is not possible to say, but one way that this can arise is if there are 11 duplicate image pairs, all of which are detected in both directions. In fact, because of artifacts arising from the musical similarity measure applied to pages with no musical content, it turns out that two of the identity matches are misclassified, and there are in fact 12 duplicate image pairs detected by this measure, which we publish on the semantic web (retrievable using `curl -H 'Accept: text/n3'` http://duplicate-pages.emo.data.t-mus.org/). Figures 1 and 2 illustrate some of the duplicate image pairs found using this method.

142b   199b

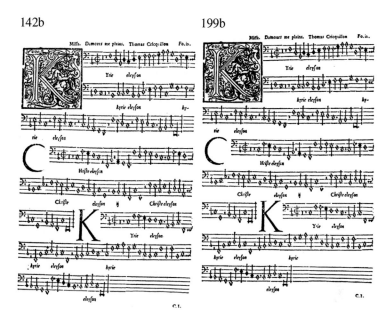

Fig. 2 Two pages with high image and musical similarity, from Susato (1546b): not shots of the same physical page, but most likely a misbound gathering

Thirdly, there are some interesting cases to investigate: in particular, the large number of high-image/medium-melodic cases; the seven cases of high-image similarity and low-melody similarity; and the low-image/low-melodic similarity cases. Since these are not in fact exact duplicates, it is apparent that combining the outlier judgments of the two similarity measurements was necessary for the basic task; the cases with one or other measure (but not both) showing high similarity reveal other relationships between the material on each page.

Figure 3 shows a pair of pages with high image similarity, but a melodic similarity between the pages that is no higher than expected according to the fit. Note the reuse of decorated initial capitals, a feature of the printing technology and resources of an individual printer in the sixteenth century—an individual printer (Tielman Susato, in this case) would not have a wide repertoire of type for decorated capitals, and so would reuse one of the appropriate size each time there was a call for one. Since we are here dealing with mass settings, there will be many examples of initial "K"s and "C"s for *Kyrie* and *Christe* movements.

Figure 4 highlights another feature of this set of works: many of the mass settings are "parody masses": settings based on musical material of another work, which gets reused throughout the mass setting. In this case, we have the ending of the *Gloria* and the start of the *Credo* from Thomas Crequillon's *Missa Kein in der Welt so schn*, both using the material from the song for a substantial fraction of the page.

Finally, Fig. 5 illustrates that this consideration of outliers also catches non-musical material: Aruspix will attempt to perform OMR on images that it is given,

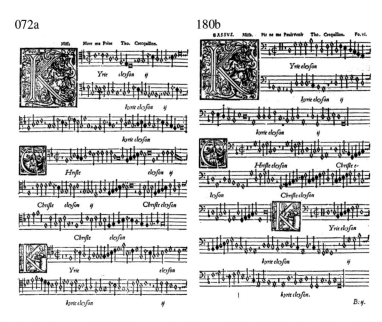

Fig. 3 A pair of pages with high image and medium melodic similarity, from Susato (1546a)

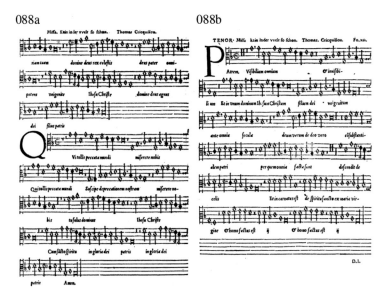

Fig. 4 A pair of pages with high melodic and medium image similarity, from Susato (1545)

076a   017b

Fig. 5 A pair of pages with low melodic and low image similarity, from Susato (1545)

and there is no metadata accompanying the set of images to indicate which contain musical material and which do not. However, the essentially random output from OMR on title pages will be dissimilar to most of the detected content, as will the image features compared with image features from pages which do contain musical material; this also explains the seven cases with high image similarity and low melodic similarity, in which one or both of the pages contain substantial amounts of text.

4   Conclusions

We have shown that a combination of image and music similarity measures can be used to identify duplicates and near-duplicate photos in digitized archives, and also to identify pairs of pages of possible interest falling short of being considered duplicates. Even though the similarity measures themselves are simple, their combination is sufficient to identify all the duplicates with no false positives, in this particular dataset. Analysis of other outlier cases shows potential to identify reuse of musical material, reuse of type, and classification into music-containing and non-musical pages.

4.1 Further Work

The distance measures between items we have used in this investigation are very simple; we have used SIFT image features without attempting to detect higher-level objects, and musical features with no attempt to consider perceptual similarity or even duration of individual notes. We could improve the image distance measure to take into account the coherence of groups of matches, as is done in pose estimation, though this would not address the most obvious false-positive of reuse of type for decorated initials. We could also attempt to deal with this by considering image features only on those regions which are detected as music by the Optical Music Recognition program. We would also like to make our approach scale. At present, the method is workable on datasets of this size, 475 pages, corresponding to individual books or restricted sets of books, and in practice there are already interesting duplicates present in sets of that size. In principle, we would like to run our method on larger datasets as a whole to investigate whether there is any contamination or other connections between sources; however, the pairwise comparison leads to $O(N^2)$ time complexity, and so building a feature index is a necessary step to apply this to larger collections.

We have published our similarity judgments from this investigation as Linked Data at http://duplicate-pages.emo.data.t-mus.org/, and we will expand this resource as we generate more data. As well as publishing individual duplicate pairs, we aim to publish higher-level judgments, such as the presumed cause of the duplication from the photographic process as in Susato (1545) or the binding in Susato (1546b). Finally, in the *Transforming Musicology* project, we aim to apply a similar method to similarity judgments of more general musical artifacts, such as musical recordings and editions of musical works.

Acknowledgements This work was supported by the *Transforming Musicology* project, AHRC AH/L006820/1.

References

Kassler, M. (1972). Optical character-recognition of printed music: A review of two dissertations. *Perspectives of New Music, 11*(1), 250–254.

Lowe, D. G. (1999). Object recognition from local scale-invariant features. In *International Conference on Computer Vision, 1999* (pp. 1150–1157).

Prerau, D. S. (1970). *Computer Pattern Recognition of Standard Engraved Music Notation.* MIT Libraries, Cambridge, MA.

Pruslin, D. H. (1966). *Automatic Recognition of Sheet Music.* MIT Libraries, Cambridge, MA.

Pugin, L. (2006). Aruspix: An automatic source-comparison system. In: W. B. Hewlett, & E. Selfridge-Field (Eds.), *Music analysis east and west* (pp. 49–60). Cambridge, MA: MIT Press.

Pugin, L., & Crawford, T. (2013). Evaluating OMR on the early music online collection. In *Proceedings of ISMIR 2013* (pp. 439–444).

Rebelo, A., Fujinaga, I., Paszkiewicz, F., Marcal, A. R., Guedes, C., & Cardoso, J. S. (2012). Optical music recognition: State-of-the-art and open issues. *International Journal of Multimedia Information Retrieval, 1*(3), 173–190.

Rose, S. (2011). Introducing early music online. *Early Music Review, 143*, 14–16.

Susato, T. (Ed.) (1545). Missarum quatuor vocum: Liber secundus / a prestantissimis musicis Nempe Ioan. Lupo hellingo. & Thomas Cricquillione. Compositarum catalogus hic infra designatur. Antwerp.

Susato, T. (Ed.) (1546a). Missarum quinque vocum: Liber primus/a diversis musicis compositarum, quarum nomina catalogus indicabit. Antwerp.

Susato, T. (Ed.) (1546b). Missarum quatuor vocum: Liber tertius / a diversis musicis compositarum. Antwerp.

Swetz, F. J., & Katz, V. J. (2011, January). Mathematical Treasures - Billingsley Euclid. *Convergence*. http://www.maa.org/press/periodicals/convergence/mathematical-treasures-billingsley-euclid

van der Loo, M. P. J. (2010). extremevalues, an R package for outlier detection in univariate data. R package version 2.1.

Wu, S., & Manber, U. (1994). A fast algorithm for multi-pattern searching. TR-94-17, Department of Computer Science, University of Arizona.

Fast Model Based Optimization of Tone Onset Detection by Instance Sampling

Nadja Bauer, Klaus Friedrichs, Bernd Bischl, and Claus Weihs

Abstract There exist several algorithms for tone onset detection, but finding the best one is a challenging task, as there are many categorical and numerical parameters to optimize. The aim of this task is to detect as many true onsets as possible while avoiding false detections. In recent years, model-based optimization (MBO) has been introduced for solving similar problems. The main idea of MBO is modeling the relationship between parameter settings and the response by a so-called surrogate model. After evaluating the points of an initial design—each point represents here one possible algorithm configuration—the main idea is a loop of two steps: firstly, updating a surrogate model, and secondly, proposing a new promising point for evaluation. While originally this technique has been developed mainly for numerical parameters, here, it needs to be adapted for optimizing categorical parameters as well. Unfortunately, optimization steps are very time-consuming, since the evaluation of each new point has to be performed on a large data set of music instances for getting realistic results. Nevertheless, many bad configurations could be rejected much faster, since their expected performance might appear to be very low after evaluating them on just a small partition of instances. Hence, the basic idea is to evaluate each proposed point on a small sample and only evaluate on the whole data set if the results seem to be promising.

N. Bauer (✉) • K. Friedrichs • B. Bischl • C. Weihs
Chair of Computational Statistics, Faculty of Statistics, TU Dortmund, Dortmund, Germany
e-mail: bauer@statistik.tu-dortmund.de; friedrichs@statistik.tu-dortmund.de;
bischl@statistik.tu-dortmund.de; weihs@statistik.tu-dortmund.de

© Springer International Publishing Switzerland 2016
A.F.X. Wilhelm, H.A. Kestler (eds.), *Analysis of Large and Complex Data*, Studies
in Classification, Data Analysis, and Knowledge Organization,
DOI 10.1007/978-3-319-25226-1_39

1   Introduction

A tone onset is the time point of the beginning of a musical note or other sound. Onset detection is an important step for music transcription and other applications frequently encountered in music processing. Although several approaches have been developed for onset detection, neither of them works well under all circumstances—instrumentation, tempo, or music genre. It is hence essential to find an optimal onset detection algorithm for a desired music data set. This task has two main problems: an optimization strategy which can handle with many categorical and numerical parameters as well as computational resources for optimizing on large data sets.

To find an optimal onset detection algorithm we use the **mlrMBO** R-Package as a comprehensive tool for MBO.[1] We apply an instance based MBO by handling each music piece as a problem instance and propose a fast variant (FMBO) where a small subset of instances is used to predict the onset detection performance. Section 2 introduces the onset detection algorithm and parameters we aim to optimize. In Sect. 3 the main MBO procedure is presented while Sect. 4 describes our proposed FMBO approach. The data set and the experimental settings are given in Sect. 5. Section 6 presents the results in respect of the optimization strategy and the best found algorithm parameter setting. Finally, Sect. 7 summarizes the work and provides ideas for future research.

2   Onset Detection Algorithms

This section presents the classical onset detection procedure and introduces the parameters for further optimization with corresponding regions of interest. Onset detection is usually performed in three stages: pre-processing (filtering the ongoing signal), computing the onset detection function (reduction), and localizing the tone onsets (peak selection). A tutorial on basic onset detection approaches is given by Bello et al. (2005). Here, we ignore the pre-processing. In general, reduction and peak selection consist of six steps:

step 1: splitting the signal into small windows,
step 2: computing in each window an onset detection function (ODF),
step 3: applying a low-pass filter (optional),
step 4: normalizing the ODF,
step 5: thresholding the ODF,
step 6: localizing tone onsets.

At first, the ongoing signal is splitted into small windows with the *window size* of N samples (step 1). In order to profit from the fast discrete Fourier transformation

---

[1] https://github.com/berndbischl/mlrMBO

(FFT), N should be assigned just powers of two. In the onset detection literature the 46 ms window (2048 samples for a sampling rate of 44.1 kHz) is usual (Dixon 2006; Rosão et al. 2012). However, other settings also occur: Holzapfel et al. (2010) use $N = 4096$ samples (100 ms). We consider, in contrast, a wide region of interest for N: 512, 1024, 2048, 4096, and 8192 samples as small windows allow a good time representation while large windows provide a high spectral resolution. A further important parameter is the *hop size h*: distance in samples between windows' starting points. The lower h the more windows are produced. In case of $N = h$ there is no overlap between the windows. There is no broad agreement in the literature in regard of the hope size. Just to illustrate this, Dixon (2006) uses 10 ms (441 samples), Rosão et al. (2012) 23 ms (1024 samples), and Holzapfel et al. (2010) 5.6 ms (250 samples). We define the region of interest for h between 128 and N samples where every value in this interval is allowed.

After splitting the signal, an ODF is computed in each window (step 2). Many functions have been proposed in recent years. The 8 ODFs used here can be divided into four groups: amplitude based (*Amplitude Increase* as defined by Bauer et al. (2013)), spectral magnitude based (*High Frequency Content* and *Spectral Flux*), and spectral magnitude and spectral phase based (*Phase Deviation* (PD), *Weighted PD*, *Normalized WPD*, *Complex Domain* (CD), and *Rectified CD*). Detailed definitions of the last seven functions can be found in Rosão et al. (2012) and Dixon (2006). For illustration purposes we define here the *Spectral Flux* feature:

$$SF(n) = \sum_{j=1}^{N/2} H(|X[n,j]| - |X[n-1,j]|) \text{ with } H(x) = (x + |x|)/2.$$

$X[n,j]$ is the *j*th frequency bin of the *n*th window and the filter H ensures that only the rise of the spectral magnitude is considered (for avoiding the tone offset detection). The main idea of steps 3 and 4 is bringing the vector of ODF-values to a more common form. We denote this vector by $\mathbf{odf} = (odf_1, \ldots, odf_m)^T$ were m is number of windows. In step 3 a low-pass filter can be applied to $\mathbf{odf}$ in order to get rid of a possible winding structure (Holzapfel et al. 2010). This filter is equal to the exponential smoothing operator with parameter α, which is fixed here to 0.8. The categorical parameter *filter* has two possibilities: *yes* and *no*.

Rosão et al. (2012) mention two approaches for normalization of the $\mathbf{odf}$ vector: subtracting *mean*($\mathbf{odf}$) from $\mathbf{odf}$ and then dividing the result either by *standard.deviation*($\mathbf{odf}$) or by *max*($\mathbf{odf}$). In the first case the normalized vector $\mathbf{n.odf}$ has a mean of 0 and a standard deviation of 1. However, min($\mathbf{n.odf}$) and max($\mathbf{n.odf}$) are unknown. The second method guarantees min($\mathbf{n.odf}$) $= -1$ and max($\mathbf{n.odf}$) $= 1$. Depending on the denominator, the parameter *norm* can be set either to *sd* or to *max*. One of the most essential issues is the thresholding of $\mathbf{n.odf}$ (step 5):

$$T_i = \delta + \lambda \cdot th.fun(|n.odf_{i-w_T}|, \ldots, |n.odf_{i+w_T}|), i = 1, \ldots, m.$$

Here δ and λ are positive constants with the following regions of interests: [0,0.8] and [0,3], respectively. w_T represents the threshold window size: number of windows left and right of the current window for computing the "moving function." We consider the possible range for w_T between 0 and 20. The kind of the moving function is recorded by the categorical parameter *th.fun* with settings *mean* and *median*.

In the last step the tone onsets are localized according to T and **n.odf**:

$$O_i = \begin{cases} 1, & \text{if } n.odf_i > T_i \text{ and } n.odf_i = max(n.odf_{i-w_O}, \ldots, n.odf_{i+w_O}) \\ 0, & \text{otherwise.} \end{cases}$$

w_O is an additional parameter—local maximum window size—with the region of interest between 0 and 12. $w_O = 0$ means that just the threshold criterion is applied. The starting time points of the windows with $O_i = 1$ compose a vector of onset times which is then compared to the vector of the true onset times. An estimated tone onset is assumed to be correctly detected, if it matches to one true onset with a tolerance of ± 50 ms (Dixon 2006).

The goodness of the onset detection is usually measured by the F-value: $F = \frac{2c}{2c+f^++f^-}$, $F \in [0, 1]$, where c is the number of correctly detected onsets, f^+ is the number of false detections, and f^- represents the number of undetected onsets.

3   Model Based Optimization and Algorithm Configuration

The aim of MBO is minimization of a (possibly non-linear or multimodal) expensive black-box function $f : \mathcal{X} \subset \mathbb{R}^d \to \mathbb{R}, f(\boldsymbol{x}) = y, \boldsymbol{x} = (x_1, \ldots, x_d)^T$. Each x_i is a parameter with region of interest $[\ell_i, u_i]$, $\mathcal{X} = [\ell_1, u_1] \times \ldots \times [\ell_d, u_d]$ is the parameter space of $\boldsymbol{x}$, and y is the target value. MBO is a sequential procedure: after evaluating an initial design of parameter settings $\mathcal{D}$ by f, a so-called meta-model or surrogate $\hat{f}$ is fitted on the data and used to propose a new point to evaluate $\boldsymbol{x}^*$ in each iteration. A detailed outline is presented in Algorithm 4. We refer the interesting readers, exemplary, to Jones et al. (1998), Hutter et al. (2011), and Bischl et al. (2014).

Beside considering the appropriate size of $\mathcal{D}$ and the complete evaluation budget, one of the most important issues is the choice of the surrogate model and the infill criterion, which map points of $\mathcal{X}$ to numerical values and allow hence the comparability between them. A popular combination is kriging as surrogate model and expected improvement (EI) as infill criterion (Jones et al. (1998)).

Kriging models consist of two terms: a polynomial term (linear model) and an error term, which is assumed to be a realization of a stationary stochastic Gaussian process. This assumption allows the calculation of model uncertainty. Furthermore, kriging needs the specification of the so-called spatial covariance function, or kernel,

---

Algorithm 4: Sequential model-based optimization

---

1 Generate an initial design $\mathscr{D} \subset \mathscr{X}$;
2 Compute $y = f(\mathscr{D})$;
3 **while** *total evaluation budget is not exceeded* **do**
4   Fit surrogate on $\mathscr{D}$ and obtain $\hat{f}(x)$ and $\hat{s}(x)$;
5   Get new design point x^* by optimizing an infill criterion ;
6   Evaluate new point $y^* = f(x^*)$;
7   Update: $\mathscr{D} \leftarrow (\mathscr{D}, x^*)$ and $y \leftarrow (y, y^*)$;
8 **return** $y_{min} = \min(y)$ and the associated x_{min}.

---

for assessing the influence of already evaluated points on the new point to be predicted. Defining such kernels is much easier for the numerical parameters than for the categorical ones. $EI(x)$ combines model prediction $\hat{f}(x)$ and uncertainty $\hat{s}(x)$ for a point x in a certain way and is the higher the lower $\hat{f}(x)$ (exploitation of $\hat{f}$) and the higher $\hat{s}(x)$ (exploration of the response area).

There are two essential drawbacks of classical MBO: (1) kriging and EI optimizer operate just with numerical parameters and (2) it has no concept of problem instances and related noisy optimization. Hutter et al. (2011) introduced instance based optimization for Sequential Model-based Algorithm Configuration (SMAC): each parameter vector represents an algorithm configuration, whose performance is measured on a set of problem instances $\mathscr{I}$. Regarding the first drawback, they proposed to use random forest as surrogate model, where $\hat{f}(x)$ and $\hat{s}(x)$ are mean and variance of x predictions among the single trees. In SMAC, the EI is optimized as follows: First, EI is computed for all already evaluated parameter settings and after that the 10 best settings are chosen as starting points for a randomized one-exchange neighborhood search. EI of the search output as well as of further 10,000 randomly chosen points are used to identify p most promising settings.

SMAC shows essential differences to Algorithm 4: Firstly, the initialization step consists merely in evaluation of one parameter setting on a randomly chosen instance. Secondly, not only one but p promising points (candidates) are proposed by EI optimizer and, lastly, an additional *intensify* step is conducted. By the intensify step the current best parameter setting x_{best} is evaluated on an additional randomly chosen instance (for ensuring its performance) and each candidate point is iteratively evaluated on a certain subset $I_{iter.nr} \subseteq \mathscr{I}$ of instances until its performance is proved to be worse than that of x_{best}, otherwise the candidate is the new x_{best}. As each parameter vector is evaluated on a subset of instances, the surrogate model should consider the information about which vector is evaluated on which subset.

In this work we aim to compare kriging and random forest surrogates for our application. When using kriging we naively handle the categorical parameters of onset detection as numerical ones by assigning an integer number to each level in order to study the behavior of this simple technique for comparison purposes.

However, we use a different uncertainty estimator $\hat{s}(x)$ for the random forest which is based on bootstrap mechanisms as introduced in Sexton and Laake (2009). A novel infill criterion optimizer—focus search—is proposed in the **mlrMBO** R-Package and used here. Focus search shrinks the interesting parameter region in many iterations to a promising section, where different shrinking mechanisms are applied depending on the parameter type. This procedure is replicated many times and the overall best point is proposed as x^*.

4 Fast Model Based Optimization by Instance Sampling

Our FMBO proposal is based on the observation that "bad" settings of the target algorithm perform weakly (i.e., cause low F-measures) on the main part of problem instances (music pieces of a data set). So we could detect such "bad" points x^* just by looking at a small subset of selected instances $I_{\text{sel}} \subseteq \mathscr{I}$. In such cases we can avoid the expensive complete evaluation and just estimate $f(x^*)$ (mean F-measure over $\mathscr{I}$). The novelty of our approach lies, on the one hand, in the kind of I_{sel} selection and, on the other hand, in the performance estimation for "bad" points. There exist many possibilities how to select k_{sel} instances from $\mathscr{I}$. Here we implemented a simple approach: after evaluating all points of the initial design $\mathscr{D}$ on all instances, the latter are clustered according to their F-measures in k_{sel} clusters. One representative of each cluster is then chosen randomly for I_{sel}. In this manner we aim to achieve high diversity within the selected instances and hope to get reliable prediction of the selection model M_{sel}: a linear model with mean F-measures of all instances as response and individual F-measures of I_{sel} instances as impact.

In the sequential steps, x^* is first evaluated on I_{sel} and then classified to be either a "good" or a "bad" point. Also here several approaches are possible for the decision. We implemented an M_{sel}-based decision: If the upper confidence bound for model prediction is greater than the best achieved F-measure, x^* seems to be a promising point which is afterwards evaluated on $\mathscr{I}$. In this case, the selection model M_{sel} is updated after the evaluation in order to profit from the new information. Otherwise, x^* is expected to be a "bad" point and merely its estimated mean F-measure (according to M_{sel}) is used as response. In the following, we compare 95 and 99 % confidence bounds. Obviously, the first one will cause more "bad" points as its value is smaller than the value of the second one.

Algorithm 5 summarizes our FMBO procedure. Note, two models are applied here: the surrogate model M_{sur} as introduced in the previous section and the above mentioned selection model M_{sel}. In contrast to Algorithm 4, we consider a maximization problem.

Algorithm 5: Fast model-based optimization (FMBO)

---

1   Generate an initial design $\mathcal{D} \subset \mathcal{X}$;
2   Compute $y = f(\mathcal{D})$;
3   Cluster all instances according to its F-measures in k_{sel} clusters (by *kmeans*);
4   Choose one instance from each cluster randomly for $I_{sel} \subseteq \mathcal{I}$;
5   Fit M_{sel}: mean F-measure (over $\mathcal{I}$) $\sim$ individual F-measures of I_{sel};
6   **while** *total evaluation budget is not exceeded* **do**
7   Fit surrogate M_{sur} on $\mathcal{D}$ and obtain $\hat{f}(x)$ and $\hat{s}(x)$;
8   Get new design point x^* by optimizing the EI infill criterion;
9   Evaluate x^* on I_{sel};
10   Predict mean F-measure for x^* according to M_{sel}: $\hat{y^*} = \hat{g}(x^*)$;
11   Compute UCB: the upper $X\%$ confidence bound for $\hat{y^*}$ prediction;
12   **if** $UCB \geq max(y)$ **then**
13   Evaluate x^* on all instances $\mathcal{I}$: $y^* = f(x^*)$;
14   Update: $\mathcal{D} \leftarrow (\mathcal{D}, x^*)$ and $y \leftarrow (y, y^*)$;
15   Update: M_{sel};
16   **else**
17   Update: $\mathcal{D} \leftarrow (\mathcal{D}, x^*)$ and $y \leftarrow (y, \hat{y^*})$;
18 **return** $y_{max} = max(y)$ and the associated x_{max}.

---

5   Data Set and Experimental Settings

As for our experiments a large data set is meaningful and the well-known data sets of hand labeled real music pieces are unfortunately relatively small (e.g., 23 pieces by Bello et al. 2005 or 90 pieces by Holzapfel et al. 2010), we decided to use MIDI data. Our data set consists of 200 freely available MIDI recordings which first 60 s were converted to WAV files using the MIDI to WAVE Converter 6.1. The pieces are heterogeneous both in regard of genre and of music instruments. A special characteristic of the data is that the neighboring onsets are separated from each other by at least 50 ms. This seems to be reasonable when considering the ± 50 ms tolerance interval for onset detection.

The main focus of our experiments is comparing MBO vs. FMBO. vs. random search (with the same evaluation budget). Additionally, we studied the effect of the surrogate model—kriging (**km**) vs. random forest (**rf**)—and of the upper confidence bound UCB for M_{sel} prediction—99 vs. 95 %. The initial design consists of 50 points, the evaluation budget is limited by 300 iterations and $k_{sel} = 10$ (5 % of the data). For each strategy 100 replications were carried out. Note that replications are comparable along all strategies as the same set of 100 start designs were used. We parallelize our experiments using the **BatchJobs** R-Package (Bischl et al. 2015).

6 Results

6.1 Comparison of Strategies

Figure 1 presents the main results of this work. *F*-measures of 100 replications for each optimization strategy are illustrated in the associated boxplots. The first (well expected) conclusion is that all optimization strategies achieve better results than random search. The second important point is the fact that kriging surrogates perform better than random forest. In view of the naive handling of categorical parameters as numeric ones, it is a very worthwhile finding. As, by the fixed number of sequential steps, MBO strategies need much more function evaluations than FMBO, they show fewer variation in the results and also better *F*-measures.

According to Fig. 3a FMBO strategies need, on average, at least 60 % less time for 300 sequential iterations than MBO. Here, the time is measured by number of instance evaluations. The more time is saved the more points were considered to be "bad" and the worse are the results of a strategy according to Fig. 1. Random forest surrogate model with UCB = 99 % leads, in contrast to kriging versions, to more "bad" points. Hence, we abandoned the UCB = 95 % option for random forest.

Let us demonstrate the working flow of FMBO based on the kriging surrogate with UCB = 99 %. Figure 2 shows the first 150 sequential steps. In order to verify the goodness of the selection model, the "bad" predictions, which are marked here as triangle points, were evaluated on all instances separately from the main FMBO run. The true performance values for these points are signed with crosses. As can be seen, in almost all cases, it was a correct decision, not to evaluate the "bad" points as

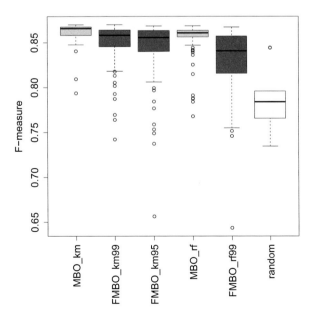

Fig. 1 Comparison of MBO and FMBO strategies. **km** means kriging model and **rf** random forest. 99 and 95 correspond to UCB values of selecting model

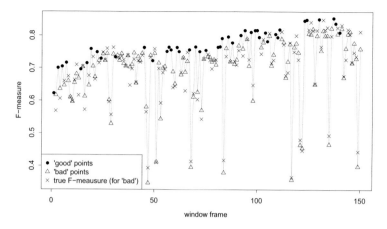

Fig. 2 First 150 points of FMBO_km99 optimization path

(a)   **(b)**

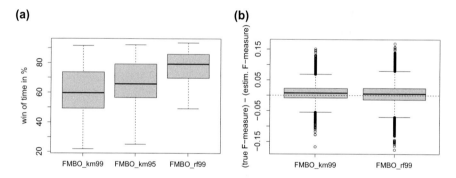

Fig. 3 (**a**) Time saving by FMBO strategies. (**b**) Error of the selecting model M_{sel}

their true mean F-measures are below the actual best value. We also see a continuing increase of the target during the optimization.

Figure 3b shows differences between the true and the predicted mean F-measures for both kriging and random forest based FMBO. As expected, the two distributions seem to be equal (because of the same selecting model M_{sel}). Furthermore, a slight underestimation of the true F-measure can be observed. The absolute error of the most points is less than 0.05 which can be seen as a positive result.

Lastly, we compared MBO and FMBO under the same budget conditions. For this reason, in each replication we noticed the number of instance evaluations conducted by the FMBO strategy and compared it to MBO with the equivalent budget. Figure 4 compares both kriging based FMBO strategies with the comparable MBO runs. Slightly better mean F-measures and considerable smaller variance can be observed for the FMBO approaches.

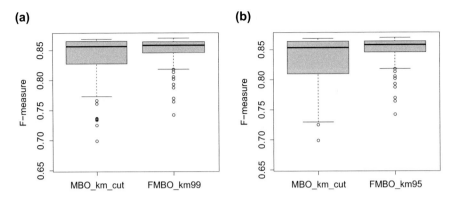

Fig. 4 Comparison of kriging based MBO and FMBO with equivalent budget for (**a**) FMBO_km99 (average budget for seq. steps = 119) and (**b**) FMBO_km95 (average budget for seq. steps = 100) strategies

6.2   Best Parameter Setting

The best achieved F-measure is 0.870. We define arbitrarily 0.865 as a threshold for a satisfying F-measure and conduct a one-factorial stability analysis around the optimal parameter setting. The best found window size is $N = 2048$ samples (46 ms) which is in accordance with the usual setting in the literature. Also the hope size $h = 602$ samples (ca. 14 ms or overlap of 60 %) appears to match with the values mentioned in Sect. 2. Spectral Flux detection function shows the best performance for our data, similar to Dixon (2006) and Rosão et al. (2012). Unfortunately, no reference values for δ and λ are mentioned in other studies. Our best settings are: $\delta = 0.038$ and $\lambda = 1.18$. For thresholding, moving median smoothing seems to perform better than the mean one. Rosão et al. (2012) use $w_T = 10$ and $w_O = 3$, while our optimal values are $w_T = 16$ and $w_O = 6$. According to the stability analysis all other detection functions and window sizes are much worse while the hop size could lie in [532, 662], δ in [0.026, 0.054], λ in [0.85, 1.4], w_T in [7, 20], and w_O in [5,6]. Low-pass filtering and moving median by thresholding yield to better results. In contrast, the kind of normalization has no impact regarding our stability analysis.

7   Conclusion and Further Research

In this paper we proposed a novel instance based FMBO approach whose main idea is to detect "bad" parameter settings early and avoid nonessential function evaluation. Kriging surrogate model based FMBO achieved a saving of more than 60 % of function evaluation in comparison with the classical MBO strategy. For the random forest surrogate this saving amounted even to 80 %. As expected, the savings are reflected in a worse performance. However, by considering the same evaluation budget FMBO strategies showed more reliable results as the corresponding classical optimization. One of the interesting findings is the fact that kriging models can by successfully applied to categorical parameters by handling them as integer ones. The best found onset detection configuration was presented and compared to the corresponding settings mentioned in the literature.

For the further research other point selection strategies should be proposed and compared with SMAC and other state-of-the-art proposals. One possibility would be a cascade of selection models with a small number of instances for initial performance prediction. However, if the uncertainty of the target prediction is too high, more instances should be evaluated. This procedure should be iteratively repeated until the uncertainty becomes low. A further concept could be a screening strategy: early identification of not significant algorithm parameters or parameter spaces. In this way the region of interest could be restricted to an especially interesting area during the optimization.

Acknowledgements This paper is based on investigations of the projects B3 and C2 of SFB 823, which are kindly supported by Deutsche Forschungsgemeinschaft (DFG).

References

Bauer, N., Schiffner, J., & Weihs, C. (2013). Comparison of classical and sequential design of experiments in note onset detection. In B. Lausen, D. V. Poel & A. Ultsch (Eds.), *Algorithms from and for nATURE AND life. Studies in classification, data analysis, and knowledge organization* (pp. 501–509). New York: Springer International Publishing.

Bello, J. P., Daudet, L., Abdallah, S., Duxbury, C., Davies, M., & Sandler, M. B. (2005). A tutorial on onset detection in music signals. *IEEE Transactions on Speech and Audio Processing, 13*(5), 1035–1047.

Bischl, B., Lang, M., Mersmann, O., Rahnenfuerer, J., & Weihs. C. (2015). BatchJobs and BatchExperiments: Abstraction mechanisms for using R in batch environments. *Journal of Statistical Software, 64*(1), 1–25. doi:10.18637/jss.v064.i11. ISSN: 1548-7660. http://www.jstatsoft.org/index.php/jss/article/view/v064i11.

Bischl, B., Wessing, S., Bauer, N., Friedrichs, K., & Weihs, C. (2014). MOI-MBO: Multiobjective infill for parallel model-based optimization. In P. M. Pardalos, M. G. C. Resende, Ch. Vogiatzis & J. L. Walteros (Eds.), *Learning and intelligent optimization.* Lecture Notes in Computer Science (pp. 173–186). New York: Springer International Publishing.

Dixon, S. (2006). Onset detection revisited. In *Proceedings of the 9th International Conference on Digital Audio Effects* (DAFx-06) (pp. 133–137).

Holzapfel, A., Stylianou, Y., Gedik, A. C., & Bozkurt, B. (2010). Three dimensions of pitched instrument onset detection. *IEEE Transactions on Audio, Speech, and Language Processing, 18*(6), 1517–1527.

Hutter, F., Hoos, H. H., & Leyton-Brown, K. (2011). Sequential model-based optimization for general algorithm configuration. In C. A. Coello Coello (Ed.), *Learning and Intelligent Optimization.* Lecture Notes in Computer Science (pp. 507–523). New York: Springer International Publishing.

Jones, D. R., Schonlau, M., & Welch, W. J. (1998). Efficient global optimization of expensive black-box functions. *Journal of Global Optimization, 13*(4), 455–492.

Rosão, C., Ribeiro, R., & Martins de Matoset, D. (2012). Influence of peak selection methods on onset detection. In *Proceedings of the 13th International Conference on Music Information Retrieval* (ISMIR 2013) (pp. 517–522).

Sexton, J., & Laake, P. (2009). Standard errors for bagged and random forest estimators. *Computational Statistics & Data Analysis, 53*(3), 801–811.

Recognition of Leitmotives in Richard Wagner's Music: An Item Response Theory Approach

Daniel Müllensiefen, David Baker, Christophe Rhodes, Tim Crawford, and Laurence Dreyfus

Abstract In this study we aim to understand listeners' real-time processing of musical leitmotives. We probe participants' memory for different leitmotives contained in a 10-min passage from the opera *Siegfried* by Richard Wagner, and use item response theory to estimate parameters for item difficulty and for participants' individual recognition ability, as well as to construct novel measurement instruments from questionnaire-based tests. We investigate the relationship between model parameters and objective factors, finding that prior Wagner expertise and musical training were significant predictors of leitmotive recognition ability, while item difficulty is explained by chroma distance and perceived emotional content of the leitmotives.

1 Introduction

1.1 Psychology of Leitmotives

The leitmotives in Richard Wagner's *Der Ring des Nibelungen* serve a range of compositional and psychological functions, including the introduction of musical structure and mnemonic devices for the listener. These leitmotives are short musical ideas that are representative of concepts in the dramatic narrative, and differ greatly in their construction, salient aspects (e.g. rhythmic, melodic, harmonic), and their usage in particular scenes and contexts. While the topic of leitmotives in Richard Wagner's music has been discussed extensively in the traditional musicological literature (Dalhaus 1979), little work has been done on the perception and psychology of real-time processing of these musical ideas. In this study, we perform

D. Müllensiefen (✉) • D. Baker • C. Rhodes • T. Crawford
Goldsmiths, University of London, New Cross, London SE14 6NW, UK
e-mail: d.mullensiefen@gold.ac.uk; d.baker@gold.ac.uk; c.rhodes@gold.ac.uk; t.crawford@gold.ac.uk

L. Dreyfus
University of Oxford, Wellington Square, Oxford OX1 2JD, UK
e-mail: laurence.dreyfus@magd.ox.ac.uk

© Springer International Publishing Switzerland 2016
A.F.X. Wilhelm, H.A. Kestler (eds.), *Analysis of Large and Complex Data*, Studies in Classification, Data Analysis, and Knowledge Organization,
DOI 10.1007/978-3-319-25226-1_40

473

a psychological experiment to attempt to understand how individuals are able to recall leitmotives, investigating both musical- and listener-based parameters. Using an item response theory (IRT) approach, we estimate difficulty parameters of the leitmotives (items) themselves, as well as parameters characterising participants' individual recognition ability.

A small number of prior studies have empirically investigated the perception of leitmotives. Initial research on the leitmotives used a learning paradigm to explore how listeners with various musical backgrounds would encode and subsequently recognise various leitmotives in real time. Using an excerpt from *Das Rheingold*, (Deliège 1992) found that musicians were able to encode musical material much more rapidly than non-musicians, and that each of the leitmotives presented different levels of difficulty in their recognition. This research was expanded upon by introducing additional visual stimuli, as well as considering listener parameters beyond musical training, finding (Albrecht 2012) that visual stimuli did not help leitmotive recognition, but that the non-musical parameter of Wagner expertise did predict an individual's recognition ability. This study explores the difficulty of encoding the leitmotives and the contributions of extra-musical factors to an individual's recognition rate.

1.2   Experimental Design and Procedure

The experiment used a within-subjects design. Participants were asked to listen actively to the same 10-min passage from Richard Wagner's opera *Siegfried* used by Albrecht (2012). This passage was chosen for its narrative qualities and high leitmotive density. The participants were told in advance that they would perform a memory recall task following the listening phase, in which they would have to indicate explicitly whether or not they recall hearing musical material from the passage, and to rate the subjectively perceived emotional qualities of the musical material, such as the level of emotional arousal and valence expressed. After the listening phase, participants were played a list of 20 excerpts, each containing a leitmotive. Ten of these leitmotives had occurred in the passage that they had heard before; the other ten were taken from a passage from the same performers' recording of Richard Wagner's *Götterdämmerung*. For each item participants had to indicate: whether they had heard the leitmotive in the 10-min listening phase or not; how confident they were in their decision; and also how emotionally arousing they perceived the leitmotive together with an emotional judgement (happy–sad) both on 7-point scales.

After completing this memory recognition task, participants filled out questionnaires assessing factors that were believed to contribute to an individual's leitmotive recognition ability: musical training, measured using the Musical Training subscale of the Goldsmiths Music Sophistication Index self-report questionnaire (Müllensiefen et al. 2014); affinity for the music of Richard Wagner and objective Wagner knowledge, measured with two novel questionnaire instruments that were constructed via factor analysis and Rasch modelling (see Sects. 2.2 and 2.3 below); and German speaking ability, on a 7-point agreement Likert scale.

1.3 Advantages of Item Response Approaches in Psychological Research

IRT (Rasch 1960; Birnbaum 1968; Lord 1980) was developed to assess individuals on attributes that are not directly observable ("latent" traits, such as aspects of intelligence or personality) using data from the individuals' performance on a suitable test. Among the most commonly cited advantages of IRT and latent trait models are: their foundation in well-established statistical theory (maximum-likelihood modelling); and their ability to quantify uncertainty via confidence intervals. In addition, Rasch models are a special class of IRT models which possess the property that item and person scores can be considered independent from the particular sample used.

Most concepts in cognitive psychology that are used to describe mental processes (such as memory capacity or attention span) are unobservable, yet item response approaches are still relatively rare within cognitive or experimental psychology. Borsboom (2006) discusses a number of reasons for the slow uptake of IRT models in most areas of psychology and also encourages its wider application. The current study represents a suitable scenario for IRT, where experimental data is generated by individuals taking a newly designed test, and where the two main research questions investigate (a) person-based factors explaining the individual's ability to perform on the test and (b) per-item factors contributing to item difficulty. We are asking what characterises listeners who perform better at encoding leitmotives in a realistic listening situation, and what musical or acoustic factors contribute to the recognisability of individual leitmotives. Compared to traditional analysis approaches in cognitive psychology, the IRT approach enables us to estimate participant ability and item difficulty within the same model and to quantify the uncertainty about both kinds of parameters through confidence intervals.

2 Data Analysis

2.1 Variables Measuring Participant Background

As described in Sect. 1.2, we collected four person-specific pieces of information: musical expertise, German speaking ability, Wagner affinity and Wagner knowledge. The experiment used a convenience sample ($N = 100$), with additional recruiting effort made to recruit participants with either familiarity or fondness of the music of Richard Wagner from across the greater London area, though more ($N = 14$)

individual's data was used in a pilot experiment and their survey and quiz response were retained for the final Rasch modelling ($N = 114$). The experimental ($N = 100$) sample was made up of 55 females (55 %) and 45 males (45 %) with a mean age of 28.7 (s.d. $= 11.82$). It is worth noting that the following analyses proceed in a step-wise fashion, where we first fit IRT and factor models to the data of the several tests and questionnaire separately and aim to establish sound measurement models for each of these novel tests. Only subsequently we combine data in a structural equation model (SEM) and a regression model. This step-wise analysis procedure allows us to check model assumptions at each stage and, where necessary, to apply adjustments to individual models (e.g. by excluding items that violate assumptions). However, the construction of the measurement models and the modelling of the structural relations between the factors of interest were carried out independently to avoid modelling bias.

2.2   Factor Analysis of Wagner Affinity Survey

To model individuals' affinity for Wagner's music we applied factor-analytic techniques to the Likert-scale data of the survey. We conducted minimal residual factor analysis on the 14 items of the affinity questionnaire using the R psych package (Revelle 2014). Parallel analysis (Horn 1965) as well as Velicer's Simple Structure (Revelle and Rocklin 1979) and the Minimum Average Partial correlation (Velicer 1976) criterion were employed to decide on the number of factors, giving ambiguous results (suggesting either 1 or 2 factors). We inspected the items for their respective factor loadings on a one-dimensional solution, finding that only one had a factor loading of less than 0.6 (with a loading of 0.482). After the removal of this item, "How often do you perform the music of Richard Wagner?", we reran the minimum residual factor analysis, and all the diagnostics suggested one-dimensional factor solution. Cronbach's α for this solution was 0.97, indicating a high internal reliability of this new Wagner affinity scale in terms of classical test theory.

2.3   Rasch Modelling of the Wagner Quiz

We designed the Wagner knowledge quiz to measure a postulated latent trait of "Wagnerism", the extent to which an individual has developed knowledge of the life and music of Richard Wagner both in terms of musicological knowledge as well as a detailed understanding of the narrative and music of his operas. The quiz had 14 multiple choice items, each with four response options, and each item was scored as either correct (1) or incorrect (0).

Because of the limited sample size ($N = 114$) we fit a Rasch model, a comparatively simple member of the family of IRT models (de Ayala 2009), requiring relatively few parameters to be estimated. The initial Rasch model was fitted using the conditional maximum likelihood criterion as implemented in eRm package in R (Mair and Hatzinger 2007) which assumes equal item difficulty as well as equal discrimination across the participant subgroups. However, Pononcy's non-parametric T_{10} (with median split sampling 1000 matrices) for subgroup invariance as well as the T_{pbis} test for equal item discrimination both indicated that the assumptions were not met. Applying a step-wise elimination procedure based on individual item fit removed 6 items and resulted in a new Rasch model containing 8 items. This resulting model passed both the T_{10} and T_{pbis} tests as well a non-parametric version of the Martin-Loef (Glas and Verhelst 1995) test for unidimensionality indicating that the main assumptions for Rasch models were met for the final 8-item model. The item difficulty parameters of the final version of the Wagner knowledge test showed a good range, from 1.04 for the item "When did Wagner die?" to -1.28 for the item "What opera is considered to be among the romantic operas that paved the way for Wagner's music dramas?".

2.4 Listening Response Analysis

The memory test contained 20 items, and participants responded with either a "yes" or "no" depending on whether they recognised the leitmotive from the 10-min listening passage or not. Each response was scored using a binary coding as either correct (1) or incorrect (0). These binary responses were then analysed by fitting using a Rasch Model for the same reasons as in Sect. 2.3. Applying the non-parametric T_{pbis} test as implemented in the R package eRm (Mair and Hatzinger 2007) to the model indicated an equal item discrimination, but the T_{10} test suggested that it was missing subgroup invariance. A graphical model check also indicated that several items differed in difficulty across the high and low performing subgroups of subjects. However, the result of the non-parametric Martin-Loef test suggested that the memory test would meet the criterion of unidimensionality.

The failure of the Rasch model to meet the criterion of subgroup invariance leaves several options. First, we explored fitting a two-parameter model with an additional guessing parameter per item (but equal discrimination) to accommodate for the possibility that participants were guessing on individual items, using the marginal maximum likelihood approach provided by the R package ltm (Rizopoulos 2006). However, the two-parameter solution appeared to be degenerate with several difficulty parameters outside the normal range. Second, we considered excluding items from the test until model assumptions are met, as was done for the Wagner knowledge test, or alternatively modelling items with a multi-dimensional IRT model. But because the leitmotive items themselves are the objects of interest in one of the subsequent analysis stages, excluding several items from the small initial pool of only 20 would leave too few to generate interesting results in terms of the memorability of different types of leitmotives. Therefore, we opted to accept the existing model, acknowledging that one of the model assumptions is not met.

This means that there is some uncertainty about the item difficulty parameters. However, as Lord (1980, p. 190) points out, the use of a Rasch model might still be justified when the sample size is small, even if assumptions do not hold. In this case, estimators derived from the Rasch model might not be optimal, but might still be more accurate than estimators derived from more flexible IRT models (e.g. the 3-parameter model).

2.5 Modelling Memory for Leitmotives with a Structural Equation Model

We specified an SEM to determine the contributions of person-specific variables to explain the memory performance in the leitmotive recognition test. The person parameters from the Rasch model for the memory test (see Sect. 2.4) served as the target variable. As predictor variables we specified the musical training and German speaking scores, and a latent Wagner expertise variable, hypothesised to influence Wagner knowledge and affinity scores (which we treated as observed variables in the context of this SEM). We also specified covariances between Gold-MSI scores and Wagner knowledge as well as Wagner affinity scores. This initial model was entirely hypothesis driven and was fit using the R package lavaan using maximum likelihood with robust standard errors (Yves 2012).

The initial model already showed an almost acceptable fit as suggested by the fit indices derived from the robust estimator (Comparative Fit Index = 0.94; Tucker–Lewis Index = 0.8, RMSEA = 0.16, SRMR = 0.07). We inspected the model parameters and removed one non-significant regression path (from German speaking ability to memory scores) and one non-significant covariance (between musical training and Wagner knowledge). We refit the model, resulting in a model with only significant path coefficients and showing almost perfect fit indices (CFI = 1; TLI = 1, RMSEA < 0.001, SRMR = 0.01). The model, depicted in Fig. 1, shows that Wagner expertise and musical training both positively influence the participants ability to recognise leitmotives in the listening test; Wagner expertise is about twice as influential as musical training.

2.6 Modelling Leitmotive Difficulty

Previous evidence in the literature (Müllensiefen and Halpern 2014) suggests that different musical features are responsible for the correct recognition of previously heard melodies ("hits") and the correct identification as novel of melodies that have not been previously heard ("correct rejections"). Therefore, we split the set of 20 leitmotives into "old" motives (that had been heard previously in the experiment) and "novel" motives (that did not occur in the passage), and ran two separate

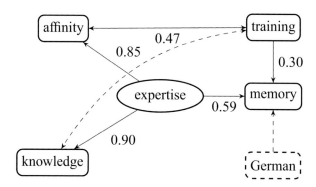

Fig. 1 Structural equation model for memory of leitmotives, incorporating Wagner knowledge and affinity, their combination into Wagner expertise, and the effect of that and generic musical training on score in the memory test. The *dashed lines* and *boxes* indicate non-significant relations removed from the final model, which contains only significant influences

Table 1 Final regression model for "novel" motives

p-value	t statistic	Error	Estimate	
0.0356*	2.597	1.5628	4.0578	Intercept
0.0480*	−2.392	1.5788	−3.7761	Chroma distance
0.0705	−2.132	0.2135	−0.4550	Valence

* denotes the significance level of $p < 0.05$

linear regression analyses with the item difficulty scores from the Rasch model as dependent variables. In both models the predictor variables were the mean of the participants' arousal and valence ratings carried out at the recognition phase as well as an acoustical distance measure based on chroma feature extraction (Mauch and Dixon 2010) and a criterion for distance thresholding (Casey et al. 2008). In addition, we used the number of occurrences of each leitmotive during the 10-min listening passage as a predictor for the regression model for "old motives".

Having only 10 observations per model, we found it necessary to reduce the number of predictor variables using step-wise backward and forward model selection using the Bayesian Information Criterion (BIC) as the model fit index, rather than using a threshold of statistical significance. The coefficients of final regression model for the "novel" motives are given in Table 1. The model has an adjusted R^2 of 0.35 but fails to reach significance overall ($F_{(2,7)} = 3.4$, $p = 0.09$). The model includes the chroma feature distance as a predictor, indicating that motives with a large chroma distance (loosely, "sounding dissimilar") to any segment within the 10-min listening passage are easier to identify as novel than motives with a small chroma distance (closer harmonically). In addition, the participants' valence ratings are selected as a predictor in the final model, albeit with a non-significant coefficient estimate. Here, motives rated as rather sad were more difficult to identify as novel motives. Neither the number of occurrences in the listening passage nor the perceived emotional arousal of the leitmotive was predictor in the regression model.

Table 2 Final regression model for "old" motives

p-value	t statistic	Error	Estimate	
0.1005	1.856	1.4217	2.6390	Intercept
0.0924	−1.911	0.2750	−0.5256	Arousal

The final regression model for the "old" motives is given in Table 2. The model has an adjusted R^2 of 0.23 and also fails to reach significance overall ($F_{(2,8)} = 3.7$, $p = 0.09$). The model includes the mean arousal ratings as the single predictor, indicating that motives that are perceived as more arousing are also recognised better as old motives. None of the other predictor variables (number of occurrences, emotional valence, harmonic distance) featured in the final regression model for old items.

3 Discussion

The decision to use an IRT approach was motivated by several factors which might generalise to similar research scenarios in empirical musicology. Firstly, we had to devise new measurement instruments for assessing very specific abilities that have not been well investigated before (e.g. Wagner expertise), and the IRT approach framework in general and Rasch modelling in particular provide a rigorous framework for constructing new tests as well as measuring the latent ability to perform on these tests. As a result the Wagner knowledge test and the Wagner affinity survey are now finished tools that can be used in any subsequent Wagner research; we have confirmed the specific objectivity of the Wagner knowledge test, and it should therefore generalise well to a new sample. Secondly, the leitmotive recognition experiment had the dual purpose of measuring the ability of participants with different backgrounds to recognise leitmotives that they had been previously exposed to in the 10-min listening passage, as well as measuring the difficulty of individual leitmotives to be recognised or identified as novel. This dual aim "to gather data simultaneously about participants as well as about items of a test" is not very common in psychological research which tends to focus on the psychological mechanisms of the participants. But this approach is well-suited to empirical music research that uses ecologically valid stimuli. The IRT framework and the Rasch model that we used provide a very elegant way of generating data characterising participants and leitmotive items at the same time and within the same model. The structural equation analysis using participants' ability coefficients demonstrates how important expertise and familiarity with a particular musical style are in order to perform well on a listening test with stimuli from this style in fact, Wagner expertise proved to be much more important than musical expertise in order to perform well on the listening test. The SEM also showed that musical training did not (directly) correlate with specific Wagner knowledge, and Wagner knowledge can be regarded as an effective type of musical expertise that is not linked to instrumental practice.

The fact that the Rasch model from the listening test did not exhibit subgroup invariance suggests some caution in interpreting the results of the subsequent regression analyses, and clearly both regression models suffer from the low item count of $N = 10$, as the coefficients of some model predictors did not reach the usual significance level. However, both regression models suggest that emotional processing of the leitmotives is linked to performance on the cognitive memory task, supporting the idea that cognitive and emotional processes during music listening are not separate but can significantly influence each other. In a forthcoming investigation, we aim to measure electrodermal activity and heart-rate data from listeners attending performances of Gergiev's production of the *Ring* and correlate those data with leitmotive occurrence.

We also found that for novel leitmotives harmonic distance in the acoustical signal was a predictor of their perceptual difficulty, indicating that harmonic distance can partially model a memory process that leads to the illusion of the familiar. However, this result should be replicated with a new set of leitmotive stimuli taken from a different passage, where the findings from the present study with regards to the influential predictor variables can serve as proper hypotheses. We also note, given Wagner's own theory of *Gefühlverständnis*, that it is unclear how much Wagner himself intended the listener to recognise leitmotive, and whether the greater difficulty we find associated with sadder motives is therefore more in line with his artistic intentions.

While the IRT approach has proved very useful for the analysis of our data, we note a few caveats. Firstly, IRT models require a substantial amount of data in order to be fit and to produce coefficients with acceptable confidence intervals. This is even more true for models with additional discrimination or guessing parameters. It is worth noting that the sample size of the memory experiment ($N = 100$) is at the lower bound of what is commonly recommended (de Ayala 2009), even for simple Rasch models.

Secondly, not all psychological or empirical music research questions can be implemented as tests where correct/incorrect answers can be scored objectively. Much music psychological work investigates the appearance of musical stimuli and can ask for subjective perceptions rather than objective ability to perform a test (Kingdom and Prins 2009). In these scenarios, IRT approaches appear to be less useful.

Finally, IRT models generally do not allow for a detailed analysis of the types of individual participants' biases. Here, techniques from signal detection theory (Macmillan and Creelman 2005) that can distinguish between e.g. "false alarms" and "misses" allow for a greater insight into the nature of the cognitive processes behind the performance on a particular test and into potentially interesting interactions between both person and item characteristics.

In sum, IRT is most useful when the main research questions target individual differences between participants and data from a large sample with good variation in test performance and related background variables can be obtained. Using an IRT approach we have been able to show how individual differences in musical training and Wagner expertise lead to differential performance on the leitmotive recognition task. Because recognising leitmotives in the constant auditory stream of Wagner's music affects a listener's musical perception, the individual differences we have identified may well influence the experience of Wagner's music, both in cognitive and emotional terms.

Acknowledgements This work was supported by the *Transforming Musicology* project, AHRC AH/L006820/1.

References

Albrecht, H. (2012). Wahrnehmung und Wirkung der Leitmotivik in Richard Wagners Ring des Nibelungen — Eine empirische Studie zur Wiedererkennung ausgewählter Leitmotive aus musikpsychologischer und musiksemiotischer Perspektive. Masters Dissertation.
Birnbaum, A. (1968). Some latent trait models and their use in inferring an examinee's ability. In F. M. Lord & M. R. Novick (Eds.), *Statistical theories of mental test scores* (pp. 395–479). Reading, MA: Edison-Wesley.
Borsboom, D. (2006). The attack of the psychometricians. *Psychometrika, 71*(3), 425–440.
Casey, M., Rhodes, C., & Slaney, M. (2008). Analysis of minimum distances in high-dimensional musical spaces. *IEEE Transactions on Audio, Speech and Language Processing, 16*(5), 1015–1028.
Dalhaus, C. (1979). Richard Wagner's music dramas. Cambridge: Cambridge University Press.
De Ayala, R. J. (2009). *Theory and practice of item response theory.* New York: Guilford.
Deliège, I. (1992). Recognition of the Wagnerian Leitmotiv. *Jahrbuch der Deutschen Gesellschaft für Musikpsychologie, 9,* 25–54.
Glas, C. A. W., & Verhelst, N. D. (1995). Testing the Rasch model. In G. H. Fischer & I. W. Molenaar (Eds.), *Rasch models: Foundations, recent developments, and applications* (pp. 69–95). New York: Springer-Verlag.
Horn, J. (1965). A rationale and test for the number of factors in factor analysis. *Psychometrika, 30*(2), 179–185.
Kingdom, F., & Prins, N. (2009). Psychophysics: A practical introduction. London: Academic Press.
Lord, F. M. (1980). Applications of item response theory to practical testing problems. New York: Routledge.
Macmillan, N. A., & Creelman, C. D. (2005). Detection theory: A user's guide (2nd ed.). Mahwah, NJ: Erlbaum
Mair, P., & Hatzinger, R. (2007). Extended Rasch Modeling: The eRm Package for the Application of IRT Models. In *R. Journal of Statistical Software, 20*(9), 1–20. doi:http://dx.doi.org/10.18637/jss.v020.i09
Mauch, M., & Dixon, S. (2010). Approximate note transcription for the improved identification of difficult chords. In *Proc. International Society for Music Information Retrieval Conference,* Utrecht (pp. 135–140).
Müllensiefen, D., Gingras, B., Musil, & Stewart, L. (2014). The Musicality of Non-Musicians: An Index for Assessing Musical Sophistication in the General Population. *PloS One, 9*(2): e89642. DOI:10.1371/journal.pone.0089642.
Müllensiefen, D., & Halpern, A. (2014). The role of features and context in recognition of novel melodies. *Music Perception, 31*(5), 418–435.
Rasch, G. (1960). *Probabilistic models for some intelligence and attainment tests.* Copenhagen: Danmarks paedagogiske institut.
Revelle, W. (2014). psych, Procedures for personality and psychological research. R package version 1.4.8. http://personality-project.org/r/psych-manual.pdf

Revelle, W., & Rocklin, T. (1979). Very simple structure - alternative procedure for estimating the optimal number of interpretable factors. *Multivariate Behavioral Research, 14*(4), 403–414.

Rizopoulos, D. (2006). ltm: An R package for latent variable modelling and item response theory analyses. *Journal of Statistical Software, 17*(5), 1–25.

Velicer, W. (1976). Determining the number of correlations components from the matrix of partial. *Psychometrika, 41*(3), 321–327.

Yves, R. (2012). lavaan: An R package for structural equation modeling. *Journal of Statistical Software, 48*(2), 1–36.

Part XI
Data Analysis in Interdisciplinary Domains

Optimization of a Simulation for Inhomogeneous Mineral Subsoil Machining

Swetlana Herbrandt, Claus Weihs, Uwe Ligges, Manuel Ferreira, Christian Rautert, Dirk Biermann, and Wolfgang Tillmann

Abstract For the new generation of concrete which enables more stable constructions, we require more efficient tools. Since the preferred tool for machining concrete is a diamond impregnated drill with substantial initial investment costs, the reduction of tool wear is of special interest. The stochastic character of the diamond size, orientation, and position in sintered segments, as well as differences in the machined material, justifies the development of a statistically motivated simulation. In the simulations presented in the past, workpiece and tool are subdivided by Delaunay tessellations into predefined fragments. The heterogeneous nature of the ingredients of concrete is solved by Gaussian random fields. Before proceeding with the simulation of the whole drill core bit, we have to adjust the simulation parameters for the two main components of the drill, diamond and metal matrix, by minimizing the discrepancy between simulation results and the conducted experiments. Due to the fact that our simulation is an expensive black box function with stochastic outcome, we use the advantages of model based optimization methods.

1 Introduction

In the building industry durable high-strength reinforced concrete is an essential component which contributes to build stable constructions. Precisely because of these characteristics, subsequent machining of reinforced concrete can be difficult

S. Herbrandt (✉) • U. Ligges
Department of Statistics, TU Dortmund, Dortmund, Germany
e-mail: herbrandt@statistik.tu-dortmund.de; ligges@statistik.tu-dortmund.de

C. Weihs (✉)
Chair of Computational Statistics, Faculty of Statistics, TU Dortmund, Dortmund, Germany
e-mail: weihs@statistik.uni-dortmund.de

M. Ferreira • W. Tillmann
Institute of Materials Engineering, TU Dortmund, Dortmund, Germany
e-mail: manuel.ferreira@tu-dortmund.de; wolfgang.tillmann@udo.edu

C. Rautert • D. Biermann
Institute of Machining Technology, TU Dortmund, Dortmund, Germany
e-mail: rautert@isf.maschinenbau.uni-dortmund.de; biermann@isf.de

© Springer International Publishing Switzerland 2016
A.F.X. Wilhelm, H.A. Kestler (eds.), *Analysis of Large and Complex Data*, Studies in Classification, Data Analysis, and Knowledge Organization,
DOI 10.1007/978-3-319-25226-1_41

even when using diamond impregnated tools. Due to substantial initial investment costs for such tools, an increase of efficiency is of special interest.

To be able to control stochastic characteristics of the multiphase material like size and position of concrete aggregate and of the multiphase tool (e.g., diamond orientation) in simulated grinding experiments, different approaches for the preliminary stage of such a simulation were derived. The actual version of the single diamond simulation is described in Sect. 2 and has, in comparison to the previous versions, the advantage of faster evaluation without loss of precision. Before proceeding with the next simulation level, the simulation parameters should be adjusted. As reference for this adjustment, data from single diamond experiments are used. In these trials, forces affecting a diamond when scratching on basalt or cement are recorded (Raabe et al. 2011; Weihs et al. 2014). Following a central composite design in cutting speed v_c and depth per revolution a_e with two centers and three repetitions results in 30 time series for each force direction and each material. The general idea is a two steps procedure of parameter adjustment and parameter modeling. In the first step the parameters are adjusted for each available (v_c, a_e)-combination. Then, regression models for the simulation parameters are derived to have the possibility of simulating processes with any (v_c, a_e)-combinations. This approach involves some difficulties. Despite data preprocessing, there are still negative forces in the time series of the conducted experiments, the simulation cannot deal with. Additionally it was not possible to eliminate the workpiece tilting that dominates the seasonal component, due to very small depths per revolution. To handle this problem, the parameters of the sine waves were estimated and the depth per iteration in the simulation was adopted, assuming that the total depth per revolution is given by a_e. The workpiece tilting in combination with noise unfortunately leads to the problem of identifying the point of first diamond workpiece contact, which is the beginning of the drilling process. Shifting of the simulated time series in time direction should help manage this missing information when comparing observed forces and simulation outcome.

2   Simulation Model

Since the optimization will be applied to the single diamond simulation, where the diamond has the shape of a pyramid with a rounded tip and almost no wear, we refrain from a detailed description of segment simulation.

2.1   Tool and Workpiece

The basic idea of diamond simulation is to build a lattice defining the diamond's shape and to apply a Delaunay tessellation which constructs three-dimensional simplices of these grid-points in a way, that each simplex contains the vertices of an

(a)   (b)   (c)

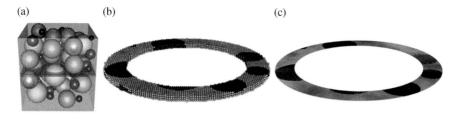

Fig. 1 (**a**) Basalt grains, (**b**) coarse grid for concrete with diameter 40 mm and 8 basalt grains (*dark gray*), and (**c**) finer grid representing the workpiece surface

irregular tetrahedron. By the no wear assumption the number of points in the lattice can and should be as small as possible. If diamond wear is allowed, the number of points and thereby also the number of simplices should be greater because the wear is simulated by removing simplices, which should not be too big. In this case we also require points inside the diamond shape which, for example, can be arranged as the diamond crystal structure (Raabe et al. 2011).

The segment simulation, as well as the concrete simulation, makes use of a subprocedure that randomly fills a part of a predefined space (e.g., given by the segment or workpiece shape) with objects (e.g., diamonds or basalt grains). For simplicity basalt grains are assumed to have the form of spheres, whose diameters follow a normal distribution with parameters estimated from grading curve of the basalt used for the concrete in the real experiments. Due to the known expected size and shape of the objects, the expected object volume and therefore the expected number n of objects that are needed to fill a desired percentage of the available space volume, can be estimated. For grains with diameters $d_1 \geq \cdots \geq d_n$ we sample in decreasing diameter order positions in the workpiece, check which positions provide enough space for a grain with this diameter and sample one of these points (see Fig. 1a).

As in the previous workpiece simulations, two point grids, a coarse grid containing information about the material heterogeneity and a finer grid which represents the workpiece, are used. The points of the coarse grid with minimal point distance δ_{coarse} fill the complete hollow cylinder. From the information about positions and sizes of the basalt grains, the material for each point is determined. Heterogeneity within the workpiece materials is simulated by sampling Gaussian random fields for each basalt grain and all cement points (Raabe et al. 2012; Schlather et al. 2014). Here material specific covariance functions with parameters estimated from the force time series' seasonality of single diamond experiments on basalt and cement are used. The result is shown in Fig. 1b where the gray graduation constitutes different heterogeneity values. To reduce simulation time, all points of the finer grid with minimal point distance δ_{fine} lie on the same level creating the surface of the workpiece. Additionally, we refrain from subdividing the point set into simplices, which is the major change to the previous workpiece simulations. Instead, the surface lattice is adapted during the process simulation (see Sect. 2.2).

In the last step the material is determined for each point of the finer grid and the heterogeneity values are interpolated from the coarse grid by material separated ordinary Kriging.

2.2 Iterative Process

After arranging the simulated diamond on the simulated workpiece surface, the iterative process starts with a first movement of the diamond along the drilling path. The next step is to determine all workpiece points inside the diamond. To simulate the cutting process these points have to move to their new positions with respect to the material m which they represent. Since cement wear leads to brittle fracture and steel is dominated by abrasive wear, a flexible model is needed that satisfies both types of wear. Consider the following point height H_m density for material m

$$P\left(H_m = h\right) = 2\xi P\left(H_m^- = h\right) I_{h \leq h_D}\left(h\right) + 2\left(1 - \xi\right) P\left(H_m^+ = h\right) I_{h > h_D}\left(h\right),$$

where $H_m^- \sim N\left(h_D, |h_W - h_D| \, s_m^-\right)$, $H_m^+ \sim N\left(h_D, |h_W - h_D| \, s_m^+\right)$, h the new height, h_W is the actual point height, and h_D is the height of the diamond's edge below (see Fig. 2).

Assuming that there is a probability ξ to sample a new height h below the diamond's edge h_D, represented by the random variable H_m^-, and a small probability $1 - \xi$ for "diamond jumps," represented by H_m^+, we can simulate different types of wear by adjusting the parameters s_m^- and s_m^+. After sampling the new point heights, the removed workpiece volume v_i (i number of iteration) is calculated. Since the chip volume would not explain higher forces on steel, in addition the sum of heterogeneity values z_i of these points is taken into consideration.

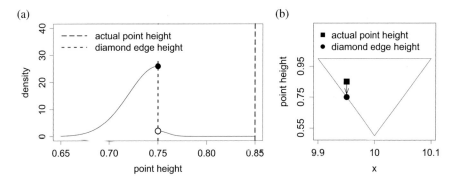

Fig. 2 (**a**) Density of point height distribution for $h_W = 0.85$ and $h_D = 0.75$, (**b**) diamond convex hull with points $h_W = 0.85$ and $h_D = 0.75$

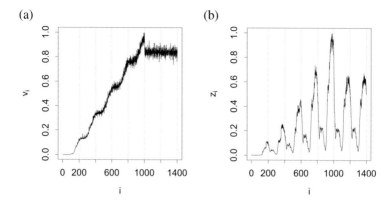

Fig. 3 (**a**) Scaled chip volume v_i, (**b**) scaled sums of heterogeneity values z_i

At the end of iteration $I = N \cdot r_U$, where r_U is the sampling rate per revolution and N the number of revolutions, the normal and radial forces are computed as a weighted sum of $v = (v_1, \ldots, v_I)^T$ and $z = (z_1, \ldots, z_I)^T$

$$F_{\text{norm,sim}} = g_{\text{het,norm}} \cdot \frac{z}{\max\{z_1, \ldots, z_I\}} + g_{\text{vol,norm}} \cdot \frac{v}{\max\{v_1, \ldots, v_I\}}$$

$$F_{\text{rad,sim}} = g_{\text{het,rad}} \cdot \frac{z}{\max\{z_1, \ldots, z_I\}} + g_{\text{vol,rad}} \cdot \frac{v}{\max\{v_1, \ldots, v_I\}},$$

where $g_{\text{het,norm}}$ and $g_{\text{vol,norm}}$ are the weights for the normal force $F_{\text{norm,sim}}$ and $g_{\text{het,rad}}$ and $g_{\text{vol,rad}}$ are the weights for the radial force $F_{\text{rad,sim}}$.

Figure 3 shows the resulting chip volume and sum of heterogeneity values for a short simulation with $I = 1400$ iterations. While v mainly explains the trend of the time series, z will dominate the seasonality. In the last two revolutions the level of removed volume is constant because the contact area between diamond and workpiece stops increasing. Since the diamond used for the simulations in the optimization has the shape of a pyramid, this phenomenon will not be observed there. Nevertheless it can be observed in time series of conducted experiments when using diamonds in the form of a truncated octahedron, as used in segments, or even in experiments with single segments.

3 Optimization

The simulation, presented above, has 11 parameters $\theta \in \Theta \subset \mathbb{R}^{11}$ which have to be adjusted to achieve realistic output. The parameters r_U, δ_{coarse}, and δ_{fine} have to fit for both force directions and both materials (basalt and cement). While the wear parameters s_m^- and s_m^+ are material specific but do not depend on the two force directions, we have the opposite case for the weight parameters $g_{\text{het,f}}$ and $g_{\text{vol,f}}$ with $f \in \{\text{norm, rad}\}$.

Due to different lengths I and J of the time series F_{sim} and the measured force time series F, we require a similarity measure that can deal with this difficulty. The aim is to minimize the expected deviation

$$O = E\left[d_{\text{DTW}}\left(F_{\text{sim}}, F\right)\right] + E\left[d_R\left(F_{\text{sim}}, F\right)\right]$$

between measured and simulated forces where

$$d_{\text{DTW}}\left(F_{\text{sim}}, F\right) = \frac{c_{p^\star}\left(F_{\text{sim}}, F\right)}{I + J}$$

is the normalized warping path distance of time series F_{sim} and F and

$$c_{p^\star} = \min\left\{c_p\left(F_{\text{sim}}, F\right) = \sum_{l=1}^{L}\left\|F_{\text{sim},i_l} - F_{j_l}\right\| \mid p\left(I, J\right) - \text{warping path}\right\}$$

are the total path costs for the best path $p^\star = \left[\left(i_1^\star, j_1^\star\right), \ldots, \left(i_L^\star, j_L^\star\right)\right]$ connecting the $i_l^\star$-th point of F_{sim} with the $j_l^\star$-th point of F ($l = 1, \ldots, L$) (see, e.g., Müller 2007; Ding et al. 2008 for dynamic time warping and Giorgino 2009 for the R package). The second part of the criterion is considered because the DTW distance based measure makes no distinction between, e.g., downward deviations and deviations caused by phase displacement. In case of downward deviation the overall force would be too small, leading to incorrect conclusions in further evaluations. To avoid this situation, the criterion is extended by a revolution based extreme value distance

$$d_R\left(F_{\text{sim}}, F\right) = \frac{1}{2}\left[\frac{\left\|F_{\text{sim,min}} - F_{\text{min}}\right\|}{\left\|F_{\text{sim,min}}\right\| + \left\|F_{\text{min}}\right\|} + \frac{\left\|F_{\text{sim,max}} - F_{\text{max}}\right\|}{\left\|F_{\text{sim,max}}\right\| + \left\|F_{\text{max}}\right\|}\right],$$

where $x_{\text{min}} = \left(\min\{x_1, \ldots, x_r\}, \ldots, \min\{x_{(N-1)\cdot r+1}, \ldots, x_{N\cdot r}\}\right)^T$ is the vector of minima for the N revolutions with r observations each of a force vector x (x_{max} analog).

For the model based optimization, calculated with the R package "mlrMBO" (Bischl et al. 2013), the criterion O is estimated by

$$\hat{O}\left(\theta\right) = \frac{1}{4K}\sum_{k=1}^{K}\sum_{m\in\{b,c\}}\sum_{f\in\{\text{norm,rad}\}}\hat{O}_{\text{min}}\left(\theta, m, f\right)$$

with K repetitions of the two force directions f for each material m. Due to the difficulty that the grinding starting point (first diamond workpiece contact) cannot be examined for the measured reference forces, only the best fitting shifted subsequence

$$\hat{O}_{\min}(\theta, m, f) = \min_{u \in U} \frac{1}{L} \sum_{l=1}^{L} \hat{O}_{l,u}(\theta, m, f)$$

with

$$\hat{O}_{l,u}(\theta, m, f) = d_{\text{DTW}}\left(F^{m,u}_{\text{sim},f}(\theta), F^{m,u}_{f,l}\right) + d_R\left(F^{m,u}_{\text{sim},f}(\theta), F^{m,u}_{f,l}\right)$$

is taken into account, where U is the set of turns. The realized force $F^m_{\text{sim},f}$ is shifted against the l-th measured force $F^m_{f,l}$ in time direction for each $u \in U$. The K repetitions should minimize the uncertainty since the simulation output is stochastic. In spite of every effort to accelerate the evaluation of the simulation, the computation of the objective function $\hat{O}$ is still very time consuming. Hence, the objective function $\hat{O}$ is approximated by a stochastic Kriging model with the Matérn correlation function (Ankenman et al. 2010).

In the first step of the optimization, the objective function is evaluated for an initial Latin hypercube design $D_0 \subset \Theta$ to estimate the Kriging parameters. To find the next point for evaluation a focus search is performed. Let denote $D_1, \ldots, D_S \subset \Theta$ random Latin hypercube designs, representing the global search, and $\theta_1^\star, \ldots, \theta_S^\star$ points evaluated with the surrogate function and fulfilling an infill criterion. These points are determined by examining the best point in D_s ($s = 1, \ldots, S$) according to the infill criterion, reducing the parameter space Θ around this point and sampling a new design D_{s1} from the reduced space Θ_1 (local search). After repeating this procedure T times, the best of the ST candidate points is evaluated with the objective function. Then the Kriging parameters are updated and the focus search repeats from the beginning.

Two different infill criteria are used. In the first optimization the lower confidence bound infill criterion, based on the difference between Kriging predictor and its standard error, is applied (Jones 2001). To manage the stochastic character of the simulation outcome leading to a stochastic objective function, $K = 25$ repetitions for each material are evaluated for each parameter vector θ. The second criterion is the augmented expected improvement which is more suitable for the optimization with stochastic output (Picheny et al. 2013).

Figure 4 shows the optimization course, with the augmented expected improvement infill criterion, of 108 objective function evaluations as boxplots of the $K = 10$ single evaluations for basalt and cement. After the evaluation of the initial design (first 50 points), the values of $\hat{O}$ converge. Furthermore, better results can be observed for evaluations on basalt, according to the choice of the objective

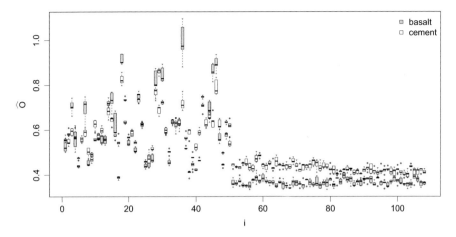

Fig. 4 Optimization course using the augmented expected improvement as infill criterion and 50 points in the initial design

function. One reason for this result could be the presence of negative observations in combination with notably strong workpiece tilting in the cement series. The reason why the values of the objective function converge to this value is not least because of the input data. Treating one of the observed time series for a (v_c, a_e)-combination as simulated and the others with this (v_c, a_e)-combination as reference for the two materials leads approximately to the same values of the objective function to which the objective function converges for basalt and cement in Fig. 4. Therefore, it is not possible to achieve better overall results. The single evaluations $\hat{O}_{l,u}(\theta, m, f)$ for the l-th reference force in f direction on material m and the u turns shifted corresponding simulated force, however, show a quite reasonable accordance (see Fig. 5).

The optimization with more repetitions and the lower confidence bound infill criterion leads to a similar but worse performance than the optimization with the augmented expected improvement criterion. Although the two infill criteria lead to almost the same best values of $\hat{O}$, the difference between the resulting optimal parameters, relative to the parameter range used for the optimization designs, is still quite large. Especially for the material specific parameters s_m^- and s_m^+, the relative difference ranges between 15 % and almost 50 %. The parameter δ_{coarse} and the weight parameters $g_{\text{het},f}$ and $g_{\text{vol},f}$ with a relative difference of less than 4 % seem to be quite similar.

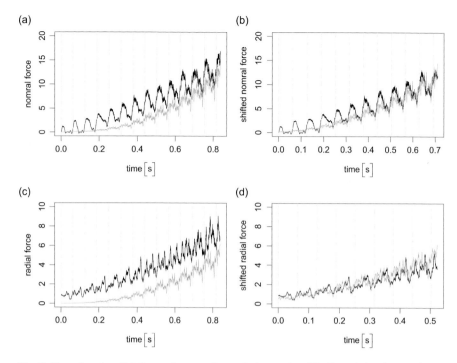

Fig. 5 Normal and radial forces for one observed time series (*black*) and the simulated time series (*gray*) for the point $\theta^\star$ with the best fit on basalt before (**a, c**) and after shifting (**b, d**): (**a**) $\hat{O}_{1,0}(\theta^\star, basalt, norm) = 0.429$, (**b**) $\hat{O}_{1,2}(\theta^\star, basalt, norm) = 0.284$, (**c**) $\hat{O}_{1,0}(\theta^\star, basalt, rad) = 0.585$, (**d**) $\hat{O}_{1,5}(\theta^\star, basalt, rad) = 0.127$

4   Conclusion and Outlook

The presented version of the simulation for the grinding process with one diamond is much faster than the previous versions. An opportunity for a speed up for the simulation with the complete drill core bit involving hundreds of diamonds is still available if necessary. The performed model based optimization with the augmented expected improvement infill criterion results in acceptable accordance between simulated and measured forces. Before deriving the regression models for the optimized simulation parameters for different process parameter combinations, attempts to validate the optimization results for basalt and cement on concrete will be done. There are two main difficulties that must be resolved. The one of not knowing the basalt grain size and position can be solved by examining the angles of entering and exiting the basalt grains in the workpiece following the drilling path. This information can be used to create a simulated counterpart, but it would not disclose information about the position of the first diamond contact. Shifting the simulated time series in time direction for subsequent generation of different starting positions, as used in the objective function of the parameter optimization, would be a time saving possibility to deal with this challenge. Since all intermediate optimization results, as well as the simulation times needed for each evaluated simulation, are available, it would be possible to compare simulations with several parameter sets with similarly good results and perhaps find a solution with maybe not the best parameters but with much faster calculation time. After the validation on concrete, the next step will be to proceed with the segment simulation. In addition to the information from the diamond simulation, follow-up requirements are the breakout rules for diamonds in segments and the linked wear rate of the metal matrix, surrounding the diamonds in the segment, for the diverse materials.

Acknowledgements This work has been supported by the Collaborative Research Center "Statistical modeling of nonlinear dynamic processes" (SFB 823) of the German Research Foundation (DFG)

References

Ankenman, B., Nelson, B. L., & Staum, J. (2010). Stochastic kriging for simulation metamodeling. *Operations Research, 58*(2), 371–382.

Bischl, B., Bossek, J., Richter, J., Horn, D., & Lang, M. (2013). mlrMBO: mlr: Model-based optimization. R package version 1.0, https://github.com/berndbischl/mlr

Ding, H., Trajcevski, G., Scheuermann, P., Wang, X., & Keogh, E. (2008). Querying and mining of time series data: Experimental comparison of representations and distance measures. In *Proceedings of the VLDB Endowment*, 1 (2), August 23–28, 2008, Auckland (pp. 1542–1552).

Giorgino, T. (2009). Computing and visualizing dynamic time warping alignments in R: The dtw package. *Journal of Statistical Software, 31*(7), 1–24. http://www.jstatsoft.org/v31/i07/

Jones, D. R. (2001). A taxonomy of global optimization methods based on response surfaces. *Journal of Global Optimization, 21*(4), 345–383.

Müller, M. (2007). Dynamic time warping. In *Information retrieval for music and motion* (pp. 69–84). New York: Springer.

Picheny, V., Wagner, T., & Ginsbourger, D. (2013). A benchmark of kriging-based infill criteria for noisy optimization. *Structural and Multidisciplinary Optimization, 48*(3), 607–626.

Raabe, N., Rautert, C., Ferreira, M., & Weihs, C. (2011). Geometrical process modeling and simulation of concrete machining based on Delaunay tessellations. In S. I. Ao, C. L. Douglas, W. S. Grundfest & J. Burgstone (Eds.), *Proceedings of the World Congress on Engineering and Computer Science 2011 (WCECS'11)*, October 19–21, 2011. Lecture Notes in Engineering and Computer Science (Vol. II, pp. 991–996). San Francisco: International Association of Engineers, Newswood Limited.

Raabe, N., Thieler, A. M., Weihs, C., Fried, R., Rautert, C., & Biermann, D. (2012). Modeling material heterogeneity by Gaussian random fields for the simulation of inhomogeneous mineral subsoil machining. In P. Dini, P. Lorenz (Eds.), *SIMUL 2012: The Fourth International Conference on Advances in System Simulation*, November 18–23, 2012, Lisbon (pp. 97–102).

Schlather, M., Malinowski, A., Oesting, M., Boecker, D., Strokorb, K., Engelke, S., et al. (2014). RandomFields: Simulation and analysis of random fields. R package version 3.0.10, http://CRAN.R-project.org/package=RandomFields

Weihs, C., Raabe, N., Ferreira, M., & Rautert, C. (2014). Statistical process modelling for machining of inhomogeneous mineral subsoil. In *German-Japanese interchange of data analysis results* (pp. 253–263). New York: Springer.

Fast and Robust Isosurface Similarity Maps Extraction Using Quasi-Monte Carlo Approach

Alexey Fofonov and Lars Linsen

Abstract Isosurface similarity maps are a technique to visualize structural information about volumetric scalar fields based on sampling the field's range by a number of isovalues and comparing corresponding isosurfaces. The result is displayed in the form of a 2D gray-scale map that visually conveys structural components of the data field. In this paper, we present a novel way to establish isosurface similarity maps by introducing a quasi-Monte Carlo approach for computing isosurface similarities. We discuss our approach and implementation details in comparison to the state of the art. We show that our method produces significantly lower computational costs, yet it is simpler and more intuitive to use, is more flexible in its applicability, and more robustly generates high-quality results.

1 Introduction

Volumetric scalar fields are generated in numerous areas of science, engineering, and medicine. They stem from running simulations, e.g., of some physical phenomena, or taking measurements, e.g., using medical imaging techniques. Isosurface extraction has established itself as one of the key analysis tools for such data fields. Looking at isosurfaces one can distinguish different structural components of underlying data, which correspond to different ranges of the scalar values. Manual inspection of all isosurfaces within the field's range would be a tedious task.

In this paper, we build upon a strategy to support interactive volumetric scalar field analysis known as isosurface similarity maps. The main idea of the approach is to build a 2D gray-scale map, which visualizes pair-wise distances between all isosurfaces from a sampling of the field's range, where dark colors correspond to small distances and bright colors to large distances. Looking at such similarity maps, one can easily distinguish ranges of field values that correspond to similar structures. Therefore, these maps help to choose proper volume visualization parameters such as a good selection of isosurfaces or a suitable transfer function.

A. Fofonov (✉) • L. Linsen
Jacobs University Bremen, Bremen, Germany
e-mail: a.fofonov@jacobs-university.de; l.linsen@jacobs-university.de

© Springer International Publishing Switzerland 2016
A.F.X. Wilhelm, H.A. Kestler (eds.), *Analysis of Large and Complex Data*, Studies in Classification, Data Analysis, and Knowledge Organization,
DOI 10.1007/978-3-319-25226-1_42

The main contribution of our paper is a new approach for fast and robust extraction of isosurface similarity maps. Our primary goal was to reduce the high computational costs of existing approaches. We replace the concept of using distance transforms as surface descriptors and mutual information estimates for isosurface similarity computations by a quasi-Monte Carlo (qMC) approach for comparing isosurfaces by looking at the enclosed volumes. We compare the two approaches by discussing implementation details, the quality of the produced similarity maps, and computation times. We can show that our method is significantly faster, more reliably produces the desired results, and is more intuitive and more flexible to apply.

2 Related Work

In addition to their property of being easily visualized and understood, isocontours have proven to be a suitable descriptor for scalar fields (Bajaj et al. 1997; Bruckner and Möller 2010; Carr et al. 2006; Duffy et al. 2013; Khoury and Wenger 2010; Scheidegger et al. 2008; Tenginakai et al. 2001). The concept of isosurface similarity maps was introduced by Bruckner and Möller (2010). They present structural information of a volume data set by depicting similarities between individual isosurfaces quantified by an information-theoretic measure. This concept can even be successfully extended to multimodal volume data (Haidacher et al. 2011). One obvious disadvantage of the approach is the considerably high costs of generating the isosurface similarity map. The reported implementation can require several hours of processing time for a single data frame.

Besides the basic purpose of data visualization, the concept has a wide area of potential applications, such as volume quantization and compression, volume segmentation, or multi-dimensional classification. However, for generating similarity maps, some requirements have to be satisfied. First, a proper distance function has to be defined to generate a similarity matrix. Second, since isocontours are defined implicitly, it is important to describe all isocontours with the same accuracy for a fair comparison. Third, the isocontours do not only differ by enclosing areas, but have different shapes, which need to be taken into account.

3 Distance Computation

3.1 Mutual Information

Mutual information is a widely used measure of similarity between two random variables. It has successfully been applied to a range of problems (Huang et al. 2006; Haidacher et al. 2008; Viola et al. 2006). A formal definition (Yao 2003) of the mutual information between two random variables X and Y is given by

$$I(X,Y) = \sum_{x \in X} \sum_{y \in Y} p_{X,Y}(x,y) \log \left(\frac{p_{X,Y}(x,y)}{p_X(x)p_Y(y)} \right),$$

where $p_{X,Y}$ denotes the joint probability distribution function of X and Y, while p_X and p_Y denote the marginal probability distribution functions of X and Y, respectively. Mutual information can also be expressed in terms of entropy by

$$I(X,Y) = H(X) + H(Y) - H(X,Y),$$

where $H(X)$ and $H(Y)$ denote the marginal entropies and $H(X,Y)$ the joint entropy of X and Y. For our purposes, it is more convenient to work with normalized values (Kvalseth 1987) given by

$$\hat{I}(X,Y) = \frac{2I(X,Y)}{H(X) + H(Y)}.$$

In order to apply the mutual information measure to isosurfaces, a proper surface descriptor is needed. Bruckner and Möller (2010) propose to use for this purpose the distance transform of the isosurfaces (Jones et al. 2006). Thus, the distances from any point of the data volume to a pair of the to-be-compared isosurfaces can be considered as random variables X and Y. The resulting value of the normalized mutual information measure can be interpreted as a similarity measure between the isosurfaces. Similar isosurfaces have a value closer to 1, while dissimilar isosurfaces are expected to have a value closer to 0.

The nature of the distance transform (DT) makes it sensitive to changes in the geometry of the isosurfaces' shapes. Generally, isosurfaces exhibit an onion peel-like structure, which means that one can expect smooth shape transitions for close-by isovalues provided that the isovalues correspond to the same structural component. Conversely, if DTs of a set of isosurfaces are very similar, we can assume that they belong to the same component.

We propose to replace the isosurface similarity computation based on mutual information with an approach that computes isosurface distances by comparing which areas in the data volume are enclosed by them. Hence, we need to integrate characteristic functions of the areas enclosed by isosurfaces. We propose to compute the integrals using a qMC approach. The main benefits of such approach are that it is faster than the mutual information extraction, it can be easily used for spaces of any dimension, and it can be used for isosurfaces defined over any spatial data structure ranging from structured grids over unstructured meshes to unstructured point-based data.

Fig. 1 Calculation of the
distance between isocontours.
Example of two isocontours
with distance 0.8

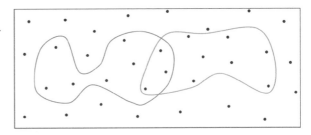

3.2 Quasi-Monte Carlo Approach

We compute random points that sample the spatial domain uniformly and evaluate
the scalar field at these random points. From the interpolated scalar function, we
can derive whether the random point lies inside (including boundary) or outside the
isosurface. This information is stored in a binary vector, where the dimensionality
of the vector is equal to the amount of random points. Hence, the binary vector
is a representative descriptor of the isocontour. Note that the descriptor does not
only allow for the estimation of the size of the area enclosed by the isocontour, but
also captures the location and shape of the enclosed area. Hence, when comparing
two isosurfaces with respect to this descriptor, one can estimate how closely the
isosurfaces' location and shape match.

Based on the introduced vectors, we can introduce a distance function between
the respective isocontours. Let A and B be isocontours, $M_{A \wedge B}$ the number of points
inside both isocontours (logical *and*), and $M_{A \vee B}$ the number of points inside of, at
least, one of the isocontours (logical *or*). Then, we define a distance $d(A, B)$ between
A and B by the Jaccard distance

$$d(A, B) = 1 - \frac{M_{A \wedge B}}{M_{A \vee B}}.$$

The idea of the distance calculation is illustrated in Fig. 1. The example shows
two isocontours with $M_{A \wedge B} = 3$ and $M_{A \vee B} = 15$, i.e., distance $d(A, B) = 0.8$.
The accuracy or complexity of the calculations can be adjusted by increasing or
decreasing the amount of random points. Due to the randomness, the error of
the approach depends only on the amount of points and not on the shapes of
the isocontours. Obviously, the algorithm can be applied to scalar fields of any
dimension and of any spatial data structure.

3.3 Sensitivity of the Measures

Due to different nature of the isosurface descriptors (DT vs. binary vector encoding
inside–outside property of random points), the resulting similarity maps may differ
substantially. To illustrate the features of the measures, we generated five synthetic
2D data sets (shown as 2D images) and corresponding similarity maps using both
methods (see Fig. 2).

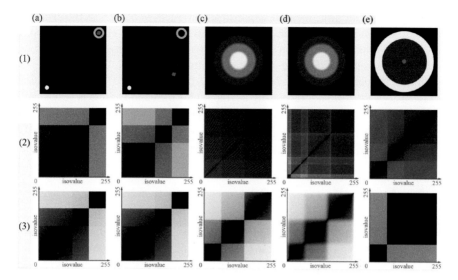

Fig. 2 Examples of differences between the distance measures. In *row (1)* five different synthetic data fields are shown. *Row (2)* depicts the corresponding similarity maps generated using mutual information measure. *Row (3)* represents the corresponding similarity maps generated using quasi-Monte Carlo approach

The first example is shown in column (a). The data field contains three objects represented by different isovalue ranges. The left-bottom corner of the 2D field contains a white circle, while the right-top corner contains dark-gray cross placed inside a light-gray ring. Looking at the similarity map (3) generated by our qMC approach, we can easily distinguish three detached dark squares on the diagonal of the map, which correspond to the three different objects. However, in the similarity map (2) generated by the mutual-information approach, one can only distinguish two squares. The reason for that is that the DTs for lower isovalues look very similar due to the close-by positions of the ring and cross.

The second example in column (b) of Fig. 2 shows the same data set, only that the cross has moved to a different location outside the ring. Now, we can observe that the mutual-information approach (2) recognizes all three objects, as the corresponding DTs now differ substantially due to big distances between the objects. The qMC approach (3), on the other hand, remains exactly the same as in column (a). Hence, we can conclude that our qMC approach is less sensitive to the spatial distribution of (non-overlapping) features and rather captures their shape differences.

The third example shown in column (c) of Fig. 2 represents three concentric circles with increasing intensity, while the fourth example in column (d) represents the same data set after adding noise. The resulting similarity map obtained by mutual information (2) for the noise-free data set (c) allows us to weakly recognize three square-like structures along the main diagonal, but there is an additional wave-like artifact covering the whole map. This artifact is caused by repeating discrete circle patterns during smooth isovalue transitions. For the noisy data (d), the similarity map (2) even exhibits another artifact in the form of artificial horizontal and vertical stripes. Our qMC measure (3), instead, produces in both cases the expected and desired result, i.e., exposing the three different objects. Hence, the second conclusion is that our approach is less sensitive to small changes and, therefore, is more robust against noise and extraneous artifacts.

Finally, in column (e) of Fig. 2, the data set also contains three concentric circles, but intensity values are not increasing anymore plus the central circle is very small. For our qMC approach (3), the difference between the big white ring with and without the small circle is negligible, while in terms of the DT this difference totally changes the picture. This confirms, again, the second conclusion from above.

4 Implementation

When using the mutual-information measure for isosurface similarity, there are many implementation details that can affect the similarity map calculation, in particular, the computation time. Hence, it is worth discussing some of the implementation parameters.

The first important parameter is the resolution of the DT. Due to high computation costs, Bruckner and Möller (2010) proposed to downsample the resolution, as even at reduced resolutions the DT captures well the characteristics of an isosurface. Since they provided computation times for all examples using a resolution of about 64^3, we decided to do the same for a fair comparison. Note that downsampling the DT is different from downsampling the data field, as a downsampled version of the original DT is used.

After DT calculation, it is possible to compute the mutual information or distances between the isosurfaces. In order to calculate marginal and joint entropy, we have to build a joint histogram. When doing so, we have to define a resolution of the histogram and its range. Like Bruckner and Möller we use a fixed histogram resolution of 128^2 throughout this paper. There are many ways to define the range, such as going from 0 to the maximum possible distance (e.g., main diagonal of data volume) or defining a certain range for each pair of isosurfaces. Depending on the data, we observed that different choices can lead to varying quality of the resulting similarity maps. We calculate the minimum and maximum distances among all DTs and use the resulting range for all pairs.

Unlike Bruckner and Möller we did not use a CUDA-based implementation of the joint and marginal entropies by Shams and Barnes (2007). The reasoning is that with a fixed number of used isovalues (256), a fixed joint histogram resolution (128^2), and using a downsampled DT resolution, the CPU implementation was actually faster than the GPU implementation in the comparisons we conducted. This is due to additional costs for memory transfer and due to the GPU initialization. The calculated DTs need to be stored in system memory or, in case of very high resolutions, on hard disk.

In contrast to the mutual information approach by Bruckner and Möller, our qMC approach has only one parameter, which is the number of random points. Intuitively, the more random points we use, the higher accuracy we achieve. In order to achieve a high performance, we propose to pass all the calculations to a GPU. There are two main steps of our similarity map computation: calculating the descriptive binary vectors for all the isosurfaces and comparing them to fill the map with distance values. Both steps are well parallelizable and can be easily implemented with high efficiency.

The amount of required memory and the time for performing the qMC approach do not depend on the data resolution (obviously, except for data loading itself). For example, there is the same amount of memory required for similarity map extraction from data of resolution 32^3 or of resolution 512^3, if the same number of random points is used. Another advantage is that no explicit isosurface extraction is required, as we only need to interpolate the values of the data field to make the inside–outside decision for each random point. In this paper, we used 32,768 random points which ensure enough accuracy for all the examples. All methods were implemented in C++, while using CUDA for GPU-based functions.

5 Results and Discussion

5.1 Similarity Map Results

To test our approach on real data, we chose three datasets with different characteristics and from different application domains. Figure 3 shows the results for the Crossed Rods dataset. In column (a), two similarity maps corresponding to qMC (1) and mutual information (2) measures are shown. Our similarity map (1) exhibits

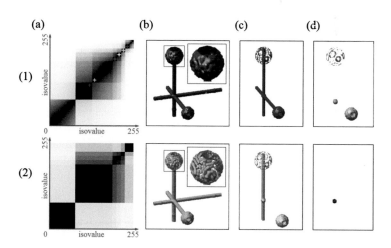

Fig. 3 Results for Crossed Rods dataset: in *column (a)* similarity maps are shown, where (*1*) is the result when using qMC and (*2*) when using mutual information measures. *Columns (b–d)* represent isosurfaces corresponding to the selections in the qMC similarity map

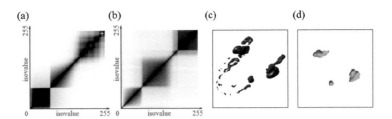

Fig. 4 Results for CT head scan: (**a**) and (**b**) show similarity maps generated using qMC and mutual information, respectively. Isosurfaces (**c**, **d**) correspond to the selection in the similarity map (**a**)

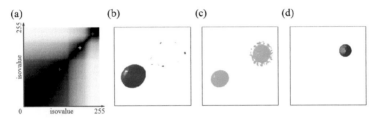

Fig. 5 Results for unstructured SPH dataset (White Dwarfs): (**a**) similarity map generated using qMC. (**b–d**) Isosurfaces corresponding to selection in (**a**)

more structures than (2). For a detailed investigation we selected six isovalues corresponding to the visible structures (marked by colored crosses) and rendered corresponding isosurfaces using the Marching Cubes approach (Lorensen and Cline 1987) [see columns (b–d)]. We observe that each of them differs from the previous one by losing a part of the construction. Hence, our similarity map (1) captured the structure somewhat better.

For a medical imaging data analysis, we used a CT scan of a man's head (see Fig. 4). The structures captured by mutual information technique (b) are also visible in our similarity map (a). Moreover, the right-top square area in (b) actually contains some more complicated structure, which is only captured in our similarity map (a). To validate our result, we picked respective isovalues and rendered the corresponding isosurfaces (c, d). Due to relatively small size of the jaws when compared to the whole head, the isosurfaces are judged to be similar when using a DT, but they are judged to be different when considering enclosed volumes.

Since our method is applicable to data of any spatial configuration, we tested it on astrophysical SPH simulation data of a two-stars system (see Fig. 5). The stars have different masses and sizes. When investigating the internal energy field, we can recognize them by occupied field ranges. The similarity map exhibits three regions (a). Figure 5b–d shows the corresponding isosurfaces using splat-based rendering (Linsen et al. 2007), which represent the expected result. The considered gridded data sets are from The Volume Library, the SPH data were provided by Marius Dan and Stephan Rosswog.

Table 1 Comparing computation times for the mutual information approach to our qMC approach

Data	Original resolution	DT resolution	Random points	Time qMC (GPU) (s)	qMC (CPU) (s)	Mut. inf. (s)
Figure 3	64 × 64 × 64	64 × 64 × 64	32,768	0.41	3.6	178
Figure 4	128 × 256 × 256	32 × 64 × 64	32,768	0.48	4.8	641
Figure 5	39,718 particles	–	32,768	0.64	–	–

5.2 Computation Times

Our approach is significantly faster than the mutual information approach by Bruckner and Möller (2010). For the presented examples, computation times are listed in Table 1. We executed the algorithms on a laptop with Intel Core i7-3630QM, NVIDIA GTX 660M, and 8 GB DDR3.

Using our approach, we achieved 430–1330 times faster similarity map computation for downsampled DT resolutions using GPU, and 50–130 times using only CPU. With increasing resolution our approach keeps the performance, as it only depends on the number of random points, while the mutual information approach has super-linearly increasing computation times leading to even higher differences between the two approaches. Note that the small difference between qMC times for the first and second example is caused by the different numbers of succeeding conditional statements and not by the different resolutions. In case of SPH data, the more complicated field interpolation leads to higher computation times.

6 Conclusion

We presented a novel approach to similarity map computation based on a qMC method. We compared our approach for calculating distances between isosurfaces with the state of the art and discussed important features as well as implementation details. For all examples, our similarity maps presented the structures in the data fields better or, at least, equally good, while being hundreds of times faster and using less system resources. Due to insignificant hardware requirements, our approach is also suitable for most devices, even those with a single core architecture. Hence, we could overcome the main drawbacks of similarity map computation based on mutual information.

References

Bajaj, C., Pascucci, V., & Schikore, D. (1997). The contour spectrum. In *Proceedings of Visualization 1997* (pp. 167–173).

Bruckner, S., & Möller, T. (2010). Isosurface similarity maps. *Computer Graphics Forum, 29*(3), 773–782.

Carr, H., Duffy, B., & Denby, B. (2006). On histograms and isosurface statistics. *IEEE Transactions on Visualization and Computer Graphics, 12*(5), 1259–1266.

Duffy, B., Carr, H., & Möller, T. (2013). Integrating isosurface statistics and histograms. *IEEE Transactions on Visualization and Computer Graphics, 19*(2), 263–277.

Haidacher, M., Bruckner, S., & Gröller, E. (2011). Volume analysis using multimodal surface similarity. *IEEE Transactions on Visualization and Computer Graphics, 17*(12), 1969–1978.

Haidacher, M., Bruckner, S., Kanitsar, A., & Gröller M. (2008). Information-based transfer functions for multimodal visualization. In *Proceedings of Visual Computing for Biomedicine* (pp. 101–108).

Huang, X., Paragios, N., & Metaxas, D. (2006). Shape registration in implicit spaces using information theory and free form deformations. *IEEE Transactions on Pattern Analysis and Machine Intelligence, 28*, 1303–1318.

Jones, M., Baerentzen, J., & Šrámek, M. (2006). 3D distance fields: A survey of techniques and applications. *IEEE Transactions on Visualization and Computer Graphics, 12*, 581–599.

Khoury, M., & Wenger, R. (2010). On the fractal dimension of isosurfaces. *IEEE Transactions on Visualization and Computer Graphics, 16*(6), 1198–1205.

Kvalseth, T. (1987). Entropy and correlation: Some comments. *IEEE Transactions on Systems, Man, and Cybernetics 17*, 517–519.

Linsen, L., Müller, K., & Rosenthal, P. (2007). Splat-based ray tracing of point clouds. *Journal of WSCG, 15*, 51–58.

Lorensen W., & Cline H. (1987). Marching cubes: A high resolution 3D surface construction algorithm. *ACM SIGGRAPH Computer Graphics, 21*, 163–169.

Scheidegger, C. E., Schreiner, J. M., Duffy, B., Carr, H., & Silva, C. T. (2008). Revisiting histograms and isosurface statistics. *IEEE Transactions on Visualization and Computer Graphics, 14*(6), 1659–1666.

Shams, R., & Barnes, N. (2007). Speeding up mutual information computation using NVIDIA CUDA hardware. In *Proceedings of Digital Image Computing: Techniques and Applications* (pp. 555–560).

Tenginakai, S., Lee, J., & Machiraju, R. (2001). Salient iso-surface detection with model-independent statistical signatures. In *Proceedings of Visualization, 2001 (VIS'01)* (pp. 231–238).

Viola, I., Feixas, M., Sbert, M., & Grller, M. (2006). Importance-driven focus of attention. *IEEE Transactions on Visualization and Computer Graphics, 12*, 933–940.

Yao, Y. (2003). Information-theoretic measures for knowledge discovery and data mining. In Karmeshu (Ed.), *Entropy measures, maximum entropy principle and emerging application* (pp. 115–136). New York: Springer.

Analysis of ChIP-seq Data Via Bayesian Finite Mixture Models with a Non-parametric Component

Baba B. Alhaji, Hongsheng Dai, Yoshiko Hayashi, Veronica Vinciotti, Andrew Harrison, and Berthold Lausen

Abstract In large discrete data sets which requires classification into signal and noise components, the distribution of the signal is often very bumpy and does not follow a standard distribution. Therefore the signal distribution is further modelled as a mixture of component distributions. However, when the signal component is modelled as a mixture of distributions, we are faced with the challenges of justifying the number of components and the label switching problem (caused by multi-modality of the likelihood function). To circumvent these challenges, we propose a non-parametric structure for the signal component. This new method is more efficient in terms of precise estimates and better classifications. We demonstrated the efficacy of the methodology using a ChIP-sequencing data set.

1 Introduction

The observations in a finite mixture model originate independently from a mixture distribution with K components that can be written as

$$f(x) = \sum_{k=1}^{K} \pi_k f_k(x; \boldsymbol{\theta}_k);$$ (1)

where $\pi_k > 0$ with $\sum_k \pi_k = 1$ is the mixing weight of component k and $f_k(x; \boldsymbol{\theta}_k)$ belongs to a given parameterized family $\boldsymbol{\theta}_k$. This model has advantages of relaxing distributional assumptions. It represents subpopulations where the population membership is not known but is inferred from the data (McLachlan and Peel 2004).

B.B. Alhaji (✉) • H. Dai • Y. Hayashi • V. Vinciotti • A. Harrison • B. Lausen
Department of Mathematical Sciences, University of Essex, Colchester, UK
e-mail: bbalha@essex.ac.uk; hdaia@essex.ac.uk; yhayasa@essex.ac.uk;
veronica.vinciotti@brunel.ac.uk; harry@essex.ac.uk; blausen@essex.ac.uk

© Springer International Publishing Switzerland 2016   507
A.F.X. Wilhelm, H.A. Kestler (eds.), *Analysis of Large and Complex Data*, Studies
in Classification, Data Analysis, and Knowledge Organization,
DOI 10.1007/978-3-319-25226-1_43

The existing literature such as Diebolt and Robert (1994) and McLachlan and Peel (2004) have demonstrated that finite mixture models can be inferred in a simple and effective way in a Bayesian estimation framework. Attention has mostly focused on parametric mixture models, when the component densities are all from the same parametric family, having different parameter values for the components. For example, all the distributions could be Poisson with different means, or all the distributions could be Negative Binomial with different parameters (even though, in practice, it is not necessary that all the densities will be of the same kind). This situation causes a persistent challenge in the diagnostic of Markov Chain Monte Carlo (MCMC) convergence due to two reasons.

The first reason is the label switching problem which results from the multi-modality of the likelihood function. Many methods exist on how to tackle the label switching problem, for example, imposing identifiability constraints (Diebolt and Robert 1994; Richardson and Green 1997; McLachlan and Peel 2004) and other methods based on relabelling algorithms (Stephens 2000b; Celeux et al. 2000; Rodriguez and Walker 2014). For a review and comparison of these methods see, for example, Jasra et al. (2005) and Sperrin et al. (2010). One limitation to the existing methods for dealing with the label switching problem is that they focus on mixture models where all components having the same type of distributions. Another drawback common to these methods is that they require heavy computational costs, which make them unsuitable for large data sets, and models with a large number of components. In practice, mixture components with different types of distributions are sometimes used, such as mixture of Poisson and Negative Binomial distributions. In such situations, the likelihood function may still have multi-modes which cause label switching problem. But the existing methods for dealing with this problem may not be applicable in this case.

The other reason is the justification of the number of components, K. Many authors have devised different strategies for estimating the number of components in Bayesian finite mixture models, for example reversible jump MCMC (Richardson and Green 1997) and Birth and Death MCMC (Stephens 2000a; Nobile and Fearnside 2007). Another approach to deal with the unknown number of components is to use a mixture of Dirichlet processes (Antoniak 1974; Escobar and West 1995), which allows an infinite number of components. This is also computationally non-trivial when a large data set with several components is involved.

This motivates our study, which we discuss in detail in the following. In certain application areas, interest may be in classifying the observations into two classes. For example, in the analysis of ChIP-sequencing (ChIP-seq) data, we are interested in whether a region of the genome is bound by the protein in question or not. For such ChIP-seq (discrete) data, although there are only two possible classes, it is inappropriate to use a mixture of two known parametric distributions (e.g. Poisson or Negative Binomial distributions). This is because such data sets usually have long tails and the tails may show multi-modal patterns.

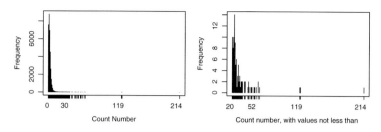

Fig. 1 Distribution of ChIP-seq data for one experiment (*left*), with zoom on the tail (*right*)

For illustration, we use ChIP-seq data generated by Ramos et al. (2010) for the experiment on CREB binding protein (CBP) for identifying the genomic regions bound by the histone acetyltransferases (see Bao et al. 2013 for a description of ChIP-seq technology and this data set). For each region (1000bp) in the genome, the data report the number of bound fragments that aligns to that region. A higher value means that the corresponding region is most likely to be bound by the protein in question. The number of regions in the data set is 33,916. The lowest count is zero and the highest count is 214, which means that some regions are tagged but with no protein of interest and a particular region is tagged with 214 counts. The mean and the variance are 2.13 and 8.76, respectively. Figure 1 shows a histogram of the count data. The left plot shows that the data set has a very long tail. If we zoom in the tail of the distribution (right plot), we see possible multi-modal patterns, suggesting that the distribution of the data is likely to consist of several component distributions. This situation has been observed also for other ChIP-seq analysis, where a two-component parametric mixture model appears to be too restrictive for the analysis of these data. An alternative approach is to use K components, with $K > 2$. In the context of ChIP-seq data analysis, this was considered by Kuan et al. (2011), who allowed the signal distribution to be a mixture of two Negative Binomial distributions (i.e. $K = 3$). However, it is very challenging to justify the true value of K. Although the reversible jump MCMC method (Green 1995) is readily available, the justification of reversible jump MCMC convergence is non-trivial and it requires heavy computational costs. Another challenge of using K components is that it is very difficult to determine exactly the component distributions. For instance, all components may be chosen as Poisson distributions, or only some components are chosen as Poisson distributions and the others are chosen as Negative Binomial distributions. As such, using a mixture distribution with K components seems unnecessary. This motivates us to consider a two-component mixture model for discrete observations, with one parametric distribution and one non-parametric distribution.

The non-parametric distribution has several advantages. It bypasses the challenges involved in the K-component mixture models, such as the label switching problem and the determination of the unknown parameter K. It does not need to justify a particular parametric distribution for the signal. In the context of ChIP-seq data, our method detects the enriched regions in the genome with higher accuracy than the mixture of parametric distributions.

2   The Model and the Posterior Distribution

Suppose that discrete observations $x_1, \ldots, x_n$ are sampled from a mixture of distributions with two components, where one component is the noise distribution and the other component is a signal distribution. We simply use the mixture density in (1) to model the data, where f_1 is the parametric distribution for the noise, f_2 is the signal distribution, and π_1 and π_2 are the corresponding mixture proportions, respectively.

Associated to each observation x_i is a latent variable z_i, i.e. $z_i = k \ (k = 1, 2)$, which represents the component from which the observation x_i originates. The complete likelihood function for $(\boldsymbol{\theta}_1, \boldsymbol{\theta}_2)$ given the full data is

$$l(\boldsymbol{\theta}_1, \boldsymbol{\theta}_2 | \boldsymbol{x}, \boldsymbol{z}) \propto \prod_{i=1}^{n} \left\{ [\pi_1 f_1(x_i; \boldsymbol{\theta}_1)]^{I[z_i=1]} [\pi_2 f_2(x_i; \boldsymbol{\theta}_2)]^{I[z_i=2]} \right\}. \tag{2}$$

The noise distribution f_1 is usually simpler to determine. For example in ChIP-seq studies (for 1000bp where the proportion of zeros is not very large), Poisson distribution is a natural choice for the noise since a genomic region not bound by the protein in question but tagged is a rare event. In contrast to this, the signal distribution can present complicated patterns. We therefore consider using a non-parametric model for the second component.

As the data are discrete, we can denote with $x_{(1)}, \ldots, x_{(L)}$ the L distinct values of the observations $x_1, \ldots, x_n$. Define

$$f_2^*(x_{(j)}) = p_j, \quad \sum_{j=1}^{L} p_j = 1; \tag{3}$$

where $p_j \text{s} \ (j = 1, \ldots, L)$ are the unknown parameters. p_j can be interpreted as the probability of $x = x_{(j)}$ given that x is drawn from the signal component. This can be viewed as a non-parametric distribution. Under this model, the distribution of x is given by

$$f(x) = \pi_1 f_1(x; \boldsymbol{\theta}_1) + \pi_2 \sum_{j=1}^{L} f_2^*(x) I[x = x_{(j)}]. \tag{4}$$

Based on the distribution (3), we have the following likelihood function given (x_i, z_i) $(i = 1, \ldots, n)$,

$$l(\boldsymbol{\theta}_1, \boldsymbol{p}, \boldsymbol{\pi} | \boldsymbol{x}, \boldsymbol{z}) \propto \prod_{i=1}^{n} \left\{ [\pi_1 f_1(x_i; \boldsymbol{\theta}_1)]^{I[z_i=1]} \left[\pi_2 \sum_{j=1}^{L} p_j I[x_i = x_{(j)}] \right]^{I[z_i=2]} \right\}$$

$$= \pi_1^{n_1} \pi_2^{n_2} \prod_{i=1}^{n} [f_1(x_i; \boldsymbol{\theta}_1)]^{I[z_i=1]} \cdot \prod_{j=1}^{L} p_j^{\sum_{i=1}^{n} I[z_i=2, x_i=x_{(j)}]} ;$$

where $n_k = \sum_i I[z_i = k]$, $k = 1, 2$.

If we choose uniform priors for $\boldsymbol{\pi}$ and $\boldsymbol{p}$ and denote the prior for $\boldsymbol{\theta}_1$ as $g_0(\boldsymbol{\theta}_1)$, we have that $\boldsymbol{\pi}, \boldsymbol{p}$ and $\boldsymbol{\theta}_1$ are independent under the posterior distributions. In particular, the posterior distribution of $\boldsymbol{\pi}$ is given by the Beta distribution.

Based on this, Gibbs sampler can be used to draw realizations from the posterior distribution and carry out a Bayesian Monte Carlo analysis. This leads to the following algorithm:

Algorithm 6: The proposed method

1 Initialization: select, $z^{(0)}, \boldsymbol{\pi}^{(0)}, \boldsymbol{p}^{(0)}$ and $\boldsymbol{\theta}_1^{(0)}$;
2 Set $m = 1$;
3 repeat
4   **for** $i = 1$ *to* n **do**
5   Update z_i with probability in

$$P(z_i = 1) \propto \pi_1 f_1(x_i; \boldsymbol{\theta}_1) ;$$
$$P(z_i = 2) \propto \pi_2 \sum_{j=1}^{L} p_j I[x_i = x_{(j)}] ;$$

6   Update $\boldsymbol{\theta}_1$ from the posterior in

$$g(\boldsymbol{\theta}_1 | \boldsymbol{x}, \boldsymbol{z}) \propto \prod_{i=1}^{n} [f_1(x_i; \boldsymbol{\theta}_1)]^{I[z_i=1]} g_0(\boldsymbol{\theta}_1) ;$$

     Update $\boldsymbol{\pi}$ from the posterior in

$$g(\boldsymbol{\pi} | \boldsymbol{x}, \boldsymbol{z}) \propto \pi_1^{n_1} \pi_2^{n_2} ;$$

7   Update $\boldsymbol{p}$ from the posterior in

$$g(\boldsymbol{p} | \boldsymbol{x}, \boldsymbol{z}) \propto \prod_{j=1}^{L} p_j^{\sum_{i=1}^{n} I[z_i=2, x_i=x_{(j)}]} ;$$

     $m = m + 1$
8 until *Enough MCMC steps have been simulated;*

---

3 Simulation Study

In the simulation study, we consider a mixture distribution with five components, where the noise component is a Poisson distribution and the signal components are Negative Binomial distributions. We sample 500 observations. Our intention is to compare our proposed method with fully parametric mixture model in terms of estimation and classification. The true model for the simulation is given by

$$f(x) = \pi_1 \text{Poi}(x; \lambda) + \sum_{k=2}^{5} \pi_k \text{NB}(x; r_k, v_k). \tag{5}$$

We chose different values for the parameters λ, r_k and v_k in order to compare our method with existing methods under different settings.

In the first scenario, we choose the set of true parameters (Set 1) as $\lambda = 2$, $\pi_1 = 0.6$, $\pi_2 = \cdots = \pi_5 = 0.1$, $r = (15, 13, 10, 8)$ and $v = (0.9, 0.7, 0.6, 0.5)$. This choice of r and v for the NB components gives the corresponding component means as $(1.68, 5.57, 6.67, 8.00)$. Such a choice implies that the means of Poisson component and all the other NB components are not too far apart. From Table 1 we can see that our method has clear posterior estimates, which approximate the true parameter value. The trace plot confirms that our method does not suffer from the label switching problem (see Fig. 2). In fact, label switching does not occur in our methodology. However, for the Poisson component and other NB components, the above situation causes some identifiability problems when traditional Gibbs

Table 1 Parameter Set 1. (i) The new method; (ii) true mixture model of five components

Model	True value										Posterior mean		Error rate
	λ	π_1	r_1	r_2	r_3	r_4	v_1	v_2	v_3	v_4	λ	π_1	e
(i)	2	0.6	15	13	10	8	0.9	0.7	0.6	0.5	2.2514 (1.8881,2.6680)	0.6987 (0.5680,0.7885)	0.31
(ii)	2	0.6	15	13	10	8	0.9	0.7	0.6	0.5	2.4371 (1.0576,4.9958)	0.2952 (0.0249,0.7433)	0.46

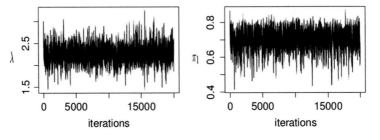

Fig. 2 MCMC trace plots for λ, π_1 for our new model for the true parameters in Table 1

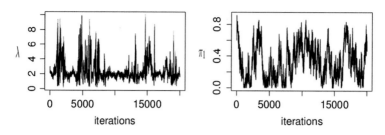

Fig. 3 MCMC trace plots for λ, π_1 for a mixture of a Poisson and four NB distributions for the true parameters in Table 1

Table 2 Parameter Set 2. (i) The new method; (ii) the true mixture model of five components

	True value										Posterior mean		Error rate
Model	λ	π_1	r_1	r_2	r_3	r_4	v_1	v_2	v_3	v_4	λ	π_1	e
(i)	7	0.6	15	20	40	30	0.4	0.3	0.3	0.2	6.8676	0.5787	0.06
											(6.4998,7.2305)	(0.5226,0.6292)	
(ii)	7	0.6	15	20	40	30	0.4	0.3	0.3	0.2	6.9622	0.5349	0.10
											(6.4599,7.4080)	(0.2279,0.6329)	

sampling method is used (see Fig. 3). The MCMC trace plots in Fig. 3 for π_1 and λ clearly show the occurrence of the label switching problem.

This issue severely distorts the posterior estimates, see Table 1. For example, the posterior mean for λ is 2.4371 (the true value is 2) and the posterior mean for π_1 is 0.2952 (the true value is 0.6). On the contrary, if we use the proposed method, the estimates for λ and π_1 are 2.2514 and 0.6987, respectively, which are closer to the true values. For simplicity, we did not provide the estimates for r and v since the main aim here is classification and under the new model r and v are not involved. Instead, we compared the misclassification rate (the ratio of the number of wrongly classified observations over the total number of observations) for the two methods. This can be easily obtained as the Bayesian approach provides the simulated z from the full posterior. From the last column of Table 1 we can see that our method has smaller misclassification rate than the parametric mixture model.

In the second set of the simulation, the choice of the true parameters are $\lambda = 7$, $\pi_1 = 0.6$, $\pi_2 = \cdots = \pi_5 = 0.1$, $r = (15, 20, 40, 30)$ and $v = (0.4, 0.3, 0.3, 0.2)$. This choice of r and v for the NB components gives the corresponding component means as $(22.5, 46.7, 93, 120)$. This gives very different component means, with the Poisson component having the smallest mean. This situation is similar to the real ChIP-seq data, in terms of long tail, and the noise component has the smallest mean value. From Table 2 we can see that our method gives posterior mean estimates for λ and π_1 with smaller bias and shorter credible intervals than the parametric mixture approach. This is because our method does not incur the label switching problem. Contrarily, the larger bias and variation in the estimates in the existing methods is due to the label switching problem, see Fig. 4. Still, the new method performs better in terms of misclassification rate. For the results, we run the

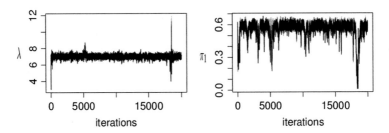

Fig. 4 MCMC trace plots for λ, π_1 for a mixture of a Poisson and four NB distributions for the true parameters in Table 2

Gibbs sampler for 20,000 steps with 10,000 steps as burn-in iterations over 100 simulations. Furthermore, we use a Metropolis-within-Gibbs sampler to simulate from the posterior distributions for the parametric mixtures, given the difficulty in simulating the parameters r and v for NB distributions.

4 Data Analysis: ChIP-seq Data

We present the application of the new method to ChIP-seq data. We consider the GBPT301.1000bp data set from the R package `enRich`. Our aim is to detect the regions in the genome that are enriched, so it is a natural two-component mixture model problem with a noise and a signal component. Several methods for the analysis of ChIP-seq data assume a parametric signal distribution mixed with a parametric noise distribution. For example, Kuan et al. (2011) propose a mixture of Negative Binomial distributions; Qin et al. (2010) adopt a generalized Poisson distribution for the signal and Bao et al. (2013) propose a Poisson distribution for noise and a Poisson distribution for the signal, and also Negative Binomial distribution for the noise and Negative Binomial distribution for the signal.

However, analysis of ChIP-seq data involving non-parametric approach focused mainly on peak calling algorithms (see Nix et al. 2008 and Zhang et al. 2008). Wang et al. (2010) employed Gibbs sampling strategy for mapping of ambiguous sequence tags. Bound regions are piled up with reads, but due to the "noise" inherent in the essay, calling "peaks" is not a straightforward task. Another demerit of non-parametric peak calling approach is that they are strongly determined by thresholds which are set heuristically in the peak calling step and the results of the analysis are compounded by the differences in the background noise (Hower et al. 2011). Therefore, we considered parametric distribution for the noise component and used non-parametric distribution to model the signal component.

Based on the posterior distribution, the posterior classification probability can be used to predict whether a region is enriched or not.

$$D_i = P(z_i = 1 | x, \theta) := \frac{\pi_1 f_1(x_i; \theta_1)}{\pi_1 f_1(x_i; \theta_1) + \pi_2 \sum_{j=1}^{L} p_j I[x_i = x_{(j)}]}.$$

Fig. 5 Number of enriched
regions identified by the
proposed model,
Poisson–Poisson mixture
model, NB–NB mixture
model on chromosome 21 at
the 0.2 % FDR

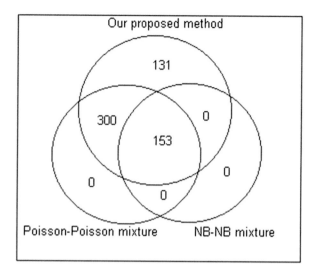

The region i will be classified as an enriched region if $D_i < c$. The threshold
value c is determined by controlling the false discovery rate (FDR) at a predefined
level (Bao et al. 2013), say 0.002. The expected FDR corresponding to the threshold
value c is given by

$$\widehat{FDR} := \frac{\sum_{i \in \text{enriched region}}(D_i)}{\sum_i I[D_i < c]}.$$

We present the result in Fig. 5, which shows a Venn diagram of the detected regions
as enriched for GBP experiment of ChIP-seq data for our proposed model, compared
with a mixture of two Poisson distributions and a mixture of two NB distributions,
at 0.2 % FDR. For the Poisson and NB mixtures we use the implementation in the
`enRich` R package. Our method detects more enriched regions than the existing
methods at the same FDR.

5 Conclusion

We developed mixture model with parametric and non-parametric components. We
achieved several advantages by using the non-parametric component. Firstly, we
neither need to specify the distributions for the signal component nor to consider
the number of components. Secondly, the method circumvents the label switching
problem. Results on simulated data verified the validity of the approach and showed
better performance in terms of estimation and classification. We illustrated the
proposed method on ChIP-seq data (GBPT301.1000bp) to detect the enriched
regions bound by proteins of interest.

Relatively large window size (1000bp) in the ChIP-seq data motivates the use of traditional mixture models that do not account for Markov dependencies. For a smaller window size, say 200bp, we expect spatial dependencies between the neighbouring windows. More elaborate models such as HMMs or Markov random fields should be considered in this case such as the method developed in Bao et al. (2014). The possible extension of this method to account for Markov dependencies is currently under investigation.

The proposed method is only valid for discrete data sets, thus a possible extension might be to develop methods able to deal with continuous data sets. In this case, a continuous distribution would be chosen for the noise component $f_1(x)$. However, new methods would need to be developed for the non-parametric component, since the posterior of z_i in Algorithm 6 will not be valid anymore. This can be explored as a future research work.

References

Antoniak, C. E. (1974). Mixtures of Dirichlet processes with applications to Bayesian nonparametric problems. *The Annals of Statistics, 2*(6), 1152–1174.

Bao, Y., Vinciotti, V., Wit, E., & 'T Hoen, P. A. C. (2013). Accounting for immunoprecipitation efficiencies in the statistical analysis of ChIP-seq data. *BMC Bioinformatics, 14*, 169.

Bao, Y., Vinciotti, V., Wit, E., & 'T Hoen, P. A. C. (2014). Joint modelling of ChIP-seq data via a Markov random field model. *Biostatistics, 15*(2), 296–310.

Celeux, G., Hurn, M., & Robert, C. P. (2000). Computational and inferential difficulties with mixture posterior distributions. *Journal of American Statistical Association, 95*, 957–970.

Diebolt, J., & Robert, C. P. (1994). Estimation of finite mixture distributions through Bayesian sampling. *Journal of the Royal Statistical Society. Series B, 56*, 363–375.

Escobar, M. D., & West, M. (1995). Bayesian density estimation and inference using mixtures. *Journal of the American Statistical Association, 90*(430), 577–588.

Green, P. (1995). Reversible jump Markov chain Monte Carlo computation and Bayesian model determination. *Biometrika, 82*(4), 711–732.

Hower, V., Evans, S. N., & Pachter, L. (2011). Shape-based peak identification for ChIP-seq. *BMC Bioinformatics, 12*(1), 15.

Jasra, A., Holmes, C. C., & Stephens, D. A. (2005). Markov chain Monte Carlo methods and the label switching problem in Bayesian mixture modeling. *Statistical Science, 20*, 50–67.

Kuan, P. F., Chung, D., Pan, G., Thomson, J. A., Stewart, R., & Kele, S. (2011). A statistical framework for the analysis of chip-seq data. *Journal of the American Statistical Association, 106*(495), 891–903.

Mclachlan, G., & Peel, D. (2004). *Finite mixture models*. New York: Wiley.

Nix, D., Courdy, S., & Boucher, K. (2008). Empirical methods for controlling false positives and estimating confidence in ChIP-Seq peaks. *BMC Bioinformatics 9*(1), 523.

Nobile, A., & Fearnside, A. T. (2007). Bayesian finite mixtures with an unknown number of components: The allocation sampler. *Statistics and Computing, 17*(2), 147–162.

Qin, Z. S., Yu, J., Shen, J., Maher, C. A., Hu, M., Kalyana-Sundaram, S., et al. (2010). HPeak: An HMM-based algorithm for defining read-enriched regions in ChIP-seq data. *BMC Bioinformatics, 11*(1), 369.

Ramos, Y. F. M., Hestand, M. S., Verlaan, M., Krabbendam, E., Ariyurek, Y., Van Galen, M., et al. (2010). Genome-wide assessment of differential roles for p300 and CBP in transcription regulation. *Nucleic Acids Research, 39*(16), 5396–5408.

Richardson, S., & Green, P. J. (1997). Bayesian analysis of mixtures with an unknown number of components (With Discussion). *Journal of the Royal Statistical Society: Series B, 59*(4), 731–792.

Rodriguez, C. E., & Walker, S. G. (2014). Label switching in Bayesian mixture models: Deterministic relabeling strategies. *Journal of Computational and Graphical Statistics, 23*, 25–45.

Sperrin, M., Jaki, T., & Wit, E. (2010). Probabilistic relabelling strategies for the label switching problem in Bayesian mixture models. *Journal of Statistics and Computing, 20*, 357–366.

Stephens, M. (2000a). Bayesian analysis of mixture models with an unknown number of components an alternative to reversible jump methods. *Annals of Statistician, 28*, 40–74.

Stephens, M. (2000b). Dealing with label switching in mixture models. *Journal of the Royal Statistical Society: Series B, 62*(4), 795–809.

Wang, J., Huda, A., Lunyak, V. V., & Jordan, I. K. (2010). A Gibbs sampling strategy applied to the mapping of ambiguous short-sequence tags. *Bioinformatics, 26*(20), 2501–2508.

Zhang, Y., Liu, T., Meyer, C., Eeckhoute, J., Johnson, D., Bernstein, B., et al. (2008). Model-based analysis of ChIP-Seq (MACS). *Genome Biology 9*(9), R137.

Information Theoretic Measures for Ant Colony Optimization

Gunnar Völkel, Markus Maucher, Christoph Müssel, Uwe Schöning,
and Hans A. Kestler

Abstract We survey existing measures to analyze the search behavior of Ant
Colony Optimization (ACO) algorithms and introduce a new uncertainty measure
for characterizing three ACO variants. Unlike previous measures, the group uncertainty allows for quantifying the exploration of the search space with respect to the
group assignment. We use the group uncertainty to analyze the search behavior of
Group-Based Ant Colony Optimization.

1 Introduction

Combinatorial optimization problems arise in many scientific disciplines such as
operations research, computer science, engineering, and commerce. Being NP-
hard, many combinatorial optimizations have an exponentially growing worst-case
runtime in the size of their problem instances for the corresponding best known
algorithms. Hence, it is desirable to compute near optimal solutions using low
computational costs with approximate algorithms like Ant Colony Optimization
(ACO) algorithms. The first ACO algorithm called Ant System (Dorigo et al.
1996) was developed in the early nineties. The main idea of ACO was inspired

G. Völkel
Institute of Theoretical Computer Science and Core Unit Medical Systems Biology,
Ulm University, Ulm, Germany
e-mail: gunnar.voelkel@uni-ulm.de

M. Maucher • C. Müssel
Medical Systems Biology, Ulm University, Ulm, Germany
e-mail: markus.maucher@uni-ulm.de; christoph.muessel@uni-ulm.de

U. Schöning
Institute of Theoretical Computer Science, Ulm University, Ulm, Germany
e-mail: uwe.schoening@uni-ulm.de

H.A. Kestler (✉)
Institute of Medical Systems Biology, Universität Ulm, Ulm, Germany
e-mail: hans.kestler@uni-ulm.de; hkestler@fli-leibniz.de

© Springer International Publishing Switzerland 2016

A.F.X. Wilhelm, H.A. Kestler (eds.), *Analysis of Large and Complex Data*, Studies
in Classification, Data Analysis, and Knowledge Organization,
DOI 10.1007/978-3-319-25226-1_44

by the foraging behavior of ants. Ants of many ant species (e.g., the Argentine ant) deposit pheromone on the paths they move along when exploring the environment. Other ants perceive the pheromone and tend to prefer paths with a larger amount of pheromone. Pheromone trails emerge due to ants choosing the same path, depositing pheromone and thus reinforcing that path. In this paper we discuss measures to analyze the development of the search behavior of ACO algorithms. The internal state of ACO algorithms determines the part of the search space that is searched effectively by the algorithm. The λ-branching factor has been used in Dorigo and Stützle (2004) to analyze the behavior of ACO algorithms. The authors suggest that the Shannon entropy can be used as a parameterless alternative to the λ-branching factor but do not demonstrate its application. The previous two measures are based on the pheromone matrix used by the ACO algorithm. Here, we investigate the development of the entropy values compared to the λ-branching factor. We introduce an uncertainty measure as the Shannon entropy of the empirical probabilities calculated from the solutions constructed by the ACO algorithm. With this uncertainty measure problem-specific heuristic values are taken into account. In contrast to the existing measures the uncertainty measure is independent of the internal state representation of the ACO algorithm. Furthermore we show that this measure can be used to quantify the uncertainty of the group assignment decisions of ACO algorithms. We use our measures to analyze ACO variants for the Vehicle Routing Problem.

2   Ant Colony Optimization

This section outlines combinatorial optimization problems and the Vehicle Routing Problem with Time Windows followed by the description of the ACO variants that are studied later.

2.1   Combinatorial Optimization Problem

Generally, for a given set of decision variables X_i ($i \in \{1, \ldots, n\}$) with finite domains $d_i = d(X_i)$ and a set of constraints $\omega \in \Omega$, $\omega : \mathcal{D}_1 \times \cdots \times \mathcal{D}_n \to \{\texttt{true}, \texttt{false}\}$ a combinatorial optimization problem (without loss of generality minimization) with respect to an objective function $f : \mathcal{D}_1 \times \cdots \times \mathcal{D}_n \to \mathbb{R}$ can be defined as follows:

$$
\begin{aligned}
f(X_1, \ldots, X_n) &\to \min \\
\omega(X_1, \ldots, X_n) &= \texttt{true}, &\quad \forall \omega \in \Omega \\
X_i &\in \mathcal{D}_i, &\quad i \in \{1, \ldots, n\}.
\end{aligned}
$$

Then the goal is to find a feasible assignment $(X_1 = c_{1j_1}, \ldots, X_n = c_{nj_n})$ that minimizes $f(X_1, \ldots, X_n)$. A feasible assignment is called a solution s of the combinatorial optimization problem and denoted briefly as $s = (c_{1j_1}, \ldots, c_{nj_n})$ with the solution components c_{ij}. A partial assignment of only a subset $\{X_i \mid i \in I\}$ of the decision variables is called a partial solution. A feasible partial solution s' is a partial solution where all constraints $\omega \in \Omega$ hold for the partial assignment. The set of feasible components $\mathscr{C}_\Omega(s')$ for a given feasible partial solution s' is the set of solution components that can be added to s' resulting in another feasible partial solution.

An example for a combinatorial optimization problem is the Vehicle Routing Problem with Time Windows (VRPTW, Bräysy and Gendreau 2005) which we study as a benchmark problem in this paper. The VRPTW is a routing problem in which the chosen routes are comprised of decisions to travel from customer v_i to customer v_j. This can be represented by assignments $X_i = j$. The domains of the decision variables are restricted such that all pairs X_i, X_j have distinct values. Two additional constraints corresponding to vehicle capacity and customer time windows limit the set of feasible solutions. More specifically, the VRPTW is a distribution problem where a given set of customers v_i ($i \in \{1, \ldots, n\}$) demands a quantity $q_i \in \mathbb{N}^+$ of a product that is available at a depot v_{n+1}. Given a limited number (m) of vehicles with limited capacity C the goal is to find a minimal number of tours and a minimal total driving distance of these tours to distribute the product. Each customer v_i must get the delivery within a given time window $[b_i, e_i]$. The delivery at customer v_i has a given service duration Δs_i. Only the begin of the service associated with the delivery must be within the time window. The pairwise distances d_{ij} and travel durations Δt_{ij} between the locations of customers and the depot are given. Each vehicle starts at the depot vertex v_{n+1}. The objective function $f(s)$ is two-dimensional and consists of the used vehicle count (or tour count) $v(s) \in \mathbb{N}^+$ and the total driving distance $d(s) \in \mathbb{R}^+$. In the newer literature a lexicographic comparison between solutions is used, where the vehicle count $v(s)$ is the primary objective and the total distance $d(s)$ is the secondary objective. Two different approaches for applying ACO to the Vehicle Routing Problem with Time Windows are detailed in the following.

2.2 ACO with Linear Solution Encoding

The construction procedure of ACO algorithms is based on a construction graph $\mathscr{G} = (\mathscr{V}, \mathscr{E})$. For the VRPTW the construction graph has a vertex v_i ($i \in \{1, \ldots, n\}$ for each customer and a vertex v_{n+j} ($j \in \{1, \ldots, m\}$) at the depot location for each vehicle. The graph contains a directed edge from vertex v_i to vertex v_j if it is possible to travel from vertex v_i to vertex v_j and to arrive within the time window of v_j. A solution component c_{ij} is an edge (v_i, v_j) in the construction graph and represents the decision to travel from vertex v_i to vertex v_j. In the context of a partial solution where the last vertex v_i is fixed implicitly we will use the vertex v_j to identify the solution component c_{ij}. The linear solution encoding represents a solution s to the VRPTW as a path through the construction graph

$$s = \Big(\underbrace{v_{n+i_1}, v_{j_1}, v_{j_2}, \ldots, v_{j_{k_1}}}_{=T_1}, \underbrace{v_{n+i_2}, v_{j_{k_1}+1}, \ldots, v_{j_{k_2}}}_{=T_2}, \underbrace{v_{n+i_3}, \ldots, v_{n+i_{k_{m+1}}}}_{T_3,\ldots,T_m} \Big),$$

where each tour T_k starts with a depot vertex v_{n+i_k} followed by the customers that are visited in the tour in that order and ends with the depot vertex $v_{n+i_{k+1}}$ of the next tour T_{k+1}. This construction graph and the linear solution encoding were first described in Gambardella et al. (1999). Let ClassicACO denote the ACO algorithm using the linear solution encoding and a corresponding sequential solution construction. ClassicACO builds a predefined number of solutions per iteration and returns the best-so-far solution after a specified number of iterations. The solutions are built incrementally by choosing the next vertex v_j to add to the current partial solution probabilistically. In construction step v for a given partial solution $s_v = (v_{i_0}, v_{i_1}, \ldots, v_{i_v})$ in linear encoding the sequential solution construction chooses the next vertex v_j with probability

$$P\left(v_j \mid s_v \right) = \begin{cases} \dfrac{\eta_j(s_v)^\alpha \cdot \tau_{i_v,j}^\beta}{\displaystyle\sum_{v_k \in \mathscr{C}_\Omega(s_v)} \eta_k(s_v)^\alpha \cdot \tau_{i_v,k}^\beta} & v_j \in \mathscr{C}_\Omega(s_v) \\ 0 & v_j \notin \mathscr{C}_\Omega(s_v), \end{cases} \tag{1}$$

where $\tau_{i_v,j} \in \mathbb{R}^+$ is the pheromone value and $\eta_j(s_v) \in \mathbb{R}^+$ the heuristic value associated with the edge (v_{i_v}, v_j) for the partial solution s_v. $\alpha, \beta > 0$ are additional weightings. The simple heuristic $\eta_j(v_{i_v}) = d_{i_v,j}^{-1}$ is used in the experiments of this paper. We call the product $r_{ij} = \eta_j(s_v)^\alpha \cdot \tau_{ij}^\beta$ the rating r_{ij} of the solution component c_{ij} (VRPTW: edge (v_i, v_j)). The pheromone matrix is initialized with $\tau_{ij} = \tau_0$ and updated using

$$\tau_{ij}(t+1) = (1-\rho) \cdot \tau_{ij}(t) + \begin{cases} \rho \cdot R(\hat{s}(n)), & c_{ij} \in \hat{s}(t) \\ 0, & \text{otherwise} \end{cases}, \tag{2}$$

where $\hat{s}(t)$ is the best solution found up to iteration t and R is a function calculating the reward based on that best-so-far solution.

2.3 ACO with Group-Based Solution Encoding

The ACO algorithm variant GB-ACO (introduced in Völkel et al. 2013) uses a group-based solution encoding and a corresponding parallel solution construction. The group-based solution encoding represents a solution s of the VRPTW as a set of paths through the construction graph—one for each tour.

$$s = \underbrace{(g_1, g_2, \ldots, g_m)}_{=(T_1, T_2, \ldots, T_m)}, \qquad g_a = (v_{n+i_a}, v_{j_{a,1}}, v_{j_{a,2}}, \ldots, v_{j_{a,k_a}}).$$

There is one group g_a per vehicle representing the tour T_a of the vehicle (if any). One group comprises the customers that are visited in the corresponding tour in the given order. The probabilistic parallel solution construction of GB-ACO starts with a solution of m empty groups which start at their respective depot node each. Solution components are added incrementally to a chosen group. For a given partial solution

$$s_v = (g_{1,v}, \ldots, g_{m,v}), \qquad g_{a,v} = (v_{i_{a,0}}, v_{i_{a,1}}, \ldots, v_{i_{a,v}})$$

and the sets of feasible components $\mathscr{C}_\Omega(s_v, g_{a,v})$ for all groups $g_{a,v}$ of s_v the probability to choose a group $g_{a,v}$ is defined as

$$P(g_{a,v} \mid s_v) = \frac{\displaystyle\sum_{v_k \in \mathscr{C}_\Omega(s_v, g_{a,v})} \eta_k(s_v, g_{a,v})^\alpha \cdot \tau_{i_{a,v},k}^\beta}{\displaystyle\sum_{k=1}^{m} \sum_{v_l \in \mathscr{C}_\Omega(s_v, g_{b,v})} \eta_l(s_v, g_{b,v})^\alpha \cdot \tau_{i_{b,v},l}^\beta}, \tag{3}$$

where the heuristic value $\eta_i(s_v, g_{c,v})$ is dependent on the group $g_{c,v}$. For the chosen group $g_{a,v}$ the probability of selecting the next vertex $v_j \in \mathscr{C}_\Omega(s_v, g_{a,v})$ for addition is defined as

$$P(v_j \mid s_v, g_{a,v}) = \frac{\eta_j(s_v, g_{a,v})^\alpha \cdot \tau_{i_{a,v},j}^\beta}{\displaystyle\sum_{v_k \in \mathscr{C}_\Omega(s_v, g_{a,v})} \eta_k(s_v, g_{a,v})^\alpha \cdot \tau_{i_{a,v},k}^\beta}.$$

For the VRPTW an empty group punishment (EGP: linear or exponential) can be applied in the group selection to support the minimization of the number of tours (non-empty groups) as described in Völkel et al. (2013). Both EGP variants reduce the rating sum of empty groups by dividing through a divisor which is linearly increasing (linear EGP) or exponentially increasing (exponential EGP) in the number of non-empty groups. The GB-ACO variant called UG-ACO uses a random selection of the next group with a pseudo-uniform distribution based on a uniform distribution modified by empty group punishment.

3 Measures

The analysis of the ACO algorithm variants can be based on their internal state per iteration t, the pheromone matrix $\tau(t)$ and the best-so-far solution $\hat{s}(t)$, and the set of generated solutions $S(t)$. In Dorigo and Stützle (2004) the λ-branching factor is used to analyze different aspects of ACO algorithms (e.g., to detect early stagnation) and it is suggested to use the entropy as a parameterless alternative of the λ-branching factor. At the end of this section we introduce the uncertainty measure which is not directly based on the pheromone matrix but is calculated from empirical probabilities.

3.1 λ-Branching Factor

Following the definition in Dorigo and Stützle (2004) for a pheromone matrix τ the minimal and maximal pheromone values for the decision variable X_i are defined as

$$\tau_i^{\min} = \min_{j \in \mathcal{V}} \tau_{ij} \quad \text{and} \quad \tau_i^{\max} = \max_{j \in \mathcal{V}} \tau_{ij} . \tag{4}$$

The λ-branching factor $b_\lambda^\tau(i)$ of vertex v_i is defined as the number of vertices v_j with a pheromone value larger than a threshold depending on λ and $[\tau_i^{\min}, \tau_i^{\max}]$, i.e.,

$$b_\lambda^\tau(i) = \left| \{v_j \mid \tau_{ij} \geq \tau_i^{\min} + \lambda(\tau_i^{\max} - \tau_i^{\min})\} \right| . \tag{5}$$

The average λ-branching factor for a given pheromone matrix τ is denoted as $\overline{b_\lambda^\tau}$. The average λ-branching factor indicates how much of the search space can be searched effectively. In Dorigo and Stützle (2004) $\lambda = 0.05$ is used for the analysis. Similarly to the branching factor $\overline{b_\lambda^\tau}$ for the pheromone values, we can define the branching factor $\overline{b_\lambda^R}$ based on the rating values of the components (provided that the heuristic values η_{ij} depend only on the previous vertex). The idea is that a λ-branching factor based on the ratings r_{ij} captures the real possibilities to explore the search space better than the one which is only based on the pheromone values.

3.2 Entropy

As described above the construction in ACO algorithms is performed on the construction graph $\mathcal{G} = (\mathcal{V}, \mathcal{E})$. The decision at vertex v_i to choose a successor vertex v_j can be modeled as random variable $X_i(s_v) \in \mathcal{C}_\Omega(s_v)$. Let $p_{i_v j} = P\left(v_j \mid s_v = (v_{i_0}, \ldots, v_{i_v})\right)$ be the probability from Eq. (1). Then the Shannon entropy of $X_i(s_v)$ is defined as

$$H(X_i(s_v)) = - \sum_{v_j \in \mathcal{C}_\Omega(s_v)} p_{i_v j} \cdot \log_2 p_{i_v j} . \tag{6}$$

This entropy can be bounded from above by the analogously defined $H(X_i)$ with $X_i \in \mathcal{V}$ [$v_j \in \mathcal{C}_\Omega(s_v)$ changes to $v_j \in \mathcal{V}$ in Eq. (6)]. The entropy $H(X_i)$ and the corresponding average entropy $\overline{H}(X)$ can only be calculated efficiently when heuristic values $\eta_j(s_v = (v_{i_0}, \ldots, v_{i_v})) = \eta_j(v_{i_v})$ that only depend on the previous vertex v_{i_v} and not on the state calculated from the whole partial solution s_v are used. Even for heuristic values $\eta_j(v_{i_v})$ the entropy $H(X_i(s_v))$ cannot be calculated efficiently because the random variables $X_i(s_v)$ are not independent from each other, e.g., choosing $X_{i_1} = v_{j_1}$ limits the sample space of X_{i_2} to $\mathcal{V} \setminus \{v_{j_1}\}$. Thus, the calculation $H(X_i(s_v))$ would require the calculation of the probability distribution for each $X_{i_1}, \ldots, X_{i_n}$ for all possible orders $(i_1, \ldots, i_n) \in S_n$.

For an efficient analysis of algorithm runs the following options remain: a calculation of $H(X_i)$ and $\overline{H}(X)$ based on probabilities proportional to $\eta_j(v_{i_v})^\alpha \cdot \tau_{i_v j}^\beta$ or a calculation of $H_\tau(X_i)$ and $\overline{H}_\tau(X)$ with probabilities proportional to $\tau_{i_v j}$. The first option captures the state of the algorithm better than the second option because it incorporates the heuristic values as well.

3.3 Uncertainty

To overcome the limitations of the previously described entropy measures, we will introduce an empirical entropy value called *uncertainty* that is intended to capture the development of the internal state of the algorithm better. Assuming the ACO algorithm runs for T iterations constructing A solutions $s_1(t), s_2(t), \ldots, s_A(t)$ per iteration t, then the sliding window $S_{w,t}$ of the solutions from w iterations starting at iteration $t - w$ is defined as

$$S_{w,t} = \left\{ s_a(k) \mid k \in \{t - w + 1, t - w + 2, \ldots, t\}, a \in \{1, \ldots, A\} \right\} \tag{7}$$

with $w \geq 2$ and $t \geq w$. Let $P_{w,t}(X_i = v_j)$ be the empirical probability calculated from the solutions of the sliding window $S_{w,t}$. The uncertainty $\mathcal{U}_{w,t}(X_i)$ within the sliding window $S_{w,t}$ to choose a successor vertex from v_i is defined as

$$\mathcal{U}_{w,t}(X_i) = - \sum_{v_j \in \mathcal{V}} P_{w,t}(X_i = v_j) \cdot \log_2 P_{w,t}(X_i = v_j) . \tag{8}$$

The mean uncertainty within the sliding window $S_{w,t}$ is denoted as $\overline{\mathscr{U}}_{w,t}(X)$. The sliding windows are motivated from the fact that ACO algorithms usually construct only a small number of solutions (e.g., $A = 10$) per iteration leading to an insufficient estimation of the probability to select a successor vertex. For problems with a group structure like the VRPTW the uncertainty of a vertex occurring in a certain group can be defined similarly to the previous approach. Let $Y_i \in \{g_1, \ldots, g_m\}$ be the random variable for vertex v_i such that $Y_i = g_j$ denotes the event that vertex v_i is contained in group g_j. The group uncertainty $\mathscr{U}_{w,t}(Y_i)$ within the sliding window $S_{w,t}$ is defined analogously to the previous uncertainty $\mathscr{U}_{w,t}(X_i)$. The mean group uncertainty is denoted as $\overline{\mathscr{U}}_{w,t}(Y)$.

4   Application

We performed an experiment with the three ACO variants ClassicACO, GB-ACO, and UG-ACO on the Solomon instances (Solomon 1987). Each algorithm had 25 repetitions on each problem instance and each algorithm used the same set of seeds for the pseudo-random number generator. The pheromone evaporation was specified as $\rho \in \{0.03, 0.06\}$, the number of constructed solutions per iteration was $A = 10$ and the three different empty group punishment strategies (none, linear, exponential) were used for GB-ACO and UG-ACO. The values were chosen such that neither early stagnation nor insufficient exploitation occurs. In the following we apply the introduced measures to the data gathered in the experiments. The figures show the data from instance c101 which is representative for the other instances of the instance class c1. Figure 1 shows the development of all discussed measures and the objective values over the iterations of the algorithms. While the λ-branching factor $\overline{b_\lambda^\tau}$ of the pheromone values distinguishes the three algorithm variants well, the λ-branching factor $\overline{b_\lambda^R}$ of the rating values cannot distinguish between GB-ACO and UG-ACO. For the two entropy measures, the algorithm variants are indistinguishable. This may be because all three algorithms use the same pheromone update rule which is dominated by the exponential evaporation. For the edge uncertainty and the group uncertainty the three algorithms show a distinct behavior. It can be observed that GB-ACO shows intermediate levels for both uncertainty measures. At the same time GB-ACO achieves the best scores for both objectives. This could indicate that GB-ACO maintains a better balance of exploration and exploitation than the other variants. Further analyses based on the uncertainty measures (data not shown) suggest that the algorithm's exploration benefits from the exponential empty group punishment strategy, whereas the exploitation is supported by the probabilistic group selection rule of GB-ACO. Based on the observations made above we examine the effects of the empty group punishment strategies on UG-ACO. For this purpose we have chosen the UG-ACO because its basic variant does not employ any explicit mechanisms for minimizing the number of used groups. Figure 2 shows a comparison of three empty group punishment (EGP)

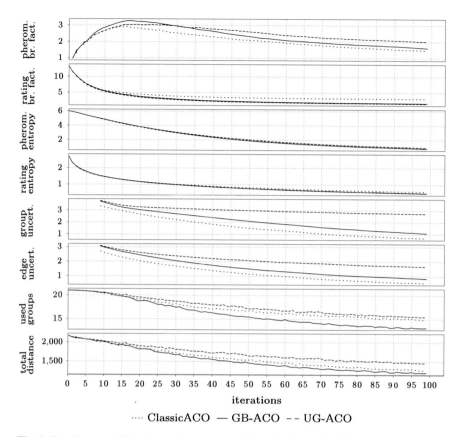

Fig. 1 Development of the discussed measures and the objective values for all three algorithms. The measures λ-branching factor (pheromone, rating), entropy (pheromone, rating), and uncertainty (group, edge) are shown. The two objective values of the generated solutions per iteration, number of used groups (tour count) and total distance, are included at the bottom. The shown values are averaged over all 25 repetitions. $\lambda = 0.05$ is used for the λ-branching factors. The uncertainty measures are calculated using a sliding window of size $w = 10$ iterations. All algorithms use $\rho = 0.03$ and the exponential group punishment is used by GB-ACO and UG-ACO (instance c101)

strategies (none, linear, and exponential) based on the uncertainty measures and the corresponding objective values. All EGP variants show a comparatively high group uncertainty. This is the expected behavior for UG-ACO, as the group selection is based on a pseudo-uniform distribution. The exponential empty group punishment generally yields the best scores for both objectives. The group uncertainty curve for this strategy decreases quickly at the beginning, but remains on a higher value than the other two strategies in the long run. The optimization progress of the algorithm can be observed through the edge uncertainty which decreases continuously. This probably means that the exponential punishment enforces a smaller number of used groups early, but the uncertainty of assigning vertices to these few groups remains

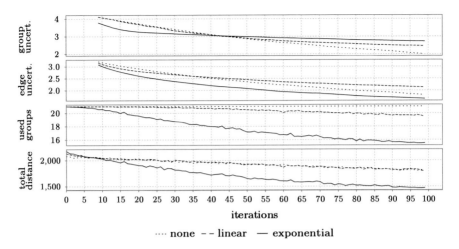

Fig. 2 Development of the mean objective values (used group count, total distance) and the mean uncertainty measures (group uncertainty, edge uncertainty) of UG-ACO on instance c101. The mean values averaged over all 25 repetitions are shown. A sliding window of size $w = 10$ is used for the uncertainty measures. The evaporation rate $\rho = 0.03$ is used

high due to the uniform sampling. The linear strategy takes longer to reduce the number of used groups, which can be seen in both the group uncertainty and the used groups score. Without empty group punishment UG-ACO is not able to minimize the number of used groups.

5   Conclusion

This paper gives an overview of measures to analyze the search behavior of ACO algorithms and introduces a new uncertainty measure. The previously proposed λ-branching factor only allows for an indirect observation of the exploration process: It describes mostly the effect of the pheromone updates but not the number of possible choices the algorithm has at a given iteration.

We introduce an uncertainty measure defined as the Shannon entropy based on the empirical probabilities of group or vertex assignments. This measure can be efficiently computed and captures the behavior of the algorithm more closely than already proposed measures. For example, the two entropy measures, previously suggested as a parameterless alternative for the λ-branching factor, cannot be used to distinguish the algorithm variants in our analysis.

Our uncertainty measure is applicable to the group assignment decisions of components (customer to tour) which is not possible with previous measures. With the group uncertainty we can observe that GB-ACO with exponential empty group punishment seems to maintain a better balance of exploration and exploitation than the other discussed ACO variants. These relations are observed when using problem-specific heuristics that do not support group minimization explicitly.

For the discussed measures the following relations can be observed (see Appendix for details). The true entropy is bounded from above by the rating entropy. The rating entropy will generally be smaller than the pheromone entropy. The true uncertainty is bounded from above by the average cross-entropy between the random variables which is equal to the sum of the average true entropy and the average Kullback–Leibler divergence between the random variables.

Acknowledgements The research leading to these results has received funding from the European Community's Seventh Framework Programme (FP7/2007–2013) under grant agreement n°602783 to HAK, the German Research Foundation (DFG, SFB 1074 project Z1 to HAK), and the Federal Ministry of Education and Research (BMBF, Gerontosys II, Forschungskern SyStaR, project ID 0315894A to HAK).

Appendix: Theoretical Relations

True Entropy Bounded by Rating Entropy Given random variable $X \in \{1, \ldots, N\}$ and assuming that without loss of generality nodes $N - k + 1, \ldots, N$ have already been visited and let r_j be the rating for choosing $X = j$, then it can be shown that the true entropy $H(X(s_v))$ is bounded from above by the rating entropy $H(X)$.

$$H(X) - H(X(s_v)) = -\sum_{j=1}^{N} \left(\frac{r_j}{\sum_{l=1}^{N} r_l} \log \frac{r_j}{\sum_{l=1}^{N} r_l} \right) + \sum_{j=1}^{N-k} \left(\frac{r_j}{\sum_{l=1}^{N-k} r_l} \log \frac{r_j}{\sum_{l=1}^{N-k} r_l} \right)$$

$$\geq -\sum_{j=1}^{N-k} \left(\frac{r_j}{\sum_{l=1}^{N} r_l} \log \frac{r_j}{\sum_{l=1}^{N} r_l} - \frac{r_j}{\sum_{l=1}^{N-k} r_l} \log \frac{r_j}{\sum_{l=1}^{N-k} r_l} \right) \geq 0$$

since $\frac{r_j}{\sum_{l=1}^{N} r_l} \leq \frac{r_j}{\sum_{l=1}^{N-k} r_l}$ and $x \log x$ is monotonically increasing.

Rating Entropy Smaller Than Pheromone Entropy The pheromone entropy $H_\tau(X)$ will generally be larger than the rating entropy $H(X)$ since the heuristic incorporated in $H(X)$ will increase the differences in probability by favoring some events. The heuristic values permitted in this case can be calculated in advance and are constant for the whole algorithm execution.

Uncertainty Relations Assume that true uncertainty $\mathbb{U}_w(X_i)$ in the sliding window of length w is calculated based on the true probabilities $p_j^{(.0)}$ of each iteration t instead of the empirical estimations. Then the average probabilities $\bar{p}_j$ in the sliding window w are known. It holds that $\mathbb{U}_1(X) = H(X^{(1)}(s_v)) = H_1(X(s_v))$ and that the true uncertainty can be bounded from above by the sum of the average entropy in the same sliding window and the average Kullback–Leibler divergence between the corresponding random variables.

$$\mathbb{U}_w(X) = - \sum_{j=1}^{N} \bar{p}_j \log \bar{p}_j = - \sum_{j=1}^{N} \left(\frac{1}{w} \sum_{t=1}^{w} p_j^{(.0)} \right) \log \left(\frac{1}{w} \sum_{t=1}^{w} p_j^{(.0)} \right).$$

Applying Jensen's inequality leads to the following:

$$\mathbb{U}_w(X) \leq - \sum_{j=1}^{N} \left(\frac{1}{w} \sum_{t=1}^{w} p_j^{(.0)} \right) \left(\frac{1}{w} \sum_{t=1}^{w} \log p_j^{(.0)} \right) = - \frac{1}{w^2} \sum_{j=1}^{N} \sum_{k=1}^{w} \sum_{l=1}^{w} p_j^{(k)} \log p_j^{(l)}$$

$$= \frac{1}{w^2} \sum_{k=1}^{w} \sum_{l=1}^{w} \underbrace{H\left(X^{(k)}; X^{(l)}\right)}_{\text{cross-entropy}} = \frac{1}{w^2} \sum_{k=1}^{w} \sum_{l=1}^{w} H\left(X^{(k)}\right) + \underbrace{D_{\mathrm{KL}}\left(X^{(k)}||X^{(l)}\right)}_{\text{Kullback-Leiblerdivergence}}$$

$$= H_w\left(X(s_v)\right) + \frac{1}{w^2} \sum_{k=1}^{w} \sum_{l=1}^{w} D_{\mathrm{KL}}\left(X^{(k)}||X^{(l)}\right).$$

References

Bräysy, O., & Gendreau, M. (2005). Vehicle routing problem with time windows, part I: Route construction and local search algorithms. *Transportation Science, 39*(1), 104–118.

Dorigo, M., Maniezzo, V., & Colorni, A. (1996). Ant system: Optimization by a colony of cooperating agents. *IEEE Transactions on Systems, Man, and Cybernetics, Part B: Cybernetics, 26*(1), 29–41.

Dorigo, M., & Stützle, T. (2004). *Ant colony optimization*. New York: Bradford Books, MIT Press.

Gambardella, L. M., Taillard, É., & Agazzi, G. (1999). MACS-VRPTW: A multiple colony system for vehicle routing problems with time windows. In D. Corne, M. Dorigo, F. Glover, D. Dasgupta, P. Moscato, R. Poli, et al. (Eds.), *New ideas in optimization* (pp. 63–76). New York: McGraw-Hill.

Solomon, M. M. (1987). Algorithms for the vehicle routing and scheduling problems with time window constraints. *Operations Research, 35*, 254–265.

Völkel, G., Maucher, M., & Kestler, H. A. (2013). Group-based ant colony optimization. In C. Blum (Ed.), *Proceeding of the Fifteenth Annual Conference on Genetic and Evolutionary Computation Conference, GECCO '13* (pp. 121–128). New York: ACM.

A Signature Based Method for Fraud Detection on E-Commerce Scenarios

Orlando Belo, Gabriel Mota, and Joana Fernandes

Abstract Electronic transactions have revolutionized the way that consumers shop, making the small and local retailers, which were being affected by the worldwide crisis, accessible to the entire world. As e-commerce market expands, the number of commercial transactions supported by credit cards—Card or Customer Not Present also increases. This growing relationship, quite natural and expected, has clear advantages, facilitating e-commerce transactions and attracting new possibilities for trading. However, at the same time a big and serious problem emerges: the occurrence of fraudulent situations in payments. In this work, we used a signature based method to establish the characteristics of user behavior and detect potential fraud cases. A signature is defined by a set of attributes that receive a diverse range of variables—e.g., the average number of orders, time spent per order, number of payment attempts, number of days since last visit, and many others—related to the behavior of a user, referring to an e-commerce application scenario. Based on the analysis of user behavior deviation, detected by comparing the user's recent activity with the user behavior data, which is expressed through the user signature, it is possible to detect potential fraud situations (deviant behavior) in useful time, giving a more robust and accurate decision support system to the fraud analysts on their daily job.

1 Introduction

The electronic commerce industry is in quick expansion at a global level. Nowadays a big majority of companies aim to get the biggest number of clients as possible, using means for the disponibilization of their services and products online. In most companies, independently of their size, it is frequent to found projects for the

O. Belo (✉) • G. Mota
University of Minho, Guimaraes, Portugal
e-mail: obelo@di.uminho.pt; pg23094@alunos.uminho.pt

J. Fernandes
Farfetch, Braga, Portugal
e-mail: joana.fernandes@farfetch.com

© Springer International Publishing Switzerland 2016   531
A.F.X. Wilhelm, H.A. Kestler (eds.), *Analysis of Large and Complex Data*, Studies in Classification, Data Analysis, and Knowledge Organization,
DOI 10.1007/978-3-319-25226-1_45

placement of their business online. The advantages are obvious. However with this proliferation of the online business a big problem arose, which one way or another, all merchants must deal with in their business area: the occurrence of fraudulent situations in payments (Kou et al. 2004; Flegel et al. 2010).

Today the majority of online payments is done through the utilization of credit cards. These transactions are considered Card or Customer Not Present (CNP). The detection of fraud in such scenarios (Delamaire et al. 2009) is often costly, either in terms of material or in human resources, with big challenges being imposed on the correct and in time identification of fraudsters. There are several traditional techniques and methods for fraud prevention (Richhariya and Singh 2012), such as identity proofing, guaranteed payments, operational management, and data quality, that combined with data mining techniques can be very effective in improving the detection rate of the prediction models. However, its efficiency is closely related to the quality and quantity of the available data, the experience and sensibility of the analyst working with the model and also even with the particular characteristics of each company. The most common data mining applications in fraud prevention/detection are based on techniques such as association rules, classification , or segmentation (Phua et al. 2005; Sanchez et al. 2009; Bhattacharyya et al. 2011).

Fraud analysts in the electronic commerce (E-Commerce) business are typically interested in customer data, such as: name, job, location, type of payment chosen, credit card information, shipping and billing country, etc. The customer behavior prior to the placement of an order is normally not taken into consideration in this analysis. For each order placed by a customer, a record is generated, containing all the steps the client made before placing the order (pages visited, items viewed and/or added to basket, time spent, number of clicks, number of payments attempted and/or credit cards used, etc). All this data constitutes the clickstream of a customer, and can pretty much describe an order completely. A clickstream record may not be by itself enough to detect fraud, therefore studying the customer behavior, and not just individual orders, is probably a better approach (Lee et al. 2001). Thus, based on these clickstream records, some kind of profiling techniques can be applied in order to reveal, with a certain accuracy, the customer behavior along time. Profile records that include a large diversity of features such as number of orders, average time spent per order, average number of cards used, etc., together with the customer data considered by fraud analysts, can be used in the construction of customer profiles. Three levels of data can therefore be considered: order, behavior, and customer. In order to capture the characteristics of a user's behavior, the concept of signature can be applied. A signature corresponds to a set of information that captures the typical behavior of a user (Cortes and Pregibon 2001; Cortes et al. 2002). For example, the average number of orders, time spent per order, number of payment attempts, and so on. In fraud and intrusion detection systems, signatures can be used in two distinct ways:

- Detection based in user profiles: the signature of the user is compared against a database of cases of known legitimate use. This kind of method fits under the class of supervised learning techniques.
- Detection based in anomalies: the signature of each customer is itself the baseline for comparison. New traffic for a user is compared against their individual signature to determine if the user's behavior changed.

In this paper the usage of a signature based method for fraud detection that implements both detection based in user profiles and in anomalies is proposed. We start the next section by presenting and discussing how signatures can be helpful on anomalous behavior detection. Then, an analysis of the data is presented (Sect. 3), followed by the method and detection techniques that resulted in the model implementation (Sect. 4). Section 5 contains the experimental setup and based on the achieved results a comparison between this model and other data mining techniques is also presented. Finally, in Sect. 6 an analysis of the obtained results is made. This paper ends with a brief set of conclusions about what was done and the effectiveness of the applied method (Sect. 7).

2 Fraud Detection Based on Signatures

Signature based methods are a fairly recent technique that has been introduced (with some success) in the detection and prevention of fraud, particularly in the telecommunications field (Cortes and Pregibon 2001; Ferreira et al. 2006; Lopes et al. 2011). This method consists in the establishment of usage patterns in a specific branch of activity, by selecting a group of features extracted from a *data stream* source. By comparing those (normal usage) patterns with outliers, it is possible the detection of fraud. Cortes and Pregibon (2001) described a transactional data stream as a dynamic continuous flow of data consisting of a set of records about some interaction processes that happened between a entities of interest. Clickstream data from E-Commerce environments are a perfect example of this definition. Therefore the usage of signature analysis for fraud detection seems very appropriate, due to the nature of the E-Commerce problems, allowing for the characterization of users almost in real-time. The features of a signature are obtained by performing statistical calculus over relevant numeric fields of the data that can be useful in explaining an user behavior. These fields in E-Commerce scenarios can be extracted from clickstream, which is a generic term to describe visitors paths through a series of web pages requested by an user in a single visit, also known as session. Clickstream data is therefore a collection of information of sessions. These fields alone can be rather uninformative, but when taken as a whole, they can be an excellent representation of an user behavior, and thus a good source for the establishment of usage profiles.

2.1 The Signature Definition

The purpose of a signature is to be adjustable and adaptable by capturing the actual user behavior and changing as it also changes, being personalized from user to user. In statistical terms, a signature can be described as an estimate of the joint probability distribution of a group of selected combined components (variables or features). If these variables consist of a unique atomic value (e.g., numeric), then they are called *simple* variables. Otherwise if they consist of two co-dependent statistical values (e.g., mean and standard deviation), they are called *complex* (Cortes and Pregibon 2001).

2.2 Signature Updating

There are two common ways for updating a signature. They are *event-driven* and *time-driven* (Cortes and Pregibon 2001; Lopes et al. 2011). In the first case with the entry of new records, the signature is updated. In the time-driven case records are collected and only after a pre-determined time period is the signature updated. The computational model for both methods is the same.

For a given temporal window w, a signature S is obtained from a function α, where $S = \phi(w)$. Ferreira et al. (2006) define a *time unit* as a certain pre-defined amount of time in which session records are accumulated and then processed, being w directly proportional to the time unit, $w = \alpha.\Delta t$. Considering at a given period of time t, a record or set of records R of a signature S_t, that should be processed with an identical format as S_t, before the signature is updated, resulting on T_R. At time $t+1$ a new signature corresponding to the update of S_t based on T_R is formed, according to the formula:

$$S_{t+1} = \beta.S_t + (1 - \beta).T_R \qquad (1)$$

in which β determines the value of the new T_R transactions in the new signature. If a certain threshold Ω, that is compared to the distance between the S and T_R vectors where $dist(S, T_R) \geq \Omega$, is met, then the signature is updated. Otherwise an alert is generated for the case detected, so it can be later analyzed. After the shift of a time unit, the signature S is updated to S' with the new clickstream information, recorded between the end of S and the updating of S' (Fig. 1).

Fig. 1 Variation of a signature—extracted from Ferreira et al. (2006)

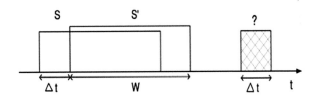

2.3   Signature Features

Features or elements of a signature are statistical values that describe a certain aspect of the behavior of a user. These are calculated through variables extracted from the clickstream data. In this concrete context of application, an element can be *simple* when it corresponds to an average value or *complex*, when it is associated with an average and standard deviation of a certain data feature. Based on some statistical analysis done over the data and on the knowledge of analysts, a group of variables considered useful for the detection of fraud was chosen and transformed into elements of the signature. The referred variables are presented in Table 1 and organized according to their respective type.

The variables chosen to be complex were those with high variability, which can be better expressed through both average and standard deviation. Simple variables, on the other hand, have very low standard deviation values and so they can be perfectly represented only by their calculated average.

Taking as an example the variable *"Number of Days To Purchase,"* which indicates the number of days that passed since a new client registration and first purchase, or the number of days since the last time an existing client made a purchase. In Table 2 it can be observed that on average fraudsters buy products in a much lower interval of days since registration or last purchase, than non-fraudsters. Also, non-fraudulent users have a much higher standard deviation value than fraudulent ones. This can be explained by the small time window fraudsters have before being caught. Thus, this variable can be better expressed through both it mean and standard deviation value.

Table 1 Elements of the signature

Simple variables	Complex variables
Time spent	Value of order
Pages visited	Order days
Time per page	Order time
Duration inactive	Number of unique actions
Number of items purchased	Number of days to purchase
Number of new visits	Number of days since last visit

Table 2 "Number of Days To Purchase" statistical analysis

Type	Min	Max	Mean	Std
Fraud	0	169	4.589	14.470
Non-fraud	0	188	19.172	34.305

2.4 Anomalies Detection

Given a record or set of records of a user, a summary is calculated and compared to its signature, to determine if there is a deviation from its typical behavior. This comparison is done by first of all extracting from the clickstream data, the set of features present in Table 1, resulting in the vector T_R, that will be compared to the user signature S, by calculating the Euclidean distance between them. If the distance is considered normal, i.e. the deviation is insignificant, another comparison is made, this time between S and a known fraudulent signature model. To perform this second comparison, the Hellinger distance formula is used. Should the distances in any of these comparisons be considered suspicious, i.e. superior to a pre-defined threshold, an alarm is generated, so that the analyst can further analyze the case.

3   The Data

In this study we used a set of real data, gathered from an E-Commerce company data warehouse. This data contained clickstream records of customers purchases ordered on the company web site. Since the data was extracted from a data warehouse, it was clean and tidy, ready for analysis, without necessity of being further treated. Each line of the data, that corresponded to a purchase placed by a customer, was labeled as fraudulent or non-fraudulent, based on previous analysis and treatment carried out by the company's analysts. Six months of data were gathered—July to December 2013—which resulted in 87,881 clickstream records. From these only 1632 were labeled as fraudulent, and the remaining 86,249 labeled as non-fraudulent. From this set of records, several smaller datasets were made in order to be used in the implementation of the signature based method. For the purpose of this study, the group of all the 87,881 records, was called DS0. From this first dataset, a subset with all the fraudulent records was created (DSF), and then from these 1000 records the subset DS1 was created, that served as the basis to define the fraudulent signature model. For the definition of the desirable values for *lim* and *lim2* thresholds used in the anomaly detection (seen later on section IV), 2000 records were extracted from DS0 and joined with the records of DS1, resulting in dataset DS2. For the creation of the signature classes, 5000 records were extracted from DS0. These were not used in any of the previous made datasets. The resulting dataset was called DS3. For testing purposes, the remaining 632 fraudulent records from DSF were joined with 4368 unused non-fraudulent records from DS0, resulting in dataset DS4, with 5000 records. To sum up, several datasets were prepared with different characteristics necessary for the implementation of the signature model. A summary of all these datasets can be seen in Table 3.

Table 3 Datasets summary

Dataset	Records	Description	Purpose
DS0	87,881	All records	
DSF	1632	All fraudulent records	
DS1	1000	Only fraudulent records	Signature model
DS2	3000	2000 non-fraudulent records + DS1	Anomaly detection
DS3	5000	Only non-fraudulent records	Signature classes
DS4	5000	632 fraudulent+4368 non-fraudulent records	Test

4 Model Implementation

4.1 Signature Processing

The signature processing starts by reading the data, which is then used to calculate the summaries for each distinct user. Then the signatures procedure is executed. As input this procedure receives a β value used in signature updating, and two other values, *lim* and *lim2*, used in the detection of anomalies. For each summary, it is verified if a signature for the user already exists. If not, a new one is created by comparing the summary with a group of pre-existent signature classes. The signature class to which the user summary has more similarities is then used established the new signature. In case the user already has a signature a comparison between it and the calculated summary is done and the previously explained process of anomalies detection, is executed.

4.2 Signature Classes

When a summary is processed it is compared to the respective user signature, although if the user hasn't a signature, a new one needs to be created. This new signature should be more complex than just the summary of a single record or a set of few records, and yet should also be a good representation of the user behavior. To address this situation, a group of non-fraudulent signatures were created, to serve as comparison for the user summary. Between these, the signature that is most resemblant with the summary is saved as the new user signature.

The creation of the signature classes was done with the dataset DS3, that contained 5000 non-fraudulent records. After calculating the summaries of those records, a clustering technique was applied in order to divide the summaries in groups with similar characteristics. The technique used was model based, assuming a variety of data models and applying the maximum likelihood estimation and the Bayes criteria to identify the most likely model and number of clusters. In particular the technique we applied selected the optimal model according to *Bayesian Information Criterion* (BIC) for *Expectation Maximization* (EM) initialized by hierarchical clustering for parameterized Gaussian mixture models, choosing the model and number of clusters based on the largest BIC.

4.3   The Signature Fraud Model

As explained previously, after a first comparison, if an alarm is not generated, then a second comparison, with a different signature is done. This signature has the particularity of representing a typical case of fraudulent behavior. Should the comparison between the summary and this signature be inferior to a certain threshold (*lim2*), then it means that the summary is very similar to the fraudulent signature and an alarm is generated.

The elaboration of the signature fraud model was done with dataset DS1, containing 1000 known fraudulent records. The summaries of these records were calculated and the mean value of each variable joined and stored resulting in a single unique signature, as it was done in the case of the signature classes.

5   Experimental Setup

5.1   Signature Based Models Tests

In the first test 18 variables (12 mean values and six standard deviation values) were used in the model signature. Then for the second test, the elements of the signature were reduced to 12 simple variables and in the third test, only the six strongest simple variables were considered, namely: *Number of Days Since Last Visit, Number of Days To Purchase, Value of Order, Number of Unique Actions, Number of Items Purchased, Number of New Visits*. The models performance was analyzed based on *Fraud Accuracy Detection, Non-Fraud Accuracy Detection*, and *Overall Accuracy*. Our focus in terms of performance went toward improving Fraud Accuracy Detection over Overall Accuracy, since to a company it is less expensive to manually analyze a non-fraudulent ordered, which was wrongly labeled, than to miss a fraudulent one. The results of the three tests, as well as the *lim* and *lim2* values, are present in Table 4. In all tests the β value used was equal to 0.7. It was chosen simply because we feel it express the right weight of new signatures' summaries, in the process of signature updating. As can be seen the third test with six simple variables, was the one that showed better results.

Table 4 Signatures models performance

Tests	% fraud acc.	% non-fraud acc.	% global acc.	lim	lim2
18 variables (complex)	53	46	48	0.3	0.3
12 variables (simple)	61	50	53	0.4	0.17
6 variables (simple)	70	47	53	0.29	0.14

Table 5 Techniques performance comparison

Technique	% fraud acc.	% non-fraud acc.	% global acc.
Random Forests	14	76	68
SVM	15	84	75
Signatures	70	47	53

5.2 Random Forests and SVM Tests

In the application of the two mining techniques, SVM and Random Forests, the DS4 dataset was used as the testing set, which was the same one used in signature based models. For the training set the DS3 and DS2 datasets were joined, resulting in a new dataset (DS5), with 8000 records of which 1000 labeled as fraudulent. From these datasets only the same six variables used previously in the signature model, were considered. The implementation of these techniques like the signature based models, was done using the R programming language.

In the case of the Random Forests and support vector machines, the libraries used were the *randomforest* and the *e1071*, respectively. In both techniques, the method chosen was classification. No fine tuning or modification of the default parameters was conducted. The results can be seen next, in Table 5.

6 Results Analysis

Table 5 shows an high overall accuracy with the SVM and Random Forests techniques, respectively, 75 and 68 %, and a lower value with the signature based model, with just over 50 % accuracy. On the other hand, the percentage of fraudulent cases is much higher in the signature method, than on the other two data mining techniques. This can be explained with the data in analysis being so much unbalanced. Since in the test data very few registries are labeled as fraudulent, the very poor detection rate of fraud in both SVM and Random Forests almost has no significance in the global accuracy. On the signature model, fraud detection is much higher, at the cost of a trade-off with overall detection, that is a little lower than in the other techniques.

Unbalanced data may not be the only reason why the classification of this dataset is so hard. In fact we applied a singular value decomposition (SVD) technique to the DS4 dataset, and this confirmed this data to be of very difficult analysis, because there aren't many differences between the behaviors of fraudsters and normal users. SVD is a well-known technique that consists in the mathematical decomposition of a matrix that splits the original matrix into three new matrices (A = U x D x V). These new decomposed matrices give insight into how much variance there is in the original matrix, and how patterns are distributed, by way of combining information from several (likely) correlated vectors, and forming basis vectors that are guaranteed to be orthogonal in higher dimensional space. This knowledge can be useful in the selection of the components that have the most variance and hence the biggest impact on our data (Vozalis and Margaritis 2005).

A good way of visualizing the distribution of data, categorized by fraud, is by plotting its left singular vectors (U matrix). In Fig. 2 the plot of two vectors from the U matrix can be seen. The data is ordered by index, with registries labeled as fraudulent appearing first. This was done to allow a proper visualization, otherwise since there are so much less cases of fraud than non-fraud, the firsts would be scattered and difficult to identify. The dark circles represent the cases labeled as fraudulent, and the light triangles are the non-fraudulent cases.

A clear division in the y-axis between the two labels would be the desirable ideally result, but as can be seen, apart from some outliers present both in fraud and non-fraud, the distribution of data is very similar. A new plot was done, but this time from the 12 variables present in the data, only the most significant ones were chosen. Vector V can be used to identify these variables, as can be seen next in Fig. 3. Variables numbered 2, 8, 9, and 12 are those showing highest impact based

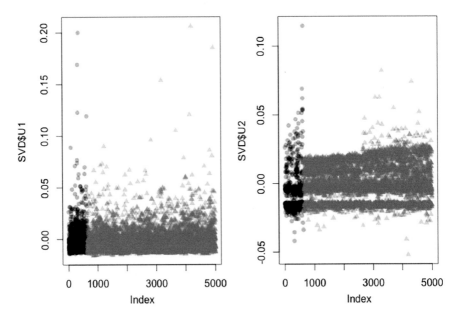

Fig. 2 SVD left singular vectors

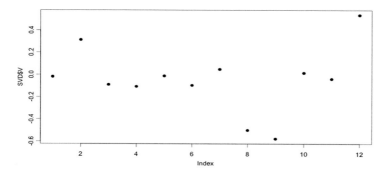

Fig. 3 Variables impact

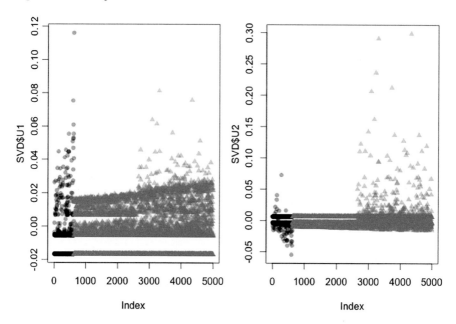

Fig. 4 SVD left singular vectors (most relevant variables)

on the application of the SVD technique. These variables correspond to *Days Since Last Visit, New Visit, Days To Purchase*, and *Order Value*. Figure 4 shows the plot of two vectors from the new SVD U matrix. Again, as can be seen there isn't a clear division of data in the y-axis. The characteristics of fraudulent cases that are so similar to the ones labeled as non-fraudulent, plus the data being so unbalanced, helps to explain the difficulty in classifying this dataset.

7 Conclusions and Future Work

In the past signature based methods were mainly applied to the telecommunication industry, and as far as we know this type of technique has never been applied in the context of E-Commerce. This fact prevents a comparison of the results. Also the scarcity of publicly available E-Commerce datasets precludes the testing of this method on different datasets. The use of Random Forests and SVM showed the high difficulty of correctly identify fraudulent cases. In that sense the signature method proved to be much more successful, with a good and better accuracy in fraud detection, traded-off by a lower value of overall accuracy. Real data is messy and hard to analyze and so it was on this particular case. At short term, in next experiments we intend to extend our datasets and the diversity of data elements and refine the method by attributing weights to feature variables (tuning its influence). We consider that the obtained results provide the basis to say that the followed method is quite useful and can detect in a satisfactory way potential cases of fraud in E-Commerce scenarios.

References

Bhattacharyya, S., Jha, S., Tharakunnel, K., & Westland, J. C. (2011). Data mining for credit card fraud: a comparative study. *Journal Decision Support Systems Archive, 50*(3), 602–613.

Cortes, C., & Pregibon, D. (2001). Signature-based methods for data streams. *Data Mining and Knowledge Discovery, (5)*, 167–182.

Cortes, C., Pregibon, D., & Volinsky, C. (2002). Communities of interest. *Intelligence Data Analysis, 6*(3), 211–219.

Delamaire, L., Abdou, H., & Pointon, J. (2009). Credit card fraud and detection techniques: a review. *Banks and Bank Systems, 4*(2), 57–68.

Ferreira, P., Alves, R., Belo, O., & Cortesao, L. (2006). Establishing fraud detection patterns based on signatures. In *Proceedings of 6th Industrial Conference on Data Mining (ICDM 2006)*, Leipzig, July 14–15.

Flegel, U., Vayssiere, J., & Bitz, G. (2010). A state of the art survey of fraud detection technology. In *Insider threats in cyber security*. Advances in Information Security (Vol. 49, pp. 73–84). New York: Springer.

Kou, Y., Lu, C., Sirwongwattana, S., & Huang, Y. (2004). Survey of fraud detection techniques. In *International Conference on Networking, Sensing, and Control* (pp. 749–754).

Lee, J., Podlaseck, M., Schonberg, E., & Hoch, R. (2001). Visualization and analysis of clickstream data of online stores for understanding web merchandising. *Data Mining and Knowledge Discovery, 5*(1), 59–84.

Lopes, J., Belo, O., & Vieira, C. (2011). Applying user signatures on fraud detection in telecommunications networks. In *11th Industrial Conference on Data Mining (ICDM 2011)*, August 30–September 3, Newark.

Phua, C., Lee, V., Smith K., & Gayler R. (2005). A comprehensive survey of data mining based fraud detection research. *Artificial Intelligence Review*.

Richhariya, P., & Singh, P. (2012). A survey on financial fraud detection methodologies. *International Journal of Computer Applications (0975 - 8887), 45*(22), 14–24.

Sanchez, D., Vila, M. A., Cerda, L., & Serrano, J. M. (2009). Association rules applied to credit card fraud detection. *Expert Systems with Applications: An International Journal, 36*(2), 3630–3640.

Vozalis, G., & Margaritis, K. (2005). Applying SVD on item-based filtering. In *ISDA '05 Proceedings of the 5th International Conference on Intelligent Systems Design and Applications* (pp. 464–469).

Three-Way Clustering Problems in Regional Science

Małgorzata Markowska, Andrzej Sokołowski, and Danuta Strahl

Abstract Three-way clustering problems have been considered since many years. They are popular specially in psychology and chemistry, but some of the propositions and methods are of more general nature. In regional science three-way data matrices consist of objects (regions), variables and time units (years). Asking which variables, in which regions and when, follow homogeneous pattern is meaningful. Three general approaches are proposed in the paper and different modes of standardization are discussed. The example on Eurostat data is also presented.

1 Introduction

Data involving three dimensions has been studied in literature (Carroll and Arabie 1980; Basford and McLachlan 1985; Pociecha and Sokołowski 1990; Kiers 1991; Smilde 1992; Vermunt 2007; Vichi et al. 2007), mainly for psychological applications. Useful information can be found on Three-Mode Company webpage.

The aim of this paper is to discuss the possibility of simultaneous classification of objects, variables and time units in regional science applications. We will use the following notation:

- Objects: $Y = y_1, y_2, \ldots, y_m$,
- Variables: $Z = z_1, z_2, \ldots, z_k$,
- Time units: $T = t_1, t_2, \ldots, t_n$.

Clustering problems can be described in two-position system, as

$$[\text{Classified items} | \text{Classification space}].$$

M. Markowska (✉) • D. Strahl
Wroclaw University of Economics, Wroclaw, Poland
e-mail: mmarkowska@ae.jgora.pl; dstrahl@ae.jgora.pl

A. Sokołowski
Cracow University of Economics, Cracow, Poland
e-mail: sokolows@uek.krakow.pl

© Springer International Publishing Switzerland 2016
A.F.X. Wilhelm, H.A. Kestler (eds.), *Analysis of Large and Complex Data*, Studies in Classification, Data Analysis, and Knowledge Organization,
DOI 10.1007/978-3-319-25226-1_46

In other words, we describe subjects for classification within some classification space. The classical problem of clustering, the Iris data is therefore [Y|Zt], since there is just one time unit in Fisher's data. Typical regional data has all three dimensions: geographical location, variables and time. Complex clustering problem for such data cube can be written as [YZT|.]. Question which is being asked within this problem sounds sensible—*Which regions, for which variables and when are similar, and can be placed in the same cluster*? But the way to answer this question is not so obvious.

In traditional numerical taxonomy (Sneath and Sokal 1963, 1973) there is a very useful term Operational Taxonomic Unit (OTU) which is defined as "the thing(s) being studied". The definition is intentionally vague. In fact everything can serve as "thing(s)" and it is possible not to stick to a word "object". For clustering variables ([Z|Yt]) we have variables as OTUs and objects forming a classification space. For dynamic clustering of regions [YT|Z], a region in a given year is an individual OTU, and classification space is defined by set of variables characterizing regions.

We discuss the possibility of solving the complex clustering problem on an example concerning the innovation of Polish economy on regional level. Poland is divided into 16 NUTS 2 regions. The following nine variables are taken into consideration:

- LLL—Participation of adults aged 25–64 in education and training by NUTS 2 regions—percentage,
- HRST—Human resources in science and technology—percentage of active population,
- HIT—Employment in high and medium high-technology manufacturing by NUTS 2 regions—percentage of total employment,
- KIS—Employment in knowledge-intensive services by NUTS 2 regions—percentage of total employment,
- HIT2—Employment in high and medium high-technology manufacturing by NUTS 2 regions—percentage of total employment in manufacturing,
- KIS 2—Employment in knowledge-intensive services by NUTS 2 regions—percentage of total employment in services,
- EDUC III—Persons aged 25–64 with tertiary education attainment by sex and NUTS 2 regions,
- YOUTH—Early leavers from education and training by NUTS 2 regions—percentage,
- EPO—Number of patents registered in a given year in the European Patent Office (EPO) per one million of workforce.

Data covers 2004–2012 period so the data cube consists of $16 \times 9 \times 9 = 1296$ data points. There are three ways of performing standardization:

- Local standardization—object-wise—each variable is standardized separately for each year on the object (province) set,
- Local standardization—time-wise—each variable is standardized over the set of time units (years) separately for each object,

- Global standardization—spatio-temporal—each variable is standardized on the set of all objects in all time units.

There are three possible approaches in dealing with the complex problem.

2 Approach 1: Total

Set of OTUs (OTU is an individual value of a province in 1 year—each unit of data cube) is subject for clustering. In our example we found six clusters. Part of Group 1 results, are presented in Fig. 1.

OTUs belonging to Group 1 are marked with 1. For the sake of clarity it would be wise to ignore individual zeros in rows, surrounded by units, or units surrounded by zeros. We name this operation smoothing. After smoothing, Group 1 (later identified as "weak innovators") is presented in Table 1.

Province	Variable	Year 04	Year 05	Year 06	Year 07	Year 08	Year 09	Year 10	Year 11	Year 12
PL11	LLL	0	0	0	1	1	0	1	1	1
PL11	HRST	0	0	0	0	0	0	0	0	0
PL11	HIT	1	0	0	0	0	0	0	0	0
PL11	KIS	0	0	0	0	0	0	0	0	0
PL11	KIS2	0	0	0	0	0	0	0	0	0
PL11	HIT2	1	1	0	1	1	1	1	0	0
PL11	EDUC	0	0	0	0	0	0	0	0	0
PL11	YOUTH	0	0	0	0	0	0	0	0	0
PL11	EPO	0	0	0	0	0	0	0	0	0
Total		2	1	0	2	2	1	2	1	1
PL12	LLL	0	0	0	0	0	0	0	0	0
PL12	HRST	0	0	0	0	0	0	0	0	0
PL12	HIT	0	0	0	0	0	0	0	1	1
PL12	KIS	0	0	0	0	0	0	0	0	0
PL12	KIS2	0	0	0	0	0	0	0	0	0
PL12	HIT2	0	0	0	0	0	0	0	0	0
PL12	EDUC	0	0	0	0	0	0	0	0	0
PL12	YOUTH	0	0	0	0	0	0	1	1	1
PL12	EPO	0	0	0	0	0	0	0	0	0
Total		0	0	0	0	0	0	1	2	2
PL21	LLL	1	0	0	0	0	0	0	0	0
PL21	HRST	0	0	0	0	0	0	0	0	0
PL21	HIT	0	1	0	0	1	0	0	0	0
PL21	KIS	0	0	0	0	0	0	0	0	0
PL21	KIS2	0	0	0	0	1	0	0	0	0
PL21	HIT2	0	1	0	0	0	0	0	0	0
PL21	EDUC	0	0	0	0	0	0	0	0	0
PL21	YOUTH	1	1	1	1	1	1	1	1	1
PL21	EPO	0	0	0	0	0	0	0	0	0
Total		2	3	1	1	4	1	1	1	1

Fig. 1 Part of Group 1 membership

Table 1 Group 1

	LLL	HRST	HIT	KIS	KIS2	HIT2	EDUC	YOUTH	EPO
PL11	07–12					04–10			
PL12			11–12					10–12	
PL21								04–12	
PL22					08–12			07–08	
PL31			04–12	04–06				04–05	04–05
PL32				04–12	04–05		04–05	04–10	
PL33	04–11	06–12	04–12	04–12		04–11			04–12
PL34			04–12			05–12			04–12
PL41				08–12	08–12			06–08	
PL42			10–12			10–11			
PL43					04–12		06–12		
PL51									
PL52		04–06					04–05		
		09–12					09–12		
PL61		05–12					04–12		
PL62	04–07		04-12			04–12	04–08		04–12
	10–12						11–12		
PL63									

Results are rather "fragmented", but definitely we can answer the question—*Which provinces, for which variables and when could be considered as weak innovators among Polish NUTS 2 regions?*

3 Approach 2: Separate

This approach has three steps:

- Step 1—Separate clustering of objects (Y), variables (Z) and time units (T). As a result we obtain so-called marginal groups.
- Step 2—Joint partition into $k(Y) \cdot k(Z) \cdot k(T)$ groups which gives us joint groups.
- Step 3—Clustering of joint groups to get final groups.

Performing Step 1 with Ward's method we clustered Polish provinces into three groups (all the others, Mazowieckie, Slaskie+Dolnoslaskie+Pomorskie), variables into three groups (YOUTH, EPO+HIT2+HIT, the rest of variables), and time span also into three periods: 2004–2006, 2007–2008, 2009–2012. The same number of groups for each dimension is just a coincidence. Each number of groups was determined on the basis of particular dendrogram (there is no place to present them here). So we have $3 \times 3 \times 3 = 27$ joint groups.

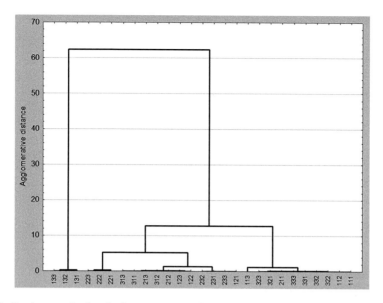

Fig. 2 Dendrogram for Step 3 of separate approach

Joint groups are coded according to marginal groups, e.g. 123 means all the others provinces characterized by EPO, HIT2 and HIT variables in 2009–2012. It is rather obvious that on the basis of dendrogram (Fig. 2) there are three homogeneous final groups.

4   Approach 3: Consecutive (Hierarchical)

We have three meta-dimensions of data cube: objects (Y), variables (Z) and time units (T). We can cluster object first, then for each object cluster we can group variables, and finally group time units for each object–variable cluster. There are six different orders of such consecutive clustering: Y-Z-T, Y-T-Z, Z-Y-T, Z-T-Y, T-Y-Z, T-Z-Y, everyone giving most probably different final result.

5   Conclusion

It can be easily noticed that there are some problems in all three approaches. In Approach 1 the number of OTUs is very large, in Approach 2—marginal partitions are preserved in the final one and there are too many versions within Approach 3.

After detailed analysis of the interpretability of results, the influence of different methods of standardization and possibilities of clear presentation we can give the following recommendations how to solve the complex clustering problem [YZT,.], with results that can be useful for regional analysis:

- Apply global spatio-temporal standardization.
- Use Approach 1 (each standardized data entry is a separate OTU).
- Smooth your results (ignoring single breaks in cluster membership).
- Present your results deciding the importance (so the order) of perspective (*which? why? when?*).

Acknowledgements The paper was prepared within the project financed by the Polish National Centre for Science, decision DEC-2013/09/B/HS4/0509.

References

Basford, K.E., & Mclachlan, G. J. (1985). The mixture method of clustering applied to three-way data. *Journal of Classification, 2*, 109–125.

Carroll, J. D., & Arabie, P. (1980). Multidimensional Scaling. *Annual Review of Psychology, 31*, 607–649.

Kiers, H. A. L. (1991). Hierarchical relations among three-way methods. *Psychometrica, 56*(3), 449–470.

Pociecha, J., & Sokołowski, A. (1990). Three-way clustering problems. *Control and Cybernetics, 19*(1–2), 179–187.

Smilde, A. K. (1992). Three-way analyses. Problems and prospects. *Chemometrics and Intelligent Laboratory Systems, 15*, 143–157.

Sneath, R. R., & Sokal, P. H. A. (1963). *Principles of numerical taxonomy*. San Francisco: W.H. Freeman and Company.

Sneath, R. R., & Sokal, P. H. A. (1973). *Numerical taxonomy. The principles and practice of numerical taxonomy*. San Francisco: W.H. Freeman and Company.

Vermunt, J. K. (2007). A hierarchical mixture model for clustering three-way data sets. *Computational Statistics & Data Analysis, 51*, 5368–5376.

Vichi, M., Rocci, R., & Kiers, H. A. L. (2007). Simultaneous component and clustering models for three-way data: within and between approaches. *Journal of Classification, 24*(1), 71–98.

Part XII
Data Analysis in Social, Behavioural and Health Care Sciences

CFA-MTMM Model in Comparative Analysis of 5-, 7-, 9-, and 11-point A/D Scales

Piotr Tarka

Abstract In this article author presents the results of comparative analysis in reference to scales based on 5-, 7-, 9-, and 11-point response categories. An attempt was made to find the optimum number of responses among these scales but in this regard to the assumptions underlying the Confirmatory Factor Model and MultiTrait-MultiMethod. For this purpose, the data was collected from a sample of young consumers ($n = 200$) studying at the universities in Poland. The specific aim of the research was focused on their attitudes, which measured different aspects of the companies' unethical behavior in the context of marketing activities. For the comparison of scales, the author has applied four models derived from the generalized CFA-MTMM model. This model allowed the recommendation of the best scale, and also helped to evaluate the effects associated with the use of particular type of scale on the CFA-MTMM alternative models and extracted, through their agency, factors.

1 Introduction

In hereby article, the author compares the four scales, i.e. based on different number of response categories, but with the same *Agree/Disagree* format of responses. A problem which is discussed relates to some methodological issues in the context of making the appropriate choices between 5-, 7-, 9-, and 11-point scale in the phenomena measurement. Choosing the right scale, before the whole research process begins, seems to play an important role, not only in the phase of data collection and ensuring information quality derived from it, but also causes further, inevitable effects in evaluation of the measurement models which take responsibility for the extraction of respective factor traits.

In order to solve the methodological problem, focused on the effects of the number of response categories in the scales, author decided to develop the four CFA-MTMM models. Thus, comparing the 5-,7-,9-, and 11-point scale, we not only

P. Tarka (✉)
Department of Marketing Research, Poznan University of Economics, Poznan, Poland
e-mail: piotr.tarka@ue.poznan.pl

© Springer International Publishing Switzerland 2016
A.F.X. Wilhelm, H.A. Kestler (eds.), *Analysis of Large and Complex Data*, Studies in Classification, Data Analysis, and Knowledge Organization,
DOI 10.1007/978-3-319-25226-1_47

obtained, through their agency, the chance to verify the construct validity of interest, but also we could analyze the level of information explained in these models due to different level of measurement in the scales. Finally, treating these scales as if they were different methods and comparing their results in CFA-MTMM models, we were able to provide the empirical proof of their possible similarities or differences.

2 The Number of Points in Scales with A/D Format

In the literature we can find information that Likert (1932) was the first who proposed 5-point scale with *Agree/Disagree* format. Much later, Dawes (2008) claimed that comparable results can also be obtained from 7- to 10-point scales, which may yield even more information than Likert's proposition. This point of view is confirmed by the information theory which states that if more response categories are added to scale, more information about the variable of interest can be obtained. For example, Alwin (1992) when considered a set of hypotheses related to the theory of information and when he tested them with panel data, he found that except for the 2-point scale, "the reliability is generally higher for measures involving more response categories" (p. 107).

Some other yet researchers when they approached to issues of increasing the number of categories, conducted comparisons of the quality of scales with different lengths, where quality referred to the strength of the relationship between the observed variable and the underlying construct of interest (Saris and Andrews 1991; Alwin 2007). Recently, Revilla et al. (2014) compared 5-point *Agree–Disagree* scale with 7- and 11-point option. In their study, they proved that quality of the measured variables decreased as the number of categories has been increased, so that the best scale, in their opinion, was a 5-point one. This contradicts the main statement of the theory of information, which as mentioned, argues that more categories mean more information about the variable of interest. However, their study had some limitations too. For example, they did not examine the scales with other alternative numbers of categories such as the 9-point scale which might confirm the tendency that using more response categories would not improve the scores. Therefore, author decided to include in his research one additional option, i.e. 9-point scale, however, in the study, only the results derived from CFA-MTMM models were considered.

3 From Campbell–Fiske to CFA-MTMM Model

As we know, in Campbell and Fiske (1959) MTMM analysis, each of two or more traits is measured with two or more methods. They suggested summarizing the correlations between all the traits measured with all the methods into a MTMM matrix, which could be directly examined for convergent and discriminant validation.

Because the MTMM approach (Campbell and Fiske 1959) faced many problems, this analytical strategy was the target of much criticism (Schmitt and Stults 1986; Marsh 1989). Even Campbell and Fiske (1959) were aware of the serious limitations of MTMM, arguing that their guidelines in the application of MTMM matrix should be carefully selected by researchers. They just wanted to (Marsh and Grayson 1995, p. 180) "provide a systematic, *formative evaluation* of MTMM data at the level of the individual trait-method unit, qualified by the recognized limitations of their approach, not to provide a *summative evaluation* of global summaries of convergent validity, discriminant validity and method effects."

The problems with MTMM led to alternative analytic solutions, mostly within the framework of covariance structure modeling which included *General CFA Model or Composite Direct Product Model* (for their review, see e.g., Browne 1984; Marsh 1989; Kenny and Kashy 1992; Marsh and Grayson 1995). Much attention was then paid to **Confirmatory Factor Analysis—CFA**, which continues to be, the best method of choice (Kenny and Kashy 1992).[1]

A generalized and complete version of CFA-MTMM model depends on the function of three components: a trait component, a method component, and a unique or error component. For each trait i, method j, person k, and measure m, where $m = i + (j - 1)I$, and I equals the total number of traits, the observed score X_{ijk} for i trait measured using j method for k person equals:

$$X_{ijk} = a_{mi}T_{ik} + b_{mj}M_{jk} + E_{ijk}. \qquad (1)$$

Here a_{mi} is the factor loading for measure m on the trait factor T_i, b_{mj} is the factor loading for measure m on the method factor M_j, and E_{ijk} is the uniqueness of the observed score. Each variable serves as an indicator on both trait and method factor.

In the CFA model, MTMM matrices can be analyzed in the context of the factors, defined by different measures of the same trait, which suggest trait effects. On the other hand, factors, which are defined by measures assessed with the same method, correspond to method effects. If the researcher wants to test the suitability of models, namely, the extent to which they fit empirical data, he can use the taxonomy of the nested models, proposed by Widaman (1985) and further developed by Marsh (1989). This taxonomy seems to be the most appropriate for all CFA-MTMM studies, because it provides a general framework in making decisions and inferences based on the effects of trait and method factors.[2]

---

[1] In fact, the beginning of these models was due to works of Werts and Linn (1970) or Jöreskog (1970), who proposed to treat the MTMM matrix as a CFA model. Much later, because there were problems with estimation of the parameters in an ordinary CFA-MTMM model requiring at least three measures for each construct, Saris et al. (2004) have developed the Split-Ballot MTMM (or SB-MTMM) model where each respondent could answer all questions only twice.

[2] In practice, a hypothesized, e.g. CFA—Correlated Traits and Correlated Methods model is set as baseline and compared with the nested more restrictive models in which specific parameters are eliminated or are constrained to zero or 1.0.

4   The Process of the Empirical Research

The subject of empirical study referred to consumers' judgment about the unethical behavior (represented by companies) in the area of marketing activities which influence the consumers' market consumption. Once the data was collected, the author generated four factor traits which were given the following names: **Poor Quality Products** (PQP); **Unfair Advertising Practices** (UAP); **Bad Approach to Client Service** (BATCS) and **Lack of Social Responsibility** (LOSR). Each factor trait was loaded with respectively observed variables as follows:

1. PQP: *In recent years, the quality of products offered by the companies has not improved*; *Companies do not take an effort to design products that meet consumers' real needs*; *Companies manufacture products that usually wear out, thus they earn money on client servicing*; *Companies manufacture products that are rich in their outer design, but are nonfunctional.*
2. UAP: *Most of the advertising contents are misleading and far away from the truth*; *Advertisements prepared by companies cannot be treated as a plausible source of information*; *Products which are often advertised, fail more frequently than products which are advertised less often*; *Companies give false color to their products.*
3. BATCS: *The quality of client service has been getting worse from year to year*; *The guarantees given with products within the client service are unfavorable*; *The way in which clients are encouraged by sellers to buy products is dishonest*; *Companies that sell clients their products, do not care to keep a good contact with them in the longer term.*
4. LOSR: *Companies do not pay much attention to the fact that the client is the most important for their activity and business*; *Companies are interested in pursuing their profits than in caring about their clients*; *Firms often make attempts to get as much of their clients' wallets as possible*; *Companies in pursuit of clients, have changed their marketing practices for worse.*

In sum, all factor traits reflected multidimensional construct that has been given the joint name: **Unethical Marketing Operations of Companies on Market**.

In the study, each factor trait was measured four times, with four variants of scales, by means of four separate questionnaires which were delivered to respondents assuming two-week interval for each questionnaire. The statements included in each questionnaire were repeated using different methods. In the description of response categories to particular scales (5, 7, 9, and 11), only the end points were labeled, as for the example of 5-point scale: 1—*totally disagree*, 5—*totally agree*.

The data was collected in 2014 year in the academic community of students (aged between 19–21) from five distinct Poznan universities in Poland.[3] The

---

[3] Adam Mickiewicz University, University of Technology, University of Economics, University of Life Sciences and University of Medical Sciences.

respondents were selected on the basis of simple random method of units selection. All sampling frames with the complete list of units were prepared and delivered by each university. During the study, the participants were asked to express their attitude to the above mentioned statements. The analysis was conducted in AMOS 21 software.

5 Selected CFA-MTMM Models in Scales Comparison

The author studied possible differences or similarities between the scales, using the four types of models. However, one of them (i.e., **CFA-CT/CM** model) stood for a basis, a reference point for comparisons of all other three models: **CFA-PCT/CM**—*Perfectly Correlated Traits/Correlated Methods*, **CFA-CT/UM**—*Correlated Traits/Uncorrelated Methods*, **CFA-NT/CM**—*No Traits/Correlated Methods*. So in general, the four CFA models were used: CT/CM; PCT/CM; CT/UM; NT/CM, but they have all originated from the one general CFA-MTMM model [see Eq. (1)].

The first, applied in the analysis of CFA-CT/CM model (see Fig. 1) was developed according to structure, which measured each of the observed variable and

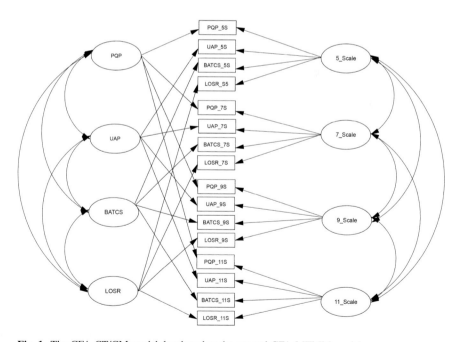

Fig. 1 The CFA-CT/CM model developed as the general CFA-MTMM model

its loading both onto a trait and a method factor. In this model, the large trait factor loadings indicated convergent validity whereas the large method factor loadings indicated the presence of method effects. Yet, in the other part, the large trait correlations, in particular, those approaching 1.0, indicated the lack of discriminant validity. In its assumptions, CFA-CT/CM model decomposed the variance of each observed variable into respective components (that were explained by traits) and simultaneously, decomposed variance explained by methods. In consequence, the researcher was able to assess the trait factor intercorrelations and the method factor intercorrelations. Unfortunately, when testing the full model, the problems of estimation may appear due to the unstable solutions of the model (Marsh 1989; Kenny and Kashy 1992). So, despite the fact that CFA-CT/CM model appears to be very attractive, in practice, even if the researcher wishes to apply it to data alone, it is rather useless. This model must be therefore used as a preliminary option for the nested models.

The next CFA-PCT/CM model, which was contrasted to CFA-CT/CM, assumed some differences due to the specification of the parameters in the factor trait correlations. In the study, they were assumed to be perfect, that is, they were set to 1.0.

The third proposed alternative, namely CFA-CT/UM model, differed from the CFA-CT/CM in the sense that it had no specified correlations between the methods. Finally, the major distinction between the CFA-CT/CM and the CFA-NT/CM model was the absence of trait factors.

The summary of fit indices related to these models is presented in Table 1 and the results of model comparisons are summarized in Table 2.

Table 1 The summary of goodness-of-fit indices for the CFA-MTMM models

Models	χ^2	df	CFI	RMSEA	TLI	GFI	AGFI
CT/CM	74.61	64	0.97	0.02	0.97	0.95	0.91
NT/CM	360.32	86	0.72	0.12	0.65	0.79	0.68
PCT/CM	149.06	69	0.86	0.08	0.78	0.89	0.81
CT/UM	116.22	67	0.93	0.03	0.91	0.92	0.89

Table 2 The CFA-MTMM model comparisons

Models	$\Delta\chi^2$	Δdf	ΔCFI	$\Delta AGFI$
The test of convergent validity				
Model CT/CM vs. NT/CM	285.71	22	0.25	0.23
The test of discriminant validity				
Model CT/CM vs. PCT/CM	75.45	5	0.11	0.10
Model CT/CM vs. CT/UM	41.61	3	0.04	0.02

5.1 The Goodness of Fit and Parameter Estimates

Having assumed the configuration between factor traits and methods and obtained, on their basis, further results, we can notice that the model CFA-CT/CM ($\chi^2_{(64)}$; RMSEA = 0.02; CFI = 0.97; TLI = 0.97; GFI = 0.95; AGFI = 0.91) reached the best overall fit in comparison to the three models. As it can be observed, the models: CFA-NT/CM ($\chi^2_{(86)}$; RMSEA = 0.12; CFI = 0.72; TLI = 0.65; GFI = 0.79; AGFI = 0.68) and PCT/CM ($\chi^2_{(69)}$; RMSEA = 0.08; CFI = 0.86; TLI = 0.78; GFI = 0.89; AGFI = 0.81) had a miserable fit (see, e.g., TLI indices which were of 0.65 and 0.78, respectively). On the other hand, the results proved that the CFA-CT/UM model obtained a good fit as compared to CFA-CT/CM.

Comparing CFA-CT/CM with CFA-NT/CM, we were able to investigate a *convergent validity*. From the scores observation it resulted that the models had substantial and statistically significant values ($\chi^2_{(22)} = 285,71; p < 0.001$), so the evidence of the convergent validity was confirmed. This assessment can be also supported by the large differences associated with $\Delta\chi^2$ value as well as sizeable ΔCFI and $\Delta AGFI$ indices (0.25 and 0.23, respectively). The results proved, that all four scales (5, 7, 9, 11) equally converged within the four extracted factor traits. However, a more precise analysis of trait- and method-related variance as well as the assessment of individual parameters estimates (see Table 3) revealed attenuation

Table 3 The CFA-CT/CM model—standardized estimates for the factor loadings

Traits and methods	PQP	UAP	BATCS	LOSC	5PS	7PS	9PS	11PS
5-Point Scale (5PS)								
PQP	0.61				0.14			
UAP		0.72			0.52			
BATCS			0.53		0.24			
LOSR				0.59	0.49			
7-Point Scale (7PS)								
POP	0.92					0.51		
UAP		0.86				0.40		
BATCS			0.77			0.68		
LOSR				0.72		0.48		
9-Point Scale (9PS)								
POP	0.35						0.37	
UAP		0.31					0.67	
BATCS			0.32				0.55	
LOSR				0.48			0.53	
11-Point Scale (11PS)								
POP	0.29							0.39
UAP		0.31						0.75
BATCS			0.26					0.64
LOSR				0.37				0.67

of the extracted factor traits caused by specific method effects. This was discernible, in particular, in the size of factor loadings calculated on the basis of 9- and 11-point scale.

Information obtained at the parameters level, associated with the convergent validity, may have additionally supported our findings if we had compared variance proportions.[4] For example, in the methods: 5- and 7-point scale (Table 3), the largest difference appeared in 5-point scale which measured trait (PQP) at the level 0.61 of variance and which was explained by this scale only at the level of 0.14. However, a close inspection of Table 3 still reveals that, a stronger convergent validity was sustained in 5- and 7-point scale as compared to weak convergence of 9- and 11-point scale.

The next phase of investigation in terms of factor traits and methods (scales) referred to the evidence of ***discriminant validity***. As Byrne argued (2010, pp. 290–291) "in testing for evidence of trait discriminant validity, one is interested in the extent to which independent measures of different traits are correlated, so these values should be negligible. When the independent measures represent different methods, correlations bear on the discriminant validity of traits and when they represent the same method, correlations bear on the presence of method effects, another aspect of discriminant validity."

Following the above assumptions, the CFA-CT/CM model was compared with CFA-PCT/CM. This analysis allowed the assessment of discriminant validity of the factor traits. Greater discrepancy and significant difference of $\Delta\chi^2$ as well as sizeable ΔCFI and $\Delta AGFI$ indicate strong discriminant validity. In our case, the compared models yielded $\Delta\chi^2$ value that was statistically significant ($\Delta\chi^2_{(5)} = 75.45$; $p < 0.001$), so the difference in ΔCFI was only at the level of 0.11 which was a modest proof of the discriminant validity.

The discriminant validity was also estimated in reference to the method effects by comparing the following models: CFA-CT/CM and CFA-CT/UM, in which the small differences in the indices of $\Delta\chi^2$; ΔCFI and $\Delta AGFI$ would indicate the discrimination between methods (scales). On the basis of the strength of both statistical ($\Delta\chi^2_{(3)} = 41.61$) and nonstatistical ($\Delta CFI = 0.04$; $\Delta AGFI = 0.02$) criteria, we may conclude that the evidence of discriminant validity for the considered scales was slightly stronger than it was for the factor traits. However, this evidence was not sufficient. It was rather the initial proof that some scales were somewhat correlated. More information can be found in Table 4 presenting correlations calculated for the factor traits and methods. Their examination reveals that 5- and 7-point scales had something in common (correlation 0.56). The same situation is present, but to a lesser degree, in 9- and 11-point scales. One possible explanation of this finding is that, the number of responses in both scales, i.e. 5 and 7, had the similar level of measurement. The same interpretation might apply to 9- and 11-point scale, however in their case, the correlation was weaker (0.48). All the

---

[4]They are computed on factor loadings either for factor traits or methods.

Table 4 The CFA-CT/CM model—correlations

Traits and methods	PQP	UAP	BATCS	LOSR	5PS	7PS	9PS	11PS
PQP	1.0							
UAP	0.69	1.0						
BATCS	0.18	0.23	1.0					
LOSR	0.37	0.58	0.49	1.0				
5PS					1.0			
7PS					0.56	1.0		
9PS					0.24	0.31	1.0	
11PS					0.19	0.21	0.48	1.0

other correlations between scales, namely, 5 and 9 (0.24); 9 and 7 (0.31); 11 and 5 (0.19); 11 and 7 (0.21) indicated rather their dissimilarity.

6 Conclusions

Empirical results showed that different scales which use different number of response categories provide inconsistent scores in terms of validity for the particular factor traits. In general, scales with many categories (9 and 11) produced worse results in CFA-MTMM models than their alternative options such as 5- and 7-point scale. It appears therefore, that wider scales (as 9 and 11) bring less information from the measurement. This may be due to the fact that more points are placed in scale. Also the requirement for a personal in-depth interpretation of these points (by the respondents), leads to more method effects and hence to the lower validity. In this regard, a human ability to differentiate from different variants of the answers is indeed very limited.

The study which was conducted had some limitations too. For example, it did not test the impact of having only the end points labeled versus having all the points labeled. However, the author decided to keep this problem open to other researchers who would like to pursue the similar subject of research. In this work, one was only interested in the effects of respective scales which affect the CFA-MTMM models quality, explained through the agency of goodness-of-fit indices as well as the obtained parameter estimates and the size of the standardized factor loadings.

References

Alwin, D. F. (1992). Information transmission in the survey interview: Number of response categories and the reliability of attitude measurement In P. V. Marsden (Ed.), *Sociological methodology* (pp. 83–118). Washington: American Sociological Association.
Alwin, D. F. (2007). *Margins of errors: A study of reliability in survey measurement.* Wiley: Hoboken.

Browne, M. W. (1984). The decomposition of multitrait-multimethod matrices. *British Journal of Mathematical and Statistical Psychology, 37*, 1–21.

Byrne, M. B. (2010). *Structural equation modeling with AMOS*. New York: Taylor and Francis.

Campbell, D.T., & Fiske, D. W. (1959). Convergent and discriminant validation by the multitrait-multimethod matrix. *Psychological Bulletin, 56*, 81–105.

Dawes, J. (2008). Do data characteristics change according to the number of points used? an experiment using 5-point, 7-point and 10-point scales. *International Journal of Market Research, 50*, 61–77.

Jöreskog, K. G. (1970). A general method for the analysis of covariance structures. *Biometrika, 57*, 239–251.

Kenny, D. A., & Kashy, D. A. (1992). Analysis of the multitrait-multimethod matrix by confirmatory factor analysis. *Psychological Bulletin, 112*, 165–172.

Likert, R. (1932). A technique for the measurement of attitudes. *Archives of Psychology, 140*, 1–55.

Marsh, H. W. (1989). Confirmatory factor analyses of multitrait-multimethod data - many problems and a few solutions. *Applied Psychological Measurement, 13*, 335–361.

Marsh, H. W., & Grayson, D. (1995). Latent variable models of multitrait-multimethod data. In R. H. Hoyle (Ed.), *Structural equation modeling: Concepts, issues and applications* (pp. 177–198). Thousand Oaks, CA: Sage.

Revilla, A. M., Saris, W. E., & Krosnick, J. A. (2014). Choosing the number of categories in agree-disagree scales. *Sociological Methods and Research, 43*, 73–97.

Saris, W. E., & Andrews, F. M. (1991). Evaluation of measurement instruments using a structural modeling approach In P. P. Biemer, R. M. Groves, L. Lyberg, N. Mathiowetz, & S. Sudman (Eds.), *Measurement errors in surveys* (pp. 575–597). New York: Wiley.

Saris, W. E., Satorra, A., & Coenders, G. (2004). A new approach to evaluating the quality of measurement instruments: The splitballot MTMM design. *Sociological Methodology, 34*, 311–347.

Schmitt, N., & Stults, D. M. (1986). Methodology review: Analysis of multitrait-multimethod matrices. *Applied Psychological Measurement, 10*, 1–22.

Werts, C. E., & Linn, R. L. (1970). Path analysis: Psychological examples. *Psychological Bulletin 74*, 194–212.

Widaman, K. F. (1985). Hierarchically tested covariance structure models for multitrait-multimethod data. *Applied Psychological Measurement, 9*, 1–26.

Biasing Effects of Non-Representative Samples of Quasi-Orders in the Assessment of Recovery Quality of IITA-Type Item Hierarchy Mining

Ali Ünlü and Martin Schrepp

Abstract Inductive Item Tree Analysis (IITA) comprises three data analytic algorithms for deriving reflexive and transitive precedence relations (surmise relations or quasi-orders) among binary items. With the help of simulation studies, the IITA algorithms were already compared concerning their ability to detect the correct precedence relations in observed data. These studies generate a set of surmise relations on an item set, simulate a data set from each of the surmise relations by applying some random response errors, and then try to recover the initial surmise relations from those noisy data. We show that, in the currently published studies however, the representativeness of sampled quasi-orders was not considered or implemented unsatisfactorily. This led to non-representative samples of quasi-orders, and hence to biased or wrong conclusions about the quality of the IITA algorithms to reconstruct the underlying surmise relations. In our paper, results of a new, truly representative simulation study are reported, which correct for the problems. On the basis of this study, the three IITA algorithms can now be compared reliably.

1 Introduction

Inductive Item Tree Analysis (IITA) is a data mining technique that tries to extract logical implications between items from binary data (Schrepp 1999, 2003; Sargin and Ünlü 2009).

An important application context of IITA algorithms is *knowledge space theory* (Doignon and Falmagne 1985, 1999). The basic idea of this approach is that in

A. Ünlü (✉)
Chair for Methods in Empirical Educational Research, TUM School of Education and Centre for International Student Assessment (ZIB), TU München, Arcisstr. 21, 80333 Munich, Germany
e-mail: ali.uenlue@tum.de

M. Schrepp
SAP AG, Walldorf, Germany
e-mail: martin.schrepp@sap.com

© Springer International Publishing Switzerland 2016   563
A.F.X. Wilhelm, H.A. Kestler (eds.), *Analysis of Large and Complex Data*, Studies in Classification, Data Analysis, and Knowledge Organization,
DOI 10.1007/978-3-319-25226-1_48

many knowledge domains, for example mathematics, knowledge is organized in a hierarchical structure. Some pieces of knowledge can only be learned successfully, if other more basic pieces of knowledge are already mastered. In that sense, pieces of knowledge, and hence the items or problems used to measure them, are organized in a reflexive and transitive relation, a *quasi-order*, which is also called a *surmise relation* in knowledge space theory.

An example with six elementary algebra problems helps to motivate the main idea.

a. A car travels on the freeway at an average speed of 52 miles per hour. How many miles does it travel in 5 h 30 min?
b. Using the pencil, mark the point at the coordinates $(1, 3)$.
c. Perform the multiplication $4x^4y^4 \cdot 2x \cdot 5y^2$ and simplify as much as possible.
d. Find the greatest common factor $14t^6y$ and $4tu^5y^8$. Simplify as much as possible.
e. Graph the line with slope -7 passing through $(-3, -2)$.
f. Write an equation for the line that passes through the point $(-5, 3)$ and is perpendicular to the line $8x + 5y = 11$.

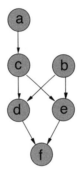

A plausible surmise relation for these items. For instance, mastery of problem *e* implies mastery of problem *b*.

A surmise relation can be used to design efficient adaptive, computer-based knowledge assessment and training procedures. For example, the *Assessment and Learning in Knowledge Spaces* (ALEKS) system is a fully automated math tutor on the Internet (e.g., see Falmagne et al. 2013).

Another typical application field of IITA are questionnaires, which contain statements that people can agree or disagree with (e.g., see Schrepp 2003). In this case, $j \rightarrow i$ (or equivalently, $i \leq j$) is interpreted as "*Each person who agrees to statement j also agrees to statement i.*"

2   Inductive Item Tree Analysis

The general scheme of IITA consists of three steps. First, a set of candidate surmise relations is generated. Second, for each of these candidate surmise relations, the fit to the data set is calculated. Third, the surmise relation yielding the best fit is selected as the solution. Currently, there are three versions of IITA available, which differ in the second step of this scheme, that is, in the measure used to calculate the fit of a surmise relation to the data.

2.1 Construction of Candidate Surmise Relations

Let I be a set of dichotomous items. For a given binary data matrix D, a set of candidate quasi-orders $QO(D)$ is generated by an *inductive construction* (for details, see Schrepp 1999).

1. We start with the surmise relation $\leq_0$ defined by: $i \leq_0 j :\Leftrightarrow b_{ij} = 0$ for all $i, j \in I$, where b_{ij} is the number of response patterns R in the data matrix D with $i \notin R$ and $j \in R$. That is, b_{ij} is the number of observed counterexamples (e.g., students solving question j but failing to solve i) to an implication $j \rightarrow i$ or relational dependency $i \leq_0 j$.
2. Assume the surmise relation $\leq_L$ for an $L \in \{0, \ldots, n\}$, with n the sample size, is already constructed. In step $L + 1$ of the process, all item pairs $(i, j) \notin \leq_L$, which fulfill the condition $b_{ij} \leq L+1$ and mutually do not cause an intransitivity together with all the dependencies contained in $\leq_L$, are added to $\leq_L$ to construct the surmise relation $\leq_{L+1}$.

2.2 Fit Measure and Selection of Best Fitting Surmise Relation

The three variants of IITA differ in this step of the general scheme.

The probability γ_L for a relational dependency $i \leq_L j$ to be violated due to random errors is estimated by

$$\gamma_L = \frac{\sum\{b_{ij}/(p_j n) \mid i \leq_L j \wedge i \neq j\}}{(|\leq_L| - m)},$$

where m is the number of items (and n the sample size). The probability γ_L is used to estimate for each pair of items the expected number of violations, t_{ij}, under the assumption that $\leq_L$ is the true surmise relation underlying the data.

The *original IITA* algorithm distinguishes two cases (see Schrepp 1999).

1. $i \not\leq_L j$: In this case, we assume that the items i and j are independent. Thus, t_{ij} equals the expected number of response patterns R with $i \notin R$ and $j \in R$, in the sense that $t_{ij} = (1-p_i)p_j n(1-\gamma_L)$. Note that under pure stochastic independence we would have $t_{ij} = (1 - p_i)p_j n$. The original IITA algorithm applies the correction factor $(1 - \gamma_L)$ for the influence of random errors even in this case, which is disputable and changed in the improved versions of IITA (see below).
2. $i \leq_L j$ and $i \neq j$: In this case, violations of $i \leq_L j$ must result from random errors. Thus, $t_{ij} = \gamma_L p_j n$.

The fit of any $\leq_L$ in the selection set $QO(D) = \{\leq_L: L = 0, 1, \ldots, n\}$ given the data matrix D is measured by the *diff coefficient*,

$$\text{diff}(\leq_L, D) := \frac{\sum_{i \neq j}(b_{ij} - t_{ij})^2}{m(m-1)}.$$

The idea is to assess the discrepancy between the observed and expected numbers of counterexamples. The smaller the diff value is, the better is the fit of the quasi-order to the data.

As already mentioned, the corrected and minimized corrected versions of IITA, which we describe next, differ from the above original IITA algorithm regarding the definition of the diff coefficient.

The first improvement corrects a problem in the estimation of the t_{ij}'s. Original IITA distinguishes only between the two cases $i \leq_L j$ and $i \nleq_L j$. This results in methodologically inconsistent estimation of t_{ij} for some item pairs, and especially when longer chains of items are present. In addition, the independence case is estimated properly. For details, see Sargin and Ünlü (2009).

A correct choice for t_{ij} for $i \nleq_L j$ depends on whether $j \nleq_L i$ or $j \leq_L i$. In other words, a third distinction for the estimation of the t_{ij}'s is introduced. The modification using these improved estimators is called the *corrected IITA* (see Sargin and Ünlü 2009).

1. If $i \nleq_L j$ and $j \nleq_L i$, then $t_{ij} = (1 - p_i)p_jn$. Independence is assumed, and the additional factor $(1 - \gamma_L)$ is omitted.
2. If $i \nleq_L j$ and $j \leq_L i$, then $t_{ij} = (p_j - (p_i - p_i\gamma_L))n$. This estimator is derived as follows. The observed number of people who solve item i is p_in. Hence, the estimated number of people who solve item i and item j is $p_in - t_{ji} = (p_i - p_i\gamma_L)n$. Note that $j \leq_L i$, and the estimator is $t_{ji} = p_i\gamma_Ln$. Thus, $t_{ij} = p_jn - (p_i - p_i\gamma_L)n = (p_j - (p_i - p_i\gamma_L))n$. This estimator is not only mathematically motivated, but also interpretable.

A further improvement is the extension of the corrected IITA algorithm to optimize the fit criterion specified by the diff coefficient, as a means to better reconstruct the correct implications from the data. This approach yields what is called the *minimized corrected IITA*.

More precisely, let the diff coefficient be based on the corrected estimators. The diff coefficient can be viewed as a function of the error probability γ_L, and we minimize that function with respect to γ_L. The fit measure then favors quasi-orders that lead to smallest minimum discrepancies, or equivalently, largest maximum matches, between the observed and expected summaries b_{ij} and t_{ij}. This optimum error rate can be expressed in closed analytical form, for details see Ünlü and Sargin (2010), and can now be used for an alternative IITA procedure, the minimized corrected algorithm, in which a minimized diff value is computed for every quasi-order (based on the estimators of the corrected algorithm and the inductive procedure of the original algorithm).

The quasi-order $\leq_{\mathrm{IITA}}$ in QO(D), which yields the smallest diff value, is selected as the solution of the IITA analysis. In other words,

$$\leq_{\mathrm{IITA}} = \arg\min_{\mathrm{QO}(D)} \mathrm{diff}(\leq_L, D).$$

2.3 Software

There are two freely available software packages that implement the discussed algorithms.

The program *ITA 2.0* developed by Schrepp (2006) implements both the predecessor classical ITA (van Leeuwe 1974) and the original IITA. It can be downloaded under www.jstatsoft.org/v16/i10.

The *R* package *DAKS* developed by Sargin and Ünlü (2010) and Ünlü and Sargin (2010) implements the original, corrected, and minimized corrected IITA algorithms. It is available from the *Comprehensive R Archive Network* CRAN.R-project.org/package=DAKS or from www.jstatsoft.org/v37/i02.

3 Simulation Design for IITA Comparison Studies

The IITA procedures were already investigated in a number of simulation studies (see the afore mentioned references). The goal of those studies was to determine whether the algorithms are able to detect the true surmise relations underlying noisy data. The simulation studies all had the following structure.

1. A true surmise relation $\leq$ is created on an item set I by a random process.
2. The *knowledge structure*

   $$\mathbf{K} = \{K \subseteq I \mid \forall i, j \in I : ((i \leq j \wedge j \in K) \implies i \in K)\}$$

   corresponding to $\leq$ is constructed. A set of n response patterns is simulated by drawing n elements from $\mathbf{K}$ randomly and, for each of these, by simulating *careless errors* (an item is in the knowledge state K, but not in the response pattern) and *lucky guesses* (an item is not in the knowledge state K, but in the response pattern) with specified probabilities α and β, respectively.
3. The simulated data set is analyzed by IITA and the best fitting surmise relation $\leq_{\mathrm{IITA}}$ is computed.
4. The true surmise relation $\leq$ and the data analytic solution $\leq_{\mathrm{IITA}}$ are compared by counting the item pairs in which they differ.

The first step is the essential and difficult part. The goal of designing sound simulation studies essentially reduces to the problem of realizing samples of quasi-orders that are *representative* for the population of all quasi-orders. The direct method to draw a representative sample of quasi-orders on a set of n items is by computing all possible quasi-orders, storing them, and then sampling from these. But this does only work for small n, since the number of quasi-orders increases very rapidly with n (e.g., see Pfeiffer 2004). On a set of six items, for instance, we already have 209,527 (labeled) surmise relations.

Previous studies tried to avoid this problem by implementing special procedures to draw random quasi-orders. The simplest method is to create a random binary relation on the n items and then to compute the transitive closure. More advanced strategies tried to compensate for the fact that this simple strategy produces samples of quasi-orders, which are far from being representative concerning quasi-order size (number of implications).

For example, in Sargin and Ünlü (2009) the following procedure was used.

1. The process starts with a relation $\mathscr{R}$ that contains exactly the reflexive item pairs (i, i) for $i = 1, \ldots, m$.
2. All other pairs are added with a probability δ to $\mathscr{R}$. The probability δ itself is drawn randomly from a normal distribution with $\mu = 0.16$ and $\sigma = 0.06$. Values less than 0 or greater than 0.30 are set to 0 or 0.30, respectively.
3. The transitive closure of $\mathscr{R}$ is the resulting random quasi-order.

This random process is already an improvement of an older procedure that draws δ based on a uniform distribution on the interval 0 to 0.40, or 0 to 1, which resulted in non-representative samples consisting of overly large surmise relations. Yet, this improved drawing process still produced non-representative samples, as can be seen in Fig. 1.

4   Representative Simulation Study

Theoretically, the corrected and minimized corrected IITA algorithms should perform better than the original IITA algorithm in general. However, in a simulation study by Sargin and Ünlü (2009), the original IITA algorithm still worked better as long as the error probabilities for careless errors and lucky guesses were small. This was due to the non-representative, or biased, random procedure used to sample the underlying quasi-orders.

A (near to) representative sampling process will produce the least biased results when generalizing the findings obtained from IITA related simulation studies to the population of all quasi-orders. Further considerations make clear as well that, if the performances of the IITA algorithms are to be investigated and compared, we ought to assure that they are tested with representative random quasi-orders.

Currently there exists no method to produce truly representative samples of surmise relations on larger item sets. Therefore, we have decided to run our comparison study for six items, the purpose of this paper is to exemplify the biases or wrong conclusions that may be induced by non-representative samples of quasi-

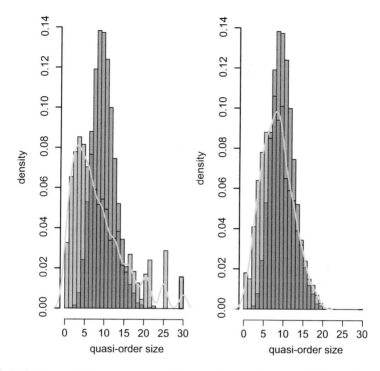

Fig. 1 In *light gray*, histogram densities of quasi-order size for 10,000 (*left panel*) and 1000 (*right panel*) δ sampled according to the absolute normal method (*left panel*) and averaged normal method (*right panel*). In addition, kernel density estimate curves of the samples are plotted, to assist visualization. In *gray*, the true distribution is shown, with the overlapping areas of "true" and "sampled" printed in *dark gray*. Averaging is over 100 quasi-orders generated for each of the 1000 δ values

orders in comparative IITA analyses, since in this case it is possible to construct the population of all surmise relations and to draw from it a true simple random sample.

4.1 Settings

In the simulation study the following settings are made.

1. Throughout the simulation study six items are used.
2. The original, corrected, and minimized corrected IITA procedures are compared based on simple random samples of the set of all 209,527 surmise relations on six items.
3. A single response error probability τ is specified and takes the values 0.03, 0.05, 0.08, 0.10, 0.15, and 0.20.
4. Sample size n is varied as 50, 100, 200, 400, 800, 1600, and 6400.
5. For each of the 42 combinations of these settings, the four steps of the general simulation design described in Sect. 3 are repeated 1000 times.

6. The three algorithms are compared and evaluated relative to each other using the symmetric differences between the derived and true quasi-orders. The criterion *dist* is used, the number of item pairs in which "derived" and "true" differ.

The important change we make in this simulation study is to base the investigation on true random samples; compared to Sargin and Ünlü (2009), for instance, where the biased absolute variant of the normal sampling was used. We will see that the difference in representativeness of corresponding quasi-order samples had a negative impact on the comparison of the IITA algorithms and biased the findings reported in the published literature.

4.2 Results

Figures 2 and 3 summarize the main differences when basing comparison on true random rather than biased absolute normal samples. The panels "rand" stand for

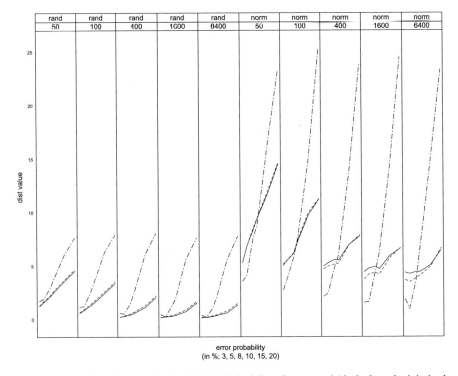

Fig. 2 Average *dist* values under the original (*dashed-dotted*), corrected (*dashed*), and minimized corrected (*solid*) IITA algorithms, as a function of error probability conditioned on sample size. Considered for true simple random and absolute normal sampling

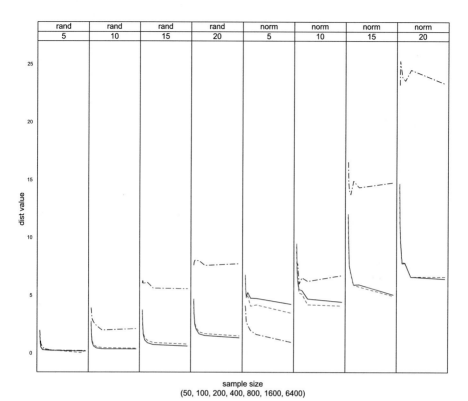

Fig. 3 Average *dist* values under the original (*dashed-dotted*), corrected (*dashed*), and minimized corrected (*solid*) IITA algorithms, as a function of sample size conditioned on error probability. Considered for true simple random and absolute normal sampling

six items and simple random sampling. The panels "norm" are for nine items and absolute normal sampling (Sargin and Ünlü 2009).

The main finding is as follows. See the panels "rand" of Figs. 2 and 3, where the three curves can be increasingly ordered from solid, dashed, to dashed-dotted, invariantly along the entire τ- and n-axis. In other words, overall, consistently the same ranking of the IITA methods can be observed. The minimized corrected version performs slightly better than the corrected, and the corrected version significantly improves on the original.

In the published simulation study, based on biased samples of quasi-orders, the original algorithm performed better than the other two algorithms for small error probabilities. This can be seen from the "norm" panel for 5 % of Fig. 3, where the dashed-dotted curve for the original algorithm lies entirely below the dashed and solid curves for the corrected and minimized corrected algorithms, respectively. Or, see the "norm" panels of Fig. 2, where the curves intersect. Moreover, albeit differences between the corrected and minimized corrected methods are negligible, ordering reversals or greater discrepancies between these two, such as in the "norm" panels for 5 % and 10 % of Fig. 3, are not observed for true random samples. See also the "norm" panels for 1600 and 6400 of Fig. 2, where the corrected algorithm yields lower *dist* values as compared to its minimized variant.

From a theoretical point of view, these ambiguous results are surprising, since the corrected or optimized estimation procedures ought to improve. We see that those biasing effects resulted from non-representative samples of quasi-orders. They are resolved, if a representative mechanism is applied.

5 Conclusion

Representativeness of random quasi-orders drawn and postulated to underlie item hierarchy mining related simulation has been seen to be an important requirement for reliable and sound comparison of IITA-type data analyses. Biasing effects of non-representative samples in previous study made authors conclude that it depends on the size of the error probability which of the three IITA algorithms performs best. However, we have been able to show that those conclusions were biased or wrong. In particular, in all conditions for the error probability, the corrected and minimized corrected algorithms outperform the original algorithm, if a representative random process for generating surmise relations is applied. Expected theoretically, however in previous study not observed unambiguously, is the following ranking of the three IITA methods, which we have now clearly obtained. The minimized corrected algorithm is slightly better than the corrected, and the corrected algorithm considerably improves on the original IITA algorithm.

References

Doignon, J.-P., & Falmagne, J.-Cl. (1985). Spaces for the assessment of knowledge. *International Journal of Man-Machine Studies, 23*, 175–196.

Doignon, J.-P., & Falmagne, J.-Cl. (1999). *Knowledge spaces*. Berlin: Springer.

Falmagne, J.-Cl., Albert, D., Doble, C., Eppstein, D., & Hu, X. (2013). *Knowledge spaces: Applications in education*. Heidelberg: Springer.

Pfeiffer, G. (2004). Counting transitive relations. *Journal of Integer Sequences, 7*, 1–11. Article 04.3.2.

Sargin, A., & Ünlü, A. (2009). Inductive item tree analysis: Corrections, improvements, and comparisons. *Mathematical Social Sciences, 58*, 376–392.

Sargin, A., & Ünlü, A. (2010). The R package DAKS: Basic functions and complex algorithms in knowledge space theory. In H. Locarek-Junge & C. Weihs (Eds.), *Studies in classification, data analysis, and knowledge organization* (pp. 263–270). Berlin: Springer.

Schrepp, M. (1999). On the empirical construction of implications between bi-valued test items. *Mathematical Social Sciences, 38*, 361–375.

Schrepp, M. (2003). A method for the analysis of hierarchical dependencies between items of a questionnaire. *Methods of Psychological Research Online, 19*, 43–79.

Schrepp, M. (2006). ITA 2.0: A program for classical and inductive item tree analysis. *Journal of Statistical Software, 16*(10), 1–14.

Ünlü, A., & Sargin, A. (2010). DAKS: An R package for data analysis methods in knowledge space theory. *Journal of Statistical Software, 37*(2), 1–31.

van Leeuwe, J. F. J. (1974). Item tree analysis. *Nederlands Tijdschrift voor de Psychologie, 29*, 475–484.

Correlated Component Regression: Profiling Student Performances by Means of Background Characteristics

Bernhard Gschrey and Ali Ünlü

Abstract Multicollinearity is one of the main problems when using regression analytic approaches to predict outcome variables. The application of traditional regression analytic approaches often provides unstable and unreliable estimates of the parameters when multicollinearity occurs. In this paper we apply a regression analytic method called correlated component regression (CCR), developed by Magidson (Correlated component regression: re-thinking regression in the presence of near collinearity. In Abdi et al. (eds) New perspectives in partial least squares and related methods, Springer, Heidelberg, pp 65–78, 2013), for characterizing student performances in PIRLS/TIMSS 2011 (Martin and Mullis, Methods and procedures in TIMSS and PIRLS 2011, TIMSS & PIRLS International Study Center, Chestnut Hill, 2013) through selected background characteristics, such as cultural and socio-economic characteristics. On the basis of various criteria, we compare the findings of CCR with the results of OLS regression regarding the prediction of student performance values. An implemented cross-validation procedure and step-down algorithm are utilized to perform a special type of variable reduction. Thus, the results of our study will provide more reliable sets of background variables for characterizing large scale educational data in the domains of reading, mathematics, and science.

1 Introduction

To analyze student performances obtained from large scale assessment studies such as *Trends in International Mathematics and Science Study* (TIMSS), *Progress in International Reading Literacy Study* (PIRLS), or *Programme for International Student Assessment* (PISA), a plenty of background characteristics have been collected. These comprise, inter alia, cultural and social background characteristics, as well as subject-specific attitudes and self-concepts (Mullis et al. 2009a,b and OECD 2012).

B. Gschrey • A. Ünlü (✉)
Chair for Methods in Empirical Educational Research, TUM School of Education and Centre for International Student Assessment (ZIB), TU München, Arcisstr. 21, 80333 Munich, Germany
e-mail: bernhard.gschrey@tum.de; ali.uenlue@tum.de

© Springer International Publishing Switzerland 2016
A.F.X. Wilhelm, H.A. Kestler (eds.), *Analysis of Large and Complex Data*, Studies in Classification, Data Analysis, and Knowledge Organization,
DOI 10.1007/978-3-319-25226-1_49

This extensive data pool offers a variety of possibilities for analyzing student performances. For instance, Bos et al. (2012a,b) could show that in groups with higher performances, the portion of students with high subject-specific attitudes and self-concepts is also significant, higher. However, characterizations of student performances by means of cultural and socio-economic background characteristics are largely limited to descriptive analyses in the published reports.

We apply a recent regression analytic method—called *correlated component regression* (CCR) and developed by Magidson (2013)—for characterizing student performances by means of selected background characteristics in regards to the domains of reading, mathematics, and science. A specific issue, which has to be considered when using traditional regression analytic approaches, is occurrence of multicollinearity, respectively, suppression effects (Lynn 2003). Pandey and Elliott (2010) distinguish between four types of suppressor variables: the *classic suppressor*, the *negative suppressor*, the *reciprocal suppressor*, and the *absolute and relative suppressor*. Basically, multicollinearity/suppression effects arise when two or more independent variables are moderately or highly correlated to each other and have no direct effect on the outcome variable. By suppressing irrelevant variance in each other, those variables cause the suppressed variables to obtain a substantial regression weight. Hence, treatment of multicollinearity/suppression effects becomes essential, since they can lead to unstable and non-significant estimates of the regression coefficients, or to increasing variances and standard errors of the estimated coefficients (Pandey and Elliott 2010 and Rawlings et al. 1998).

Especially, when applying multiple regression in the context of large scale assessments such as PIRLS/TIMSS or PISA, where the correlations among the predictors can be quite high, seemingly good predictive performance of a model may be associated with overfitting. Magidson (2013) showed that—based on a multicollinear structure of the data—the CCR approach provides more stable and reliable estimates of the regression coefficients. In this paper, we analyze the structure of our data and compare the outcomes of CCR with the results of traditional OLS regression by means of various criteria. For applications of CCR, the software *CORExpress*® is used (Magidson 2011).

The basic idea of CCR is to reduce confounding effects due to high predictor correlations, thus obtaining more reliable parameter estimates. CCR produces $C < P$ (P, number of predictors) correlated components to predict an outcome variable Y, whereat each component S_c, $c = 1, \ldots, C$, is an exact linear combination of the predictors $X_1, X_2, \ldots, X_P$, and the weights are chosen to maximize the components' ability to predict Y. Those components can be viewed as composite predictors used in an iteratively built "regularized" model.

The first component S_1 is a weighted average of all P one-predictor models of target variable Y on X_p, and a one-component CCR model is the regression of Y on S_1. That is, $Y = \alpha_p^{(1)} + \lambda_p^{(1)} X_p + \epsilon_p^{(1)}$, $S_1 = \sum \lambda_p^{(1)} X_p$, and $Y = a^{(1)} + b_1^{(1)} S_1 + \epsilon^{(1)}$. The cth component S_c is a weighted average of all one-predictor models "in which the previous components $S_1, \ldots, S_{c-1}$ are controlled for," that is, $S_c = \sum \lambda_p^{(c)} X_p$ where $Y = \alpha_p^{(c)} + \gamma_{1,p}^{(c)} S_1 + \cdots + \gamma_{c-1,p}^{(c)} S_{c-1} + \lambda_p^{(c)} X_p + \epsilon_p^{(c)}$, and a c-component CCR model is the regression of Y on $S_1, \ldots, S_{c-1}$ and S_c, that is, $Y = a^{(c)} + b_1^{(c)} S_1 + \cdots + b_c^{(c)} S_c + \epsilon^{(c)}$.

In CCR, cross-validation and step-down procedures are used to choose the optimal number of correlated components and predictors in the model. The selection process is analogous to backward variable selection and driven entirely by cross-validation sample performance. For particular values of C the process starts with all variables included and then eliminates variables with the smallest standardized coefficients one at a time, re-estimating the model at each step, with cross-validation performance criterion values computed for the selections. In particular, variables having direct effects on the outcome variable Y can be determined as those having high loadings on the first component. Those variables are called *prime predictors*. For details, see Magidson (2010) and Magidson and Bennett (2011). We want to concentrate on application.

Student performances have rarely been characterized by regression analytic approaches due to the multicollinear structure of the data. Furthermore, the application of traditional regression analytic approaches should be treated with caution when multicollinearity or suppressor variables occur, because in such cases estimates of the parameters may become inaccurate. In this regard, the present paper investigates whether CCR will outperform OLS regression when characterizing student performances by means of a set of selected (generally correlated) background characteristics.

2 Study Design

For characterizing student performances in the domains of reading, mathematics, and science, we analyzed data from the national (German) PIRLS/TIMSS 2011 study (Mullis et al. 2009a,b). We used the national data set of 3928 students with performance values in all three domains (Bos et al. 2012a,b). Pertinent background variables for characterizing student performances have been selected from the national background model, with regard to gender (one item), socio-economic factors (six items), and cultural factors (four items). We also selected German grade, mathematics grade, and science grade as additional relevant variables for characterizing performances.

2.1 Detecting Multicollinearity

Since a multicollinear structure of the data may deteriorate accuracy and predictive power of a multiple linear regression model, we analyzed the presence of multicollinearity with respect to the mentioned variables. Although regression coefficients can be estimated when multicollinearity occurs, the estimates tend to become inaccurate and misleadingly non-significant (variances and standard errors of the estimated parameters become large). As can be seen in Fig. 1, the sum of the overlaps $1' + 2'$, representing the parts of the variables that overlap the criterion,

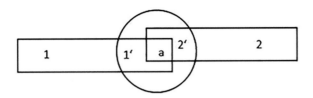

Fig. 1 Visualization of an overlapping area of common variances through a non-zero correlation among two predictors, 1 and 2. The sum of the overlaps with the criterion (*circle area*), $1' + 2'$, would overestimate the actual covered amount, $1' + (2' - a)$

would overestimate the total overlap due to a non-zero correlation between the predictors. The actual amount of the criterion covered by the two predictors is $1' + (2' - a)$.

Concerning this matter, we initially investigated pairwise correlations as well as partial correlations (representing the correlation between a predictor and the criterion after removing common variance with other predictors from both the criterion and the observed predictor) and semipartial correlations (representing the unique contribution of a predictor to the unaltered criterion after removing the contributions of the remaining predictors from the predictor of interest) among the variables and computed the overlapping area of common variances (Cohen et al. 2003 and Sheskin 2011).

However, the inspection of pairwise correlations is not sufficient, since multicollinearity can exist with pairwise correlations close to zero. Therefore, we analyzed the occurrence of multicollinearity by appropriate diagnostic fits such as *tolerance* and *variance inflation factor* (VIF), although Berk (1977) and O'Brien (2007) pointed out that in some cases these fits provide less accurate benchmarks for detecting multicollinearity. An additional and more extensive examination of multicollinearity is performed on the basis of *eigenvalues, condition indexes*, and *decomposition of variance* (Albers 2009 and Belsley et al. 2004).

2.2 Application of Correlated Component Regression

To consider the effects of multicollinearity, or respectively suppression, and to provide stable parameter estimates, we performed linear CCR analyses, considering student performances in the domains of reading, mathematics, and science as a function of the aforementioned (14) predictors, and compared the outcomes to the results of OLS regression. Additionally, we tested the assumptions of linear regression by appropriate diagnostic plots and searched for uncommon cases, respectively, outliers or influential points (Cook and Weisberg 1982 and Rawlings et al. 1998). Finally, a special type of variable reduction of the most relevant variables is conducted in order to maximize the predictive power of the particular models and to remove irrelevant predictors.

For capturing suppressor/confounding effects among the independent variables, we reduced the number of correlated components stepwise by k-fold cross-validation and computed the corresponding goodness-of-fit statistics R^2 and $CV\text{-}R^2$ (cross-validated R^2) for the particular models (for details, see Magidson and Bennett 2011 and Refaeilzadeh et al. 2008). Referring to the application of k-fold cross-validation, we ran ten rounds by selecting $k = 6$ equal and random created folds ($n = 287$ cases in each fold), since 6 is the smallest integer between 5 and 10 that divides evenly 1722 (1722 cases without missing values in each domain are available). Thus, a matrix of fit statistics (R^2 and $CV\text{-}R^2$) for models ranging from $C = 14$ to $C = 1$ correlated components for the particular domains is obtained. Regarding the optimal number of correlated components C, the model that maximizes $CV\text{-}R^2$ has to be chosen.

A step-down algorithm that works in conjunction with k-fold cross-validation is used to achieve a variable reduction of the most relevant variables ($P^* < P$) and to enhance predictive power of the particular models (for technical details, see Magidson and Bennett 2011). We initially estimated models including all predictors and evaluated the models using k-fold cross-validation. Removing stepwise a variable with the lowest standardized coefficient and repeating the evaluation of the new models yield a matrix of fit statistics (R^2 and $CV\text{-}R^2$) for models with one to fourteen predictors.

To obtain the final CCR models with respect to the estimated models, we selected the models with the highest $CV\text{-}R^2$ values. Those models contain the optimal number of correlated components as well as the most relevant predictors. The final regularized CCR models capture suppressor/confounding effects and provide more reliable parameter estimates.

3 Results

3.1 Multicollinearity Analysis

For the 14 predictors, moderate and partly high bivariate correlations ($r > 0.600$) among the variables were obtained. Moreover, when comparing the partial correlations and beta-coefficients to the corresponding pairwise correlations, the findings indicated a large overlapping area of common variance.

Table 1 provides the total explained variance (R^2) and the computed overlapping area of common variance for the three domains. As can be seen, the common variances are relatively high compared to the total explained variances. This indicates that the predictors capture some of the same variability in the criterion and do not enhance predictive power of the particular regression models. In this case, the fit statistic R^2 seems to be little meaningful.

Since the calculation of the diagnostic fits tolerance and VIF could not explicitly identify critical variables, we additionally analyzed the occurrence of

Table 1 Total explained variance (R^2) and overlapping area of common variance

Reading		Mathematics		Science	
R^2	Common var	R^2	Common var	R^2	Common var
0.489	0.390	0.525	0.384	0.445	0.340

R^2 coefficient of determination, *Common var* overlapping area of common variance

multicollinearity by examining eigenvalues, condition indexes, as well as the corresponding proportion of variance (Table 2). The findings corroborate the presence of multicollinearity. The computed eigenvalues point out that at least eight dimensions are characterized by comparatively low values, respectively, values near zero (≤ 0.05). These low values indicate collinearity among some of the predictor variables. Hence, small changes in the data values can lead to large changes in the estimates of the regression coefficients (Albers 2009 and Belsley et al. 2004). Furthermore, regarding the calculation of the condition indexes and the respective proportion of variance, four item pairs having a high condition index (>15) as well as high loadings on the same dimensions could be determined.

3.2 Correlated Component Regression Analysis

Inspecting the standard diagnostic plots (Cook and Weisberg 1982 and Rawlings et al. 1998) for the particular multiple regression models, we can affirm the assumptions of linear regression. Thus, results of the particular models can be expected to provide fairly robust estimates of the parameters disregarding the effects of possible suppression and multicollinearity.

For determining the optimal number of correlated components (C) for the particular domains, Table 3 provides the required fit statistics R^2 and $CV\text{-}R^2$, listed by increasing number of correlated components. The model that maximizes $CV\text{-}R^2$ has to be chosen.

The results show that the three domains can be described by three correlated components. Although OLS regressions yield a higher R^2 for the saturated models ($C = P$) compared to the respective CCR models (0.4890 vs. 0.4887, 0.5253 vs. 0.5244, and 0.4445 vs. 0.4439), CCR models with $C < P$ correlated components yield a higher $CV\text{-}R^2$ (0.4781 vs. 0.4793, 0.5162 vs. 0.5176, and 0.4327 vs. 0.4334), suggesting that these CCR models will outperform OLS regressions when applied to new data.

Fit statistics of the final regularized CCR models ($C < P$, $P^* < P$) can be found at the bottom of Table 3. The most predictive model for the domain of reading is characterized by a three-component model with five predictors. Thus, the predictive power of the model can be additionally improved by variable reduction ($CV\text{-}R^2_{\text{final}} = 0.4803$). Similar results were identified for the domain of science. The predictive power of the model can be slightly enhanced to $CV\text{-}R^2_{\text{final}} = 0.4357$ by removing

Table 2 Decomposition of variance: eigenvalues, condition indexes, and loadings

Model	Eigenvalue[a]	CI[b]	Constant	Gender	Language	Books	Mother birth	Father birth	Person birth	Ed. degree	Social status	Poverty	Voc. tr. mother	Voc. tr. father	German grade	Math grade	Science grade
1	12.07	1.00	0.00	0.00	0.00	0.00	0.00	0.00	0.00	0.00	0.00	0.00	0.00	0.00	0.00	0.00	0.00
2	1.34	2.99	0.00	0.00	0.00	0.00	0.00	0.00	0.00	0.06	0.08	0.13	0.00	0.00	0.00	0.00	0.00
3	0.62	4.40	0.00	0.00	0.00	0.00	0.00	0.00	0.00	0.05	0.06	0.78	0.00	0.00	0.00	0.00	0.00
4	0.27	6.68	0.00	0.00	0.00	0.00	0.00	0.00	0.00	**0.39**	**0.77**	0.00	0.00	0.00	0.00	0.00	0.00
5	0.17	8.37	0.00	0.01	0.07	0.01	0.04	0.04	0.00	0.06	0.01	0.00	0.02	0.01	0.04	0.06	0.04
6	0.15	8.86	0.00	0.08	0.02	0.16	0.02	0.02	0.00	0.07	0.04	0.06	0.00	0.00	0.01	0.02	0.03
7	0.10	11.12	0.00	**0.62**	0.00	0.17	0.00	0.00	0.00	0.01	0.01	0.00	0.00	0.00	0.00	0.00	0.00
8	**0.05**	**15.25**	0.00	0.04	0.11	**0.49**	0.06	0.05	0.00	0.06	0.00	0.00	0.17	0.06	0.01	0.00	0.01
9	**0.05**	**15.42**	0.00	0.01	**0.63**	0.09	0.12	0.08	0.02	0.02	0.00	0.01	0.02	0.02	0.00	0.03	0.14
10	**0.04**	**16.54**	0.00	0.06	0.14	0.00	0.00	0.02	0.00	0.01	0.00	0.00	0.04	0.02	0.03	**0.35**	**0.66**
11	**0.04**	**18.17**	0.00	0.02	0.00	0.00	0.18	0.29	0.01	0.00	0.01	0.00	0.24	0.17	0.16	0.01	0.06
12	**0.03**	**19.64**	0.00	0.12	0.00	0.00	0.00	0.00	0.01	0.00	0.01	0.00	0.10	0.16	**0.63**	**0.38**	0.05
13	**0.03**	**20.08**	0.00	0.01	0.00	0.00	**0.51**	**0.48**	0.00	0.00	0.00	0.00	0.15	0.15	0.10	0.05	0.00
14	**0.02**	**23.48**	0.00	0.00	0.02	0.00	0.06	0.00	**0.71**	0.09	0.00	0.00	0.11	0.17	0.01	0.00	0.01
15	**0.01**	**45.48**	0.99	0.04	0.00	0.07	0.01	0.01	0.25	0.16	0.00	0.01	0.15	0.27	0.00	0.04	0.01

CI condition index

[a]Eigenvalues near zero indicating multicollinearity

[b]Condition indexes > 15 indicating multicollinearity

B. Gschrey and A. Ünlü

Table 3 R^2 and $CV\text{-}R^2$ for respective numbers of correlated components (C) and final CCR models

Reading				Mathematics				Science			
P	C	R^2	$CV\text{-}R^2$	P	C	R^2	$CV\text{-}R^2$	P	C	R^2	$CV\text{-}R^2$
14	1	0.4312	0.4277	14	1	0.4401	0.4371	14	1	0.3911	0.3875
14	2	0.4843	0.4787	14	2	0.5126	0.5075	14	2	0.4358	0.4288
14	**3**	**0.4887**	**0.4793**	**14**	**3**	**0.5244**	**0.5176**	**14**	**3**	**0.4439**	**0.4334**
14	4	0.4889	0.4789	14	4	0.5251	0.5170	14	4	0.4445	0.4328
14	5	0.4890	0.4783	14	5	0.5253	0.5164	14	5	0.4445	0.4329
14	6	0.4890	0.4781	14	6	0.5253	0.5163	14	6	0.4445	0.4327
...	...	...	...	...	...	...	...	...	...	...	...
14	14	0.4890	0.4781	14	14	0.5253	0.5162	14	14	0.4445	0.4327
5	**3**	**0.4844**	**0.4803**	**14**	**3**	**0.5244**	**0.5176**	**10**	**3**	**0.4440**	**0.4357**

P number of predictors, C number of correlated components, R^2 coefficient of determination, $CV\text{-}R^2$ cross-validated coefficient of determination

four predictors. An exception here is the domain of mathematics, which shows no further enhancement of predictive power when the number of predictors is reduced stepwise ($CV\text{-}R^2_{\text{final}} = 0.5176$).

Comparing the results of OLS regression and CCR, Table 4 provides the estimated standardized regression coefficients for the three competence domains. Standardized regression coefficients have been separately estimated for the saturated OLS regression model, the cross-validated CCR model ($C < P$), and the final CCR model ($C < P$, $P^* < P$). As can be seen in Table 4, the estimated regression parameters and the respective constants (intercepts) show noteworthy differences related to the used regression analytic methods. These differences may become crucial when assigning students based on their performances to the competence levels, which are typically used in PIRLS, TIMSS, or PISA to characterize student proficiency and to report assessment results (e.g., see Bos et al. 2012a,b).

4 Discussion

In addition to the student performances, PIRLS and TIMSS 2011 studies provide a plenty of background information. However, current analyses for characterizing student performances by means of background characteristics are largely limited to descriptive analyses. A more extensive characterization by appropriate methods is rarely presented in the published reports. In the present paper we have applied traditional OLS regression and recent CCR for characterizing student performances through selected background variables and compared the results of the two methods on the basis of various criteria.

Since the occurrence of multicollinearity, respectively, suppressor variables, may have a negative effect on the accuracy of parameter estimation, we have initially

Table 4 Standardized regression coefficients for OLS regression and CCR

Variable	Reading			Mathematics			Science		
	Std. coefficients			Std. coefficients			Std. coefficients		
	OLS	CCR	CCR	OLS	CCR	CCR	OLS	CCR	CCR
	$C = 14$	$C = 3$	$C = 3$	$C = 14$	$C = 3$	$C = 3$	$C = 14$	$C = 3$	$C = 3$
	$P = 14$	$P = 14$	$P = 5$	$P = 14$	$P = 14$	$P = 14$	$P = 14$	$P = 14$	$P = 10$
	$\hat{\beta}*$	$\hat{\beta}*$	$\hat{\beta}*$	$\hat{\beta}*$	$\hat{\beta}*$	$\hat{\beta}*$	$\hat{\beta}*$	$\hat{\beta}*$	$\hat{\beta}*$
Gender	−0.024	−0.028	–	0.077	0.078	0.078	0.157	0.150	0.154
Language	−0.032	−0.017	–	−0.045	−0.026	−0.026	−0.038	−0.036	−0.041
Books	0.099	0.096	0.108	0.099	0.086	0.086	0.124	0.123	0.119
Mother birth	−0.006	−0.009	–	−0.009	−0.017	−0.017	−0.042	−0.052	−0.051
Father birth	−0.007	−0.009	–	−0.021	−0.014	−0.014	−0.082	−0.071	−0.074
Person birth	−0.017	−0.015	–	−0.023	−0.018	−0.018	−0.004	−0.005	–
Ed. degree	0.034	0.038	–	0.049	0.039	0.039	0.016	0.023	–
Social status	0.042	0.038	–	0.033	0.032	0.032	0.048	0.042	–
Poverty	−0.001	−0.005	–	−0.025	−0.031	−0.031	0.006	0.000	–
Voc. tr. mother	0.082	0.081	0.101	0.054	0.057	0.057	0.068	0.069	0.069
Voc. tr. father	−0.041	−0.035	–	−0.043	−0.022	−0.022	−0.015	−0.017	–
German grade	−0.326	−0.328	−0.337	−0.158	−0.147	−0.147	−0.237	−0.209	−0.218
Math grade	−0.232	−0.231	−0.221	−0.462	−0.464	−0.464	−0.243	−0.266	−0.254
Science grade	−0.128	−0.126	−0.132	−0.071	−0.083	−0.083	−0.105	−0.109	−0.115
Constant	489.8	485.9	444.4	487.7	476.1	476.1	429.8	431.6	427.1

OLS ordinary least squares, *CCR* correlated component regression, *C* number of correlated components, *P* number of predictors
*Significant ($p < 0.01$)

analyzed the structure of the data. On the basis of several statistics we have been able to confirm a multicollinear structure of the large scale assessment data, indicating that estimates of the regression coefficients may become inaccurate when using traditional OLS regression for characterizing student performances. Based on the fit statistics R^2 and $CV\text{-}R^2$, we have compared OLS regression to CCR. As the results of our study corroborate, CCR outperforms traditional OLS regression. Especially when reducing the number of correlated components and removing irrelevant predictors from the models, predictive power has been improved. We have also seen noteworthy differences regarding the estimated regression parameters, and future research can investigate the impact that may have on the classification of students to competence levels in educational comparison studies.

The application of special regression analytic approaches to characterizing student performances in the context of large scale educational assessments such as PIRLS, TIMSS, or PISA, where the number of predictors can be high (more than 400), is lacking in literature so far. Since the consideration of multicollinearity and suppression effects is crucial for the application of regression analysis, OLS regression seems not to be *the* appropriate method for this purpose. However, CCR promises to be a very useful instrument to identify those background variables that may have considerable influence on student performances.

In this regard, a large scale real application of CCR including many more background predictor variables must be systematically performed. A further issue that typically arises when applying regression analytic approaches in the context of large scale educational assessments are the occurrences of high portions of missing values. A detailed investigation of the effects of missing value structures in the data and their imputations for CCR related analyses in large scale assessments is another interesting direction for future research on this topic.

References

Albers, S. (2009). *Methodik der Empirischen Forschung*. Wiesbaden: Gabler.

Belsley, D., Kuh, E., & Welsch, R. E. (2004). *Regression diagnostics: Identifying influential data and sources of collinearity*. Hoboken: Wiley.

Berk, K. N. (1977). Tolerance and condition in regression computations. *Journal of the American Statistical Association, 72*, 863–866.

Bos, W., Tarelli, I., Bremerich-Vos, A., & Schwippert, K. (2012a). *IGLU 2011: Lesekompetenzen von Grundschulkindern in Deutschland im Internationalen Vergleich*. Münster: Waxmann.

Bos, W., Wendt, H., Köller, O., & Selter, C. (2012b). *TIMSS 2011: Mathematische und Naturwissenschaftliche Kompetenzen von Grundschulkindern in Deutschland im Internationalen Vergleich*. Münster: Waxmann.

Cohen, J., West, S. G., Aiken, L., & Cohen, P. (2003). *Applied multiple regression/correlation analysis for behavioral science*. Mahwah, NJ: Lawrence Erlbaum Associates.

Cook, R. D., & Weisberg, S. (1982). *Residuals and influence in regression*. New York: Chapman & Hall.

Lynn, H. S. (2003). Suppression and confounding in action. *The American Statistician, 57*, 58–61.

Magidson, J. (2010). Correlated component regression: A prediction/classification methodology for possibly many features. Belmont, MA.

Maidson, J. (2011). CORExpress 1.0 User's Guide. Belmont, MA.

Magidson, J. (2013). Correlated component regression: Re-thinking regression in the presence of near collinearity. In H. Abdi, W. W. Chin, V. E. Vinzi, G. Russolillo, & L. Trinchera (Eds.), *New perspectives in partial least squares and related methods* (pp. 65–78). Heidelberg: Springer.

Magidson, J., & Bennett, G. (2011). *Correlated component regression (CCR) - a brief methodological description*. Technical Paper, Logit Research, Kent. www.logitresearch.com/Library_of_technical_papers

Martin, M. O., & Mullis, I. V. S. (2013). *Methods and procedures in TIMSS and PIRLS 2011*. Chestnut Hill: TIMSS & PIRLS International Study Center.

Mullis, I., Martin, M. O., Kennedy, A. M., Trong, T. L., & Sainsbury, M. (2009a). *PIRLS 2011 assessment framework*. Chestnut Hill: TIMSS & PIRLS International Study Center.

Mullis, I., Martin, M. O., Ruddock, G. J., O'Sullivan, C.Y., & Preuschoff, C. (2009b). *TIMSS 2011 assessment framework*. Chestnut Hill: TIMSS & PIRLS International Study Center.

O'Brien, R. (2007). A caution regarding rules of thumb for variance inflation factors. *Quality & Quantity, 41*, 673–690.

OECD. (2012). PISA 2009 Technical Report. Paris: OECD Publishing.

Pandey, S., & Elliott, W. (2010). Suppressor variables in social work research: Ways to identify in multiple regression models. *Journal of the Society for Social Work and Research, 1*, 28–40.

Rawlings, J. O., Pantula, S. G., & Dickey, D. A. (1998). *Applied regression analysis: A research tool*. New York: Springer.

Refaeilzadeh, P., Tang, L., & Liu, H. (2008). *Cross-validation*. Tempe: Arizona State University.

Sheskin, D. J. (2011). *Handbook of parametric and nonparametric statistical procedures*. Boca Raton: CRC Press.

Analysing Psychological Data by Evolving Computational Models

Peter C.R. Lane, Peter D. Sozou, Fernand Gobet, and Mark Addis

Abstract We present a system to represent and discover computational models to capture data in psychology. The system uses a Theory Representation Language to define the space of possible models. This space is then searched using genetic programming (GP), to discover models which best fit the experimental data. The aim of our semi-automated system is to analyse psychological data and develop explanations of underlying processes. Some of the challenges include: capturing the psychological experiment and data in a way suitable for modelling, controlling the kinds of models that the GP system may develop, and interpreting the final results. We discuss our current approach to all three challenges, and provide results from two different examples, including delayed-match-to-sample and visual attention.

P.C.R. Lane (✉)
School of Computer Science, University of Hertfordshire, College Lane, Hatfield AL10 9AB, Hertfordshire, UK
e-mail: peter.lane@bcs.org.uk

P.D. Sozou
Centre for Philosophy of Natural and Social Science, London School of Economics and Political Science, Houghton Street, London WC2A 2AE, UK

Department of Psychological Sciences, University of Liverpool, Bedford Street South, Liverpool L69 7ZA, UK
e-mail: p.sozou@liverpool.ac.uk; p.sozou@lse.ac.uk

F. Gobet
Department of Psychological Sciences, University of Liverpool, Bedford Street South, Liverpool L69 7ZA, UK
e-mail: fernand.gobet@liverpool.ac.uk

M. Addis
Faculty of Arts, Design and Media, Birmingham City University, City North Campus, Perry Barr, Birmingham B42 2SU, UK
e-mail: mark.addis@bcu.ac.uk

© Springer International Publishing Switzerland 2016   587
A.F.X. Wilhelm, H.A. Kestler (eds.), *Analysis of Large and Complex Data*, Studies in Classification, Data Analysis, and Knowledge Organization,
DOI 10.1007/978-3-319-25226-1_50

1 Introduction

Psychology, as with other sciences, generates a large body of experimental data. Using this information to foster scientific understanding requires techniques of data analysis to find patterns, and then to produce understanding from the patterns. A standard method for finding such patterns in psychology is to construct a computational model, which is a process that generates the patterns in the data. In this paper, we describe a technique for creating process-based models from experimental data; these models represent patterns from the data, and provide a base for future analysis and understanding.

There exist a number of different approaches to develop computer models in psychology, including mathematical modelling, symbolic modelling, and connectionism (see Gobet et al. 2011, for an overview). Developing models has been notoriously difficult, as there has been no systematic methodology for doing so until recently (but see, e.g., Lane and Gobet 2012).

The research presented in this paper addresses this problem by creating a system to discover viable models in a semi-automated fashion. One way to solve this kind of problem is to search over a defined space of potential functions or programs (Langley et al. 1987). Recently, researchers have applied this foundation in many domains including psychology (Frias-Martinez and Gobet 2007) and mechanics (Schmidt and Lipson 2009).

Our proposed system also works by searching over a defined space. The first component is a generic definition of a *symbolic cognitive model*, applicable to a wide number of domains. The second component is the discovery process, which searches over a large number of candidate models, seeking models with a good fit to the results in the target domain. The third component is the domain-specific information required, such as how the model interacts with the experiment, how the experiment is conducted, and how the results are analysed; this information is extracted from the published literature.

2 Representation of Computational Models

The computational models used in our experiments are composed from three main components:

1. An input/output module, to interact with the experiment;
2. A short-term memory (STM), to store information; and
3. A control program, which determines the behaviour of the model. The control program stores its last calculated value as the *current value*.

This structure is typical of symbolic models used in psychology. The design of the following operators, settings and timing parameters are all taken from prior work, such as Cowan (2001), Frias-Martinez and Gobet (2007) and Samsonovich

(2010). The search space is intended to cover a large range of cognitively plausible architectures.

The input/output module is responsible for accessing the experimental stimuli and providing the response(s). For example, in the first case study below, the input is a series of three images: these are represented as inputs 1, 2 and 3. The output is the name of one of the images. In other experiments, the input may be more complicated. The required inputs and outputs are created for each experiment separately. We describe the specific input/output mechanisms for the two case studies separately.

The STM is a fixed-capacity working memory (Cowan 2001), and is treated as a simple buffer of three items. Items can be pushed onto the STM, with items at the bottom being forgotten. The model can also retrieve items directly by position, and compare items in two positions.

The control program defines the behaviour of the model. The program is written in a lisp-like language, using a fixed set of *operators*; this language is the *Theory Representation Language*. The operators are responsible for interaction with the input/output module, interaction with the STM, and for general control flow (such as sequential or conditional operators). The operators are shown in Table 1, and a sample program in Fig. 1.

The sample program consists of six operators. A characteristic feature of lisp-like languages is that the program structure is highlighted by the brackets: this feature is useful for the search process, as will be seen in the next section. By matching the brackets (and as shown by indentation), the `seq` operator uses two blocks: in this example, the first is an `if` operator, itself using three blocks; the second is an `output` operator, which uses no blocks.

This sample program first compares STM items 1 and 2 (`compare-1-2`). If they are equal (`if` operator), then the first stm slot is retrieved (`access-stm-1`),

Table 1 Operators for the control program

Operator	Time (ms)	Description
access-stm-1 (2 or 3)	50	Put item in STM slot 1 (2 or 3) into current value
compare-1–2 (2–3 or 1–3)	200	Current value is true/false if STM item 1/2/3 = item 2/3/1
if	200	Selects between two sub-programs based on a condition
nil	50	Set current value to 0 ("false")
putstm	50	Push given value on to STM
seq	50	Sequentially do two sub-programs

Fig. 1 A sample control program

```
(seq
    (if (compare-1-2)
        (access-stm-1)
        (access-stm-3))
    (output))
```

otherwise the third stm slot is retrieved (`access-stm-3`). The retrieved value is implicitly stored as the model's *current value*. The program finally outputs (`output`) the current value; this final operator is based on the domain-specific input/output module for the experiment.

Each operator in Table 1 has an associated time, in milliseconds, reflecting the physical processing time of the underlying cognitive process. In many psychological experiments, the data include results on reaction times: the time that passes between the input being shown and the output being generated. The times allocated to each operator have been derived from earlier work in cognitive modelling, and enable simulated reaction times to be computed for each model. For the sample program, the running time will be 50 for the `seq`, 200 for the `if`, 200 for the `compare-1-2`, 50 for one of the `access` operators, and 200 (depending on domain) for the `output`: a total of 700 ms.

A further aspect of the models provides an element of non-determinism. We allow each operator to fail with a probability of 2 %. When an operator fails, the control program simply moves to the next step in the program. For example, if the `if` operator failed in the program above, the next operator is `output`, which would output whatever the current value happened to be.

For our experiments, we keep the architecture of the computational models fixed. In each case study, all the models have the same domain-specific input/output module, the same STM, and the same set of operators within the control program. A wide variety of behaviours can be achieved by changing the control program. The next section describes an automated way of adapting the control program to attempt to fit a target experiment.

3 Search Process: Genetic Programming

Genetic programming (GP) (Koza 1992; Poli et al. 2008) is one of the number of techniques for searching large spaces of candidate solutions; other related techniques include Monte Carlo techniques (Metropolis and Ulam 1949) and simulated annealing (Kirkpatrick et al. 1983). The advantages of GP for our application include its long history of previous successes, that it constructs a population of many candidate solutions, and there is a simple method of controlling the search process, through a domain-specific fitness function.

The fitness function is how the quality of a model is determined. For our case studies, the fitness function is determined partly from the performance of the model (how often the model's output agrees with a human's, for the same inputs), and partly from its reaction times or the size of control program. The fitness function is used to compare the models, guiding the search towards those models which have better values of fitness.

The GP algorithm (with parameters: the size of population, the number of elite models to retain in each generation, and the target number of iterations):

1. Create an initial current population from a randomly generated set of candidate models (of varying size, constructed recursively up to a restricted depth).
2. Rank the models using the *fitness function*.
3. Construct an empty next population.
4. Add the best N models to the next population, where N is the number of elite models to retain.
5. Repeat until the next and current populations are the same size:

    a. Select two models randomly from the current population, probabilistically preferring those with better fitness values.
    b. Use crossover and mutation (see below) to generate two new models.
    c. Add the two new models to the next population.

6. Replace the current population with the next population.
7. Repeat from (2) until target number of iterations is complete.

The GP algorithm uses two processes to generate new models from current ones: *crossover* and *mutation*. The processes are illustrated in Fig. 2, with the original programs shown on the left of the figure. Crossover selects two blocks at random, one from each program, and makes two new programs by replacing the

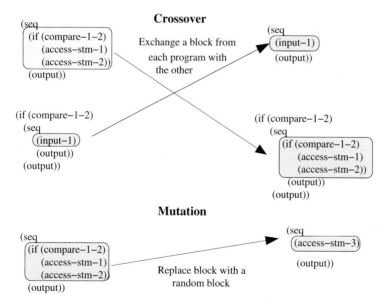

Fig. 2 Illustration of crossover and mutation

selected block in the first with the selected block in the second, and vice-versa. Mutation replaces a randomly selected block with a new, randomly created block, of arbitrary size.

4 Case Study 1: Delayed Match to Sample

The Delayed Match to Sample (DMTS) task explores a number of cognitive processes, including object recognition, categorisation and STM. The experiment consists of a number of trials. In each trial, a target image is initially presented. After a delay, two images are presented, one of which is the target image; the participant must identify the target image. Various studies have used this task, some looking at brain activity whilst the experiment is conducted, including Chao et al. (1999) and Habeck et al. (2003).

In this case study, we follow Frias-Martinez and Gobet (2007) and Lane et al. (2014), and simulate one set of results from Chao et al. (1999): Table 2.

4.1 Domain-Specific Model Details

The models for the DMTS study use the general framework presented in Sect. 2, adding processes to obtain the input stimuli and the output response. The stimuli are presented as a set of three inputs, with each stimulus being a number from 1 to 6, each number representing a distinct image. The first stimulus is the *target*, and is presented for 1.0 s. Then there is a gap of 0.5 s, before the *left* and *right* paired inputs are shown for a further 2.0 s. The response is obtained by the model outputting either *left* or *right* to indicate which of the two inputs is the same as the original target; a random "guess" is made if no output is made. The theory language is thus augmented with five further operators. These operators read the input from the target, left or right position, or output the response "left" or "right"; each operator takes 200 ms.

The models are tested on all 60 combinations of input values: for each of six inputs, there are five possible distractors, and each pair can be presented with the target on the left or the right.

Table 2 Accuracy and reaction time obtained by humans in DMTS task (Chao et al. 1999)		Accuracy		Time	
	Stimulus	Mean	Standard error	Mean	Standard error
	Tools	95 %	± 1.2 %	767 ms	± 27.5 ms

4.2 Fitness Function and Search Parameters

The fitness function is determined by observing the accuracy of the model and its reaction time, before comparing these figures with those in Table 2.

The *performance* is measured by recording each correct response obtained by the model as a 1.0, and each incorrect response as a 0.0. The *reaction time* is measured by recording the simulated time that the model takes to produce a response; this time is measured from after the paired inputs are removed. The mean of the model's responses (its accuracy) and the mean of its reaction times are subtracted from the values in Table 2, reporting the absolute values. The smaller these values are, the "more fit" the model is.

The *fitness function* ranks the models and is computed from the mean performance and the mean reaction time. In order to make the reaction time values of approximately equal weight to the performance values, the mean difference is divided by the target time of 767.0, and any difference greater than 1.0 is capped at 1.0. In this way, the two numbers are each between 0 and 1.0, and can be combined by averaging. The final fitness value will be a number between 0 and 1.0, with 0 representing the best performing, fastest possible model, and 1.0 the worst performing, slowest possible model. Note that any output made before the paired inputs are removed is an error, and that trial's reaction time does not appear in the measurements of fitness.

For example, if the model produces a mean accuracy of 88 %, and a mean reaction time of 600 ms, then the fitness value is computed as:

$$0.5 \times |0.95 - 0.88| + 0.5 \times |767 - 600|/767 = 0.144$$

For the GP algorithm, the population size is 10,000, with 20 elite individuals held between each generation, and the search is run for 50 iterations.

4.3 Results

Table 3 gives mean figures for accuracy, reaction time, and program size. In brackets is given the mean absolute deviation from the target values in Table 2. For space reasons, we do not show an example of the models. The means were obtained by running the GP algorithm 10 times, independently, and combining the results for the best model from each run.

Table 3 Results for DMTS case study, mean (deviation from target) over ten runs

Accuracy (%)	Reaction time (ms)	Size
93.8 (1.5)	786.0 (28)	135.9

The results demonstrate that the GP search algorithm is reliably locating high quality models. The models are close to the target accuracy and mean reaction time results obtained in humans. The models obtained are quite complex, and differ in many ways, illustrating the potential diversity in explanations that may be possible.

For a more detailed discussion of this case study, including a comparison of different fitness functions, see Lane et al. (2014); the results reported here follow the original protocol more closely, use a different fitness function, and show a substantially better fit to the human data.

5 Case Study 2: Visual Attention

For the second case study, we apply our system to a visual-attention task, used by Kornblum (1969). In this task, participants had to watch for one of four lights, and then press one of four corresponding buttons. The experiment explored the impact that repeated lights had on reaction times and error rates.

Each trial consisted of a light being lit, waiting for a button to be pressed, with the response and reaction times recorded. Each set of trials was divided into eight blocks of 150 trials. Each of the eight blocks used a different "probability of non-repetition" (pnr), which is the probability that the next lit light will not be a repeat of the previous lit light. Kornblum (1969) presents results for the mean error rate, and reaction times separately for each of the eight blocks, and separate analysis for when lights are repeated or not.

In this case study we look only at the error rate for the different pnr values: these values have been estimated from Fig. 2 of Kornblum (1969): Table 4.

5.1 Domain-Specific Model Details

The models for the Kornblum study use the general framework presented in Sect. 2, adding processes to obtain the input stimuli and the output response. The stimuli are represented as a set of four inputs, corresponding to the four lights. The lit light will be represented by an input of 1.0 on the corresponding input to that light, and unlit lights by inputs of 0.0. The outputs are similar. Four outputs are provided, which must be set to 1.0 by the model as appropriate; a random "guess" is made

Table 4 Human error rates on Kornblum attention task

PNR value	0.75	0.39	1.00	0.47	0.97	0.56	0.92	0.88
Error rate (%)	2.9	3.6	3.1	4.0	4.4	4.1	4.2	4.0

if no output is made. The theory language is thus augmented with eight further operators. Four operators query whether each light is lit, and are called `input-1`, etc. Four operators are used to press one of the four buttons. Each operator takes 200 ms. The models are tested using a sequence of 1200 numbers: 150 for each pnr value. The 150 numbers are generated randomly, using the pnr value to determine repeats.

5.2 Fitness Function and Search Parameters

The fitness function is determined from the error rate of the model, compared with the human accuracy, and also a measure of model complexity.

The *performance* is the absolute difference between the model's error rate for each pnr value compared with the target error rate, from Table 4. The smaller this value is, the "more fit" the model is.

The *complexity* is the size of the control program. To make the complexity a number between 0 and 1, the size of the control program is capped at 1000, and then divided by 1000. Hence, 1 is the largest value for the complexity and 0 (for an empty program) the smallest. The assumption is that smaller programs will be better than larger ones, and is a widely used heuristic in GP systems (Poli et al. 2008). The overall fitness function is computed as the mean of the performance and complexity measures. For the GP search algorithm, the population size is 500, with 50 elite individuals held between each generation, and the search is run for 50 iterations.

5.3 Results

Table 5 gives mean figures for mean difference in performance, and total program size. The standard deviation is shown in brackets for the performance values. The means were obtained by running the search 10 times, independently, and combining the results for the best model from each run.

The results demonstrate that the GP search algorithm is reliably locating high quality models, based on the performance measure. The general findings from case study 1 still apply: a range of viable models is discovered by the system, illustrating the potential diversity in possible explanations.

Table 5 Results for Kornblum case study, means obtained over ten runs

Performance	Size
0.004 (0.001 std dev)	15.9

6   Conclusion

The case studies presented in this paper demonstrate how a generic definition of a class of symbolic cognitive models can be used to create viable models in two different areas of psychology: decision making and visual attention. The cognitive models' behaviour is determined by a control program. The genetic-programming search algorithm is able to work through a small subset of the range of possible models, and locate high quality candidate models. Analysis demonstrates that the models are a good fit to the target data used here.

Future work will improve the quality of the modelling, by including more of the target behaviours explained in the psychological literature. The fitness functions used here combined different objectives into a single measure; we plan to use multiple-objective techniques to improve the comparisons. Also, we will extend the application of the models to more domains, such as categorisation and risk assessment. Finally, the main aim of this research is to create models which improve psychological theories in these domains; in future work, the generated models will be analysed to produce this understanding.

Acknowledgements This research was supported by ESRC Grant ES/L003090/1.

The implementation was written for the Java 7 platform in the Fantom language, and used the ECJ evolutionary computing library (Luke 2013).

References

Chao, L., Haxby, J., & Martin, A. (1999). Attribute-based neural substrates in temporal cortex for perceiving and knowing about objects. *Nature Neuroscience, 2,* 913–920.

Cowan, N. (2001). The magical number 4 in short-term memory: A reconsideration of mental storage capacity. *Behavioral and Brain Sciences, 24,* 87–114.

Frias-Martinez, E., & Gobet, F. (2007). Automatic generation of cognitive theories using genetic programming. *Minds and Machines, 17,* 287–309.

Gobet, F., Chassy, P., & Bilalić, M. (2011). *Foundations of cognitive psychology.* London: McGraw Hill.

Habeck, C., Hilton, J., Zarahn, E., Flynn, J., Moeller, J., & Stern, Y. (2003). Relation of cognitive reserve and task performance to expression of regional covariance networks in an event-related fMRI study of non-verbal memory. *NeuroImage, 20,* 1723–1733.

Kirkpatrick, S., Gelatt, C. D., & Vecchi, M. P. (1983). Optimization by simulated annealing. *Science, 220*(4598), 671–680.

Kornblum, S. (1969). Sequential determinants of information processing in serial and discrete choice reaction time. *Psychological Review, 76*(2), 113–131.

Koza, J. R. (1992). *Genetic programming: On the programming of computers by means of natural selection.* New York: MIT Press.

Lane, P. C. R., & Gobet, F. (2012). A theory-driven testing methodology for developing scientific software. *Journal of Experimental and Theoretical Artificial Intelligence, 24,* 421–456.

Lane, P. C. R., Sozou, P. D., Addis, M., & Gobet, F. (2014). Evolving process-based models from psychological data using genetic programming. In R. Kibble (Ed.), *Proceedings of the 50th Anniversary Convention of the AISB* (pp. 144–149).

Langley, P., Simon, H. A., Bradshaw, G., & Zytkow, J. (1987). *Scientific discovery: Computational explorations of the creative processes*. Cambridge, MA: MIT Press.

Luke, S. (2013). The ECJ owner's manual. http://cs.gmu.edu/~eclab/projects/ecj/docs/manual/manual.pdf

Metropolis, N., & Ulam, S. (1949). The Monte Carlo method. *Journal of the American Statistical Association, 44*(247), 335–341.

Poli, R., Langdon, W. B., & McPhee, M. F. (2008). *A field guide to genetic programming*. Raleigh, USA, Lulu Books.

Samsonovich, A.V. (2010). Toward a unified catalog of implemented cognitive architectures. In *Proceedings of the 2010 Conference on Biologically Inspired Cognitive Architectures* (pp. 195–244). Amsterdam: IOS Press.

Schmidt, M., & Lipson, H. (2009). Distilling free-form natural laws from experimental data. *Science, 324*(5923), 81–85.

Use of Panel Data Analysis for V4 Households Poverty Risk Prediction

Lukáš Sobíšek and Mária Stachová

Abstract One of the main approaches to tracking causality between income, social inclusion and living conditions is based on regression models estimated using various statistical methods. This approach takes into account quantitative and qualitative information about individuals or households that is collected in different periods of time (years in particular), thus allowing it to be transformed into multidimensional data sets, called panel data. Regression models based on panel data are able to describe the dynamics over time periods, so that the patterns can be related to changes in other characteristics.

This paper utilises one of these approaches to panel data analysis RE-EM trees which are used to predict the risk-of-poverty rate of households located in the four "Visegrad" countries. The risk-of-poverty rate of individual households is computed on the basis of cluster analysis results, and it takes into account household living conditions as well as income. Subsequently, the risk-of-poverty rate is used as the outcome for the prediction model above. Certain household characteristics were chosen as predictors including: information about the "head" of the household (age, education level, marital status, etc.) and information about the number of members in the household.

The results show slight differences in poverty determinants among Visegrad countries. The determinants with the highest impact on the risk-of-poverty rate are: number of household members (Czech Republic, Hungary and Slovakia) and education level (Poland).

L. Sobíšek (✉)
Faculty of Informatics and Statistics, University of Economics, Prague, W. Churchill Sq. 4,
130 67 Prague 3, Czech Republic
e-mail: lukas.sobisek@vse.cz

M. Stachová
Faculty of Economics, Matej Bel University, Tajovského 10, 975 90 Banská Bystrica, Slovakia
e-mail: maria.stachova@umb.sk

© Springer International Publishing Switzerland 2016   599
A.F.X. Wilhelm, H.A. Kestler (eds.), *Analysis of Large and Complex Data*, Studies
in Classification, Data Analysis, and Knowledge Organization,
DOI 10.1007/978-3-319-25226-1_51

1 Introduction

The measurement of the risk of household poverty is an integral part of the living condition monitoring. To understand the poverty means to consider different aspects of living situation (Fusco et al. 2010; Ward et al. 2009). It can be looked upon from different angles. Perry (2002) divides them into the following three groups.

The first and most widespread approach restricts the assessment indicators measuring poverty to economic or financial ones (current income in particular). Methods using this concept incorporate some means that connect income with minimum acceptable living standards in a given society, thus determining where to draw a line of the risk of poverty. The threshold is usually defined at 50–60 % of the mean or median (Eurostat 2014).

Another way of assessing poverty—by direct measuring of current living conditions—focuses on the outcome dimension and indicators of material deprivation. It takes into account whether a particular entity (e.g. a household) owns some possessions (e.g. a car) or is able to participate in certain activities (e.g. going on vacation abroad). However, no attempts are made to find the reason why somebody does not own or do something, be it financial straits or different preferences.

The third approach is based on feedback of respondents in a survey. It is rather subjective because the result depends on respondents' preferences and demands.

There are many studies available which consider the household poverty of European countries from different points of view. Some only take into account household income (Andriopoulou and Tsakloglou 2011), others include macro factors (e.g. unemployment rate or GDP) of the region to which the household belongs, (Reinstadler and Ray 2010) or use a principal components analysis to include proxy variables for community and household social capital endowment, and a set of variables describing household economic well-being (Santini and De Pascale 2012).

We believe that a combination of all factors mentioned in Perry (2002) are to be considered when studying poverty. This approach is supported by Nolan and Whelan (2011), for instance, who give reasons why the measurement of income has to be supplemented by that of material deprivation.

In this paper, we present a "mosaic" of all the three approaches to assessing whether particular households belong to worse- or better-off ones. For this reason we apply a cluster analysis, the households being divided into two within-group homogeneous clusters. Those belonging to the same cluster are on a similar level of poverty risk.

In the next step of our analysis we fit a prediction model based on the RE-EM tree method, predicting the classification of the household into the respective poverty risk cluster. The estimated model is based on the training data set that covers the period between 2007 and 2009. The model is then used to make new prediction entries in the database from 2010 onwards.

For further details on the data and methodology, see the following section below. The results we achieved are shown in the penultimate section, a summary of the research concluding the paper.

2 Data and Methodology

Eurostat data, namely the EU statistics on income and living conditions (EU-SILC survey) were employed in the analysis. Since the data were collected over a 4-year period, they could be converted into a multidimensional panel format. The data set contains information on European households, their members, social status and housing conditions, covering the period from 2007 to 2010. In this paper, we only deal with households located in the four Visegrad countries (the Visegrad Four, V4), i.e. the Czech Republic (CZ), Hungary (HU), Poland (PL) and Slovakia (SK), due to their economic and social proximity.

First of all, we carried out a latent class cluster analysis to define poor households, having used commercial software Latent GOLD (Vermunt and Magidson 2005). The clustering method implemented in the software allows searching for K-latent classes that represent case segments, analysing variables of mixed scale types (nominal, ordinal, continuous and counts) simultaneously and specifying continuous and discrete covariates that predict class membership. The clustering variables in this study are data on households, such as information affecting social exclusion (e.g. household insolvency, indebtedness, enforced lack of basic necessities), data on total income, type of dwelling, housing conditions, etc. They were chosen taking into account special characteristics and living standards of each V4 country. We obtained two clusters and values for each household that indicates the probability of membership in the "poorer" cluster. This classification served as a new response variable for the prediction.

In the second phase of the research, we split the data into a training set (covering the period 2007–2009) and a test data set (from 2010). The former data set was used for fitting the prediction model based on the RE-EM tree method, the function *REEMtree()* from R package "REEMtree" (Sela and Simonoff 2011b) having been applied. (The structure of mixed effects models for longitudinal data is being combined with the flexibility of tree-based estimation methods; see Sela and Simonoff 2011a). As predictors for this model we take information about the "head" of the household (age, education level, marital status, etc.) and information about the number of members in the household. Reinstadler and Ray's (2010) study employs a similar set of determinants when calculating a risk-of-poverty rate.

The latter data set was used for testing the prediction ability of the model.

3 Results

The above methodology was applied to the respective data sets. The first stage of the analysis utilised latent class clustering, with the resulting percent of households classified in the "poorer" cluster as shown in Table 1 which also shows the variables used for the clustering. The resulting profile plots of clusters for each V4 country are shown in Figs. 1, 2, 3, and 4.

Table 1 Clustering summary for all V4 countries

Variables used for the clustering	CZ	HU	PL	SK
HS040—capacity to afford paying for one week annual holiday away from home (nominal variable)	+	+	+	+
HS050—capacity to afford a meal with meat every second day (nominal variable)	−	+	+	+
HS060—capacity to face unexpected financial expenses (nominal variable)	+	−	+	+
HS090—do you have a computer? (nominal variable)	+	+	+	+
HS100—do you have a washing machine? (nominal variable)	−	+	−	−
HS110—do you have a car? (nominal variable)	+	+	+	+
HS120—ability to make ends meet (ordinal variable)	−	−	+	−
HS140—financial burden of the total housing cost (ordinal variable)	−	−	+	+
HS130—lowest monthly income to make ends meet (continuous variable in hundreds, local currency)	+	+	+	−
HY020—total disposable household income (continuous variable in thousands, local currency)	+	+	+	+
Percent of households classified in the "poorer" cluster1 (%)	51.7	50.0	54.2	42.3

It displays variables used for clustering and the cluster1 (the "poorer") membership in percentage

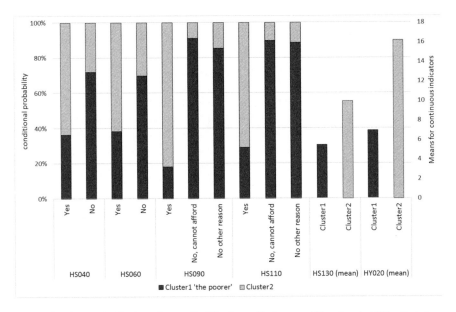

Fig. 1 Czech households clustering profile. The figure displays conditional probabilities that show how the clusters are related to the nominal indicator variables. For continuous variables means are displayed instead of probabilities

The following factors were chosen as the most appropriate variables for clustering of Czech households: the capacity to afford paying for a one-week annual holiday away from home, the ability to cope with unexpected financial expenses,

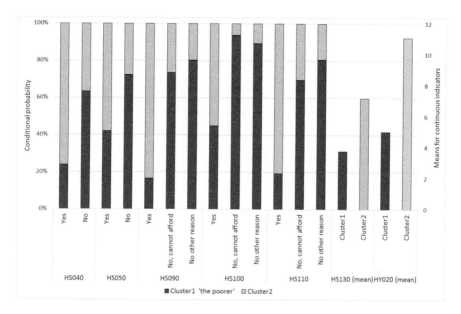

Fig. 2 Hungarian households clustering profile. The figure displays conditional probabilities that show how the clusters are related to the nominal indicator variables. For continuous variables means are displayed instead of probabilities

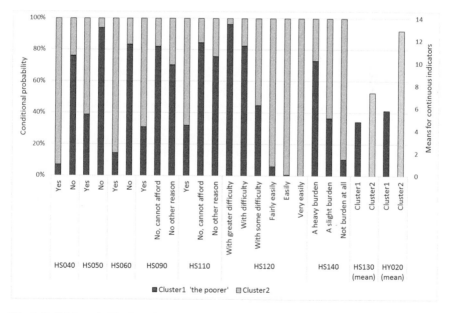

Fig. 3 Polish households clustering profile. The figure displays conditional probabilities that show how the clusters are related to the nominal (or ordinal) indicator variables. For continuous variables means are displayed instead of probabilities

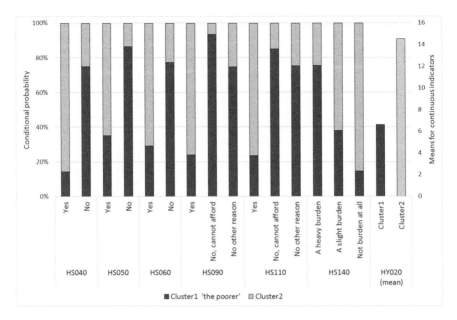

Fig. 4 Slovak households clustering profile. The figure displays conditional probabilities that show how the clusters are related to the nominal (or ordinal) indicator variables. For continuous variables means are displayed instead of probabilities

the ownership of a computer and/or a car, the lowest monthly income to make ends meet and the total disposable household income; see Table 1 and Fig. 1. According to the above criteria, 51.7 % of households belong to the segment (cluster) of the "less well off" (i.e. poor ones).

The key variables chosen for clustering of Hungarian households are as follows: the capacity to afford paying for a one-week annual holiday away from home, the capacity to afford a meat dish every second day, the ownership of a computer, a washing machine and/or car, the lowest monthly income to make ends meet and the total disposable household income. It is apparent from Table 1 and Fig. 2 that half of Hungarian households can be labelled as "poorer" ones.

The following variables for clustering of Polish households were employed: the capacity to afford paying for a one-week annual holiday away from home, the capacity to afford a meat dish every second day, the ability to cope with unexpected financial expenses, the ownership of a computer and/or a car, the lowest monthly income to make ends meet and the total disposable household income. Table 1 and Fig. 3 indicates that 54.2 % of Polish households are classified in the cluster with a higher risk of poverty.

The most appropriate variables for clustering of Slovak households are: the capacity to afford paying for a one-week annual holiday away from home, the capacity to afford a meat dish every second day, the capacity to cope with unexpected financial expenses, the ownership of a computer and/or a car, the lowest monthly income to make ends meet, total housing costs and total disposable

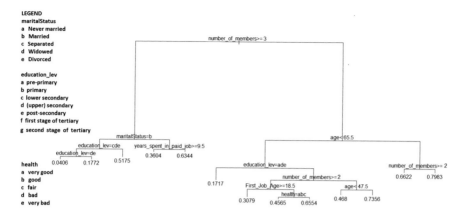

Fig. 5 RE-EM tree estimated on Czech household data

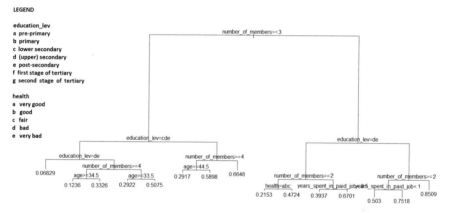

Fig. 6 RE-EM tree estimated on Hungarian household data

household income. As shown in Table 1 and Fig. 4, 42.3 % of Slovak households rank among the poorer ones.

In the second phase of the analysis we estimated the RE-EM tree model on training data set. The outcome variable is the probability of belonging to the cluster with higher risk of poverty. The same predictors were employed for each country, particularly information on the person responsible for accommodation and the number of the members of the household. We obtained four RE-EM tree models as indicated in Figs. 5, 6, 7, and 8.

Figure 5 shows the RE-EM tree containing all households in the Czech Republic. The root node is split according to the number of household members. If it is higher than two, the respective households are placed in the right branches, all the others being on the other side. The households on the lower nodes are divided according to their members' marital status, educational attainment, age, etc., until the splitting stop condition is satisfied. The terminal node (leaf) is ended with mean of our

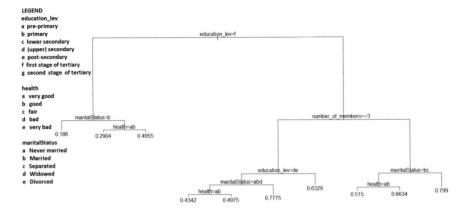

Fig. 7 RE-EM tree estimated on Polish household data

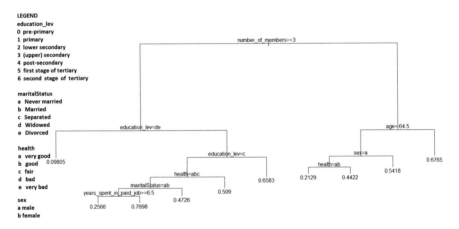

Fig. 8 RE-EM tree estimated on Slovak household data

response (in our case it is the probability of membership into the "poorer" cluster) for all observations assigned to this node.

The RE-EM tree in Fig. 6 illustrates the data from Hungary. The division of the root node is based on the number of household members, the other key nodes being split by education levels, age, years spent in employment, etc.

Figure 7 displays the RE-EM tree estimated on Polish household data set. The splitting variable of the root node is the level of education in this case, the other nodes branching out according to the number of household members, their marital status, state of health, etc.

Figure 8 presents the RE-EM tree of Slovak households. Again, the predictor in the root node is the number of members, the sub-nodes ramifying according to age, education levels, gender, state of health, etc.

Table 2 Predictive abilities of RE-EM tree models for all V4 countries

	CZ	HU	PL	SK
Error rate (%)	21.63	20.57	30.5	25.9
Number of object in training data	6792	5367	8440	3483
Number of object in test data	2264	1789	2819	1161

The predictive abilities of RE-EM tree models are evaluated on the test data set and expressed by error rates in Table 2. The numbers of households in both the training and test data sets for each V4 country are displayed in the table as well.

4 Conclusion

Predicting the risk of poverty for individuals and households is an aspect of responsible social policy.

In the paper, we present the results of the analysis which is aimed at the factors determining the risk of poverty. Another objective is to estimate a predictive model that can be employed to find out if there is a risk of poverty or social exclusion for particular households.

We performed a latent class cluster analysis to determine the level of poverty risk of households in the fours Visegrad countries, using a combination of indicators that had been shown in previous studies to be useful in defining vulnerability to poverty. In addition to household income, the living conditions and material property of the household were used to set the threshold of poverty risk. As the results show, these factors combine differently in different countries. Probability indication of poverty risk became the basis for the construction of the predictive model. It was built with the use of the RE-EM tree method, which enables the avoidance of restrictive parametric assumptions and provides the flexibility to use time-varying covariates. A bimodal non-linear distribution of the outcome did not allow us to use other well-established models, such as mixed effects ones (Baltagi 2012; Pinheiro and Bates 2009). It is apparent from the research results that, whereas in the Czech Republic, Hungary and Slovakia the parent node in the RE-EM tree is split based on the number of the members of the household, in Poland, the factor impacting the most on the splitting variable is the level of education completed. These results should be considered when designing anti-poverty policies. For example, in Poland, where the low education level implies higher risk-of-poverty, the social policy promotes and emphasises on education by, e.g., scholarships.

In the future research project, we plan to investigate whether resampling methods enhance the estimation accuracy. Further, we will compare the RE-EM model with an alternative approach. We intend to apply a classification method designed for cross-sectional data to a data set that would include both the values from a given year and auxiliary variables representing the dynamics in the data.

Acknowledgements The present paper was supported by VEGA1/0127/11 grant project entitled
"The poverty distribution within the EU countries". EU-SILC microdata were provided for
research purposes under EU-SILC/2011/33 agreement concluded between the European Commission, Eurostat and the Technical University in Košice, Slovakia. Eurostat bears no responsibility
for the results and conclusions reached by the authors. Data were analysed by Lukáš Sobíšek who
cooperates in the project. This work was also supported by projects Mobility—enhancing research,
science and education at Matej Bel University, ITMS code: 26110230082, under the Operational
Program Education co-financed by the European Social Fund.

References

Andriopoulou, E., & Tsakloglou, P. (2011). The determinants of poverty transitions in Europe and
    the role of duration dependence. IZA Discussion Paper No. 5692. Bonn: The Institute for the
    Labor.
Baltagi, B.H. (2012). *Econometric analysis of panel data*. Chichester: Wiley.
Eurostat (2014). *Income distribution statistics*. Luxembourg: Eurostat.
Fusco, A., Guio, A. C., & Marlier, E. (2010). *Income poverty and material deprivation in European
    countries*. Luxembourg: Office for Official Publication of the European Communities.
Nolan, B., & Whelan, C. T. (2011). Poverty and deprivation in Europe. Oxford: Oxford University
    Press.
Perry, B. (2002). The mismatch between income measures and direct outcome measures of poverty.
    Social Policy Journal of New Zealand, 19, 101–127.
Pinheiro, J., & Bates, D. (2009). *Mixed-effects models in S and S-PLUS*. New York: Springer.
Reinstadler, A., & Ray, J.-C. (2010). *Macro Determinants of Individual Income Poverty in 93
    Regions of Europe*. CEPS-INSTEAD Working Paper no. 2010–13.
Santini, I., & De Pascale, A. (2012). *Social Capital and Household Poverty: The Case of European
    Union*. Working Paper no. 109 from Sapienza University of Rome.
Sela, R. J., & Simonoff, J. S. (2011a). RE-EM trees: A data mining approach for longitudinal and
    clustered data. *Machine Learning, 86*, 169–207.
Sela, R. J., & Simonoff, J. S. (2011b). REEMtree: Regression trees with random effects. R package
    version 0.90.3.
Vermunt, J. K., & Magidson, J. (2005). *Technical guide for latent GOLD 4.0: Basic and advanced*
    [online]. Belmont, MA: Statistical Innovations Inc.
Ward, T., Lelkes, O., Sutherland, H., & Tóth, I. G. (Eds.) (2009). *European inequalities: Social
    inclusion and income distribution in the European Union*. Budapest: Tárki.

Part XIII
Data Analysis in Library Science

Collaborative Literature Work in the Research Publication Process: The Cogeneration of Citation Networks as Example

Leon Otto Burkard and Andreas Geyer-Schulz

Abstract In educational and scientific publishing processes scientists and prospective scientists (students) in their different roles (author, editor, reviewer, production editor, lector, reference librarians) invest a large amount of work into the proper handling of scientific literature in the widest sense. In this contribution we introduce the LitObject middleware and its combination with the popular open-source tool Zotero. The LitObject middleware supports the exchange of sets of scientific objects (literature objects) consisting of bibliographic references and documents (e.g. PDF-documents) by scientists. In our contribution we emphasize several process improvements with a special focus on the cogeneration of citation networks.

1 Introduction

In every publication literature is cited. To be able to cite literature the authors of a publication have to search, find, possibly evaluate and read the cited publications, file the fulltext document and, later, have to retrieve the publications filed. This time-consuming process of literature work has to be done by every researcher. To facilitate literature management scientists can use literature management software such as EndNote, Mendeley or Zotero that are compared and described in more detail by Hensley (2011). Based on the list of expectations from Gilmour and Cobus-Kuo (2011), management of literature also includes, for example, the import of references and fulltext documents from digital libraries, gathering metadata from documents as well as the organization in a database and annotation of literature. Additionally, properly formatted citations should be provided in various styles by the software. Literature management tools support the work with literature objects by providing functions to arrange them logically, for example, in a hierarchic folder structure, linking literature objects and in most cases provide a fulltext search for all organized literature objects and options to annotate them. In this paper the bundle

L.O. Burkard (✉) • A. Geyer-Schulz
Information Services and Electronic Markets, Karlsruhe Institute of Technology,
Karlsruhe, Germany
e-mail: mail@leon-burkard.de; andreas.geyer-schulz@kit.edu

© Springer International Publishing Switzerland 2016   611
A.F.X. Wilhelm, H.A. Kestler (eds.), *Analysis of Large and Complex Data*, Studies
in Classification, Data Analysis, and Knowledge Organization,
DOI 10.1007/978-3-319-25226-1_52

of reference and fulltext documents will be called a *literature object*. Literature objects are managed with the help of a PDF-manager as introduced by Mead and Berryman (2010). A PDF-manager is a software application that manages not only a reference but also mainly fulltext documents in PDF-format. Unfortunately, as described by Hull et al. (2008), metadata support for retrieving correct references into PDF-manager software is error-prone: There is no "universal method to retrieve metadata. For any given publication, it is not possible for a machine or human to retrieve metadata using a standard method" (Hull et al. 2008, p. 8). Also there are various options how metadata can be represented. The PDF-format itself only offers insufficient and limited metadata fields to embed metadata and as a consequence this feature is used only rarely according to Howison and Goodrum (2004).

Apart from new ways of organizing and managing literature the way of publishing changes: Glänzel and Schubert (2004) examined that in the 1980s about 25 % of all publications had only one author. This percentage decreased to 11 % until the year 2000. The average journal publication in 2011 had more than four authors. In the field of computer science single author publications represented only about 15 %, the majority was written by three and more authors according to Solomon (2009). However, the most common used literature management tools operate as user desktop applications with proprietary and restricted sharing and collaboration features. Even worse is the situation on the tool support side for extended use-cases with requirements such as the circulation of literature within the scope of fair use. One application for this requirement could be the temporary access to the authors' cited literature by the lector, reviewer or editor of the publication within the review process. Another use case is the support of collaborative literature work to facilitate the activity of writing a multi-author publication.

In order to improve the part of collaborative literature work we developed the LitObject middleware. The LitObject middleware serves as a foundation for various extended services that require a structured access to literature objects. As an example for an extended service we present the utilization of the LitObject middleware as a cogenerated data basis for citation networks.

2 A Simplified Publication Process

For a better motivation and understanding of the necessity for a middleware for literature objects we introduce a simplified publication process as depicted in Fig. 1. A detailed conceptual description of the publication process with an emphasis on the author's and editor's tasks can be found in University of Chicago Press (1982). The high-level perspective process consists of four main subprocesses:

In the first subprocess *Creation* the article for the first submission is prepared. This subprocess includes literature work, particularly retrieving literature objects and using references in the written document. After submission of the article the subprocess *scientific quality management (SQM)* starts. *SQM* includes several subprocesses such as the whole review process including the prior selection of

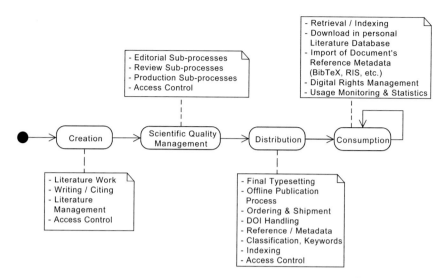

Fig. 1 A simplified publication process

appropriate reviewers as well as returning the submission to the author(s) for improvement, correction and submission of a camera ready version, editorial decisions about the orientation and corresponding selection of submissions and, last but not least, all tasks of a production editor as key person for the timely completion and coordination of the volume. The subprocess *Distribution* concentrates on the technical publication steps starting with the final typesetting of the submission for print and online publication followed by tasks for the correct classification, indexing as well as DOI handling and generation, respectively, as well as provision of reference metadata for usage in libraries, online catalogs, search engines and literature management environments. Also traditional ordering and shipment processes in combination with logistic, payment and accounting services are affected. Tasks in the *Distribution* subprocess are commonly executed by personnel in publishing companies, *SQM* usually by academic volunteers and writing articles in the subprocess *Creation* by authors. Finally, in the *Consumption* subprocess the corresponding publication is found, retrieved, (delivered and payed), read, filed and cited. This subprocess also includes usage monitoring as feedback for authors and the editorial board.

2.1   Challenges in the Publication Process with Regard to Literature

Challenges in the *Creation* subprocess are: How can past literature searches be rediscovered? How can literature objects be shared in multi-author publication scenarios and how, in general, can the result of literature searches executed by scientists or students be stored permanently? How can process information be gathered and used for process improvements through extended services (e.g. cogeneration of citation graphs, machine-learning, etc.)?

Reviewing as well as editorial work in the *SQM* subprocess includes checking and at least partially reading the author's cited literature. This leads to a repetition of work as authors, reviewers and editors have to search, find and retrieve the same document for a proper citation check or further discussion about the content of the publication. The subprocess *SQM* also implies, strictly speaking, the documentation of scientific research (e.g. lab books, videos, recordings, printouts of measurement instruments, software, data sets) and preservation of all work that the publication refers to and is based upon. Because of different subscription contracts not all scientists involved in the publication process have access to the same literature bundle needed in the *SQM* process. Therefore, the option to submit a bundle of written document and used literature in combination with a (time-limited) access permission for all affected roles is attractive for the subprocess *SQM*.

The third subprocess *Distribution* deals with enhancements in the generation and provision of the several types of metadata. To possess an increasing data basis of submissions and corresponding literature that already might be classified can support applications and research in the fields of automatic classification, indexing, clustering and linking of related literature.

In the last subprocess *Consumption* questions arise how literature management tools can be integrated in the cycle of retrieving literature and submitting it to the LitObject middleware. Especially current approaches to import literature objects in literature management tools such as Zotero or Mendeley are mostly based on individual web-crawlers for each publication website system that have to be updated after every minor update in the website structure of the publisher. The idea of individual web-crawlers has already been pursued in the research project UniCats (presented by Lockemann et al. 2000) where a "wrapper generator" supported the development.

A more appealing approach would be to embed information directly into the website such as DublinCore tags, ContextObjects in Spans (CoIns) or Highwire Press Tags to name a few. However, in the current development state they either have no support to link a reference to fulltext documents, are lacking proper transformations to common reference data formats such as RIS or BibTeX, have ambiguous fields or do not support the description of various documents on one page such as the description of a website for a collection of papers. They are good to promote information about the element in question on the homepage but, at the moment, they should not serve as a solid metadata basis for usage in scientific publications. The main problems in this process are the error-prone and often irreversible transformations between the different metadata formats as stated by Hull et al. (2008, p. 7).

3   The LitObject Middleware

In response to the challenges of Sect. 2.1 we propose the LitObject middleware as a system component which automates the transport and transformation of literature objects as shown in Fig. 2. The key idea is that the LitObject middleware acts as a central system with webservices between the various literature management tools and digital libraries as well as a repository for the publishing process (Fig. 1).

The system as presented in Fig. 2 has three main system boundaries: The first subsystem boundary is the local literature management software (illustrated by two literature management software applications). The second subsystem is the middleware, the third subsystem is an extended service. In this paper we describe a system for the cogeneration of graphs in Sect. 4 as an example for an extended service.

Our solution to the problems of Sect. 2.1 is to keep the organization of literature on an individual level with local literature management software. The local literature management is extended by a plugin that adds an export of literature objects to the middleware (as presented in Fig. 3). The exported literature objects can be acted on (e.g. display or edit) by a website that accesses the LitObject middleware. Through the website literature objects can be imported and exchanged by local literature management systems, digital libraries and extended services.

To avoid the creation of duplicates with every import from the website and to provide not only an import, but also a notification mechanism that automatically detects changes on the side of literature management tools as well as the server side, the plugins could be extended by a synchronization mechanism as indicated by the dotted line in Fig. 3.

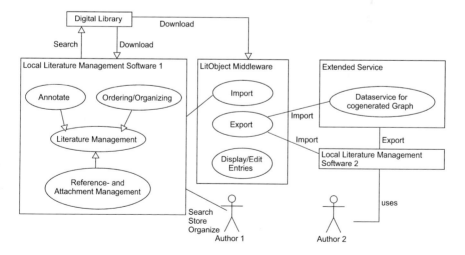

Fig. 2 System diagram of a middleware for literature objects

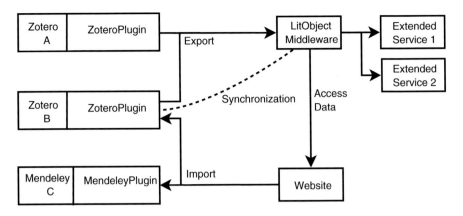

Fig. 3 Component structure for the LitObject middleware

The LitObject middleware follows the design pattern of a resource oriented architectural style as described by Fielding (2000). The main concepts of a resource oriented architecture (RESTful architecture) are a client server model with a separation between a client that requests data from a server utilizing the HTTP protocol (Fielding and Reschke 2014b), statelessness that means not to handle any session context on the server side, resource identification, a uniform interface, self describing messages and hypermedia. According to Pautasso and Wilde (2011) a resource is everything that is "relevant for an application (and its state)". Resources have an identifier resolved by a uniform resource identifier (URI) (Fielding and Reschke 2014a). Resources have a representation for their data format. Usually the Java Script object notation (JSON) or the extended markup language (XML) are used for this purpose. A client interacts with a resource through its representation by using the methods of the HTTP protocol: GET for retrieving, PUT for updating, POST for creating and DELETE for deleting resource elements. The application state is handled by using links within the representation that guide the usage of the webservice.

The LitObject middleware follows the introduced constraints. We have identified two main resources (see Fig. 4 for a detailed overview), the items resource that hosts literature objects and the collections resource which serves as a named collection for a set of corresponding literature objects. Both the items and the collections resource have instances (with identifiers) that are depicted by <item-id> and <collection-id> in Fig. 4. The item resource has three subresources refersto, bibtex and attachments. The refersto resource is itself a link list to other literature objects. The literature object is described by its reference at the resource bibtex. Fulltext documents that are part of the literature object are described and hosted at the attachments subresource. The collections resource, on the other hand, is a list of links to one or more literature objects identified by a URL. A plugin for a local literature management

```
http://<host>:<port>/<api_user>        - base URL of the LitObject middleware

/items                                  - list of all literature objects
    /<item-id>                          - URL of one single literature object
        /refersto                       - list of links to other literature objects
        /bibtex                         - reference of the literature object in the BibTeX format
        /attachments                    - list of all documents of the literature object
        /attachments/<id>               - URL of one single (fulltext) document of the literatur object
        /attachments/<id>/content       - content/download of the fulltext document

/collections                            - list of all collections
    /<collection-id>                    - URL of one single collection which is a list of links to
                                          URLs of <item-id> resources
```

Fig. 4 URL structure of the LitObject middleware

application, like Zotero or an extended service, interacts with the URL structure via a JSON representation utilizing the HTTP verbs GET, POST, PUT and DELETE.

4 Cogenerated Citation Networks with the Help of the LitObject Middleware

In Sect. 3 we have introduced the LitObject middleware that serves as a generic service to interconnect various literature management applications and extended services. One example of an extended service for the LitObject middleware is the provision of a technical infrastructure for cogenerated citation networks.

There exist various citation link networks such as the Arxiv HEP-PH citation graph (Stanford University 2003) and also service interfaces to request link databases, for example, the ArnetMiner system as presented by Tang et al. (2008) The approach for building up the data set in the ArnetMiner system is to extract researcher profiles at first, then querying databases with the researchers' name as identifiers and storing these information in a database. An alternative approach is followed by the LitObject middleware by utilizing the exported literature objects of the (locally) organized literature. Three use cases can be distinguished that are supported by a naming convention for collections following the scheme `<collection/paper-name>:<linktype>`.

1. The author's own paper: What did he cite: `<collection-name>:cited`
2. The author's retrieval process: What literature did he find relevant for the topic? This is a latent construct: `<collection-name>:relevant`
3. The author follows the citation structure in papers. How does he record the citations he followed: `<paper-read>:citationfollowed`

As depicted in Fig. 5, at first authors organize their literature with the help of a literature management software such as Zotero following the naming convention

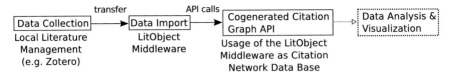

Fig. 5 Steps for utilizing the LitObject middleware as data basis for citation networks

introduced above. In the second step, they transfer their literature objects to the LitObject middleware. As shown in Fig. 4 there are two main resources in the LitObject middleware: one for the literature objects (`items`) and the second one for collections (`collections`). The extension for the literature management software exports at first all literature objects to the `items` resource and in a second step creates the same link structure as the local managed collections at the `collections` resource on the LitObject middleware side. As these imported literature objects as well as their arrangement in collections are saved in a database in a structured manner their structure is exposed by an API and—following the REST paradigm of various representations of resources—in different data formats, for example, the pajek data format or as comma separated values (csv) data.

5   Discussion

Other extended services—for example, as depicted in Fig. 3 on top of the component "Website"—are at least technically possible, e.g. the generation of an organization-wide search index or an internal repository. Also the LitObject middleware can serve as foundation for scientific projects in the fields of information retrieval, machine learning or clustering.

   Key reasons to continue the development on the LitObject middleware are that the usage of the commercial sharing and collaboration features of Mendeley, Zotero, etc., limits research capabilities, forces vendor lock-in and exposes research activities permanently as well as reduces interoperability between various literature management tools.

   Additionally there are legal as well as technical challenges: Within the sphere of "fair use" the exchange of full literature objects is allowed. However the restrictions of fair use are ambiguous, differ between countries and are non-uniform between publishers. Already the exchange of (digital) literature objects within one organization unit is sketchy as some Libraries advice against the sharing of literature as for instance the Health Science Library of the University of North Carolina (2014). In this technically focused publication a deeper exploration is beyond the scope, however, an exhaustive evaluation of these topics by country specific lawyers would be desirable.

   On the technical side as stated by Hull et al. (2008), there still exist issues how to identify literature objects globally as not every object has a digital object identifier

(DOI), international standard book number (ISBN) or a uniform resource name (URN). The questions how to get metadata and in which representation is solved technically, for example, by the unAPI specification (Chudnov et al. 2006b) that is described and motivated more in detail in an article by Chudnov et al. (2006a). Unfortunately, unAPI is lacking the linkage to fulltext documents, however, only a small adaptation is necessary to support this requirement as well. Finally, citation network analysis and visualization services (e.g. Pajek) can utilize this interface.

6 Summary

In this paper we have introduced the LitObject middleware that is a RESTful web service to interconnect various literature management software tools. The aim of the LitObject middleware is to improve collaborative literature work for advanced publishing processes. The LitObject middleware offers the possibility to import and export literature objects. Literature objects are the combination of a reference in a data format such as BibTeX and one or more fulltext document that belong to the reference. Additionally, literature objects can also be grouped together in collections. On top of the LitObject middleware various extended services are possible. As one possible extended service we presented an application for the cogeneration of citation networks. With the help of many authors who import literature objects into the LitObject middleware it is possible to build up a network of citations and clusters of literature that belong to a particular topic.

References

Chudnov, D., Back, G., Binkley, P., Celeste, E., Clarke, K., D'arcus, B., et al. (2006b). unAPI version 1. Accessed September 23, 2014, http://unapi.info/specs/unapi-version-1.html

Chudnov, D., Binkley, P., Summers, E., Frumkin, J., Giarlo, M. J., Rylander, M., et al. (2006a). Introducing UnAPI. *Ariadne*, (48). http://www.ariadne.ac.uk/issue48/chudnov-et-al/

Fielding, R. T. (2000). *Architectural Styles and the Design of Network-Based Software Architectures*. Ph.D. thesis, University of California, Irvine.

Fielding, R. T., & Reschke, J. (2014a). Hypertext transfer protocol (HTTP/1.1): Message syntax and routing. *RFC 7230 (Proposed Standard)*.

Fielding, R. T., & Reschke, J. (2014b). Hypertext transfer protocol (HTTP/1.1): Semantics and content. *RFC 7231 (Proposed Standard)*.

Gilmour, R., & Cobus-Kuo, L. (2011). Reference management software: A comparative analysis of four products. *Issues in Science and Technology Librarianship, 66*. doi:10.5062/F4Z60KZF.

Glänzel, W., & Schubert, A. (2004). Analyzing scientific networks through co-authorship. In H. Moed, W. Glänzel, & U. Schmoch (Eds.), *Handbook of quantitative science and technology research* (pp. 257–276). Dordrecht: Kluwer Academic Publishers. doi:10.1007/1-4020-2755-9_12.

Hensley, M. K. (2011). Citation management software: Features and futures. *Reference & User Services Quarterly, 50*(3), 204–208.

Howison, J., & Goodrum, A. (2004). *Why Can't I Manage Academic Papers Like MP3s? The Evolution and Intent of Metadata Standards*. Working Paper, Institute for Software Research, Carnegie Mellon University. http://www.repository.cmu.edu/isr/494/

Hull, D., Pettifer, S. R., & Kell, D. B. (2008). Defrosting the digital library: Bibliographic tools for the next generation web. *PLoS Computational Biology, 4*(10). doi:10.1371/journal.pcbi.1000204.

Lockemann, P., Christoffel, M., Pulkowski, S., & Schmitt, B. (2000). UniCats: ein System zum Beherrschen der Dienstevielfalt im Bereich der wissenschaftlichen Literaturrecherche. *IT - Information Technology, 42*(6), 34–40.

Mead, T. L., & Berryman, D. R. (2010). Reference and PDF-manager software: Complexities, support and workflow. *Medical Reference Services Quarterly, 29*(4), 388–393. doi:10.1080/02763869.2010.518928.

Pautasso, C., & Wilde, E. (2011). Introduction. In E. Wilde & C. Pautasso (Eds.), *REST: From research to practice* (pp. 1–18). New York: Springer.

Solomon, J. (2009). Programmers, professors, and parasites: Credit and co-authorship in computer science. *Science and Engineering Ethics, 15*, 467–489. doi:10.1007/s11948-009-9119-4.

Stanford University. (2003). High-energy physics citation network. Accessed September 23, 2014, https://www.snap.stanford.edu/data/index.html#citnets

Tang, J., Zhang, J., Yao, L., Li, J., Zhang, L., & Su, Z. (2008). ArnetMiner: Extraction and mining of academic social networks. In *Proceedings of the 14th ACM SIGKDD International Conference on Knowledge Discovery and Data Mining (Las Vegas)* (pp. 990–998). New York: ACM.

University of Chicago Press. (1982). *The Chicago manual of style* (13th ed.). Chicago: The University of Chicago Press.

University of North Carolina Health Sciences Library. (2014). Copyright basics: Sharing articles - LibGuides at University of North Carolina Chapel Hill. Accessed September 23, 2014, http://www.guides.lib.unc.edu/c.php?g=9031&p=45264

Subject Indexing of Textbooks: Challenges in the Construction of a Discovery System

Bianca Pramann, Jessica Drechsler, Esther Chen, and Robert Strötgen

Abstract The purpose of this paper is to present the steps on the way of finding a suitable classification system for subject indexing of textbooks and to integrate this into a resource discovery system. Textbooks are usually not indexed at great length. Yet, the research library of the Georg Eckert Institute has a unique status regarding the range of collected textbooks. This makes a detailed subject indexing indispensable. The currently used local solution shall be replaced by a more standardised solution to serve greater dissemination and compatibility of our data and to simplify the search process for the user.

1 Introduction: The Georg Eckert Institute and the Research Library

The Georg Eckert Institute for International Textbook Research (GEI), a member of the Leibniz Association, conducts application-oriented, multidisciplinary research into textbooks and educational media, informed primarily by history and cultural studies. The central mission of the Institute is to conduct research into

B. Pramann (✉) • J. Drechsler
Georg Eckert Institute for International Textbook Research - Member of the Leibniz Association,
Braunschweig, Germany
e-mail: pramann@gei.de; drechsler@gei.de

E. Chen
Georg Eckert Institute for International Textbook Research - Member of the Leibniz Association,
Braunschweig, Germany

Max-Planck-Institut für Wissenschaftsgeschichte - Max Planck Institute for the History
of Science, Berlin, Germany
e-mail: echen@mpiwg-berlin.mpg.de

R. Strötgen
Georg Eckert Institute for International Textbook Research, Braunschweig, Germany

Stiftung Wissenschaft und Politik - German Institute for International and Security Affairs,
Berlin, Germany
e-mail: robert.stroetgen@swp-berlin.org

© Springer International Publishing Switzerland 2016
A.F.X. Wilhelm, H.A. Kestler (eds.), *Analysis of Large and Complex Data*, Studies
in Classification, Data Analysis, and Knowledge Organization,
DOI 10.1007/978-3-319-25226-1_53

historical, political and geographically significant depictions in textbooks and other educational media used for schools in order to further the understanding of the past, the present and the future. The GEI equips and supports its global partners through knowledge transfer resources and its research infrastructure provides an impetus to research into textbooks. At the heart of the Institute is our research library, around which much of the digital research infrastructure is built. It contains the most comprehensive collection in the world of textbooks and curricula for history, geography, social studies/politics and religion/philosophy/ethics from 160 countries, in addition to primers and early readers. The library inventory currently lists close to 253,000 items; of which the textbook collection comprises the majority with 178,000 volumes, the remaining 75,000 volumes being academic literature. In recent years the library has developed into an acclaimed, modern research facility, with researchers and academics from around the world attracted by its accessibility and high quality provision of services. The next chapter describes the challenges concerning the indexing of textbooks—for the librarian and the researcher.

2 The Challenges of Subject Indexing Textbooks and Curricula

The condensed and canonical character of textbooks gives them central significance in academic, political and educational respects. Such a specialised collection presents its own particular challenges, not only in terms of international search and acquisition but also with regards to cataloguing and indexing the material. Without this extremely laborious process, researchers would not be able to find a specific textbook among our vast collection.

Why is the indexing of educational media according to content so vital? With textbooks conventional search parameters are of little use: textbooks and curricula frequently share the same titles, and those titles are not always revealing. Textbooks, in particular, are typically published in series, and the precise age group for which a textbook has been designed is often not immediately apparent. Using the publisher or author as a search criterion is not a valid alternative as these are commonly not known to the user. Universal numbers, such as ISBN numbers, are only rarely assigned to materials such as curricula (or known to the researcher) so that they cannot be used as search parameters either. An additional complicating factor is the absence of any similarly comprehensive or international collection of textbooks and curricula in Germany, meaning that the cataloguing regulations of the German National Library do not contain detailed subject cataloguing instructions for such material. Here, textbooks are in the same category as, for example, calendars of events or annual reports of activities. They are only indexed very generally by subject group. Yet, we prefer a comprehensive approach as we believe that the broad scope of our textbook collection makes detailed subject indexing of this material essential, especially with regards to the exacting and differentiated theories to be investigated with its help.

By virtue of its extensive collection and detailed indexing of educational media for schools the GEI's research library has acquired a unique status. Indexing textbooks and curricula according to country, school subject and target age group enables the material to be collated to an internationally comparable standard according to fundamental and substantive criteria. As very few institutions submit this material to such a detailed subject indexing process, there is currently no suitable classification scheme available that encompasses the standardised entry of the criteria mentioned above such as country, school subject and education level. As a result, the GEI currently uses a classification system designed in-house, which can only assign records at a local level. The desire for greater standardisation and improved metadata compatibility, as well as the need for a sustainable and expandable system of textbook classification, has led to the search for similar, compatible standards for country/region, educational level and school subjects. The overall objective is to continue to improve usability in terms of searching in and working with the collection and to achieve a greater degree of internationalisation.

3 Reviewing Classification Systems

How did we look for a new classification system? What were our criteria? The new classification system should be internationally used and widely disseminated; it should be extendable and compatible with other systems; and last but not least it should be user-friendly and be available in different languages. Concerning the content, we were looking for compatible standards for country/region, educational level and school subjects. After formulating these requirements and determining the criteria for a new system of classification, existing classification systems were reviewed and assessed for their suitability to detailed textbook indexing. In addition to well-established standard systems such as Dewey decimal classification (DDC) or systems widely used in Germany, such as Regensburger Verbundklassifikation (RVK) and Basisklassifikation (BK) (Balakrishnan 2013), specialised educational classification systems and thesauri, like ISCED fields of education and training or UNESCO IBE education thesaurus, were consulted. A review of the different models quickly made apparent that no existing classification system would be applicable for the GEI; there were none that fulfilled all prerequisites equally well or allowed the degree of detail required for the criteria of country, school subject and education level. Educational studies classifications and thesauri are well suited to mapping academic subjects, but do not contain the desired geographical specification, which is necessary to identify the country or region in which a textbook or curricula is approved or applicable. On the other hand, the well-established standard systems were very complex, yet not suited to index textbooks in the needed detail.

These observations prompted the decision to subdivide the subject classification into different categories. This allowed those components from other classification systems that best served the index specifications to be combined, whilst also fulfilling the requirements of a standardised, internationally distributed and user-friendly classification system.

Table 1 Mapping of possible thesauri

Local classification	ISCED (UNESCO 2013; UNESCO Institute for Statistics 2012)	UNESCO-IBE (UNESCO 2007)
u030 Geography	0521 Environmental sciences	680 Social studies: Geography instruction
u050 History	0222 History and archaeology	680 Social studies: History instruction
u070 Social/political studies	0312 Political sciences and civics	680 Social studies: political education/social studies
u091 Values education	0223 Philosophy and ethics	682 Values education: moral education

The decision was made to code the geographical units according to a standard based on ISO 3166-2 (Scheven 2013) as defined by the International Organisation for Standardisation (ISO). This ISO standard allows further subcategories of countries, which had not been possible with the standards applied thus far or the internal notation. Curricula, in particular, need to be classified according to the area for which they are valid (in Germany, for example, each of the 16 Länder has a different curricula), but many textbooks also require detailed classification of geographical validity. In federal countries the validity of curricula and textbooks frequently only applies to regional subunits of the country. Such information must be precisely indexed in order to make it available to researchers as a search option. This ISO standard was the only system that fulfilled this specific requirement.

The International Standard Classification of Education (ISCED) (Schneider 2008a, b) proved to be a suitable tool with which to define education level. The ISCED was developed by UNESCO with the aim of classifying and defining school systems and types of schools; it endeavours to enable comparisons between international qualifications, making it particularly suitable for comparing the international educational levels of textbooks and curricula. Corresponding classifications from the mappings produced by UNESCO, to which countries refer when compiling the Report of National Education Data, can be of assistance for subject indexing and provide a suitable framework for the classification of education systems in the various countries.

Defining terms for school subjects is a more difficult process because of the differences in subject classification and assignation of curricular material between countries/federal states. Table 1 uses two educational studies thesauri to demonstrate the complexities of this issue.

The process of deciding on terms for school subjects is still in progress. In order to find a solution that meets all requirements and is internationally acceptable, European partners—at the beginning EDISCO (Italy), Emmanuelle (France) and MANES (Spain)—are invited to contribute to the process. These partner institutions also collect textbooks, albeit restricted to specific countries, and therefore have similar requirements and encounter corresponding challenges in the classification of the school subject to which a textbook relates.

4 Examples of Implementation and Application

The decision to use different classifications allows a subject indexing of textbooks and curricula which is both custom-made, and yet still standardised. Equally, classifying media in this way enables its straightforward and user-friendly integration into diverse research contexts. Standard classification allows, therefore, for a structured search, according to applicable country, school subject and educational level, in the "Curricula Workstation" (Georg Eckert Institute for International Textbook Research 2015a), a database of curricula information. The Curricula Workstation was developed by the Georg Eckert Institute as part of a project sponsored by the German Research Foundation (DFG). It makes a very important contribution to textbook and educational research infrastructure by making a valuable, yet largely inaccessible, research source such as curricula permanently accessible and searchable.

The fundaments of the classification system also provide for structured search options in the GEI-DZS database (Georg Eckert Institute for International Textbook Research 2015b), a directory of the textbooks approved in each of the German federal states, which simplifies the search for textbook approval data.

However, this model of subject indexing is of particular importance in the presentation of media in the GEI's discovery system (Drechsler and Strötgen 2014). The TextbookCat research tool (Georg Eckert Institute for International Textbook Research 2015c), which has been developed through the DFG-funded project "The Promotion of Outstanding Research Libraries", is intended to supplement the library's OPAC and to considerably improve the search options available for the library's textbook collection. Unlike the OPAC, it allows users to browse holdings. It is based on the open source discovery software VuFind and enables over 178,000 titles from the library's unique textbook collection to be searched through, according to variable criteria such as country, federal state, education level and subject. The research tool's search interface is available in English as well as German, in order to accommodate the library's international target group. TextbookCat was developed to meet the specific needs of researchers in this field and is carefully tailored to textbook searches.

Our intent is to amalgamate our current holdings of data and subsequently enable researchers to conduct detailed international searches for textbooks. At present our solutions are primarily based on proprietary solutions and are not yet available for external application. We plan to enable the integration of other databases through the development and adoption of international standards.

5 Outlook: Standardisation as the Route to Becoming a "Global Textbook Resource Centre"

The application of standardised and internationally recognised classification systems results not only in our own infrastructure (Strötgen 2014) better matching the research requirements of our users but also facilitates the Institute's plan to integrate textbook metadata from other institutions into a comprehensive, international catalogue with the help of the "TextbookCat" which is based on the discovery system software VuFind. The standardised indexing and classification of textbooks is a preliminary step in the long-term goal of consolidating metadata, and enables the necessary commutability of data. These plans and preparations are rooted in the Georg Eckert Institute's key objective of creating a Global Textbook Resource Centre (GLOTREC); a vision that would provide an international research centre for textbook resources and support researchers around the world. The project in its finished form will comprise three parts; the first to include the construction and operation of a textbook centre. The aim of this will be the standardisation and the virtual amalgamation of existing textbook collections, as mentioned above. This will represent a major advance for this field of research as in many libraries textbooks are not recorded separately and the generalised nature of their titles renders them laborious to locate. The Georg Eckert Institute also envisions a methodical examination of the collated textbook collections of its varied partners in order to detect any gaps and will supervise targeted acquisition, particularly of non-European textbooks, to rectify them. A workshop was held in autumn of 2014 with representatives from relevant textbook collections (EDISCO/Italy, Emmanuelle/France and MANES/Spain) who discussed the likely design and organisation of this priority field.

The "Global Textbook Resource Centre" will not be confined to a central catalogue of textbooks; its second key area will be a "Centre for Digitisation and Research" through which the availability of full text versions will be intensively developed in response to research needs. The Institute will contribute its own digitised full text versions and continue with its digitisation programme, but as a centre of competence for textbook digitisation it will also support the process in other countries and use its own programme to drive the digitisation of particularly important textbook examples in other countries and collections.

GLOTREC's infrastructure project will be the third key area and will further underpin the Georg Eckert Institute's role as a central hub and forum for textbook research. Guest academics will be invited to Braunschweig to work with the Institute collection, but also with its researchers, and to give regular feedback on the user-friendliness of the research infrastructure. Findings of research conducted by, and with, guest academics will be discussed at conferences and workshops. Commensurate with contemporary academic desideratum, these plans will fortify and further improve the position of the Georg Eckert Institute as a significant international social research facility.

6 Summary

The paper tried to formulate the various steps taken on the road to a standardised classification system for textbooks. At the beginning, looking for an integral solution, we thoroughly examined the more commonly used classification systems.

Yet, we soon realised that a combination of several standardised systems seemed more future-oriented. This combined classification system is already effectively in use in some applications at the GEI. One example is the GEI's "TextbookCat" which is based on the open source discovery software VuFind and enables the GEI's textbook collection to be easier searched, considering the specific needs of researchers in this field. The next years will see further extensions of the classification systems and the reduction of deficits especially in the field of subject classification, a further internationalisation and an extended exchange with partners from all over the world. The major goal is to build an international textbook catalogue with integrated data provided by institutes from other countries that also collect textbooks.

References

Balakrishnan, U. (2013). Das VZG-Projekt "coli-conc"-Brückenbildung zwischen DDC und RVK. Paper presented at the RVK-projekt workshop, Göttingen. Accessed January 9, 2015, http://www.gbv.de/Verbundzentrale/Publikationen/publikationen-der-vzg-2013/pdf/Balakrishnan_131120_RVK_WS_Konkordanz.pdf

Drechsler, J., & Strötgen, R. (2014). TextbookCat - Das neue Rechercheinstrument für die Schulbuchsammlung des Georg-Eckert-Instituts. Paper presented at the 3. Deutsches VuFind-Anwendertreffen, Frankfurt am Main. Accessed January 09, 2015, http://www.hebis.de/de/1ueber_uns/projekte/portal2/vufind-treffen-2014_praes/Strtgen_TextbookCat.pdf

Georg Eckert Institute for International Textbook Research: Curricula Workstation. (2015a). Accessed January 09, 2015, http://www.curricula-workstation.edumeres.net/

Georg Eckert Institute for International Textbook Research: GEI-DZS: Datenbank zugelassener Schulbücher. (2015b). Accessed January 09, 2015, http://www.gei-dzs.edumeres.net/

Georg Eckert Institute for International Textbook Research: TextbookCat. (2015c). Accessed January 09, 2015, http://www.tbcat.edumeres.net/

Scheven, E. (2013). Der Ländercode nach ISO 3166 und seine Nutzung. Paper presented at the Bibliothekskongress, Leipzig. Accessed January 09, 2015, http://www.opus-bayern.de/bib-info/volltexte/2013/1513/pdf/B_Kongress_LC.pdf

Schneider, S. L. (Ed.) (2008a). *The International standard classification of education (ISCED 97) an evaluation of content and criterion validity for 15 European Countries.* Mannheim: MZES.

Schneider, S. (2008b). Anwendung der Internationalen Standardklassifikation im Bildungswesen (ISCED 97) auf deutsche Bildungsabschlüsse. Paper presented at the expert-workshop at the Forschungsdatenzentrum Bundesinstitut für Berufsbildung (BIBB-FDZ). Accessed January 09, 2015, http://www.bibb.de/dokumente/pdf/Slides.Silke_.Schneider.pdf

Strötgen, R. (2014). New information infrastructures for textbook research at the Georg Eckert Institute. *History of Education and Children's Literature, IX*(1), 149–162.

UNESCO. (2013). Revision of the international standard classification of education: Fields of education and training (ISCED-F). Accessed January 09, 2015, http://www.uis.unesco.org/Education/Documents/isced-37c-fos-review-222729e.pdf

UNESCO. (2007). UNESCO-IBE education thesaurus. Accessed January 09, 2015, http://www.ibe.unesco.org/en/services/online-materials/unesco-ibe-education-thesaurus.html

UNESCO Institute for Statistics. (2012). International standard classification of education: ISCED 2011. Accessed January 09, 2015, http://www.uis.unesco.org/Education/Documents/isced-2011-en.pdf

The Ofness and Aboutness of Survey Data: Improved Indexing of Social Science Questionnaires

Tanja Friedrich and Pascal Siegers

Abstract In this paper we adopt a user-centered indexing perspective to propose a concept for the subject indexing of social science survey data. Operationalization processes in survey development mean that the constructs being studied are mostly hidden in the verbalization of the questionnaire (latent subject content). Indexable concepts are therefore found at two different semantic levels that, inspired by research on the indexing of pictures (Shatford, Cat Classif Q 6(3):39–62, 1986), we treat as the *ofness* and *aboutness* of survey data. We apply a syntax of term linking and role indicators, combining directive terms (e.g., attitude, experience, perception) with subject terms (e.g., corruption, foreigners). Each directive and subject term combination represents a retrievable unit of interest to the secondary researcher.

1 Introduction

While data documentation is a core responsibility of specialized institutions such as data centers and data archives, principles of subject indexing have largely been developed in the library and information science (LIS) community, and thus are highly adapted to textual material. As research data are mostly numerical it is all but impossible to index this kind of information according to established subject indexing principles. There is a common understanding that research data (or datasets) represent a type of information which is distinct from books or journals (Gold 2007) and that their nature is highly discipline-specific (Borgman 2012). Therefore, any attempt to develop documentation standards for this kind of information should consider the characteristics of research data on a discipline basis.

We investigate specifics of social science survey data drawing on Shatford's theory for the indexing of pictures, we propose a concept of syntactic indexing for this type of data. To familiarize the reader with this information type and to identify aspects that are relevant to its indexing, the following section deals with

T. Friedrich (✉) • P. Siegers
GESIS, Cologne, Germany
e-mail: tanja.friedrich@gesis.org; pascal.siegers@gesis.org

© Springer International Publishing Switzerland 2016   629
A.F.X. Wilhelm, H.A. Kestler (eds.), *Analysis of Large and Complex Data*, Studies
in Classification, Data Analysis, and Knowledge Organization,
DOI 10.1007/978-3-319-25226-1_54

the characteristics and usage of survey data. In Sect. 3 we outline Shatford's theory, adapt it to the indexing of survey data, and present our concept of syntactic indexing on these grounds. The section closes with a few remarks on possible retrieval scenarios. Finally, we draw some conclusions and make suggestions for future work.

2 Survey Data: Its Characteristics and Its Usage

Crucial to user-centered indexing (Fidel 1999) is that indexers need to know and investigate users' information needs and seeking practices. We assume that a user orientation is all the more important when it comes to the indexing of non-text material or, in our case, research data. A mere document-oriented approach to indexing[1] would, in contrast, pose several problems, not least the fact that most research data are numerical data. Consequently, our point of departure is the question: How do users look for data?

Social science survey data[2] are distributed for purposes of secondary use by data archives. In these archives, field experts (social scientists) as well as experts in data documentation give support to secondary researchers who are looking for data that are suited to their research interests. From their experience in advising researchers, these experts know the cornerstones of survey data seeking. With regard to indexing, this knowledge gives valuable insights that should help us identify indexing rules that promise to provide optimal retrieval results.

Simply put, survey research is about measuring the existence of or the relationship between social phenomena. As a distinct type of information, survey data have several characteristics that have a direct influence on how these particular data should be treated in documentation. These characteristics are best understood by envisioning the process of survey research. Basically, this research process consists of the formulation of a theoretical problem and hypothesis, data collection, data analysis, and support or rejection of the initial hypothesis or theory.[3] For our purposes, the first half of this process is the most interesting one, that is, the theoretical evolvement of a research problem and the development of an instrument for data collection. This phase of research design bears the content inherent in the data, because the topics of surveys are not contained in the datasets, but in the surrounding materials, like study descriptions, primary publications, and—most importantly—the questionnaire. The questionnaire contains all the measurements

---

[1]A comparison and detailed description of the two opposed indexing approaches of "user orientation" and "document orientation" is given by Fidel (1999).

[2]Survey data are only one of several forms of data that are used in empirical social research. Other social science data are, for example, data from observation, from depth interviews, focus groups, or experiments (cf. Blair et al. 2014).

[3]This description of the research process follows Bernard (2013). There are more detailed accounts of the process available, for instance in Bryman (2012).

that have been used in the study. Thus, if we assume that conceptual analysis in subject indexing determines "what a document is about—that is, what it covers" (Lancaster 1998), the key to the aboutness of survey data is the questionnaire. Taking the secondary data user's perspective, we know from archival practice that even though survey data are archived and retrieved study-wise, users are very rarely looking for entire surveys, but rather for particular measurements of constructs that have been employed in a study (i.e., one or a few questions that have been asked in a survey). For successful retrieval and thus subject indexing this means that the appropriate indexing level is not given by the complete study itself but by the individual measurements of the studied constructs. Since these are reflected in the questionnaire, the latter should be the primary information source for the indexing.

What a study is about or what it covers can be traced back to the social phenomena that the primary investigator intended to study. But often they will not appear literally in the questionnaire, because only simple social phenomena can easily be measured by directly asking respondents about them (examples are *age*, *gender*, or *household size*). Most other social phenomena (or *constructs*, cf. Bernard 2013) are more complex and have to be broken down into simpler variables in order to be measurable (Bernard 2013). This process, referred to as *operationalization*, translates the constructs into a measurement. For example, a researcher would be ill-advised to collect data on xenophobia by asking respondents "How xenophobic would you say you are?" because either the respondents would not understand the term or, for reasons of social desirability, might refrain from giving an accurate self-description (Bernard 2013). This example shows that a construct which is as complex as *xenophobia* needs to be operationalized in order to become measurable. For instance, one suitable question within a measurement of xenophobia might be "Do you think that there is more crime due to foreigners living in our country?" Survey questions of this kind have a clear content that is easily accessible through the question wording (i.e., *crime, foreigner*), but they also refer to an additional level of meaning that is the construct (i.e., *xenophobia*).

How does all this affect the data seeking practices of secondary researchers? Researchers who do not collect their own data still develop their research questions and hypotheses from a theoretical problem. They then look for individual data that refer to specific variables of interest for their own research. Therefore, we can assume that secondary data users search using terms that depict complex constructs (like *xenophobia*) as well as terms that depict measurements (like *crime* and *foreigners*).[4] These two cases resemble the process of operationalization in survey design. We view them as two subject layers that relate to the same data. For indexing purposes this means that, while the questionnaire is the primary information source, it is crucial to realize that the questions usually do not literally contain the studied constructs but rather verbal representations or paraphrases. The challenge for subject indexing is to capture both subject levels in indexing, the construct level and the measurement level. A theoretical foundation for the problem of indexing different

---

[4]Exemplary tests that we ran with log files from the GESIS data catalogue support this assumption.

subject levels can be found in the work of Sara Shatford Layne on the indexing of pictures.

3   Indexing Survey Data

3.1   Ofness and Aboutness According to Sara Shatford Layne

In order to capture the two subject levels of survey data analytically, we draw on an indexing theory that has been developed for indexing pictures (Shatford 1986; Shatford Layne 1994). The underlying motivation of Shatford's theory was that "[t]he subjects of pictures have essential qualities that make them different from the subjects of textual works, and it should be possible to improve access to those subjects if their qualities are clearly understood" (Shatford 1986). As we elaborated earlier, this is also true for survey data and, as our analysis shows, Shatford's ideas can to some extent be applied to our indexing problem.

Building on earlier work by art historian Erwin Panofsky, Shatford developed a theory of indexing pictures, considering different aspects and attributes that she identified to be particularly important. She named four classes of attributes that needed to be considered in picture indexing: biographical, subject, exemplified, and relationship attributes (Shatford Layne 1994). Not all of them are applicable in our context, but the *subject attributes* are. Shatford also calls them *aspects* of a picture and defines three of these aspects as of interest in indexing: *ofness* and *aboutness* of a picture, *generic* and *specific identity* of a picture, and four *subject facets* of a picture (time, space, activities, events). The aspects that are of interest in the case of survey data indexing are *ofness* and *aboutness*. According to Shatford, it is possible to distinguish analytically (at least) two subject levels or meanings of pictures. The first level refers to the concrete and objective subject or *factual meaning* of a picture, that is to say, the objects that are depicted. Shatford calls this subject level the *ofness* of the picture. She then uses the familiar term *aboutness* to refer to another level which is abstract and subjective and which can also be called the *expressional meaning* of a picture. Shatford uses examples to illustrate her theory: "[...] an allegorical image might be *of* a man and a lion, but be *about* pride [...]; or an image *of* a person crying might be *about* sorrow" (Shatford Layne 1994). In other words, ofness refers to what is visibly depicted in the picture, while aboutness refers to an intended meaning that is not visible in the picture, but identifiable on the grounds of world knowledge. We found an impressive similarity to the subject levels of survey data in this theory and adapted it for our purposes.

3.2   The Ofness and Aboutness of Survey Data

Subject indexing of survey data can benefit from Shatford's theory because here too the indexer is dealing with two subject levels. The first level of meaning in survey data refers to the literal question wording, i.e., the manifest content of the questionnaire. This corresponds to the *ofness* of a picture. The second level of meaning is the construct that the primary investigator wishes to measure. As described above, in many cases the construct is not directly accessible from the question wording—it remains latent. This corresponds to the *aboutness* of a picture. The major challenge for subject indexing of survey data is to capture the latent content of the questions, i.e., aboutness.

For these reasons, we suggest an approach to indexing aboutness of survey data that is based on the questionnaire but which approximates the latent content of the data. We mentioned above that operationalization is the process that translates the construct into a *measurable unit*. A closer look at this measurable unit reveals that it has two components that have been deduced from the construct: the *topic* and the *attribute* of the measurement. To illustrate this point somewhat further: It is not possible to "measure" a table. The attributes of a table that are of interest (i.e., the height, width, weight) have to be specified (Schnell et al. 2011). Conversely, it is not possible to measure the height, etc. if the object which the measurement addresses (e.g., tables or chairs, etc.) has not been specified. Thus, a useful operationalization defines a measurable unit corresponding to a theoretical construct by providing information on the *topic* of the measurement (e.g., foreigners) as well as on the *attributes* of the measurement (e.g., attitude, behavior, feeling, etc.). The measurable unit as a combination of topics and attributes represents the subject content of the construct. We argue that indexing the topics and the attributes of the measurement with combinations of thesaurus terms[5] enables a good approximation of the aboutness of survey data (e.g., ATTITUDE, FOREIGNERS) whereas indexing only the topic of the content (e.g., FOREIGNERS, CRIME) would merely capture the ofness of the data.

Identifying the aboutness of a measurement often requires more than simply an evaluation of the text of one particular question. For instance, the European Values Study (EVS 2011) asks respondents whether they agree with the statement "Both the husband and wife should contribute to household income" (see v164, Table 1).

The topic represented in the question wording, i.e., the ofness of the question, might be indexed using the thesaurus terms HOUSEHOLD INCOME and MARRIED COUPLE. This would be the literal manifestation of the operationalization of a construct, as it is found in the questionnaire. The disadvantage of indexing the ofness of this question is that the selected terms are ambiguous in that they suggest an indicator of household income for married couples more than an opinion

---

[5]In order to provide satisfactory recall, we use a social science thesaurus that allows us to control for synonyms and term relationships.

Table 1 Question Q48 from EVS (2011)

		Agree strongly	Agree	Disagree	Disagree strongly	Don't know	No answer
v160	A pre-school child is likely to suffer if his or her mother works	1	2	3	4	8	9
v161	A job is alright but what most women really want is a home and children	1	2	3	4	8	9
v162	Being a housewife is just as fulfilling as working for pay	1	2	3	4	8	9
v163	Having a job is the best way for a woman to be an independent person	1	2	3	4	8	9
v164	Both husband and wife should contribute to household income	1	2	3	4	8	9

Q48 People talk about the changing roles of men and women today. For each of the following statements I read out, can you tell me how much you agree with each. Please use the responses on this card

on who should contribute to the income of the household. But the question format and the context contain more information to help uncover the latent content of the question. First of all, the fact that for this question answers are to be given on a four point scale ranging from 1 = I strongly agree to 4 = I strongly disagree shows that the question is not about a behavior (i.e., earning money), but about an evaluation. Secondly, the statement is one of several statements about the roles of men and women (see Table 1). This shows that, in fact, the construct *attitudes towards gender roles* is measured. This aboutness would best be indexed using the thesaurus terms ATTITUDE and GENDER ROLE. The first term, ATTITUDE, refers to the attributes of the measurement whereas the second, GENDER ROLE, stands for the topic. To index the more specific ofness of the first statement, one could additionally assign the term HOUSEHOLD INCOME.

As the example shows, aboutness can be revealed by systematically assessing topics and attributes of the measurement. Identifying aboutness requires careful examination of the context of the question; it also requires field knowledge, and data literacy (Gray 2004). These requirements correspond to the knowledge needed to index the aboutness of a picture according to Shatford (1986).

3.3 How to Index Ofness and Aboutness

Indexing surveys at the level of the measurement unit, as is proposed here, is very exhaustive. That is to say, it results in a large number of assigned terms per study, which leads to high recall in retrieval. We would also expect low precision as our in-depth indexing should result in multiple assignment of terms. To avoid this problem, we propose an indexing syntax (cf. Lancaster 1998) that allows for assignment of terms to single measurement units within one survey (*term linking*, Lancaster 1998) and that combines terms according to their different roles (*role indicators*, Lancaster 1998). To define the role indicators we revert to the already mentioned constitutive components of a measurement unit: the topic and its attribute. First, we use the idea of ofness to assess the topic of the measurement. Indexing terms that represent the topics are derived from the question wordings. We call these terms *subject terms*. A subject term can be any subject area that is relevant for the social sciences. In line with the general indexing rule of specificity (Lancaster 1998), we choose the most specific term for each measured subject. If necessary, we allow combinations of subject terms. Secondly, we introduce *directive terms* to index the attributes of the measurements. Terms in this role can be assigned to one of four broad classes of attributes: (1) cognition, (2) evaluation, (3) emotion, and (4) action.

These classes are based on similarities of constructs used in social sciences. Theories of action explain (1) individual or collective behavior (i.e., actions) as a function of (2) individuals' perception of the social world (i.e., cognitions) and (3) positive or negative values that individuals assign to specific objects or behaviors (i.e., evaluations). Examples for this structure of explanations are rational choice theory (Coleman 1986), the theory of planned behavior (Ajzen 2012), and theories of values (for instance Schwartz 1994). Particularly in (social) psychology, (4) emotions also have a central role in the study of individual behavior (e.g., Festinger's 1957 theory of cognitive dissonance). The classes have a primarily heuristical value because they link directive terms if they are related to the same class of attributes (e.g., evaluations) and the same topic (e.g., immigration). For instance, ATTITUDES and PREJUDICE describe different forms of evaluation. Users searching for questions on attitudes towards immigrants might also be interested in questions on prejudice towards immigrants (see Sect. 3.4).

It is important to note that the four classes are neither exhaustive nor exclusive. In some cases assigning directive terms to the four classes might be ambiguous depending on the specific discipline. The terms in Table 2 are preliminary suggestions for directive terms with possible assignments to our four broad attribute classes. In the first step the selection of directive terms and the assignment to attribute classes will be based on theoretical reasoning. However, these decisions have to be preliminary, because they have to be evaluated in the course of the evaluation of the indexing. Evaluation will, however not lead to a firm and finalized

Table 2 Examples for directive terms from four attribute classes

Cognition	Evaluation	Emotion	Action
PERCEPTION	ATTITUDE	MOOD	BEHAVIOR
KNOWLEDGE	PREFERENCE	FEAR	USE/UTILIZATION
AWARENESS	JUDGMENT	ANGER	CHOICE
INTEREST	PREJUDICE	HAPPINESS	INTERACTION
BELIEF	SATISFACTION	HATE	COMMUNICATION

list of keywords, but only to a fair starting point, because "[o]nce a vocabulary has grown to a leveling-off point we can expect continued gradual growth, to increase specificity and to accommodate new topics [...] new terms will be derived from the indexing and searching operations" (Lancaster 1972, 103). This means that throughout our indexing work with the thesaurus, we will continue evaluating and updating our directive term choices and the assignment to the classes of attributes.

We argue that we reach the level of aboutness in subject indexing of survey data by linking directive with subject terms. The directive terms that we use are chosen from an initially limited but extendable pool of terms that are assigned to one of the four broad attribute classes (see Table 2) in order to facilitate faceted retrieval (see Sect. 3.4).

The following examples show how this syntactic indexing works. The Eurobarometer 76.1 survey includes questions about corruption in Europe. The first sample statement is "There is corruption in the in the national public institutions in Germany." (European Commission 2014). The ofness of the item would be indexed using the subject terms CORRUPTION and PUBLIC INSTITUTIONS. In addition, the statement refers to the subjective perception of social reality which is a special form of cognition. We would assign the directive term PERCEPTION here. In this way we index the aboutness of the question by assigning three thesaurus terms that have two different roles. The second sample statement is "Are you personally affected by corruption in your daily activities?" (European Commission 2014). The subject terms are again derived from the question wording: CORRUPTION and EVERYDAY LIFE. In this case, the question is about personal involvement in corruption. This is part of the broader class of action (because corruption involves two interacting parties). The suitable thesaurus term would be EXPERIENCE. This example shows that by adding a directive term to subject terms we can reveal the aboutness level of survey data in subject indexing (see Fig. 1).

In practice there are some special cases that merit discussion. First, there might be cases where there is a one-to-one correspondence between construct, question wording, and thesaurus terms. For example, the construct *life satisfaction* is usually measured by one direct question: "All things considered, how satisfied are you with your life as a whole these days?" (EVS 2011). In this case ofness and aboutness are not distinct, and we will most likely find the precombined term LIFE SATISFACTION in the thesaurus. Second, some questions measure objective or

Fig. 1 Indexing measurable units with subject and directive terms

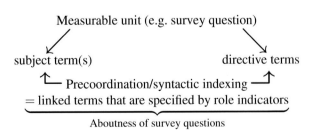

Measurable unit (e.g. survey question)

subject term(s)   directive terms

Precoordination/syntactic indexing
= linked terms that are specified by role indicators

Aboutness of survey questions

ascriptive characteristics of the respondents (for instance *age*, *gender*, and *marital status*). These objective characteristics cannot be classified with one of the four broad classes of attributes in survey research. In many cases single thesaurus terms will be available for indexing these questions (i.e., AGE, HOUSEHOLD INCOME, etc.).

To sum up: we capture the ofness of each measurable unit by assigning one or more subject terms. Linking these with directive terms reveals the aboutness of the data. Each directive and subject term combination represents a measurable unit of interest to the secondary researcher and makes the data retrievable.

3.4   Ofness and Aboutness in Retrieval

The system of syntactic indexing allows for various faceting mechanisms in retrieval. For example, refinement of search results by the two role operators (subject and directive terms) is possible. Also, a retrieval of measurements ("refine by concept/construct") seems interesting.

Syntactic indexing at the variable level and faceting mechanisms in retrieval are particularly useful for question databases, where they enable users to search for specific measurements for questionnaire design. As mentioned above, most users in data catalogues search for very general terms like youth, crime, corruption. Precision, however, is low for these queries because many specific measurements that relate to the terms. We suggest two ways of post-query faceting possibilities to the user. First, subject terms that are frequently related to the search term can be suggested to increase the precision of the results. With respect to corruption such terms might be PUBLIC INSTITUTIONS or EVERYDAY LIFE (see example in Sect. 3.3). Second, the directive terms associated to a topic term can be suggested as facets so that users can then refine the results according to whether they search for data about accepting or offering a bribe (BEHAVIOR) or how respondents perceive corruption in their country (PERCEPTION). Moreover, the classification of directive terms with the classes of attributes links similar directive terms. This allows for suggesting alternative queries. A search for attitudes towards immigrants might be extended to the directive term PREJUDICE if both are also frequently linked to the topic term IMMIGRANT. Thus, the classification of directive terms improves retrieval because similarities in the measurement are revealed.

4   Conclusion

Different subject levels need to be considered when indexing survey data. These different levels are suitably captured by the theory of ofness and aboutness. The use of a syntax with term linking and role indicators allows for systematic indexing at the aboutness level and enables us to make measurable units searchable for secondary research. Syntactic indexing of survey data permits sophisticated faceted searching and can be a promising component of an efficient, user-oriented question database. The indexing concept outlined here should work for surveys of individuals as well as for surveys of organizations. It could also be adjusted to data collected by lab experiments in social psychology or economics, or to data from observational studies.

References

Ajzen, I. (2012). The theory of planned behavior. In P. A. M. Lange, A. W. Kruglanski, & E. T. Higgins (Eds.), *Handbook of theories of social psychology* (pp. 438–459). London: Sage.

Bernard, H. R. (2013). *Social research methods. Qualitative and quantitative approaches* (2nd ed.). Thousand Oaks, CA: Sage.

Blair, J., Czaja, R. F., & Blair, E. A. (2014). *Designing surveys. A guide to decisions and procedures* (3rd ed.). Thousand Oaks, CA: Sage.

Borgman, C. L. (2012). The conundrum of sharing research data. *Journal of the American Society of Information and Technology, 63*(6), 1059–1078.

Bryman, A. (2012). *Social research methods* (4th ed.). Oxford: Oxford University Press.

Coleman, J. S. (1986). Social theory, social research, and a theory of action. *American Journal of Sociology, 91*(6), 1309–1335.

European Commission (2014). *Eurobarometer 76.1 (2011)*. TNS Opinion & Social, Brussels (producer). GESIS Data Archive, Cologne. ZA5565. Data file Version 4.0.0, doi:10.4232/1.11847.

EVS (2011). *European Values Study 2008: Integrated Dataset (EVS 2008)*. GESIS Data Archive, Cologne. ZA4800. Data file Version 3.0.0. doi:10.4232/1.11004.

Festinger, L. (1957). *A theory of cognitive dissonance*. Stanford: Stanford University Press.

Fidel, R. (1999). User-centered indexing. *Journal of the American Society for Information Science, 45*(8), 572–576.

Gold, A. (2007). Cyberinfrastructure, data, and libraries. Part 1: A cyberinfrastructure primer for librarians. *D-Lib Magazine, 13*(9/10).

Gray, A. (2004). Data and statistical literacy for librarians. *IASSIST Quarterly, 28*, 24–29. http://www.iassistdata.org/downloads/iqvol282_3gray.pdf

Lancaster, F. W. (1972). *Vocabulary control for information retrieval*. Washington: Information Resources Press.

Lancaster, F. W. (1998). *Indexing and abstracting in theory and practice* (2nd ed.). Champaign: University of Illinois.

Schnell, R., Hill, P. B., & Esser, E. (2011). *Methoden der empirischen Sozialforschung* (9th ed.). München: R. Oldenbourg Verlag.

Schwartz, S. H. (1994). Are there universal aspects in the structure and contents of human values? *Journal of Social Issues, 50*(4), 19–45.

Shatford, S. (1986). Analyzing the subject of a picture: A theoretical approach. *Cataloging and Classification Quarterly, 6*(3), 39–62.

Shatford Layne, S. (1994). Some issues in the indexing of images. *Journal of the American Society for Information Science, 45*(8), 583–585.

Subject Indexing for Author Name Disambiguation: Opportunities and Challenges

Cornelia Hedeler, Andreas Oskar Kempf, and Jan Steinberg

Abstract Author name disambiguation is becoming increasingly important due to the prevalent availability of publications in digital libraries. Various approaches for author name disambiguation are available, utilising a variety of information, e.g., author name, affiliation, title, journal and conference name or venue, citation, co-author, and topic information (Ferreira et al., SIGMOD Rec 41(2), 2012). Topics can be obtained, e.g., using subject information captured in various controlled vocabularies, classifications and mappings between them used to index publications (Torvik et al., J Am Soc Inf Sci Technol 56(2):140–158, 2005). Research interests of authors, evident in topics, might change over time though (Ferreira et al., SIGMOD Rec 41(2), 2012), and thus limit their usefulness for author name disambiguation. Here we present a longitudinal analysis of topics with respect to their suitability for author name disambiguation. We analyse the distribution of subject headings and classification notations taken from the Thesaurus (TSS) and the Classification for the Social Sciences (CSS) (http://www.gesis.org/en/services/research/thesauri-und-klassifikationen/) for research projects and literature (available in sowiport—http://sowiport.gesis.org maintained by GESIS) and the changes in distribution over time. To assess the suitability of subject information for author name disambiguation more closely, we then analyse the changes in the annotation over time for a selection of authors and author groups at different stages in their career, also taking into account the hierarchical organisation of the applied controlled vocabularies.

1 Introduction

There is an increased demand for the evaluation and assessment of the impact of research carried out by individual researchers, research communities and institutions (Mahieu et al. 2014; D'Angelo et al. 2011). The information gathered during such

C. Hedeler (✉)
School of Computer Science, The University of Manchester, Manchester, UK
e-mail: chedeler@cs.manchester.ac.uk; chedeler@gmail.com

A.O. Kempf • J. Steinberg
GESIS - Leibniz Institute for the Social Sciences, Cologne, Germany
e-mail: andreas.kempf@gesis.org; jan.steinberg@gesis.org

© Springer International Publishing Switzerland 2016   639
A.F.X. Wilhelm, H.A. Kestler (eds.), *Analysis of Large and Complex Data*, Studies
in Classification, Data Analysis, and Knowledge Organization,
DOI 10.1007/978-3-319-25226-1_55

assessments is utilised by funding bodies as well as policy makers, and plays a role for accreditations and university rankings or promotion decisions for individuals. As part of the overall assessment, bibliometric and scientometric measurements, such as the h-index (Hirsch 2005), are used as one way of measuring the impact of research. Such measures tend to rely on the number of citations of the papers written by a particular author. For these measures to perform accurately and provide a realistic measure of the impact, the citations need to be attributed to the correct person, otherwise, the impact measure of a person or institution could appear to be better or worse than it actually is. For this reason, author name disambiguation, one of the most difficult problems in the context of digital libraries (e.g., Santana et al. 2014; Ferreira et al. 2012; Torvik et al. 2005), is a relevant and core issue for information specialists which affects the quality of services and content in library and research information systems.

The remainder of the paper is structured as follows: Sect. 2 reviews existing approaches for author name disambiguation and topic evolution analysis. Section 3 introduces CSS and TSS, followed by Sect. 4 with a description of the bibliographic data set and the method we used to analyse the distribution of the terms from the classification and the thesaurus. We present the results of our analysis in Sect. 5 and conclude our paper with a discussion of future work in Sect. 6.

2   Background

2.1   Author Name Disambiguation

In addition to the assessment of the impact of research, more frequently users of digital libraries are interested in literature written by a particular author (Islamaj Dogan et al. 2009). To support research impact assessments and author-centred searches, author name ambiguity needs to be resolved, i.e., synonyms (authors that publish under different names, e.g., Ulrike and Uli Sattler, as resolved in   DBLP—http://dblp.uni-trier.de—http://dblp.uni-trier.de/pers/hd/s/Sattler:Uli. html), and polysemes [different authors with the same name, e.g., Wolfgang Schluchter: de.wikipedia.org/wiki/Wolfgang_Schluchter and de.wikipedia.org/wiki/Wolfgang_Schluchter_(Cottbus)] need to be identified to be able to attribute publications to the correct person.

Recent years have seen a number of new approaches driven by increased use of publication records and in particular of citations to assess researcher impact (Mazloumian 2012; Acuna et al. 2012). Several digital libraries are undertaking their own efforts to disambiguate authors in their data sets and manage the quality of their author and publication data using the metadata available to them (Liu et al. 2013; Reuther et al. 2006). However, as digital libraries tend to integrate information from various sources, they suffer from inconsistencies in representation of, e.g., names (e.g., initial vs. forename, different spellings, e.g., due to typos or "ue" or "u" instead of "ü"), or venue titles (abbreviation vs. full title) despite best efforts to maintain a high data quality. These differences in representation and spelling variations can have an effect on the performance of author name disambiguation approaches.

For the actual disambiguation process, a wide variety of metadata are used, including journal or conference names, affiliations of the authors, co-author networks, keywords or topics either explicitly available as metadata, or estimated from the text using, e.g., Latent Dirichlet Location (LDA) (Blei et al. 2003), abstracts, and references to other papers, including self-citations (Ferreira et al. 2012; Smalheiser and Torvik 2009; Liu et al. 2013). Examples of the use of keywords or topics along with other information for author name disambiguation include Torvik and Smalheiser (2009) and Santana et al. (2014), the former using Medical Subject Headings (MeSH) in the author name disambiguation in MEDLINE and the latter using terms extracted from the publication and venue titles. However, in some digital libraries the available metadata can be quite sparse, providing insufficient amount and detail of information to disambiguate authors efficiently. For example, in the social sciences a large number of publications have a single author (Borgman 2007) which limits the use of co-author networks. Also, the publications are more likely to be books rather than conference or journal papers, which means only publisher information is available with less information content suitable for author name disambiguation than venues, such as journals or conferences, which tend to have similar names if they cover similar research areas. There are also very few established, long running conferences, unlike in other research communities. The sparseness of the available metadata in the social sciences makes it even more important to utilise the available metadata, which includes keywords or subject topics, appropriately.

2.2 Topic Evolution Analysis

Topic models, such as LDA, are frequently used to determine the topics of publications, and are utilised as additional evidence for author name disambiguation (Ferreira et al. 2012). However, LDA does not take into account the evolution of terms representing topics over time, nor does it take into account a change in research interests of authors when applied to author name disambiguation. A number of approaches for author name disambiguation acknowledge that a change in research interest of authors makes accurate author name disambiguation harder (e.g., Liu et al. 2013) and try to account for such changes in their similarity function (e.g., Santana et al. 2014) or account for errors that could be caused by such a change in a post-processing step to improve the results of the author name disambiguation (e.g., Liu et al. 2013). In addition, efforts are being undertaken to analyse and model the evolution of topics over time (e.g., Wu et al. 2010; Blei and Lafferty 2006; Wang et al. 2008), which could provide further insights into the use of keywords or subject topics when applied to author name disambiguation.

3   Subject Indexing for Author Name Disambiguation

3.1   Classification for the Social Sciences

The effectiveness of subject indexing for author name disambiguation is tested on the GESIS literature database Social Science Literature Information System (SOLIS). Publications are annotated with terms from GESIS' proprietary domain-specific classification system Classification for the Social Sciences (CSS). CSS was developed in its current form and mapped to previous classification systems in 1995, and has been in use in practice since 1996. It consists of 159 classes organised into four different hierarchical levels. There are five major classes including 26 subclasses, such as sociology or political science. For these classes in turn, there are 102 subclasses, representing subfields of, e.g., sociology, such as sociology of religion. For two of the disciplines as represented by the subclasses there exists a further lower hierarchical level (i.e., subclasses for mass communication as part of communication sciences and economics and business economics for economics). The core classes include all those classes or disciplines that are considered to be part of the first major class, which is Social Sciences. In practice one so-called main class and one or more so-called auxiliary classes are assigned to a publication or research project by domain experts. Recently, a mapping between CSS and the Dewey Decimal Classification (DDC) has been built up manually.

3.2   Thesaurus for the Social Sciences

In addition to the classification (CSS) GESIS also maintains and uses its own discipline-specific thesaurus TSS, which is the core thesaurus for subject indexing in the German-speaking social sciences. It has been translated from German into English and French and consists of about 8000 subject headings and about 4000 non-descriptors or synonyms. The TSS has its own classification scheme, which assigns each subject heading to one or more subject categories, such as, "Fundamentals and Manifestations of Social Behavior" or "Interdisciplinary Application Areas of Social Sciences". In practice approximately between 10 and 15 subject headings are assigned to each publication. For the annotation process, a so-called geographical up-posting policy is followed for non-European regions, making geographical subject headings mandatory as soon as there is a geographical reference in the text, resulting in a potential over-representation of geographical subject headings in the assigned descriptors. Over the past few years mappings to other thesauri have been established, e.g., to the Integrated Authority File (IAF) of the German National Library and the Thesaurus of Sociological Indexing Terms.

4 Analysis of Subject Indexing

4.1 Data Set

To analyse the suitability of subject indexing for author name disambiguation, we analysed publications in the SOLIS database maintained by GESIS in terms of the distribution of classification notations of the CSS and subject headings from the TSS. SOLIS is part of the social science portal sowiport which provides an integrated search space for more than 7 million social science research publications and projects from 12 different databases. It was founded in 1978 and includes social science research literature since 1945. SOLIS captures the full name of a person, including middle name, rather than just the initials and the surname, and contains mostly German-speaking social science researchers with predominantly European rather than Asian names, which reduces the chance of two people having exactly the same full name significantly. For this reason and based on feedback from users of SOLIS, for the purpose of the analysis presented here, we perceive the data set in SOLIS to be fully disambiguated, even though no author name disambiguation has taken place in SOLIS or sowiport.

To limit spurious analysis results due to limited use of classification notations from the CSS and terms from the TSS, we limited the publications to those published in the 60-year period between 1954 and 2013. As we are interested in the change of topics of individual and groups of authors over time, we excluded authors with a single publication, leaving us with 63,683 different author names (for about 340,000 publications) for the analysis presented here. An analysis of single- and co-authorship revealed that more than 80 % of the papers are single-author papers, resulting in a limited use of co-authorship for author name disambiguation, and potentially making subject indexing a more important source of evidence for author name disambiguation.

4.2 Multilevel Analysis Approach

To answer our research question on whether topic information, i.e., classification notations or subject headings, can help to distinguish between different authors with the same name, we analysed the longitudinal distribution of annotations on multiple levels.

On a macro-level we calculated the average value of discrimination for subject headings of the TSS and classes of the CSS. It served as an approximation of the expressiveness of CSS classes and TSS descriptors.

On a meso-level we analysed the frequency of classes and subject headings with regard to three different time spans of authors' research activity.

On a micro-level we looked at individual authors from each of these three different time spans.

5 Results

5.1 Macro-Level of Study

Our macro-level analysis has shown that an author has on average 6.57 classes and subclasses from the entire CSS, and 6.02 from the core classes. Aggregation of the sub-disciplines (subclasses) shows that on average an author only has 3.98 aggregated classes, and only 3.47 when considering only the aggregated core classes.

The same analysis on the whole thesaurus (TSS) has shown that on average an author has 48.11 subject headings. The mean value of authors per subject heading is 375.46 and the frequency of subject headings ranges from one author to 29,599 authors.

The average number of authors per class of the whole CSS is 2631. This number changes when we consider only the core classes (2655), aggregated classes for the whole classification (7908) and aggregated classes of the core classes (10,421). This illustrates that the coverage of the core areas of the CSS is higher than for the more marginal areas of the classification. To gain an impression of the divergent use of classification notations of the CSS and subject headings taken from different notations of the TSS classification scheme see Fig. 1.

5.2 Meso-Level of Study

As a second step, we dealt with our research question on a meso-level. For the purpose of this analysis, we formed three different groups of authors based on the total number of years of their research activity, allowing us to analyse the topic distributions of research interests over the duration of a researcher's career. The first group of authors with a time span of research activity between 5 and 10 years consists of 16,108 authors. The second group of authors with a time span of research activity between 20 and 30 years includes 7953 authors. The third group of authors, covering between 40 and 50 years of research activity consists of 482 authors.

Our analysis shows that on average an author of the first group has 5.45 number of classes when considering all classes of the CSS, and 3.61 when considering the aggregated classes. Authors of the second group publish on average 10.98 of all classes, and 5.79 of the aggregated classes, and authors of the third group publish on average 21.50 classes and 9.31 aggregated classes, respectively.

With regard to TSS the average number of different subject headings for the first group of authors is 37.87, for the second group it is 84.48, and for the third group it is 191.92.

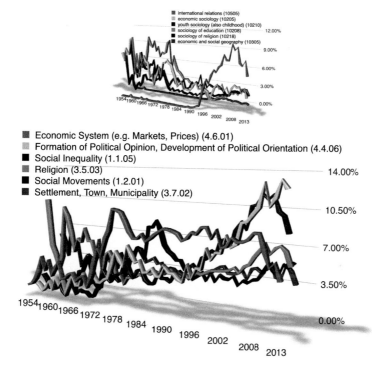

Fig. 1 Temporal distribution of a selected number of classification notations (*left*) of the CSS and notations of the TSS classification scheme (*right*)

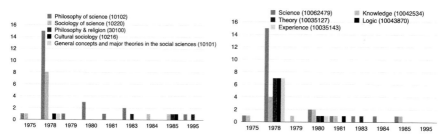

Fig. 2 Temporal distribution of a selected number of classification notations (*left*) and subject headings (*right*) for first author

5.3 Micro-Level of Study

Finally, on the micro-level we looked at the frequency of classes and subject headings for individual anonymised authors from each of these three different group-specific time spans. The first example shown in Fig. 2 published 29 publications over 10 years of research activity. Looking at the distribution of the two most common notations "Philosophy of Science" and "Sociology of Science" it

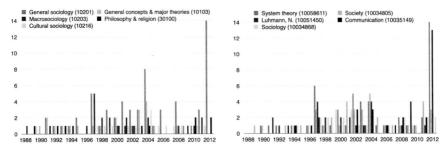

Fig. 3 Temporal distribution of a selected number of classification notations (*left*) and subject headings (*right*) for second author

can be observed that these two occur constantly throughout the entire period of activity. Even notations considered to represent research areas that tend to be less frequently studied, such as "Philosophy & Religion" and "Cultural Sociology", can be observed continuously throughout the publication activity. Looking at the distribution of the most frequently indexed subject headings the situation is much less clear. Here, however, it must be considered that the distribution of subject headings represents a much more granular description of the content of a publication than the distribution of notations of a classification system. Only the two most frequently indexed subject headings occur twice or more and can be observed throughout the whole publication period.

The second example shown in Fig. 3 has accumulated 146 publications over 24 years of publication activity. Connected to the overall significant increase in the number of publications compared to the first example, all of the five most frequently indexed classification notations continuously occur throughout the entire period of publication activity. Even the class "Cultural Sociology" appears continuously in the 1990s as well as in the 2000s. The distribution of subject headings also appears clearer than in the first example. Again linked to the overall significant increase in the number of publications even less frequent subject headings, such as "Luhmann, N." (for publications discussing the work of N. Luhmann), appear throughout the whole publication period.

The third example shown in Fig. 4 has published 241 publications in 41 years of publication activity. This large number of publications allows the creation of a fairly representative research profile based on the distribution of the most frequent classification notations as well as the most frequent subject headings in which this author publishes. On the one hand, and most likely related to the long period of publication activity it can be observed that sometimes research disciplines, such as "Education & Pedagogics", could be taken up again after a long break of inactivity of nearly 20 years. On the other hand, with regard to the distribution of subject headings, the emergence of new research interests (e.g., "Health"), that are then continuously studied over a number of years can be observed for individual authors even after a significant research activity of more than 15 years.

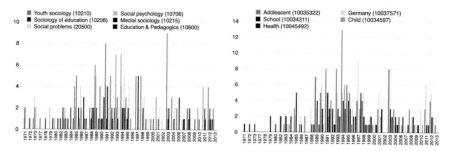

Fig. 4 Temporal distribution of a selected number of classification notations (*left*) and subject headings (*right*) for third author

6 Conclusion and Future Work

The results of the analysis presented here seem to suggest that, at least for the social sciences, subject indexing could provide useful information content for author name disambiguation. This result, however, is only apparent when analysing the distributions of classifications and subject headings at the micro-level. Analysis results at the macro-level only provide an impression of the different frequency of use of classification notations and subject headings. For this reason, speaking of an "average author" could only serve as an approximation to answer the research question. The frequency of use differs significantly between different classification notations and subject headings. Similarly, a look at the mere number of classification notations and subject headings for the different time periods of publication seems to provide little insight. Perhaps unsurprisingly, the average number of classification notations and subject headings increases continuously over the duration of a research career.

The analysis results suggest that using the most frequently occurring classification notations and subject headings for publications of an author could help create a profile for the author and aid author name disambiguation. The creation of such topic profiles can lead to the identification of topic continuity for authors. However, the analysis at the different levels of detail, in particular the analysis of authors with different durations of research activity suggests that different approaches for different author groups might be advisable. For example, for authors with a large number of publications, the subject headings might be suitable to generate an author profile, whereas for authors with a limited number of publications the classification notations representing the disciplinary assignment rather than the allocation of subject headings which represent the topic of a publication on a much more granular level might be more suitable for generating an author profile.

We aim to incorporate the generation of topic-based author profiles at appropriate level of detail to aid the author name disambiguation planned at GESIS. The author name disambiguation process is seen as a multi-step process consisting of initial author name disambiguation using existing standard algorithms that take into account publication and author-centred information, such as co-authors and year of publication. Classification and subject heading information will then be considered as part of the second step to resolve cases in doubt. Sowiport does not only contain publications annotated with CSS and TSS, but also publications annotated with entries from other controlled vocabularies. Existing cross-concordances between these controlled vocabularies and CSS as well as TSS can be utilised and enabled to use the same approach for author name disambiguation on those publications, too. As an additional step to enrich the existing data with person records, this will be followed by a third step that utilises external person-centred reference information in the form of individualised person records of the IAF that are then linked to the authors in sowiport.

References

Acuna, D. E., Allesina, S., & Kording, K. P. (2012). Future impact: Predicting scientific success. *Nature, 489*(7415), 201–202.

Blei, D. M., & Lafferty, J. D. (2006). Dynamic topic models. In *ICML '06: Proceedings of the 23rd International Conference on Machine Learning* (pp. 113–120).

Blei, D. M., Ng, A. Y., & Jordan, M. I. (2003). Latent Dirichlet allocation. *The Journal of Machine Learning Research, 3*, 993–1022.

Borgman, C. L. (2007). *Scholarship in the digital age: Information, infrastructure, and the Internet.* Cambridge, MA: MIT Press.

D'angelo, C. A., Giuffrida, C., & Abramo, G. (2011). A heuristic approach to author name disambiguation in bibliometrics databases for large-scale research assessments. *Journal of the American Society for Information Science and Technology, 62*(2), 257–269.

Ferreira, A. A., Gonçalves, M. A., & Laender, A. H. F. (2012). A brief survey of automatic methods for author name disambiguation. *SIGMOD Record, 41*(2), 15–26.

Hirsch, J. E. (2005). An index to quantify an individual's scientific research output. *Proceedings of the National Academy of Sciences of the United States of America, 102*(46), 16569–16572.

Islamaj Dogan, R., Murray, G. C., Neveol, A., & Lu, Z. (2009). Understanding PubMed(R) user search behavior through log analysis. *Database, 2009*, bap018.

Liu, W., Islamaj Doğan, R., Kim, S., Comeau, D. C., Kim, W., Yeganova, L., et al. (2013). Author name disambiguation for PubMed. *Journal of the Association for Information Science and Technology, 65*, 765–781.

Mahieu, B., Arnold, E., & Kolarz, P. (2014). Measuring scientific performance for improved policy making. Technical report.

Mazloumian, A. (2012). Predicting scholars' scientific impact. *PLoS ONE, 7*(11), e49246.

Reuther, P., Walter, B., Ley, M., Weber, A., & Klink, S. (2006). Managing the quality of person names in DBLP. In *ECDL 2006* (pp. 508–511).

Santana, A. F., Gonçalves, M. A., Laender, A. H. F., & Ferreira, A. (2014). Combining domain-specific heuristics for author name disambiguation. In *IEEE/ACM Joint Conference on Digital Libraries (JCDL)* (pp. 1–10).

Smalheiser, N. R., & Torvik, V. I. (2009). Author name disambiguation. *Annual Review of Information Science and Technology, 43*(1), 287–313.

Torvik, V. I., & Smalheiser, N. R. (2009). Author name disambiguation in MEDLINE. *ACM Transactions on Knowledge Discovery from Data, 3*(3), 1–29.

Torvik, V. I., Weeber, M., Swanson, D. R., & Smalheiser, N. R. (2005). A probabilistic similarity metric for Medline records: A model for author name disambiguation. *Journal of the American Society for Information Science and Technology, 56*(2), 140–158.

Wang, C., Blei, D., & Heckerman, D. (2008). Continuous time dynamic topic models. In *Uncertainty in Artificial Intelligence (UAI)*.

Wu, H., Wang, M., Feng, J., & Pei, Y. (2010). Research topic evolution in "Bioinformatics". In *2010 4th International Conference on Bioinformatics and Biomedical Engineering (iCBBE)* (pp. 1–4).

Index

Active learning, 91, 93, 96–101
Addis, M., 587
Adjusted control limit, 241
Adler, W., 395, 411
Ahrens, W., 385
AIC, 57, 58, 60, 61, 63, 65, 67, 201, 319
Alaoui Ismaili, O., 147
Amelynck, D., 425
Analytical CRM, 311, 312
Ant Colony Optimization, 519–521, 524, 528
Audio features, 437–439, 441, 443–445, 447
Author name disambiguation, 639–643, 647, 648

Baier, D., viii, ix, 301
Baker, D., 473
Bal-Domańska, B., 287
Bankrupts and non-bankrupts sampling, 345
Bartel, H.-G., 125
Baryła, M., 345, 369
Batch learning, 241
Bauer, N., 461
Bayesian mixture model, 507
Bayesian procedures, 301–304, 310
Belo, O., 531
Bessler, W., 323
Beta regression, 207, 209, 210, 214, 216
Biermann, D., 487
Big data, viii, 29, 43, 51, 113–115, 120, 141, 185, 187, 193
Bihn, M., ix
Biplots, 207
Bischl, B., 113, 461

Blanck, A., ix
Bone stiffness, 385, 386, 394
Bonnin, G., 437
Bootstrap, 11, 54, 57, 125, 128, 130, 131, 146, 379, 381, 396, 397, 399, 408, 411, 413, 421, 466
Brier, 395, 397–399, 405–407, 411, 413–417, 419, 420
Bukar, B.A., 507
Burkard, L.O., 611
Burkovski, A., ix

Care system, 17, 29
CART, 213, 215, 312, 313, 349–351
    logit, 311
Category-ordered data, 31, 35, 36
Causality, 8–10, 221, 223, 225–228, 599
Ccr, 575–580, 582, 584
CFA model, 553, 555
Chen, E., 621
Chernyak, E., 103
Chip-seq, 507–511, 513–517
Choice-based conjoint analysis, 301, 302, 310
Citation networks, 611, 617
Classification, 6, 7, 25, 59, 66, 67, 79–84, 87, 88, 92, 93, 96, 99–101, 113–122, 135, 137, 147, 148, 153, 156, 157, 166, 168, 173, 174, 180, 182, 198, 207–216, 241–245, 250, 254, 259, 260, 275, 276, 278–285, 288, 312, 318, 320, 335–339, 341–345, 349–352, 354, 355, 369, 374–378, 386, 397, 408, 409, 412,

421, 437–447, 457, 507, 513, 514,
532, 539, 545, 546, 584, 585, 600,
601, 607, 621, 623–627, 637–640,
642–648
models, 345
multi-label, 207
music, 437
Clustering, 31, 32, 37, 39, 40, 43–46, 49–52,
67, 86, 91, 93, 101, 125, 126,
128–131, 133, 134, 137, 139–150,
152, 153, 155–157, 166, 183, 265,
269–271, 425, 428, 429, 433, 449,
537, 538, 545–550, 601–604, 614
divisive, 43
ensemble, 137, 141, 144, 145
hierarchical, 43, 46, 49, 50, 52, 126, 128,
130, 538
mean-shift, 137, 140
problems, 545
spectral, 137, 139, 145
supervised, 147–149, 152, 153, 155, 156
Cognitive modelling, 587
Collaboration, 611, 612, 618
Concept drift, 241–243, 245, 247–250
Conjoint analysis, 301–304, 310
choice-based , 302, 310
Consumers, 9, 531, 553
Contreras, P., 43
Control charts, 241, 242, 250
Conversano, C., 207
Convex hull, 207, 208, 212–216, 490
Cornuéjols, A., 147
Corporate bankruptcy, 345, 347, 349, 351, 353,
355, 369, 370, 374, 380–382
Correlated component regression, 575, 580,
585
Crawford, T., 449, 473
Cross-validation, 80–84, 95, 182, 238, 355,
575, 577, 579
C&RT-logit, 311
Curricula, 621–625, 627

Dai, H., 507
Data
analytics, 43
category-ordered, 35, 36
multivariate functional, 173
ordinal, 54
retrieval, 629
science, viii
sharing, 629
social science, 630

symbolic, 139–141, 145
three-way, 545, 550
Decision tree, 78, 311, 312, 314, 316–319, 335,
443
Decker, R, viii, ix
Design product, 301, 303, 305, 307, 309, 310
Diatta, J., 71
Digital library, 639
Dimensionality reduction, 51, 185
D'Inverno, M., 449
Discovery system, 621, 626
Dividend policy, 335, 339
Domenach, F., 159
Dorfleitner, G., viii, ix
DPM, 425
Drechsler, J., 621
Dreyfus, L., 473
Dyadic distance, 195, 197
Dynamic weighted majority, 241

Eckardt, M., 221
E-commerce, 531, 533, 535–537, 539,
541–543
Economic environment, 369, 370, 379, 380
Efficient computation, 231
Eigenvalue, 115, 186, 431
EM algorithm, 53, 56, 65, 67
Embodiment, 425
Ensemble clustering, 137, 141, 144, 145
Ensemble methods, 141, 241, 250, 311, 395,
411
EU NUTS-2 regions, 287
European Union Regional Space, 253
Event history, 221, 225, 227, 228
Excess takeover premium, 323, 326–329

Features
audio, 437–439, 441, 443–445, 447
non-informative, 396, 411, 412, 416, 417,
419, 420
selection, 79–83, 85–89, 117–119, 122,
275, 276, 279, 282, 283, 285, 344,
408, 409, 412, 421, 438, 443, 446,
447
Fernandes, J., 531
Ferreira, M., 487
Financial planning, 357, 358, 366, 367
Finite, 8, 54, 55, 61, 67, 68, 118, 129, 143, 168,
222, 225, 228, 507–509, 511, 513,
515–517, 520
Fofonov, A., 497

Formal concept analysis, 159, 160, 168
Fraud detection, 531, 533, 535, 537, 539, 541, 543
Fraud detection and prevention, 531
Friedrich, T., 629
Friedrichs, K., 461
Frigau, L., 207
Functional data analysis, 173
Fürstberger, A., ix

GAMLSS, 236, 238, 239, 385–388, 393
Gaul, W., viii, ix
Genetic programming, 587, 590
Genre recognition, 437
Geyer-Schulz, A., viii, ix, 611
Gobet, F., 587
Górecki, T., 173
Graphical models, 221–229
Grinding process, 487, 495
Groenen, P., viii, ix
Gschrey, B., 575
Gul, A., 275, 395, 411

Harrison, A., 275, 507
Hayashi, Y., 507
Hedeler, C., 639
Hellmann, M., 265
Hennig, Ch., viii, ix, 31
Herbrandt, S., 487
Herrmann, D., 385
Hierarchical clustering, 43, 46, 49, 50, 52, 126, 128, 130, 538
High dimensional data, 44, 51, 52, 226, 575
Hildebrandt, H., 17
HMMS, 288–293, 295, 296, 516
Horn, D., 113
Household finance, 357
Households, 362, 367, 599–605, 607
Huang, C.-L., 31
Hybrid predictive models, 311
Hüllermeier, E., viii, ix

IDEFICS consortium, 385
Incremental methods, 185, 188, 194
Independence, 8–10, 115–118, 199–201, 221, 223, 225, 229, 290, 565, 566
Inductive item tree analysis, 563
Instance sampling, 461
Intemann, T., 385
Iodice D'Enza, A., 185

Isosurface similarity, 497–499, 501, 503, 505, 506
Item hierarchy mining, 563, 572
Item response theory, 3, 13, 473–475, 482, 483

Jannach, D., 437
Jostschulte, K., 265

Kempf, A.O., 639
Kestler, H.A., viii, 79, 519
Khan, Z., 275, 395, 411
K-means, 32, 45, 46, 149, 152, 153, 156, 157, 271, 428
K-nearest neighbour, 408, 411
Köppl, S., 265
Kramer, I., vii
Kraus, J., ix
Krolak-Schwerdt, S., viii, ix
Krzyśko, M., 173

Label switching, 507–509, 512, 513, 515–517
Łapczyński, M., 311
Lane, P.C.R., 587
Large scale educational assessment, 575
Latent variables, 3–5, 7–15, 56, 225, 228
Lattice, 159–165, 221, 488, 489
Lausen, B., viii, ix, 275, 395, 411, 507
Lausser, L., ix, 79
Leave-one-out, 95, 182, 231, 232, 235, 236
Leitmotives, 473–475, 477–483
Lemaire, V., 147
Leman, M. , 425
Life-length risk, 357, 359–361, 363, 365, 367
Ligges, U., 487
Limam, M., 241
Lin, C.-J., 31
Linsen, L., 497
Literature management, 611–615, 617–619
Löffler, S., 301
Logit, 7, 10, 303, 311–313, 315–320, 335–337, 339, 340, 342, 343, 351, 352, 354, 369–371, 373–382, 585
    model, 369
    multinomial, 303, 335–337, 339, 340, 342, 343
Log-likelihood, 53–63, 65, 66
Log-linear model, 195, 198, 199, 201, 204
Longevity, 357, 360–363, 367
Luebke, K., 335

Maes, P.-J., 425
Mahmoud, O., 275, 395, 411

Manufacturing firms, 369
Many features, 113, 114, 122
Many observations, 113, 114, 122
Marketing paradigms, 3, 7, 14
Markos, A., 185
Markov, 9, 39, 221, 223–226, 228, 229, 312,
        508, 509, 516
Markowska, M., 545
Martens, J.-P., 425
MDS, 195–197, 201–203
Mean-shift clustering, 140
Mejri, D., 241
Memory, 185, 186, 473–475, 477–479, 481,
        502, 503, 589
Mergers and acquisitions, 323, 328
Metodiev, M.V., 275
Microarrays, 79, 80, 84
Miftahuddin, M., 395, 411
Misclassification, 83, 119, 121, 207–210,
        242–244, 318, 336–338, 342, 513
Mixture models, 53, 68, 428, 507, 508, 516,
        517
Model based optimization, 461–467, 469–471,
        487, 495
Model selection, 53–63, 65, 67, 238, 387, 389,
        390, 392, 393, 479
Mola, F., viii, ix, 207
Mota, G., 531
Mucha, H.-J., 125
Müllensiefen, D., 473
Müller, M., viii, ix
Müssel, C., ix, 519
Multicollinearity, 575, 577–581, 584
Multi-instance multi-label, 91, 92, 96,
        101
Multinomial logit, 303, 335–337, 339, 340,
        342, 343
Multitrait-multimethod (MTMM), 553
Multivariate functional data, 173
Murtagh, F., 43
Music, 425–427, 430, 431, 434, 435, 437–439,
        441–447, 449–451, 453, 457–459,
        461, 462, 466, 467, 471–477,
        480–482, 496
Music classification, 437
Mutual, 5, 6, 9, 117–119, 142, 285, 498, 499,
        501–506

Nakayama, A., 195
Noisy, 80, 125, 126, 134, 144, 145, 276, 421,
        465, 496, 501, 563, 567
Non-informative features, 396, 411, 412, 416,
        417, 419, 420

Nonparametric, 54, 57, 88, 182, 183, 238, 409,
        421, 509, 510, 515, 516, 585

Objectification, 265, 267
Onset detection, 461–465, 471
Optical music recognition, 449
Ordinal data, 53, 54
Outliers, 34, 278, 291, 312, 454, 455, 540, 578
Overlapping, 208, 275–279, 282, 284, 285,
        399, 409, 421, 428, 569, 578–580

Pairwise likelihood, 53, 54, 56, 60, 68
Panel data, 287, 297, 599, 601, 603, 605, 607
Panel data analysis, 287
Pawełek, B., 369
P-dissimilarity, 31, 35–37, 40
Pełka, M., 137, 253
Penalized regression, 231, 232, 239
Penalty estimation, 231
Percentile curves, 385, 386, 392, 393
Perperoglou, A., 231, 395, 411
Pfeffer, M., ix
Pheromone, 520, 522, 524, 526, 528, 529
Pietrzyk, R., 357
Pigeot, I., viii, ix, 385
Pimperl, A., 17
Pitou, C., 71
Playlist features, 437
Pociecha, J., viii, ix, 345, 369
Pohlabeln, H., 385
Portides, G., 159
Pramann, B., 621
Pre-processing
        supervised, 152
        supervised and unsupervised, 147
Pricing, 301, 306, 309, 310
Principal component analysis, 173, 186, 187
Probability estimation trees, 395, 397
Product design, 301, 303, 305, 307, 309, 310
Proportional overlapping scores, 275
Psychology, 15, 78, 435, 473, 475, 545, 562,
        587, 588, 638
Publication process, 611–614

Quadtree, 71–74, 77, 78
Quadtree decomposition, 71

Ranalli, M., 53
Random forest, 84, 87, 215, 282, 335, 338,
        339, 341, 342, 344, 395–397, 399,

400, 404, 407, 408, 411, 412, 416, 419, 465–470, 472
Random projection, 43–46, 48–51
Rand's index, 125, 128
Rautert, C., 487
RE-EM tree, 599–601, 605–607
Recognition, optical music, 449
Regional science, 545
Regression
    analysis, 173, 209, 239, 256, 265, 273
    beta, 207, 209, 210, 214, 216
    penalized, 231, 232, 239
Representative random quasi-orders, 563
Research data, 629, 630
Retz, R., 91
Rhodes, C., 449, 473
Richard Wagner, 473–481
Risk aversion, 357, 358, 361–363, 366, 367
Risk of poverty, 599, 607
Rocci, R., viii, ix, 53
Rohlfing, I., viii, ix
Rojahn, J., 335
Rokita, P., 357

Sagan, A., 3
Schirra, L.-R., 79
Schmid, F., ix
Schmid, M., viii, ix
Schneck, C., 323
Scholze, F., viii, ix
Schrepp, M., 563
Schulte, T., 17
Schwenker, F., viii, ix, 91
Schöning, U., 519
Scientific discovery, 588
Selection
    feature, 79–83, 85–89, 117–119, 122, 275, 276, 279, 282, 283, 285, 344, 408, 409, 412, 421, 438, 443, 446, 447
    model, 53–63, 65, 67, 238, 387, 389, 390, 392, 393, 479
    variable, 125, 126, 128, 129, 131, 134, 336, 577
Semantic, 11, 103, 104, 159, 161, 163, 168, 169, 438, 454, 629
Sensitivity to 2008 crisis, 287
Shannon entropy, 519, 520, 525, 528
Siegers, P., 629
Signature based methods, 531, 533
Similarity, 31, 32, 35–37, 40, 44, 50, 51, 103–110, 139, 145, 147, 151, 152, 159, 161–169, 193, 199, 201–203, 207, 208, 211–216, 449–455, 457,

458, 487, 492, 497–506, 632, 641, 649
isosurface, 497–499, 501, 503, 505, 506
measure, 103, 449, 487
Simulation, 53, 54, 61–63, 65, 116, 143–145, 159, 165, 167, 229, 231, 232, 234, 236–238, 267, 306, 310, 320, 379, 396, 401, 404, 408, 411, 418, 419, 487–489, 491, 493, 495, 496, 504, 512, 563, 567–572
Singular value decomposition, 185–187, 539
Smart growth sectors, 253–255, 257, 259, 261–263
Sobczak, E., 253
Sobíšek, L., 599
Social science data, 629, 630
Social sciences research, 553
Sokołowski, A., 545
Sozou, P.D., 587
Spanning path, 43, 46
Spectral clustering, 137, 139, 145
Spiliopoulou, M., viii, ix
Stachová, M., 599
Standardisation, 621, 623–626
Statistical learning, 89, 207
Statistical simulation, 487
Steinberg, J., 639
Stochastic processes, vector valued, 221
Strahl, D., 545
Sträng, E., ix
Strötgen, R., 621
Subject indexing, 621–625, 627, 629, 631, 633, 636, 639, 641–643, 645, 647, 649
Subjective evaluation, 265
Suffix tree, 103, 105–107, 110–112
Supervised clustering, 147–149, 152, 153, 155, 156
Supervised and unsupervised pre-processing, 152
Surrogate, 53, 54, 312, 461, 464–468, 470, 493
SVM, 91–93, 95, 97–99, 101, 117, 120, 121, 173, 214, 283, 284, 342, 343, 397, 404–408, 539, 542
Symbolic data, 137, 139–141, 145

Takeover contest, 323, 325, 327, 330
Tarka, P., 553
Textbooks, 336, 621–626
Text information retrieval, 71
Text localization, 71, 72, 78
Textrank, 103–106, 109, 110
Text summarisation, 103, 111
Thesauri, 621, 623, 624, 642

Three-way data, 545, 550
Tillmann, W., 487
Time factor, 369, 370, 374
Time series, 31, 35, 40, 180, 183, 221, 222,
        224–229, 238, 435, 487, 488, 491,
        492, 494–496
Tree selection, 395
Triadic distance, 195–197, 199–205

Ünlü, A., viii, ix, 563, 575
Usage profiling, 531

Vance, C., viii, ix
Variable selection, 125, 126, 128, 129, 131,
        134, 336, 577
Vatolkin, I., 437
Vector valued stochastic processes, 221
Vermunt, J.K., viii, ix
Vertex, 160, 224, 521–526, 528

Vinciotti, V., 507
Visualization, 43, 50, 197, 209, 216, 221, 497,
        506, 542, 578, 619
Völkel G., 519
Volumetric data, 497

Weihs, C., viii, ix, 113, 241, 461, 487
Wiesenmüller, H., ix
Wilhelm, A.F.X., viii
Wöhler, C., 265
Wolf-Ostermann, K., viii, ix
Wołyński, W., 173
Workforce, 253–256, 258–263, 292, 296

Yakovlev, M., 103

Zuliana, S.U., 231

Printed in the United States
By Bookmasters